T0255915

Undergraduate Lecture Notes in Physics

Undergraduate Lecture Notes in Physics (ULNP) publishes authoritative texts covering topics throughout pure and applied physics. Each title in the series is suitable as a basis for undergraduate instruction, typically containing practice problems, worked examples, chapter summaries, and suggestions for further reading.

ULNP titles must provide at least one of the following:

- An exceptionally clear and concise treatment of a standard undergraduate subject.
- A solid undergraduate-level introduction to a graduate, advanced, or non-standard subject.
- A novel perspective or an unusual approach to teaching a subject.

ULNP especially encourages new, original, and idiosyncratic approaches to physics teaching at the undergraduate level.

The purpose of ULNP is to provide intriguing, absorbing books that will continue to be the reader's preferred reference throughout their academic career.

Series editors

Neil Ashby
Professor Emeritus, University of Colorado, Boulder, CO, USA

William Brantley
Professor, Furman University, Greenville, SC, USA

Matthew Deady
Professor, Bard College Physics Program, Annandale-on-Hudson, NY, USA

Michael Fowler
Professor, University of Virginia, Charlottesville, VA, USA

Morten Hjorth-Jensen
Professor, University of Oslo, Oslo, Norway

Michael Inglis
Professor, SUNY Suffolk County Community College, Long Island, NY, USA

Heinz Klose
Professor Emeritus, Humboldt University Berlin, Germany

Helmy Sherif
Professor, University of Alberta, Edmonton, AB, Canada

More information about this series at http://www.springer.com/series/8917

Costas Christodoulides

The Special Theory of Relativity

Foundations, Theory, Verification, Applications

Springer

Costas Christodoulides
Department of Physics, School of Applied Mathematical and Physical Sciences
National Technical University of Athens
Athens
Greece

The book is a revised version of the Greek edition. Copyright for all languages except Greek is owned by the author. Original Greek edition *I eidiki theoria tis sxetikotitas kai oi efarmoges tis* or *Η ΕΙΔΙΚΗ ΘΕΩΡΙΑ ΤΗΣ ΣΧΕΤΙΚΟΤΗΤΑΣ ΚΑΙ ΟΙ ΕΦΑΡΜΟΓΕΣ ΤΗΣ*,
© Tziolas Publications, 2014.

ISSN 2192-4791 ISSN 2192-4805 (electronic)
Undergraduate Lecture Notes in Physics
ISBN 978-3-319-25272-8 ISBN 978-3-319-25274-2 (eBook)
DOI 10.1007/978-3-319-25274-2

Library of Congress Control Number: 2015951793

Springer Cham Heidelberg New York Dordrecht London
© Springer International Publishing Switzerland 2016
This work is subject to copyright. All rights are reserved by the Publisher, whether the whole or part of the material is concerned, specifically the rights of translation, reprinting, reuse of illustrations, recitation, broadcasting, reproduction on microfilms or in any other physical way, and transmission or information storage and retrieval, electronic adaptation, computer software, or by similar or dissimilar methodology now known or hereafter developed.
The use of general descriptive names, registered names, trademarks, service marks, etc. in this publication does not imply, even in the absence of a specific statement, that such names are exempt from the relevant protective laws and regulations and therefore free for general use.
The publisher, the authors and the editors are safe to assume that the advice and information in this book are believed to be true and accurate at the date of publication. Neither the publisher nor the authors or the editors give a warranty, express or implied, with respect to the material contained herein or for any errors or omissions that may have been made.

Printed on acid-free paper

Springer International Publishing AG Switzerland is part of Springer Science+Business Media (www.springer.com)

Preface

The book at hand has developed from the lecture notes used in the teaching of the Special Theory of Relativity for many years by the author, originally to engineering students and finally to students of Physics and Mathematics at the School of Mathematical and Physical Sciences of the National Technical University of Athens. About half the material contained in the book was covered in 25 hourly lectures during the second semester of the first year of studies.

As a textbook, the book has some special characteristics: The proofs of the theorems are given in adequate detail, many figures are used, many examples are used for the comprehension of the theory and many problems are suggested for solution by the reader. The detailed solutions of all the problems are given at the end of the book. However, it must be understood that, for maximum benefit, the reader should really try to solve a problem before resorting to the solution given. As assistance in this direction, the answers to all the problems are given, as well as appropriate hints or suggestions on how to solve the problems.

Special attention has been paid to presenting the historical approach of Physics to the Special Theory of Relativity, to its experimental foundation and to the experiments performed in order to test the validity of the theory. For this reason, a large number of experiments are described in detail.

The Special Theory of Relativity found applications in a large number of problems, mainly in Physics, and an attempt is made to present the most important of them, which every physicist should be familiar with. Many of the applications are developed in the Examples and Problems and the reader should pay particular attention to them.

In order that the reader should take advantage of the ease of access to the scientific literature made possible by today's technology, references are given to the original articles and review articles concerning each topic.

The book gives a fairly complete presentation of the Special Theory of Relativity for a first approach to the subject at the undergraduate university level. A basic course on the Special Theory of Relativity could consist of Chaps. 2, 3, 6 and 9, supplemented by appropriate applications of the theory from Chaps. 4, 5 and 7.

Additional, more demanding topics may be found in the rest of the chapters and in the appendices. The mathematics needed is that of the first year of a degree in Physics. Whenever necessary, a brief presentation of additional mathematics is given in the text.

It is hoped that the book will be a useful addition to the existing literature on this fascinating subject.

Athens
November 2015

Costas Christodoulides

Contents

General Bibliography

A. Einstein, *Relativity* (Methuen, London, 1962).
A simple presentation, by Einstein himself in 1916, of the Special and the General Theories of Relativity.

A. Einstein, H.A. Lorentz, H. Weyl and H. Minkowski, *The Principle of Relativity* (Dover Publications).
The main publications which established the Theory of Relativity, Special and General.

W. Rindler, *Relativity-Special, General and Cosmological* (Oxford University Press, 2006, 2nd ed.).
Includes the presentation of the Special Theory of Relativity, with emphasis on the use of four-vectors and tensors.

M. Born, *Einstein's Theory of Relativity* (Dover Publications, 1964).
A thorough presentation of the Theory of Relativity, Special and General, at pre-university level.

V.A. Ugarov, *Special Theory of Relativity* (Mir Publishers, Moscow, 1979).
An excellent and comprehensive book with a good historical review of the formulation of the theory and the discussion of the paradoxes of relativity.

A.P. French, *Special Relativity* (W.W. Norton and Co, New York, London, 1968).
A general exposition of the Special Theory of Relativity, at an introductory university level.

E.F. Taylor and J.A. Wheeler, *Spacetime Physics* (W.H. Freeman and Co, San Francisco, 1966).
A very good presentation of the Special Theory of Relativity, with solved problems and the discussion of paradoxes.

C. Møller, *The Theory of Relativity* (Clarendon Press, Oxford, 1972, 2nd ed.).
A classical book at an advanced level for the Special and the General Theories of Relativity.

A. Pais. *'Subtle is the Lord...' The Science and the Life of Albert Einstein* (Clarendon Press, Oxford, 1982).
Einstein's biography.

The Equations of the Special Theory of Relativity

The transformations, which are given in Cartesian coordinates, correspond to two frames of reference, S and S′, whose axes coincide at $t = t' = 0$ and which are in relative motion to each other, such as the velocity of S′ relative to S is $\mathbf{V} = V\hat{\mathbf{x}}$.

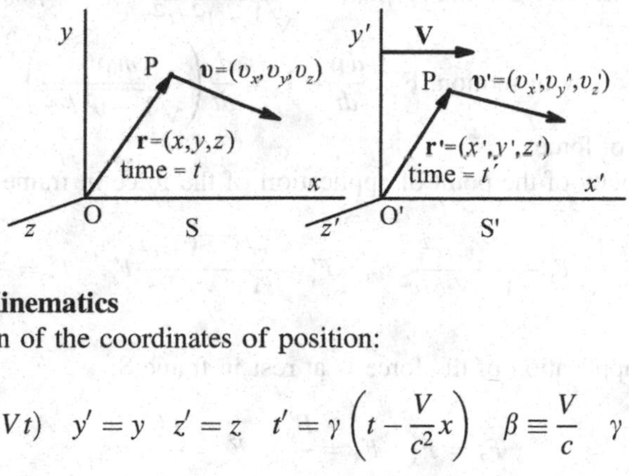

Relativistic Kinematics

Transformation of the coordinates of position:

$$x' = \gamma(x - Vt) \quad y' = y \quad z' = z \quad t' = \gamma\left(t - \frac{V}{c^2}x\right) \quad \beta \equiv \frac{V}{c} \quad \gamma \equiv \frac{1}{\sqrt{1-\beta^2}}$$

Contraction of length: $\Delta L = \Delta L_0/\gamma \quad \Delta L_0 =$ proper or rest length

Dilation of time: $\Delta t = \gamma\,\Delta\tau \quad \Delta\tau =$ proper or rest time

Memory aid *The advantages of exercise:* *He who runs relative to us,* *appears to us to be thinner and* *to live longer.*

Transformation of velocity: $\upsilon'_x = \dfrac{\upsilon_x - V}{1 - \dfrac{\upsilon_x V}{c^2}} \quad \upsilon'_y = \dfrac{\upsilon_y}{\gamma\left(1 - \dfrac{\upsilon_x V}{c^2}\right)} \quad \upsilon'_z = \dfrac{\upsilon_z}{\gamma\left(1 - \dfrac{\upsilon_x V}{c^2}\right)}$

Transformation of speed: $\upsilon' = c\sqrt{1 - \frac{(1-\upsilon^2/c^2)(1-V^2/c^2)}{(1-V\upsilon_x/c^2)^2}}$

Transformation of the Lorentz factor: $\gamma'_P = \gamma\gamma_P\left(1 - \frac{V\upsilon_x}{c^2}\right)$

Doppler effect: $f = f_0\frac{\sqrt{1-\beta^2}}{1+\beta\ \cos\theta} \quad \lambda = \lambda_0\frac{1+\beta\ \cos\theta}{\sqrt{1-\beta^2}}$

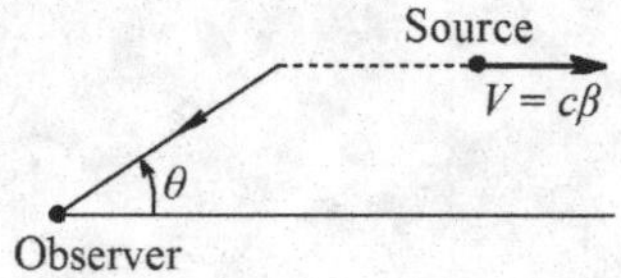

Relativistic Dynamics

Relativistic mass: $m(\upsilon) = \frac{m_0}{\sqrt{1-\upsilon^2/c^2}} = \gamma m_0 \quad m_0 = \text{rest mass}$

Relativistic momentum: $\mathbf{p} = m\boldsymbol{\upsilon} = \gamma m_0\boldsymbol{\upsilon} = \frac{m_0\boldsymbol{\upsilon}}{\sqrt{1-\upsilon^2/c^2}} \quad p = \beta\gamma m_0 c$

Newton's second law of motion: $\mathbf{F} = \frac{d\mathbf{p}}{dt} \quad \mathbf{F} = \frac{d}{dt}\left(\frac{m_0\boldsymbol{\upsilon}}{\sqrt{1-\upsilon^2/c^2}}\right)$

Transformation of force
($\boldsymbol{\upsilon}$ is the velocity of the point of application of the force in frame S):

$$F'_x = F_x - \frac{V\upsilon_y/c^2}{1-V\upsilon_x/c^2}F_y - \frac{V\upsilon_z/c^2}{1-V\upsilon_x/c^2}F_z, \quad F'_y = \frac{\sqrt{1-V^2/c^2}}{1-V\upsilon_x/c^2}F_y, \quad F'_z = \frac{\sqrt{1-V^2/c^2}}{1-V\upsilon_x/c^2}F_z$$

If the point of application of the force is at rest in frame S:

$$F_x = F'_x \quad F_y = \frac{F'_y}{\gamma} \quad F_z = \frac{F'_z}{\gamma}$$

Total relativistic energy (Kinetic energy + Rest energy):

$$E(\upsilon) = E_0 + K = m_0c^2 + K$$

$E = mc^2 = \gamma m_0c^2 \quad E = \frac{m_0\ c^2}{\sqrt{1-\upsilon^2/c^2}}$. Rest energy: $E(0) = E_0 = m_0c^2$

Kinetic energy: $K = E - m_0c^2 \qquad K = \frac{m_0c^2}{\sqrt{1-\upsilon^2/c^2}} - m_0c^2 = m_0c^2(\gamma - 1)$

Momentum: $\mathbf{p} = \frac{E}{c^2}\boldsymbol{\upsilon}$

Conservation of momentum: $\sum_{i=1}^{n}\mathbf{p}_i = \text{const.}$

Conservation of mass-energy: $\sum_{i=1}^{n}E_i = \text{const.}$

$E^2 - p^2c^2 = m_0^2c^4$ invariant

$E^2 = m_0^2c^4 + p^2c^2$

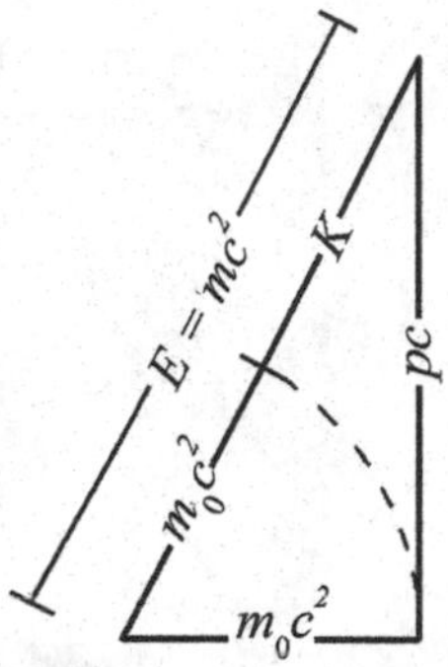

Transformation of momentum and energy:

$$p'_x = \gamma\left(p_x - \frac{\beta}{c}E\right) \quad p'_y = p_y \quad p'_z = p_z \quad E' = \gamma\,(E - c\beta\,p_x)$$

Particles of zero rest mass: $E = pc \quad p = \dfrac{E}{c} \quad \upsilon = c$ always

Photons: $E = h\nu \quad p = \dfrac{h}{\lambda} = \dfrac{E}{c}$ (Planck's constant: $h = 6.626 \times 10^{-34}$ J · s = 4.136×10^{-15}eV · s)

Equivalence of mass and energy: $\Delta E = \Delta m\, c^2$

Mass defect of the nucleus of the isotope ${}^A_Z\mathrm{X}$: $\Delta M = Zm_\mathrm{H} + (A - Z)m_\mathrm{n} - M_\mathrm{X}$ (m_H = mass of hydrogen atom, m_n = neutron mass, M_X = mass of the atom of the isotope ${}^A_Z\mathrm{X}$)

Binding energy of the nucleus of the isotope ${}^A_Z\mathrm{X}$:

$$B.E. = \Delta M\, c^2 = [Zm_\mathrm{H} + (A - Z)m_\mathrm{n} - M_\mathrm{X}]c^2$$

Electromagnetism

The electric charge is invariant.

Lorentz force: $\mathbf{F} = q\mathbf{E} + q\boldsymbol{\upsilon} \times \mathbf{B}$

Maxwell's equations:

$$\nabla \cdot \mathbf{E} = \frac{\rho}{\varepsilon_0}, \qquad \nabla \cdot \mathbf{B} = 0, \qquad \nabla \times \mathbf{E} = -\frac{\partial \mathbf{B}}{\partial t}, \qquad \nabla \times \mathbf{B} = \varepsilon_0\mu_0\frac{\partial \mathbf{E}}{\partial t} + \mu_0\mathbf{J}.$$

Wave equation: $\nabla^2\mathbf{E} = \dfrac{1}{c^2}\dfrac{\partial^2\mathbf{E}}{\partial t^2} \quad \nabla^2\mathbf{B} = \dfrac{1}{c^2}\dfrac{\partial^2\mathbf{B}}{\partial t^2} \quad c = \dfrac{1}{\sqrt{\varepsilon_0\mu_0}} \equiv 299\,792\,458$ m/s.

Transformation of the electromagnetic field:

$E'_x = E_x \quad E'_y = \gamma(E_y - VB_z) \quad E'_z = \gamma(E_z + VB_y)$

$B'_x = B_x \quad B'_y = \gamma(B_y + VE_z/c^2) \quad B'_z = \gamma(B_z - VE_y/c^2)$

$$\mathbf{E}'_\parallel = \mathbf{E}_\parallel \quad \mathbf{E}'_\perp = \gamma(\mathbf{E} + \mathbf{V} \times \mathbf{B})_\perp \quad \mathbf{B}'_\parallel = \mathbf{B}_\parallel \quad \mathbf{B}'_\perp = \gamma(\mathbf{B} - \mathbf{V} \times \mathbf{E}/c^2)_\perp$$

Chapter 1
Historical Introduction

The history of the Theory of Relativity is a very interesting subject, which is discussed in detail in many texts. In this chapter, we will refer only to the main events that led to the establishment of the Special Theory. These events will be presented initially in a concise manner in chronological order, while the more important of them will be described in more detail in the rest of the chapter. For more information, the reader may refer to specialized books which examine the historical development of the theory [1]. Some important experiments that were performed after the publication of the Special Theory of Relativity will be presented in the chapters to follow and, more systematically, in Chap. 10.

1.1 The Main Landmarks in the Development of the Special Theory of Relativity

In chronological order, the main theoretical developments and the experiments which were important in the gradual approach of Physics to the Special Theory of Relativity are listed below (marked with an asterisk are those events that will be discussed in more detail in the rest of the chapter):

1632* Galileo (Galileo Galilei, 1564–1642) publishes his book entitled *Dialogue Concerning the Two Main World Systems—Ptolemaic and Copernican*. In it he formulates the law of inertia and the principle of the invariance of the physical laws for inertial frames of reference in relative motion to each other, which led to the mathematical formulation of what is known today as the *Galilean transformation*. Galileo was the first to try to measure, without success, the speed of light.

1644 Descartes (Rene Descartes, 1596–1650) suggests that the *aether*, known to ancient philosophers, possesses mechanical properties and carries the forces exerted between material bodies. Hooke (Robert Hooke, 1635–1703) upgrades the aether to *luminiferous*, or the medium in which light propagates. Later, to the properties of the aether, Faraday (Michael Faraday, 1791–1867) will also add the transmission of

© Springer International Publishing Switzerland 2016
C. Christodoulides, *The Special Theory of Relativity*,
Undergraduate Lecture Notes in Physics, DOI 10.1007/978-3-319-25274-2_1

magnetic forces, while Gauss (Karl Friedrich Gauss, 1777–1855) and Riemann (Bernhard Riemann, 1826–1866) will add to the aether's properties the ability to carry gravitational and electric forces.

1676* The first measurement of the speed of light by Rømer (Ole Christensen Rømer, 1644–1710). In his efforts to perfect the use of the system of planet Jupiter and its satellite Io in the measurement of time, he observes the (then unknown) Doppler effect in the frequency of the satellite's eclipses. From these observations, the first evaluation of the speed of light is achieved. Historically, it is the first measurement of a universal constant.

1687* Newton (Isaac Newton, 1642–1727) publishes his book *Philosophiae Naturalis Principia Mathematica*. The three laws of motion are formulated. The concepts of absolute space and absolute time are put forward. The inertial properties of matter are also examined.

1725* Discovery of the phenomenon of the aberration of light by Bradley (James Bradley, 1693–1762). In his efforts to measure the annual parallax in the positions of the stars due to the motion of the Earth on its orbit, Bradley discovers the phenomenon of the aberration of light —the stars describe, in the duration of a year, ellipses on the celestial sphere. From measurements and the recognition of the fact that the phenomenon is due to the motion of the Earth, a fairly accurate estimate of the speed of light in vacuum is derived. The observation of the aberration of light will be used later to conclude that the medium in which light is assumed to propagate, the aether, is not dragged by the Earth in its motion through it.

1810* Arago (François Jean Dominique Arago, 1786–1853) tries to detect possible differences in the speed of light from stars, as the Earth, moving on its orbit, changes its speed relative to the source. The results, published in 1853, were negative. To interpret these results, Fresnel (Augustin Fresnel, 1788–1827) proposed the idea that the aether is partially dragged by the Earth, as the latter moves.

1842 Discovery of the Doppler effect (Christian Andreas Doppler, 1803–1853). The possibility is presented, for the first time, of the measurement of the speeds of bright objects, among which stars, galaxies etc. The classical theoretical analysis of the effect was accurate enough for these purposes.

1849*–62 Accurate measurements of the speed of light by Fizeau (Armand Hippolyte Louis Fizeau, 1819–1896) and Foucault (Jean Bernard Léon Foucault, 1819–1868). With ingenious systems of rotating wheels and prisms, the accurate measurement of the speed of light in the laboratory becomes possible.

1851 Foucault's pendulum. The rotation of the plane of oscillation of a vertical pendulum on the surface of the Earth is used to demonstrate the rotation of the Earth about its axis, as well as the fact that the Earth is only approximately an inertial frame of reference.

1851* Measurement of the speed of light in moving water by Fizeau. The speed of light in a moving medium is found to differ from its speed in the same medium, when this is at rest. The difference is of the form $k\upsilon$, where υ is the speed of the fluid relative to the laboratory and k is the *drag coefficient* of the aether by the fluid, as suggested by Fresnel, different for each fluid.

1868* M. Hoek used an interferometer in an effort to measure the drag of the aether by an optical medium, namely water, in its relative motion with respect to the aether. The results agree with Fresnel's theory of partial aether drag.

1861–2* Maxwell (James Clerk Maxwell, 1831–1879) completes and summarizes the laws of electromagnetism in four equations, which today bear his name, thus formulating the classical theory of the electromagnetic field and proving theoretically the existence of electromagnetic waves, which move with a speed equal to $c = 1/\sqrt{\varepsilon_0 \mu_0} = 3 \times 10^8$ m/s, in agreement with experiment. He proves that visible light is in fact a portion of the spectrum of electromagnetic waves. The production and detection of electromagnetic waves in the laboratory is achieved for the first time by Hertz (Heinrich Rudolf Hertz, 1857–1894). The work of Maxwell and other contemporary scientists upgrades the aether to a central element in the interpretation of electromagnetic phenomena.

1881 The first experiment of Michelson (Albert Abraham Michelson, 1852–1931) is performed, in an effort to detect the movement of the Earth relative to the aether, using an interferometer. The results showed no relative motion, within the limits of the sensitivity of the experiment.

1883 Publication of the book of Mach (Ernst Mach, 1838–1916) *Die Mechanik in ihrer Entwicklung*. Stated in it are Mach's views on space and time, as well as the origin of the inertial properties of matter, due to the interaction of bodies with the matter in the whole of the universe.

1887* Improved experiment of Michelson and Morley (Edward Williams Morley, 1838–1923). An improved interferometer, with longer paths for light, was used for the detection of the motion of Earth relative to the aether. The apparatus was 10 times more sensitive than that of the 1881 experiment. The results were again negative, setting a maximum limit of 8 km/s for the possible speed of the Earth relative to the aether.

1892* *The contraction hypothesis*. In their effort to explain the negative result of the Michelson and Morley experiment, FitzGerald and Lorentz (George Francis FitzGerald, 1851–1901, Hendrik Antoon Lorentz, 1853–1928) made the ad hoc hypothesis that bodies moving relative to the aether with a speed υ, contract in the direction of their relative motion by a factor of $\sqrt{1-\upsilon^2/c^2}$. This hypothesis actually interprets the negative result of the experiment.

1892–1904 Publications of Lorentz on the electrodynamics of moving bodies.

1894–6 Discovery of the electron by Thomson (Joseph John Thomson, 1856–1940).

1895–1905 Publications of Poincaré (Jules Henri Poincaré, 1854–1912) on relativity.

1896 Discovery of radioactivity by Becquerel (Antoine Henri Becquerel, 1852–1908). For the first time, particles with relatively high energies become available for experiment.

1902* Study by Kaufmann (Walter Kaufmann, 1871–1947) of the motion of fast electrons in parallel electric and magnetic fields. A variation is observed of the inertial mass with speed, for electrons with speeds 0.8 to 0.95 times the speed of light in vacuum.

1904* Lorentz publishes his theory of the electron, which is based on the contraction hypothesis for moving bodies, and proves, for the first time, the relation $m = m_0 \big/ \sqrt{1 - \upsilon^2/c^2}$ which describes the variation of the inertial mass with the speed υ of a body. He compares the predictions of this equation with the values of m/m_0 measured by Kaufmann.

1904* Lorentz derives the exact transformation which leaves Maxwell's equations and the wave equation in vacuum invariant, in the absence of free charges and currents, i.e. when $\rho = 0$ and $\mathbf{J} = 0$. This transformation is known today as the *Lorentz transformation*. Poincaré, in 1905, completed the transformations of coordinates and of electromagnetic magnitudes, by finding the correct transformation for ρ and $\mathbf{J}$.

1905* The *Special Theory of Relativity* is formulated by Einstein (Albert Einstein, 1879–1955) in his paper entitled '*On the electrodynamics of moving bodies*'.

1.2 The Principle of Relativity of Galileo. Galileo's Invariance Hypothesis. The Law of Inertia. Inertial Frames of Reference

"Shut yourself up with some friend of yours in the main cabin below decks on some large ship, and have with you there some flies, butterflies and other small flying animals. Have a large bowl of water with fish in it; hung up a bottle that empties drop by drop into a wide vessel beneath it. With the ship standing still, observe carefully how the little animals fly with equal speed to all sides of the cabin. The fish swim indifferently in all directions; the drops fall into the vessel beneath; and, in throwing something to your friend, you need throw it no more strongly in one direction than another, the distances being equal; jumping with your feet together, you pass equal spaces in every direction. When you have observed all this carefully (though there is no doubt that when the ship is standing still everything

> must happen in this way), have the ship proceed with any speed you like, so long as the motion is uniform and not fluctuating this way and that. You will discover not the least change in all these effects named, nor could you tell from any of them whether the ship was moving or standing still."

With these words, Galileo introduces the *principle of relativity* [2]. He had observed in experiments and deduced from logical syllogisms, that the uniform relative motion between two observers does not affect the phenomena of Mechanics, as they are observed by them. This was expressed in *Galileo's principle of invariance*.

> The laws of Physics are identical for all observers moving with constant velocities relative to each other.

The principle referred only to mechanical phenomena (being the only ones then known) and constitutes a very important contribution of Galileo to the formulation of the laws of Physics.

Another great contribution of Galileo to Physics is the *law of inertia*. It states that: A body on which no external forces are applied, will continue to move with a constant velocity or remain at rest. The law of inertia was later stated as Newton's first law of motion. It initially applied to mechanical forces, but it now covers all forces. Knowing that the fundamental forces fall off at least as fast as the inverse square of the distance, we say that a body that is 'far enough' from other bodies, may be considered isolated to an approximation good enough for the law of inertia to hold.

1.2.1 The Galilean Transformation

In Physics we examine events and their development with time. Events are localized in regions of space and time which are sufficiently small. Our experience shows that we live in a universe with three spatial dimensions and one temporal. In order to define the point at which an event occurs, we need enough points of reference, relative to which to give its position in a unique way. For three spatial dimensions, four reference points are necessary, not lying on the same plane. Using these, it is possible to define a coordinate system in space. An example of a simple coordinate system is the Cartesian system with three mutually orthogonal axes. We also need a clock, with which to measure time at every point of the coordinate system. These constitute a *frame of reference*. A frame of reference in which an observer is stationary is his own frame of reference. Also, a frame of reference in which the law of inertia applies is called an *inertial frame of reference*. In Classical Physics, space and time are absolute. We will see in the next chapter how these concepts fail in the Special Theory of Relativity. We will also explain how we calibrate a frame of reference using light signals, as proposed by Einstein.

The *position* of an event in a frame of reference S is defined by giving its coordinates in space and time (x, y, z, t) in this frame (Fig. 1.1). We will assume that another frame of reference, S′, moves relative to S with a constant velocity $\mathbf{V}=V\hat{\mathbf{x}}$

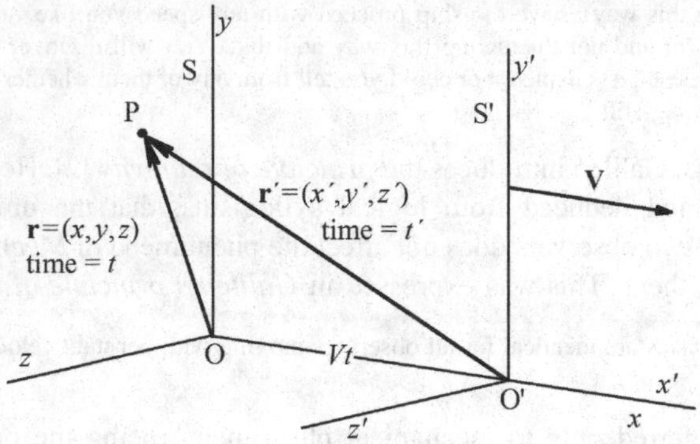

Fig. 1.1 The Galilean transformation changes the coordinates (x, y, z, t) of an event in the frame of reference S, to the coordinates of the same event (x', y', z', t') in the frame of reference S′, which is moving with a constant velocity $\mathbf{V} = V\hat{\mathbf{x}}$ relative to S

($\hat{\mathbf{x}}$ is the unit vector in the x-direction), and that at the moment $t = t' = 0$ the axes of these two frames of reference coincide. In Classical Mechanics, time is absolute and universal to all observers, so that the clocks of the two frames of reference, as set, will always show $t' = t$. Since the relative motion of the two frames takes place along the x-axis only, the distance of the origins of the axes of the two frames, O′ and O, will be $OO' = Vt$. It will, therefore, be $x' = x - Vt$ and $y' = y$, $z' = z$.

The transformation

$$x' = x - Vt, \quad y' = y, \quad z' = z, \quad t' = t. \tag{1.1}$$

is called *Galilean transformation*. It is a linear transformation and, hence, it leaves Newton's laws of motion invariant, as we will see below.

1.3 Rømer and the Speed of Light

During the Age of Explorations, between the 15th and the 17th centuries, the problem of determining geographic longitude became of paramount importance. In the north hemisphere, geographic latitude could be determined easily, by measuring the angle above the horizon of the position of Polaris, the pole star. A similar method could be used in the south hemisphere, with suitably selected stars. The determination of longitude, however, presented considerable difficulties. If there could be, on a ship, a reliable clock that gave the time at some reference point on Earth, whose geographic longitude was defined, the difference in longitudes of the ship and the reference point could be found by determining the local time of the ship, by observing, for example, the exact moment when the Sun or a star rises or reaches its maximum elevation above the horizon. The difference between the local times would give the difference in geographic longitudes. Problems arose, however, due to the difficulties in carrying on a ship, given the conditions of its motion, a clock that would show accurate time for long periods. The problem was so important that the British parliament established a prize of a substantial sum (£20 000, equivalent today to, approximately, £2 million or €3 million) for the construction of such a reliable clock. The English clockmaker John Harrison did manage to construct and perfect the *nautical chronometer*, during the period of 1735–1772, and to collect, after great difficulties, the prize.

Galileo had, in 1616, claimed a similar prize of king Philip III of Spain, by suggesting the use, as a clock, of the system of planet Jupiter and its then known satellites, which he himself had discovered using his telescope. The observation of the eclipses of one of Jupiter's satellites, as it entered the planet's shadow, would give, using tables, the local time at a reference point on the Earth. The longitude of the observer could then be determined as described above. However, the method proved to be unreliable, due to the difficulties in observing the eclipses on board a ship and the inaccuracies in Galileo's tables for the times of the eclipses.

Rømer, a Danish astronomer, worked on the same problem at the Uraniborg observatory, which had been established by Tycho Brahe on the island of Hven. He and Jean Picard observed, in 1671, a total of 140 eclipses of Io, a satellite of Jupiter. Comparing the times of the eclipses with those measured by Cassini (Giovanni Domenico Cassini, 1625–1712) in Paris, they determined the difference in the geographic longitudes of the two observatories. By the way, Cassini had observed some discrepancies in his observations of the satellites of Jupiter, which he had attributed to the finite speed of light. In one of his publications, in 1675, he actually mentions that "light takes 10–11 min to travel a distance equal to the radius of the Earth's orbit"! This would give for the speed of light a value between 2.3 and 2.5×10^8 m/s. Cassini, however, did not follow up this matter.

In Paris now, Rømer continued to observe Jupiter's satellite Io and he actually did establish the fact that the finite speed of light influenced the times of observation of the eclipses. In a dramatic move, in September 1676, he announced to the French Academy of Sciences that the eclipse of Io which was due, according to the observations he had made in August, to happen at 45 s after 5:25 a.m. on November 9th, would be delayed by exactly 10 min! Despite their disbelief, the astronomers of the Royal Observatory in Paris ascertained that the eclipse referred to was delayed by 10 min. Rømer explained to the members of the Academy, a few days later, that the delay was due to the fact that between August and November the distance between Jupiter and Earth had increased, and so the light needed 10 more minutes to reach the Earth. The geometry of the phenomenon is shown in Fig. 1.2, from the work of an unknown writer concerning the announcement of Rømer. The Earth moves around the Sun (A) passing, successively, through the points EFGHLKE. Jupiter moves at a considerably smaller speed on its orbit and may, to a first approximation, be considered stationary. Io is shown at point C to be entering the shadow of Jupiter (B), at the moment that marks the beginning of the eclipse. If the August observations were performed when the Earth was at point such as L on its orbit, in September the Earth would be at point K, which is at a distance from

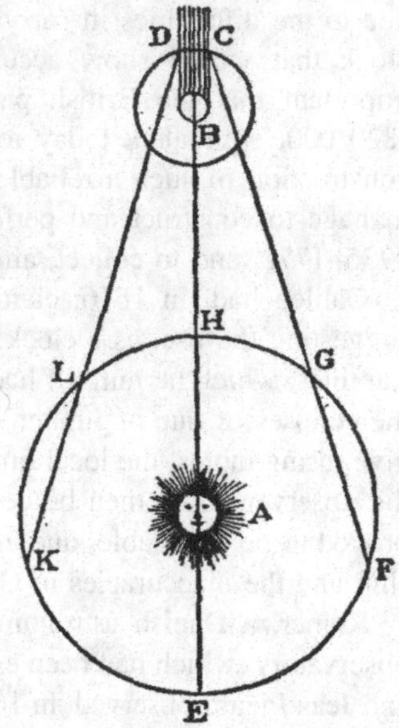

Fig. 1.2 The relative positions of the Earth and the planet Jupiter (*B*), as the former moves in its orbit around the Sun (*A*). (From a reference of an unknown writer to Rømer's observations, in 1676)

Jupiter greater by a length equal to (LK), and the delay of 10 min was the time needed for light to travel this additional distance.

What Rømer had actually observed, was a manifestation of the phenomenon later to be known as *the Doppler effect*. In this case, the role of the periodic source was played by the Jupiter-Io system and the observer was on the moving Earth. He found that the frequency of the eclipses was greater when the Earth was approaching Jupiter (e.g. section FG of the orbit) compared to when the Earth was receding from Jupiter (e.g. section LK). The orbital period of Io was found to be equal to 42.5 h from observations made when the Earth was at the point nearest to Jupiter, H, and the relative motion of the two planets had no radial component. Rømer observed that 40 revolutions of Io on its orbit lasted for 22 min less when observed when the Earth was approaching Jupiter, compared to the duration of the same number of revolutions when the Earth was moving away from it. From these measurements, Rømer concluded that light travels in 22 min the same distance that the Earth travels during 80 times the period of Io (he obviously ignored the difference between the length of the arc FG and that of the chord FG).

Rømer himself never used his measurements to evaluate the speed of light. Had he done so, he would have found the ratio c/υ of the speed of light c to the orbital speed υ of the Earth as equal to

$$\frac{c}{\upsilon} = \frac{80 \times 42.5 \times 60}{22} = 9300. \tag{1.2}$$

With the then accepted value of the distance of the Earth from the Sun of 140×10^6 km, the orbital speed of the Earth would be found to be 28.3 km/s and, therefore, $c = 260\,000$ km/s. Instead of Rømer, the speed of light was calculated by Huygens (Christiaan Huygens, 1629–1695), who, however, after corresponding with Rømer, was left with the erroneous impression that he claimed that light needed 22 min to travel the *diameter* of the Earth's orbit, thus finding a speed of light equal to 220 000 km/s. Newton, in his book *Opticks*, in 1704, possibly based on Rømer's measurements, mentions that light needs 7–8 min in order to travel a distance equal to the Sun-Earth distance, instead of 8 min and 20 s which we know today to be the correct value. With the Sun-Earth distance being, according to Newton, equal to 70×10^6 miles, the speed of light would have a value between 235 000 and 270 000 km/s.

Independently of the accuracy in the value of the speed of light calculated using Rømer's measurements, the significance of these measurements may be summarized as follows:

1. A universal constant was measured for the first time.
2. The speed of light was indeed found to be very large, but finite.
3. It was established that astronomers should take into account the finite speed of light in the timing of celestial phenomena.

Epilogue

Harrison's nautical chronometer, although useful, was very expensive, initially costing approximately as much as 30 % of a ship! From evidence of the time, it appears that it was not used extensively. Today, one may purchase, without much expense, a GPS (Global Positioning System) unit, which gives one's position anywhere on Earth with an accuracy of better than 5 m (Fig. 1.3). This system is based on the accurate timing of the arrivals of signals from atomic clocks orbiting the Earth inside 24–32 artificial satellites. It is interesting, however, to mention that,

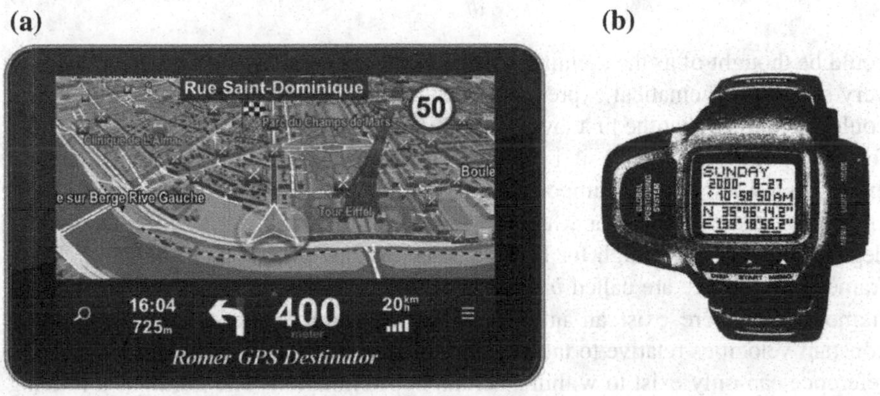

Fig. 1.3 **a** The use of GPS in a portable destinator with a city map and **b** a wristwatch positioning system

for this system to work, corrections are required, due to relativistic effects. One correction is predicted by the Special Theory of Relativity and is due to the fact that the atomic clocks in the satellites lag behind similar clocks on the surface of the Earth by about 7.2 μs per day, because of their speed relative to the clocks on the Earth. A second correction is predicted by the General Theory of Relativity and is due to the fact that the clocks in the satellites gain about 45.9 μs per day, compared to those on the surface of the Earth, due to the difference in the gravitational potential at which these two groups of clocks are situated. A total relativistic correction of 38.6 μs per day or 0.45 parts in a billion is needed.

Rømer's observations, which were intended to solve a problem in navigation, led to the determination of the speed of light, which played an important role in the development of the Theory of Relativity, without which it would be impossible for the modern system of positioning on every point on the planet to work!

1.4 Newton's Laws of Motion. Inertia and Inertial Frames of Reference

In 1687, Newton formulated his three *laws of motion* as follows:

1. A body on which the total external force applied is zero, remains at rest or moves at a constant velocity.
2. The rate of change of the momentum of a body with time is equal to the sum of the external forces exerted on it.
3. At every point, for every force applied (action), an equal and opposite force (reaction) is developed.

The first law is *the law of inertia*. The second law,

$$\frac{d\mathbf{p}}{dt} = \mathbf{F} \tag{1.3}$$

could be thought of as the definition of force, which, defined in this manner, leads to very simple mathematical expressions for the laws describing the natural forces. It could be claimed that the first law is superfluous, because it follows from the second for the special case when it is $\mathbf{F} = 0$. However, Newton's first law of motion expresses something very important, namely that it is possible for frames of reference to exist in nature for which the condition of zero external force holds to a degree that is good enough for Newton's second law of motion to be valid. These frames of reference are called *inertial* or *Galilean frames of reference*. If one such frame exists, there exist an infinite number of such frames, which move with constant velocities relative to the first and to each other. Obviously, such frames of reference can only exist to within a certain approximation. The Earth, as a rotating and revolving frame of reference, is not inertial. It can be considered to be inertial if we include, next to the real forces exerted on a body, forces which are called fictitious (or d'Alembert or inertial forces), such as the centrifugal and the Coriolis force. The center of the Sun is subject to a very small acceleration, due to the gravitational forces of neighboring stars but, mainly, due to its revolution around the center of the Galaxy. These accelerations are, however, extremely small and impossible to measure and may be neglected. The heliocentric frame of reference is, to a very good approximation, inertial. We usually speak of motion *relative to the fixed stars* to refer to motion in an inertial frame of reference, implying the motion with reference to the matter in very distant galaxies in the visible universe.

1.4.1 The Invariance of Newton's Second Law of Motion Under the Galilean Transformation

The Galilean transformation for the coordinates of position was stated above as

$$x' = x - Vt, \quad y' = y, \quad z' = z, \quad t' = t.$$

To find the Galileo transformation for velocities, we take the differentials of these equations

$$dx' = dx - Vdt, \quad dy' = dy, \quad dz' = dz, \quad dt' = dt, \tag{1.4}$$

from which there follow the relations

$$\upsilon'_x = \frac{dx'}{dt'} = \frac{dx}{dt'} - V\frac{dt}{dt'} = \frac{dx}{dt} - V\frac{dt}{dt}, \quad \upsilon'_y = \frac{dy'}{dt'} = \frac{dy}{dt'} = \frac{dy}{dt}, \quad \upsilon'_z = \frac{dz'}{dt'} = \frac{dz}{dt'} = \frac{dz}{dt} \tag{1.5}$$

or, finally,

$$\upsilon'_x = \upsilon_x - V, \quad \upsilon'_y = \upsilon_y, \quad \upsilon'_z = \upsilon_z. \tag{1.6}$$

Vectorially, for a general velocity of the frame of reference S′ relative to frame S equal to **V** and the origins of the two frames of reference coinciding when $t = t' = 0$, the transformation is:

$$\mathbf{r}' = \mathbf{r} - \mathbf{V}t, \quad t' = t. \tag{1.7}$$

Differentiating with respect to time, we have the transformation of velocity

$$\boldsymbol{\upsilon}' = \boldsymbol{\upsilon} - \mathbf{V}. \tag{1.8}$$

Differentiating again, we have the transformation for the acceleration

$$\mathbf{a}' = \mathbf{a}. \tag{1.9}$$

Writing Eq. (1.3) as

$$\mathbf{F} = m\mathbf{a}, \tag{1.10}$$

it follows that

$$\mathbf{F} = m\mathbf{a}'. \tag{1.11}$$

If we assume that

$$\mathbf{F}' = \mathbf{F} \tag{1.12}$$

and that the mass remains invariant, then

$$\mathbf{F}' = m\mathbf{a}'. \tag{1.13}$$

Thus, the Galileo transformation leaves Eq. (1.10) unchanged (invariant).

1.5 The Aberration of Light

Aristarchus of Samos put forward his *heliocentric theory* in the 3rd century B.C., possibly influenced by Heraclides Ponticus (4th century B.C.). According to his theory, the Sun is at the center of the universe and around it revolve the Earth, the Moon and the then known five planets. This system, coupled with the rotation of the Earth about its axis, simplified the interpretation of the motions of celestial bodies. The main objection from the scientific community of the time was the lack of an annual periodic change in the positions of the 'fixed' stars. The motion of the Earth about the Sun once a year, would mean that the stars would appear at different directions at different seasons of the year, with reference to the *ecliptic coordinate system*, based on the plane of the Earth's orbit, i.e. the plane of the *ecliptic* (Fig. 1.4). A star A_1 situated at the pole of the ecliptic, should describe a *parallactic orbit* on the celestial sphere, similar in shape to that of the orbit of the Earth around the Sun (i.e. almost circular). The angle α subtended at the star by the mean radius of the Earth's orbit, is called its *parallax*. Measured in radians, the parallax of a star is given by $\alpha = \arctan(R/D)$, where R is the radius of the Earth's orbit and D is the distance of the star from the Sun. For stars at smaller ecliptic latitudes (A_2), the parallactic orbits will be elliptical, with an angular magnitude of the semi-major axis equal to α, while for stars on the plane of the ecliptic (A_3), the parallactic orbits will be straight lines with angular magnitude equal to 2α.

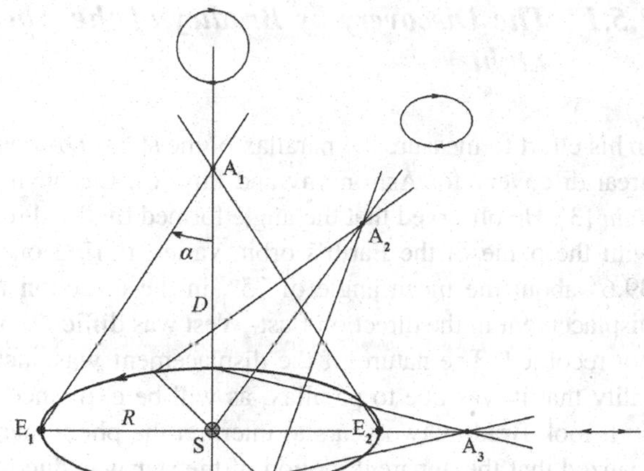

Fig. 1.4 The phenomenon of stellar parallax

The impossibility of measurement of parallax with the methods then available, led Aristarchus to suggest that the fixed stars must lie at distances very much greater than the radius of the Earth's orbit. We know today that even for the nearest to the Sun star, *Proxima Centauri,* the parallax is equal to only 0.786″, which corresponds to a distance of 1/0.786 = 1.27 pc (parsec, from the words ***par***allax and ***sec***ond) or 4.2 light years from the Sun, i.e. 260 000 times the radius of the orbit of the Earth around the Sun (astronomical units, ua). Aristarchus' conclusion was, therefore, correct, and constitutes the first, albeit qualitative, estimation of the vast distances to the stars.

With the reintroduction of the heliocentric theory by Copernicus, in 1543, the problem of measuring the annual parallax of the stars returned. Thomas Digges suggested, in 1573, that the observation of parallax would confirm Copernicus' theory. There were reports from astronomers who claimed to have observed the parallax of stars: of Jean Picard, in 1680, who measured an annual variation in the position of the pole star by 40″, of John Flamsteed, in 1689, with similar observations on the same star and of Robert Hooke, in 1674, who, on observing the position of γ *Draconis,* which passes virtually through the zenith at London and may be observed without needing to correct for the refraction of light by the atmosphere, found that the star lies further to the North in July by 23″ compared to its position in October. However, the details of these motions did not agree with those expected due to parallax and they were not considered as observations of the phenomenon. For the sake of history, we mention that Bessel (Friedrich Bessel, 1784–1846) was the first astronomer to achieve the measurement of the parallax of a star, by measuring for the star 61 of the constellation of Cygnus a parallax of 0.294″, corresponding to a distance equal to 3.4 parsec or 11.1 light years.

1.5.1 The Discovery by Bradley of the Aberration of Stellar Light

In his effort to measure the parallax of the star γ *Draconis*, Bradley made, in 1725, a great discovery for Astronomy and Physics: the phenomenon of the *aberration of light* [3]. He observed that the angle formed by the direction of the radius to the star with the plane of the Earth's orbit, varies, during one year, with an amplitude of 39.6″ about the mean angle of 75°, in the direction North–South (Fig. 1.5). The displacement in the direction East–West was difficult to measure accurately and was not recorded. The nature of the displacement was such that it excluded the possibility that it was due to parallax, as will be explained below.

It took Bradley two years to interpret the phenomenon. This happened when he realized that the apparent motion of the star was due to the movement of the Earth on its orbit around the Sun and the finite speed of light. Thomson [4] mentions a story, according to which the inspiration for the interpretation of the effect came to Bradley during a pleasure trip by boat on the river Thames. The company went up and down a distance on the river and Bradley noticed that the wind-indicator on top of the mast changed direction every time the boat changed direction, although the

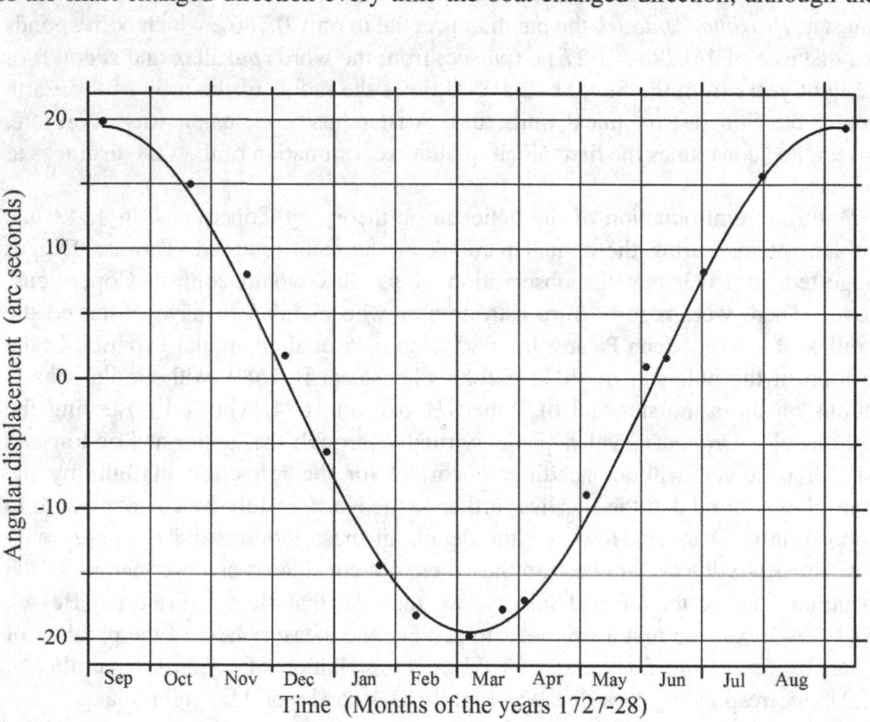

Fig. 1.5 The variation of the position of the star γ *Draconis*, in the direction North–South, around the mean value of 75° relative to the plane of the Earth's orbit. These measurements of Bradley cover the period of one year, between 1727 and 1728. The curve fitted to the experimental points is sinusoidal, with a period of 12 months and an amplitude of 39.6/2 = 19.8 arc seconds

mild wind kept blowing in the same direction. In Bradley's writings, however, there is no reference to such an incident, so we must conclude that we are dealing with yet another myth of the history of science.

In order to understand the phenomenon of the aberration of light, we will first examine what happens in the simpler case of a star in a direction perpendicular to the plane of the orbit of the Earth (Fig. 1.6). We will examine the propagation of a light pulse (or of a photon) coming from the star and moving along the axis of a telescope used for its observation (Fig. 1.6a). Let the light pulse be, at some moment, at point P_1 on the objective lens of the telescope and on the telescope's axis. For the pulse to remain on the telescope's axis, given that this is moving with the Earth with velocity $\boldsymbol{\upsilon}$ normal to the direction of propagation of the pulse, the axis of the telescope must form an angle, say α, with this direction. To evaluate the angle we reason as follows: If the light pulse needs time Δt in order to cover the length $P_1P_2 = c\Delta t$, where P_2 is another point on the telescope's axis, for the pulse to remain on the axis of the telescope, the latter must move during this time by a distance $T_1T_2 = \upsilon\Delta t$, with the Earth. T_2 is a point on the telescope's axis at which the pulse is now found (P_2) and T_1 was the position of this point of the axis a time Δt earlier. Because it is $\tan\alpha = T_1T_2/P_1P_2$, it follows that

$$\tan\alpha = \upsilon/c. \tag{1.14}$$

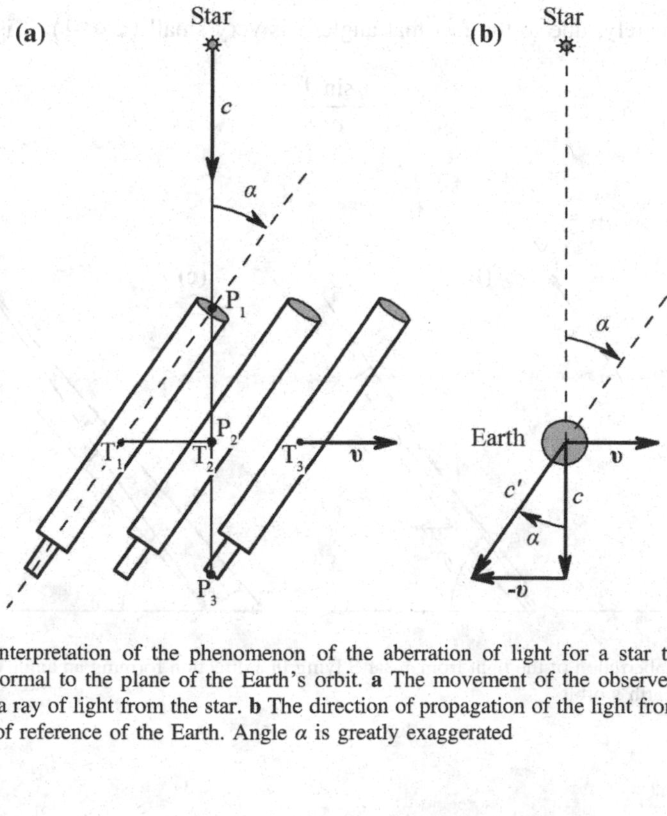

Fig. 1.6 Interpretation of the phenomenon of the aberration of light for a star that lies in a direction normal to the plane of the Earth's orbit. **a** The movement of the observer's telescope relative to a ray of light from the star. **b** The direction of propagation of the light from the star, in the frame of reference of the Earth. Angle α is greatly exaggerated

For this value of the angle, when the point on the telescope's axis considered reaches T_3, the light pulse will be coming out of the telescope's eyepiece, at P_3. The ratio $\beta = \upsilon/c$ is called *aberration constant*.

In the frame of reference of the Earth and the telescope, the light falls on the telescope forming an angle α with the direction in which the star lies (Fig. 1.6b). It should be noted that Bradley's analysis leads to the conclusion that the observed speed of the light pulse relative to the telescope is $c' = c/\cos\alpha$, which is greater than c.

In the case of the star studied by Bradley, which is not in a direction normal to the plane of the Earth's orbit but forms an angle of $\theta = 75^\circ$ with it, the velocities of the light and of the Earth on its orbit are not always perpendicular to each other. Figure 1.7 shows the observation of the star with a telescope, when the two velocities are at an angle θ to each other: (a) for stationary Earth and (b) for Earth moving with velocity $\boldsymbol{\upsilon}$ relative to the star or, better, relative to the medium in which light propagates. The aberration angle is now α relative to the line which forms an angle θ with the plane of the Earth's orbit. The velocity of the light relative to the Earth is found by the vectorial addition of the velocity of light relative to the star with the velocity $-\boldsymbol{\upsilon}$ (Fig. 1.7c). The aberration angle α will, therefore, be given by:

$$\tan\alpha = \frac{\upsilon_\perp}{c} = \frac{\upsilon \sin\theta}{c}. \tag{1.15}$$

Approximately, due to the fact that angle α is very small ($\alpha \ll 1$), it is

$$\alpha = \frac{\upsilon \sin\theta}{c}. \tag{1.16}$$

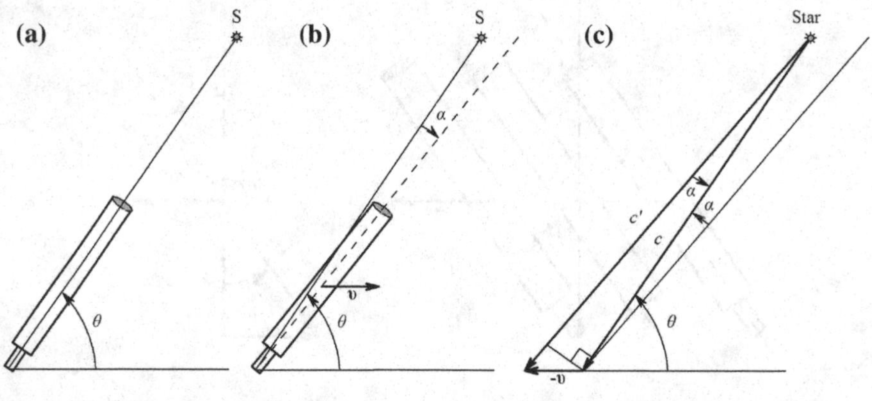

Fig. 1.7 The aberration of the light from a star S lying in a direction forming an angle θ with the plane of the Earth's orbit

As the Earth moves on its orbit (Fig. 1.8a), the angle formed by its velocity with the line towards the star varies and so does the angle of aberration. For the successive positions of the Earth A, B, C and D in Fig. 1.8a, the star appears to be moving on the celestial sphere on the *ellipse abcd* of Fig. 1.8b. The vectorial addition of the two velocities for the four positions of Fig. 1.8a is shown in Fig. 1.8c. At positions B and D, the two velocities are perpendicular to each other and the aberration angle is maximum and equal to $\alpha = \upsilon/c$, while at points A and C it is minimum and equal to $\beta = (\upsilon/c)\sin\theta$. By the way, had the movement of the star on the celestial sphere been due to the variation of the Earth's position on its orbit, the angle formed by the apparent direction of the star with the plane of the Earth's orbit would have been maximum when the Earth was at points A and C, and minimum when the Earth was at points B and D. In Bradley's observations, the

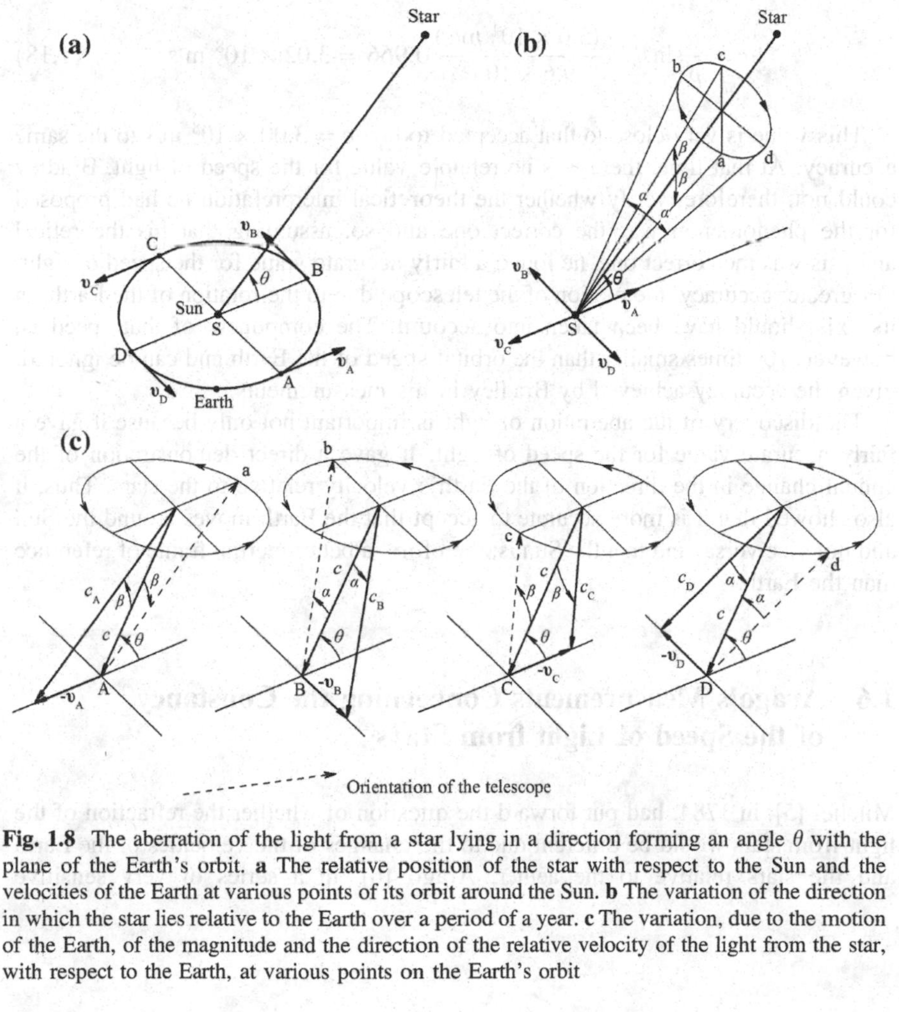

Fig. 1.8 The aberration of the light from a star lying in a direction forming an angle θ with the plane of the Earth's orbit. **a** The relative position of the star with respect to the Sun and the velocities of the Earth at various points of its orbit around the Sun. **b** The variation of the direction in which the star lies relative to the Earth over a period of a year. **c** The variation, due to the motion of the Earth, of the magnitude and the direction of the relative velocity of the light from the star, with respect to the Earth, at various points on the Earth's orbit

opposite is seen, as shown in Fig. 1.8b. Also, the direction in which the star moves on its elliptical orbit (abcd in Fig. 1.8b) is opposite to that expected due to parallax. These made it clear that the effect observed was not that of parallax.

Bradley measured the transverse component of the displacement, i.e. the minor axis of the elliptical path traced by the star on the celestial sphere (ac in Fig. 1.8b). He found that

$$2\beta = 39.6'' \quad \text{and, therefore,} \ \beta = 19.8'' = \frac{19.8}{3600} \times \frac{2\pi}{360} \ \text{rad} = 9.6 \times 10^{-5} \ \text{rad}. \tag{1.17}$$

This is approximately the angle subtended by a soccer ball at a distance of 1500 m. Substituting $\theta = 75^{\circ}$ and $\upsilon = 30$ km/s $= 3 \times 10^4$ m/s in the equation $\beta = (\upsilon/c)\sin\theta$, we find that

$$c = \frac{\upsilon}{\beta}\sin\theta = \frac{(3.0 \times 10^4 \ \text{m/s})}{9.6 \times 10^{-5}} 0.966 = 3.02 \times 10^8 \ \text{m/s}. \tag{1.18}$$

This value is very close to that accepted today, $c = 3.00 \times 10^8$ m/s to the same accuracy. At that time, there was no reliable value for the speed of light. Bradley could not, therefore, verify whether the theoretical interpretation he had proposed for the phenomenon was the correct one and so, assuming that his theoretical analysis was the correct one, he found a fairly accurate value for the speed of light. For greater accuracy, the motion of the telescope due to the rotation of the Earth on its axis should have been taken into account. The component of that speed is, however, 100 times smaller than the orbital speed of the Earth and can be ignored, given the accuracy achieved by Bradley in his measurements.

The discovery of the aberration of light is important not only because it gave a fairly accurate value for the speed of light. It gave a direct demonstration of the annual change in the direction of the Earth's velocity relative to the stars. Thus, it also showed that it is more accurate to accept that the Earth moves around the Sun and not vice versa, and that the Sun is, therefore, a better inertial frame of reference than the Earth.

1.6 Arago's Measurements Concerning the Constancy of the Speed of Light from Stars

Mitchel [5], in 1784, had put forward the question of whether the refraction of the light from stars would be different due to the changes of the velocities of the Earth and the stars relative to the aether. Arago [6], in a series of very sensitive

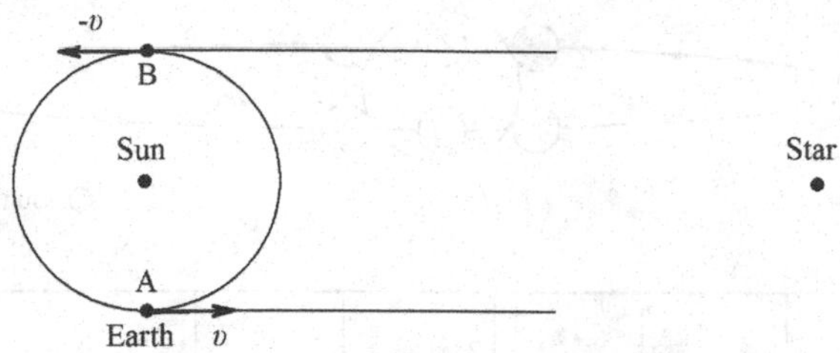

Fig. 1.9 Arago investigated the possibility that the speed of light from a star is different when the Earth, moving in its orbit, approaches the star (*point A*), compared to that when it moves away from the star (*point B*)

measurements, showed that the refraction of light is not influenced by the motions of the stars relative to the Earth. If the orbital speed of the Earth is υ (Fig. 1.9), then, for a star on the plane of the Earth's orbit, the relative speed of approach of the star and the Earth would be greater by 2υ when the Earth moved towards the star (point A) compared to the value it had 6 months earlier, when the Earth moved away from the star (point B).

The image of the star, formed by the objective lens of a refracting telescope, will be formed at the focus of the lens. If the change in the relative speed of the star and the Earth leads to a change of the speed of light relative to the telescope, then the refractive index will change by a proportion of $\pm\upsilon/c$ in the duration of a year. This is of the order of 10^{-3}. Arago found no displacement of the focus of the telescope's lens, concluding that the relative motion of the star and the Earth does not influence the speed of the light from a star, as measured by an observer on the Earth, at least to a measurable degree.

This is in agreement with other observations, such as the variation with time of the light intensity of eclipsing binary variable stars. These systems consist of two stars which revolve around their common center of mass. At some parts of its orbit, one of the stars will be approaching the observer while the other will be receding from him. Half a period later the directions of the motions will be reversed. The intensity curve of the light from the two stars as a function of time has a characteristic shape, such as the one shown in Fig. 1.10. From this curve, useful conclusions can be deduced for the stars forming the system. For the system of the figure, for example, the luminosities of the two stars per unit area are different. Thus, when the less luminous star is in front of the brighter one (point 2), the decrease in the total brightness of the system is greater than when it is behind it (point 6). Such a curve is observed for the eclipsing binary variable star Algol.

De Sitter (Willem de Sitter, 1872–1934) suggested a method of testing the independence of the speed of light from the speed of the source, based on measurements of the brightness curves of eclipsing binary variable stars. If the system is at a distance D from the Earth, the light, travelling with a speed of c, will reach the

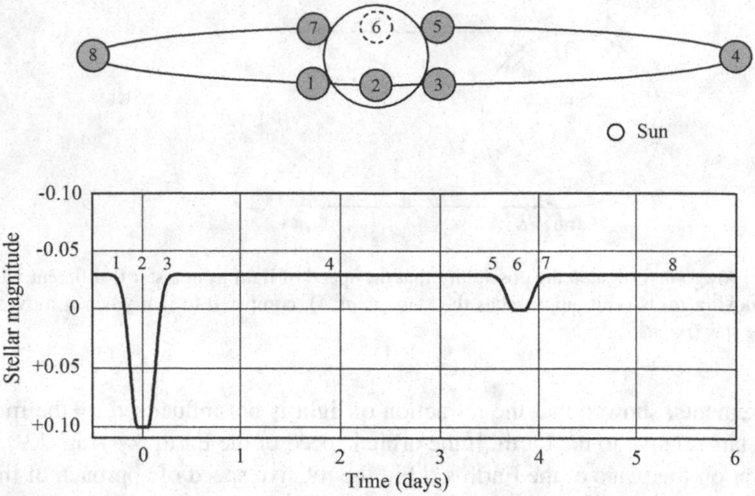

Fig. 1.10 The *brightness curve* of an eclipsing binary variable star. The binary variable star Algol exhibits such a variation in brightness. The smaller star virtually revolves around the bigger star and has a smaller luminosity per unit area than the latter. The size of the Sun is given for comparison

Earth a time D/c after its emission. A small change Δc in c, would lead to a change equal to $\Delta t = -(D/c^2)\Delta c$ in the time of arrival of the light. This would cause visible anomalies in the brightness curve of the two stars. No such anomalies were observed in practice and we are obliged to conclude that the motion of the two members of the binary star system does not influence the speed of the light emitted. It should also be stressed that the Doppler effect gives us the possibility to separate the light from each star. We are thus in a position to test whether the intensities of light from the two stars vary in agreement with the changes in their positions.

Since then, objections have been raised against both the measurements of Arago and the use of binary star brightness curves as described above. Fox [7] noted that, according to the theory of the propagation of light in matter, when an electromagnetic wave is incident on a medium, the field excites its electrons to oscillate and they, in their turn, re-emit the wave. The transmitted wave is a result of the superposition of the incident and the re-emitted waves. When the light enters from vacuum to a certain depth in the medium, which depends on the density of the electrons in the medium, we may assume that the light is propagating with its speed in that medium and all information concerning its speed in vacuum is lost. This is the Ewald-Oseen extinction theorem [8]. According to this, both Arago's and the measurements made in accordance with de Sitter's suggestion, only detect the speed of light in air or in the material of the telescope's lens. Experiments performed later, took care to minimize this effect (see Chap. 10, Sect. 10.1.2).

1.7 Measurements of the Speed of Light in the Laboratory

The first recorded reference to the speed of light is found in the work of Aristotle, who mentions—and disagrees with—the opinion of Empedocles, who states that light from the Sun must need some time to reach the Earth. Measurements of the speed of light actually begin with Galileo's experiments. In his *Dialogues*, he has Simplicio observe that, when a canon at a great distance is fired, we first see the flash and hear the sound later. Galileo concluded that the speed of light is greater than that of sound and he offers a method of measuring it. He suggests that two people, A and B, at a great distance from each other, each hold a lantern, initially covered. At some moment A uncovers his lantern. B waits till he sees the light from A's lantern and immediately uncovers his own lantern. A must see the light from B's lantern some time interval after he had uncovered his own and this is the time needed for light to travel twice the distance between A and B. Galileo actually also suggests that they practice at a small distance from each other, so that they determine their reaction time! He also mentions that he had performed the experiment, with a distance of about one kilometer between the lanterns, without noticing any time delay. From these observations he concludes that the speed of light must be at least ten times that of sound in air.

The first measurement of the speed of light 'in the laboratory' was performed by Fizeau, in 1849. The apparatus he used is shown schematically in Fig. 1.11.

The principle of operation of the apparatus was based on the measurement of the time needed for light to travel a known distance. Light from a source S was focused by a lens and then, after partial reflection by a glass plate M_1, it passed through a gap in a rotating toothed wheel W (see figure). The light beam then passed through a lens which turned it into a parallel beam. This travelled a distance L to a mirror M_2 where it was totally reflected so that it returned to the toothed wheel. If in the meantime the wheel had turned by an angle large enough for the returning light to meet a gap, the light would pass through and be seen by an observer O behind the

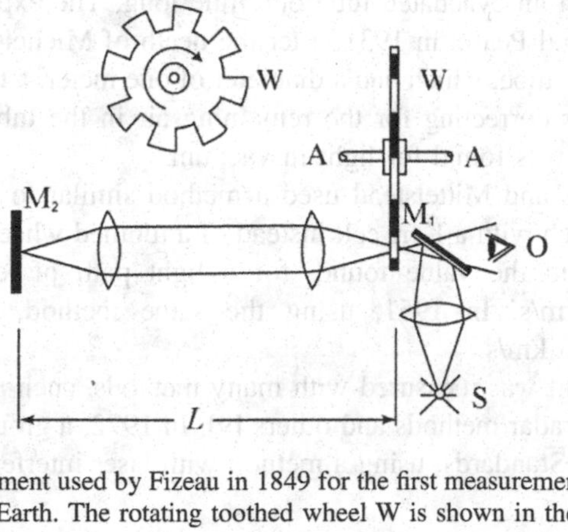

Fig. 1.11 The arrangement used by Fizeau in 1849 for the first measurement of the speed of light using a source on the Earth. The rotating toothed wheel W is shown in the figure

plate M_1. If we assume that the wheel has N teeth and N gaps and is rotating about its axis with a frequency of f revolutions per unit time, the wheel produces light pulses at a frequency of Nf, each of duration $1/2Nf$. The time needed for a pulse, from the moment it passes through the gap, to return to the toothed wheel is $2L/c$. The wheel's gaps are placed in the light's path with a frequency Nf or one every time $1/Nf$. If the time needed for the light pulse to travel to the mirror M_2 and back is an integral multiple of $1/Nf$, then the observer will see the light pulse. The condition for this is, therefore,

$$\frac{2L}{c} = k\frac{1}{Nf} \qquad (k = 1, 2, 3, \ldots) \tag{1.19}$$

from which c is determined. If we assume that the wheel starts rotating and accelerates from zero frequency, when the frequency reaches the value of $c/2NL$, for $k = 1$, the observer will see the first maximum in the light's intensity. The next maximum will be observed when the frequency becomes c/NL, for $k = 2$, and so on.

Fizeau used a toothed wheel with 720 teeth and a distance $L = 17\,266$ m and found that the first maximum in the light's intensity appeared at $f = 12.6$ revolutions per second. The value he found for the speed of light in *air* was $c = 315\,300 \pm 500\,\text{km/s}$. From this value of c and the value of the refractive index of light in air, $n = 1.000\,293$, we find that the speed of light in vacuum is greater than that in air by 88 km/s. Cornu, in 1874, used Fizeau's method, with a greater distance, and was able to observe maxima in the light intensity for up to $k = 13$.

In 1850, Foucault used a similar method, but replaced the toothed wheel by a rotating prism with faces that acted as reflectors. In his initial experiment, he used a distance L of only 4 m. Later, in 1862, he used $L = 20$ m and found $c = 298\,000 \pm 500$ km/s. In an improved Foucault arrangement, Michelson, in 1927, using a distance of $L = 22$ miles $= 35$ km between mount Wilson and mount San Antonio in California, measured the more accurate value of $c = 299\,796 \pm 4$ km/s for the speed of light in vacuum. He was planning an experiment to measure the speed of light in an evacuated tube one mile long. The experiment was completed by Pearson and Pease, in 1931, after the death of Michelson. With multiple reflections inside the tube, which had a diameter of one meter, a total path of 13 km was achieved. After correcting for the remaining air in the tube, a speed of $c = 299\,774 \pm 11$ km/s was found for light in vacuum.

In 1925, Karolus and Mittelstead used a method similar to that of Fizeau but with an optical switch with a Kerr cell instead of a toothed wheel. The method had greater accuracy and the value found, for a light path of length 300 m, was $c = 299\,778 \pm 10$ km/s. In 1951, using the same method, Bergstrand found $c = 299\,793.1 \pm 0.3$ km/s.

The speed of light was measured with many methods, such as with microwave resonance cavities, radar methods and others [9]. In 1972, a group at the American National Bureau of Standards, using a method with laser interferometry, found for the speed of light in vacuum the value of $c = 299\,792\,456.2 \pm 1.1$ km/s, which has an uncertainty at least 100 times smaller than those of previous measurements.

By 1975, the speed of light was known to be 299 792 458 m/s with an uncertainty of only 4 parts per billion. Since 1983, by international agreement, the relationship between the unit of length (meter, m) and time (second, s) is defined so that the speed of light in vacuum is, by definition, equal to

$$c \equiv 299\,792\,458 \text{ m/s}.$$

This makes easier and more accurate the measurement of distances by using electromagnetic waves (light and radio waves), as the measurements of time are much more accurate than the direct measurements of length.

1.7.1 The Possibility of Dependence of the Speed of Light in Vacuum on Its Frequency

It is well known that the speed of light in materials depends on its color or its frequency (phenomenon of dispersion). The question arises of whether something like that is also true for vacuum. Measurements performed on electromagnetic waves with frequencies from 10^8 Hz (radio waves) up to 10^{22} Hz (γ rays), showed no dependence of their speed on frequency. Naturally, measurements at high frequencies are not as accurate as those at radio and at optical frequencies. The relevant experiments are examined in Chap. 10, Sect. 10.1.3.

1.8 Attempts to Measure the Dragging of Aether by Moving Media

The phenomenon of the aberration of light could be interpreted, using the corpuscular model for light, as being analogous to the observation of the oblique fall of raindrops by a moving observer. With the wave model of light, aberration may be interpreted if it is assumed that the aether, the medium in which the light waves propagate, is not disturbed (dragged) by the moving Earth. The dragging of aether by the optical medium in which light propagates, when this medium is moving relative to the aether, was proposed by Fresnel [10] in order to explain Arago's experiments, with which he did not detect any difference in the speed of stellar light, when the Earth moved towards a star and when it receded from it. Fresnel used his theory of the elasticity of the aether to prove that the aether is partially dragged by an optical medium which is moving relative to it. If the motion of the medium relative to the aether occurs with a speed υ, Fresnel assumed that the aether is dragged with a speed which is a fraction f of this speed, i.e. with speed $f\upsilon$, where the coefficient f is the *drag coefficient* of the aether by the particular medium, and he found that

$$f = 1 - \frac{1}{n^2}, \tag{1.20}$$

where n is the index of refraction of the optical medium. The question immediately arose of whether the dragging of the aether by a moving optical medium could be measured. The experiments described below were performed to this end.

1.8.1 The Experiment of Fizeau

In 1851, Fizeau [11] used an interferometer (Fig. 1.12) in order to measure the drag coefficient f in moving fluids.

The fluid under investigation may be forced to move with a large speed υ, in a pipe, in the direction ABDC. We will assume that the apparatus moves with a speed V relative to the aether. The light from a monochromatic source S is split into two beams, 1 and 2, after partial transmission and partial reflection by a partially-silvered plate of glass G, at an angle of 45° to the direction of the incident light beam. With reflections on the mirrors M_1, M_2 and M_3, beam 1 traverses the path $GM_1M_2M_3G$, in a direction opposite to that in which the fluid is moving, while beam 2 traverses the path $GM_3M_2M_1G$, in the same direction as the fluid.

We first examine beam 2.

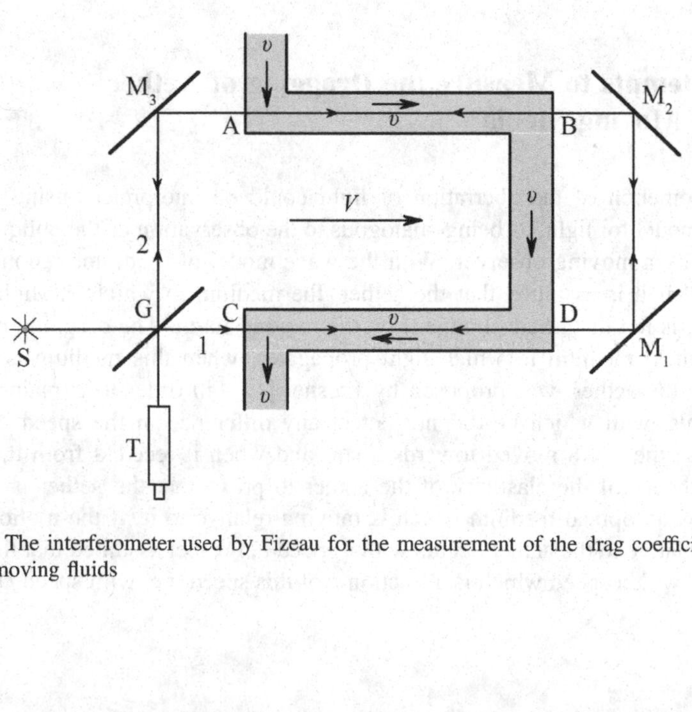

Fig. 1.12 The interferometer used by Fizeau for the measurement of the drag coefficient of the aether in moving fluids

For the path AB:

The speed of the fluid relative to the apparatus is υ.

The speed of the fluid relative to the aether, without aether dragging, is $V+\upsilon$.

The speed of the fluid relative to the aether, with aether dragging, is $(1-f)(V+\upsilon)$.

The speed of the light relative to the fluid, without aether dragging, is c/n.

The speed of the light relative to the fluid, with aether dragging, is $c/n-(1-f)(V+\upsilon)$.

The speed of the light relative to the laboratory, with aether dragging, is

$$c/n-(1-f)(V+\upsilon)+\upsilon \qquad \text{or} \qquad c/n-(1-f)V+f\upsilon.$$

For the path DC, from the last result we find by replacing υ with $-\upsilon$ and c with $-c$:

The speed of the light relative to the laboratory, with aether dragging, is $-c/n-(1-f)V-f\upsilon$.

The total time needed for beam 2 to travel the distance (AB) + (DC) is

$$t_2=\frac{l}{c/n-(1-f)V+f\upsilon}+\frac{l}{c/n+(1-f)V+f\upsilon}=2l\frac{c/n+f\upsilon}{[c/n+f\upsilon]^2-(1-f)^2V^2} \tag{1.21}$$

If the speeds υ and V are small compared to c, the denominator may be written, approximately, as $(c/n)^2$ and, therefore,

$$t_2=2l\frac{n^2}{c^2}\left(\frac{c}{n}+f\upsilon\right). \tag{1.22}$$

The corresponding time for beam 1 is found, by placing $-\upsilon$ instead of υ, to be equal to

$$t_1=2l\frac{n^2}{c^2}\left(\frac{c}{n}-f\upsilon\right). \tag{1.23}$$

The difference in the two times is

$$\Delta t=t_2-t_1=4l\frac{n^2}{c^2}f\upsilon. \tag{1.24}$$

If the light from the source S enters the interferometer having passed through a narrow vertical slit, from the interference of the two beams of light an interference pattern will result at the telescope, as shown in Fig. 1.13. We will have bright fringes at those points at which the two beams arrive with a phase difference which is an integral multiple of 2π (or the difference in the optical paths will be an integral multiple of the wavelength λ). We will have dark fringes at those points at which

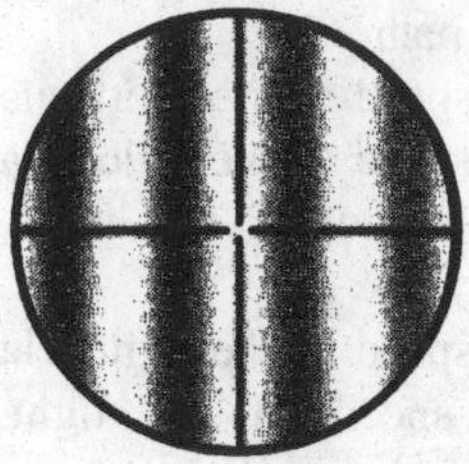

Fig. 1.13 The fringes observed through the telescope of the interferometer used in Fizeau's experiments

the two beams arrive with a phase difference which is an odd integral multiple of π (or the difference in the optical paths will be an odd integral multiple of $\lambda/2$).

The positions of the fringes could be observed when the fluid (e.g. water) was stationary and when it was moving with considerable speeds υ. The displacement of the fringes from the one case to the other would correspond to

$$\Delta k = \frac{c}{\lambda}\Delta t = 4l\frac{n^2}{c\lambda}f\upsilon \tag{1.25}$$

complete oscillations of light, with an accuracy to the first power of the ratio υ/c. This would lead to a shift of the interference pattern by a number of fringes. For the drag coefficient f found by Fresnel and given by Eq. (1.20), the last equation gives

$$\Delta k = \frac{4l}{\lambda}\frac{\upsilon}{c}(n^2 - 1). \tag{1.26}$$

Typical magnitudes in the experiments performed by Fizeau were: $l = 1.5$ m, $\lambda = 530$ nm, $n = 1.333$ and water speed of $\upsilon = 7$ m/s, for which a shift of the interference pattern was observed, equal to $\Delta k = 0.23$ of the width of a fringe. From Eq. (1.25), the drag coefficient

$$f = \frac{\lambda c}{4n^2 l\upsilon}\Delta k \tag{1.27}$$

is derived, which, for the numerical values given, is equal to $f_{\text{exp}} = 0.48$. Fresnel's formula [Eq. (1.20)] gives, for $n = 1.333$, the theoretical value of $f_{\text{theor}} = 0.43$ for the drag coefficient of water. The agreement between experiment and theory is good, at the level of 10 %.

Equation (1.26) was confirmed completely by Fizeau's experiments. The experiment was repeated by Fizeau [12] again in 1853 and by Michelson and Morley in 1886 [13].

1.8.2 The Experiment of Hoek

In 1868, M. Hoek [14] used the interferometer shown in Fig. 1.14 in an attempt to measure the dragging of the aether by an optical medium in relative motion to it. The medium was water with a refractive index n, in a tube AB, with length l. A beam of monochromatic light from a source S is split in two beams, after partial transmission and partial reflection at a partially-silvered glass plate G, at an inclination of 45° relative to the direction of the incident beam. The transmitted beam (beam 1) traversed the path along the rectangular parallelogram GCDBAG, with total reflections, successively, at the mirrors M_1, M_2 and M_3. The reflected beam (beam 2) traversed the path GABDCG. Both beams ended up at the telescope T, by which they could be observed, beam 1 after passing through the glass plate G, and beam 2 after partial reflection at the plate.

Assuming that the apparatus is stationary with respect to the aether, and that the two paths of the light beams have the same (optical) length, the two beams of light will reach the center of the optical field of the telescope with zero phase difference and will interfere constructively. The light intensity at the center of the optical field will be maximum. For light moving slightly off center compared to the central path shown in the figure, there will be a phase difference, which will grow as the distance from the center increases. A pattern similar to that shown in Fig. 1.13 will, therefore, be produced, with bright and dark fringes.

If we assume that there is no dragging of the aether by the medium and that initially the interferometer moves with a speed V relative to the aether, which we will take to be parallel to the tube AB and in the direction from A to B, for simplicity, and then we rotate the apparatus by 180°, we find, using arguments

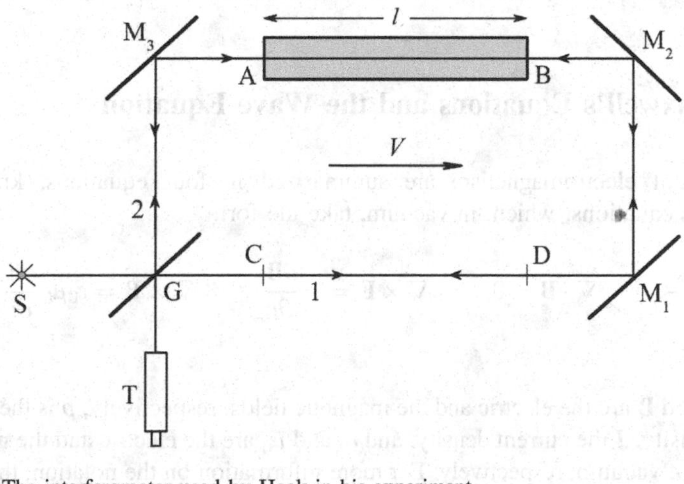

Fig. 1.14 The interferometer used by Hoek in his experiment

similar to those of the experiment of Fizeau, that a shift will be observed in the fringe pattern by $2\Delta k$ fringes, where

$$\Delta k = \frac{2l}{\lambda}\frac{V}{c}(n^2 - 1). \tag{1.28}$$

Such a displacement would be easily observable given the speed of the Earth on its orbit, $V = 30$ km/s. For $\lambda = 400$ nm, $l = 1$ m and $n = 1.333$, it would have been equal to $2\Delta k = 777$ fringes. No displacement of the fringe pattern was observed, however, during the rotation of the apparatus.

If we consider that the aether is partially dragged by the water in the pipe AB, the fact that a shift in the fringes does not occur on rotating the interferometer implies that the drag coefficient of the aether for the particular medium, f, is equal to

$$f = 1 - \frac{1}{n^2}. \tag{1.29}$$

This is exactly the relation suggested by Fresnel in the interpretation of Arago's observations. Hoek's experiment verifies Fresnel's formula, at least to an accuracy of first order in the ratio V/c.

The experiments just described are interpreted by the hypothesis that the aether is dragged by optical media. However, the dragging has the strange property of depending on the refractive index and therefore on the wavelength (color) of the light, something that is very difficult to justify. The question arises of whether it is possible to detect the motion of the Earth through the aether. Michelson, and later Michelson and Morley, tried to answer this question with their very important experiments. These experiments we will describe below.

1.9 Maxwell's Equations and the Wave Equation

The laws of electromagnetism are summarized in four equations, known as Maxwell's equations, which, in vacuum, take the form

$$\nabla \cdot \mathbf{E} = \frac{\rho}{\varepsilon_0}, \qquad \nabla \cdot \mathbf{B} = 0, \qquad \nabla \times \mathbf{E} = -\frac{\partial \mathbf{B}}{\partial t}, \qquad \nabla \times \mathbf{B} = \varepsilon_0 \mu_0 \frac{\partial \mathbf{E}}{\partial t} + \mu_0 \mathbf{J}, \tag{1.30}$$

where **E** and **B** are the electric and the magnetic fields, respectively, ρ is the volume charge density, **J** the current density, and ε_0 and μ_0 are the electric and the magnetic constants of vacuum, respectively. For more information on the notation, the reader is referred to Appendix 4 or to a book on vector analysis. The physical meaning of these equations is as follows:

$\nabla \cdot \mathbf{E} = \frac{\rho}{\varepsilon_0}$	This equation is derived from Coulomb's law and is a formulation of Gauss' law in differential form. The expression on the left is the flux of the electric field per unit of infinitesimal volume, which is shown to be proportional to the charge enclosed by this volume
$\nabla \cdot \mathbf{B} = 0$	This is Gauss' law for the magnetic field. Its flux per unit of infinitesimal volume is equated to zero due to the non-existence of magnetic monopoles, which would correspond to the positive and negative electric charges
$\nabla \times \mathbf{E} = -\frac{\partial \mathbf{B}}{\partial t}$	This is the differential form of Faraday's law of electromagnetic induction or the fact that a magnetic field varying with time produces an electric field
$\nabla \times \mathbf{B} = \varepsilon_0 \mu_0 \frac{\partial \mathbf{E}}{\partial t} + \mu_0 \mathbf{J}$	The complete form of Ampère's law, according to which an electric field varying with time or an electric current, produce a magnetic field. The first term on the right is due to the displacement current, suggested by Maxwell

In vacuum and without free charges, $\rho = 0$, or currents, $\mathbf{J} = 0$, the equations simplify to

$$\nabla \cdot \mathbf{E} = 0 \qquad \nabla \cdot \mathbf{B} = 0 \qquad \nabla \times \mathbf{E} = -\frac{\partial \mathbf{B}}{\partial t} \qquad \nabla \times \mathbf{B} = \varepsilon_0 \mu_0 \frac{\partial \mathbf{E}}{\partial t} \tag{1.31}$$

which may be combined to give an equation for the electric field $\mathbf{E}$

$$\frac{\partial^2 \mathbf{E}}{\partial x^2} + \frac{\partial^2 \mathbf{E}}{\partial y^2} + \frac{\partial^2 \mathbf{E}}{\partial z^2} = \frac{1}{c^2}\frac{\partial^2 \mathbf{E}}{\partial t^2} \quad \text{or} \quad \nabla^2 \mathbf{E} = \frac{1}{c^2}\frac{\partial^2 \mathbf{E}}{\partial t^2}, \tag{1.32}$$

and an equation for the magnetic field $\mathbf{B}$

$$\frac{\partial^2 \mathbf{B}}{\partial x^2} + \frac{\partial^2 \mathbf{B}}{\partial y^2} + \frac{\partial^2 \mathbf{B}}{\partial z^2} = \frac{1}{c^2}\frac{\partial^2 \mathbf{B}}{\partial t^2} \quad \text{or} \quad \nabla^2 \mathbf{B} = \frac{1}{c^2}\frac{\partial^2 \mathbf{B}}{\partial t^2}. \tag{1.33}$$

These equations describe an electromagnetic wave moving with a speed of

$$c = \frac{1}{\sqrt{\varepsilon_0 \mu_0}} = 3 \times 10^8 \text{ m/s}, \tag{1.34}$$

which, it should be noted, does not depend on the speeds of the source of the wave or of the observer.

The flow of electromagnetic energy per unit time and per unit area, at a point, is given by the Poynting vector:

$$\mathbf{S} = \frac{1}{\mu_0}\mathbf{E} \times \mathbf{B}. \tag{1.35}$$

Necessary for the propagation of electromagnetic waves was considered to be the existence of the aether, although its properties were impossible to explain using mechanical models.

If, however, the ignorance concerning the structure of aether could be tolerated, there was another more serious problem with the laws of electromagnetism, which could not be ignored. Up till then, it was taken for granted that the laws of Physics satisfied the *Galilean principle of (mechanical) relativity*, i.e. they had the same form in all inertial frames of reference. As we have already demonstrated in the case of Newton's second law of motion, the transition from one inertial frame of reference to another, based on the Galilean transformation, leaves the law invariant. This is not true, however, for the laws of electromagnetism. Maxwell's equations and, thus, the wave equation also, are not invariant under the Galilean transformation. We will return to this subject in Chap. 9. For the moment, it is enough to mention that the transformation of the wave equation, Eq. (1.32), from one inertial frame of reference $S(x, y, z, t)$ to another, $S'(x', y', z', t')$, which is moving with a constant velocity $\mathbf{V}=V\hat{\mathbf{x}}$ relative to S, based on the Galilean transformation given by Eqs. (1.1), transforms the wave equation for the electric field to

$$\nabla'^2\mathbf{E} = \frac{1}{c^2}\frac{\partial^2\mathbf{E}}{\partial t'^2} + \frac{1}{c^2}\left(V^2\frac{\partial^2\mathbf{E}}{\partial x'^2} - 2V\frac{\partial^2\mathbf{E}}{\partial x'\partial t'}\right). \tag{1.36}$$

Obviously, it is not possible for the wave equation to have the form of Eq. (1.32) in one inertial frame of reference and the form of Eq. (1.36) in another, unless we accept that the frame of reference of the Earth is privileged, as it is from experiments in this frame of reference that the laws leading to the 'simple' form of Eq. (1.32) for the propagation of electromagnetic waves were found. Lorentz played the leading role in the adoption of the view that there is a stationary aether, which defines a privileged frame of reference in which the laws of electromagnetism assume their 'simpler' form. This would define Newton's absolute space, in which no fictitious forces appear. The question immediately arises of whether it is possible to detect experimentally the motion of a frame of reference, such as the Earth, relative to the aether.

1.10 The Experiment of Michelson and Morley

The most serious investigation of whether it is possible to detect the motion of the Earth relative to the aether, was performed by a series experiments by Michelson in 1881 [15] and, in an improved form, by Michelson and Morley in 1887 [16].

The experiment of Michelson and Morley, is one of the historic experiments of Physics. Although Einstein, who must have been aware of the results of this experiment, does not appear to have been influenced by these in his formulation of the Special Theory of Relativity, their consequences for Physics should not be underestimated. Its negative results changed the way we view electromagnetic radiation today. The main consequence of the failure to detect such a motion is that it abolished from Physics the absolute frame of reference relative to which all phenomena could be described.

The experimental apparatus used in 1887 was an optical interferometer, of the kind known today as a Michelson interferometer (Fig. 1.15, from the original publication of Michelson and Morley). It had been assembled on a large granite block which floated in a pool of mercury (with a density of 13.6 gr/cm^3 as compared to granite's 2.7 gr/cm^3). In this way, the whole interferometer could be rotated about a vertical axis with minimized friction and disturbances. The general idea of the experiment was the comparison of the speed of light in two mutually perpendicular directions. If there was a difference in the two speeds, due to the relative motion of the interferometer in the aether, this would be observable.

A simplified diagram of the interferometer is shown in Fig. 1.16a. Monochromatic light from a source S, passes through a slit s and is turned into a parallel beam by a converging lens. The light meets a half-silvered mirror, a, which splits it in two beams of about equal intensities. One of these will pass through the half-silvered mirror a and will continue its path towards a mirror c, where it suffers total reflection and returns to a. From that point it will be reflected towards a telescope of observation, df. The other beam of light is diverted normally to its initial direction, is totally reflected by a mirror b, passes through the half-silvered mirror a, and also arrives at the telescope. In the actual experiment, as shown in Fig. 1.15, the lengths of the optical paths were greatly increased by multiple reflections on mirrors suitably placed. By this method, the paths of the light beams had in the second experiment of Michelson and Morley ten times the lengths they

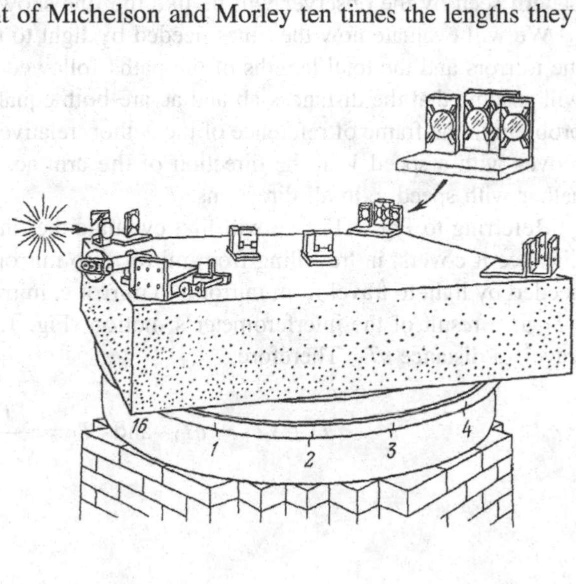

Fig. 1.15 The arrangement used in the experiment of Michelson and Morley

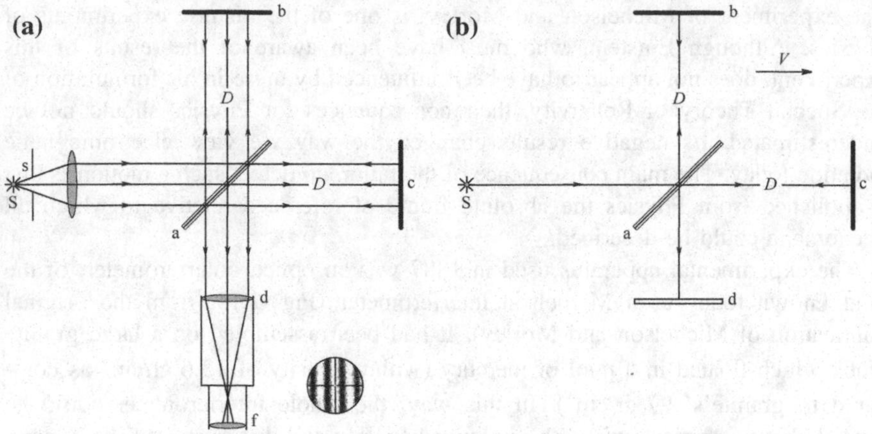

Fig. 1.16 The principle of operation of the Michelson interferometer

had in the first experiment of Michelson. This lead to a proportional increase in the sensitivity of the experiment.

If the two beams travel their paths in the same time, they will arrive at the telescope in phase and they will interfere *constructively*. In this case, at the center of the telescope's optical field there will be a maximum in light intensity. The same will happen if the times needed by the two beams to travel their corresponding paths differ by an integral multiple of the period of the light used. If slit s is vertical, then at the center of the optical field there will appear a narrow strip of light or a *bright interference fringe*. At a little distance from the bright fringe, there will appear a dark fringe, because at those points the beams arrive with a time difference of half a period (or a semi-integral multiple of it) and interfere *destructively*. The pattern seen by the observer will be like the one shown in Fig. 1.16a.

We will evaluate now the times needed by light to travel the distances between the mirrors and the total lengths of the paths followed by light. For simplicity, we will assume that the distances ab and ac are both equal to D. We will examine the problem in the frame of reference of the aether, relative to which the interferometer moves with a speed V in the direction of the arm ac. Light moves relative to the aether with speed c, in all directions.

Referring to Fig. 1.17, we will first evaluate the time light needs, and the total distance it covers, in travelling from mirror a to mirror c and back. In the time T_1 needed by light to travel from mirror a to mirror c, mirror c will move by a distance VT_1 as a result of the interferometer's motion (Fig. 1.17b). In this time, the light travels a distance cT_1. Therefore,

$$D + VT_1 = cT_1 \quad \text{and} \quad T_1 = \frac{D}{c - V}. \tag{1.37}$$

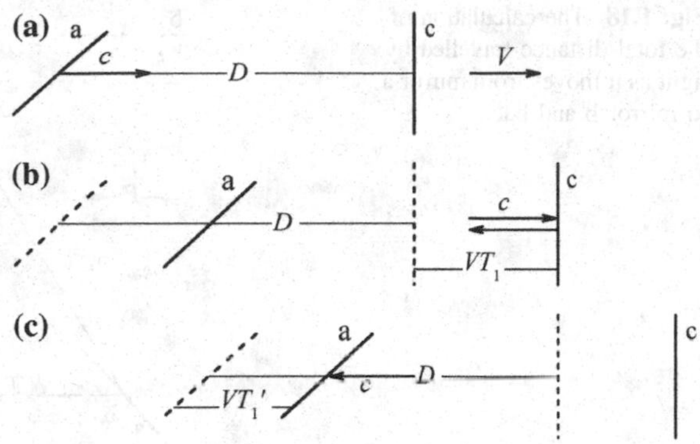

Fig. 1.17 The calculation of the total distance travelled by light as it moves from mirror a to mirror c and back

In the return of light from mirror c to mirror a, in time T_1', mirror a moves by a distance VT_1' (Fig. 1.17c). Therefore,

$$D - VT_1' = cT_1' \quad \text{and} \quad T_1' = \frac{D}{c+V}. \tag{1.38}$$

The total time needed for light to travel from mirror a to mirror c and back, is

$$T_1 + T_1' = \frac{D}{c-V} + \frac{D}{c+V} = 2D\frac{c}{c^2 - V^2}. \tag{1.39}$$

The distance travelled by light in this time is

$$L_1 = c(T_1 + T_1') = 2D\frac{c^2}{c^2 - V^2}. \tag{1.40}$$

Referring to Fig. 1.18, we will evaluate now the time needed by light, and the total distance it covers, in travelling from mirror a to mirror b and back. In the time T_2 needed by light to travel from mirror a to mirror b, mirror b will have moved by a distance VT_2 as a result of the interferometer's motion. In this time light travels a distance cT_2. This length is the hypotenuse of a right-angled triangle, the other sides of which have lengths D and VT_2. Therefore,

$$(cT_2)^2 = D^2 + (VT_2)^2 \quad \text{and} \quad T_2 = \frac{D}{\sqrt{c^2 - V^2}}. \tag{1.41}$$

The time T_2' needed by light to travel from mirror b back to mirror a is equal to T_2. Thus, the total time needed by light to travel from mirror a to mirror b and back is

$$T_2 + T_2' = \frac{2D}{\sqrt{c^2 - V^2}}. \tag{1.42}$$

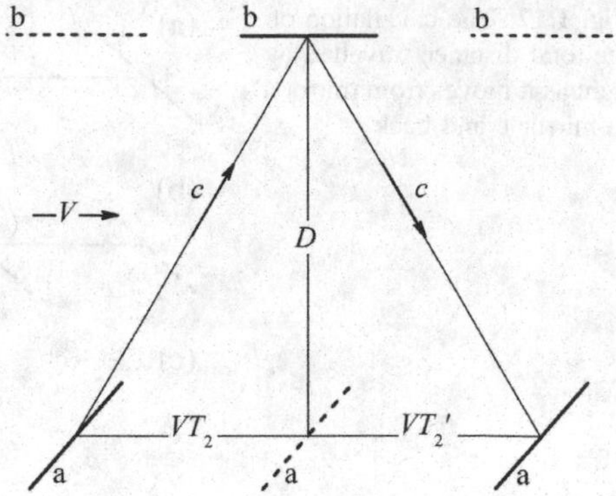

Fig. 1.18 The calculation of the total distance travelled by light as it moves from mirror a to mirror b and back

The distance travelled by light in this time is

$$L_2 = c\left(T_2 + T_2'\right) = 2D\frac{c}{\sqrt{c^2 - V^2}}. \tag{1.43}$$

The length difference of the two paths is

$$\Delta L = L_1 - L_2 = 2D\left(\frac{1}{1 - V^2/c^2} - \frac{1}{\sqrt{1 - V^2/c^2}}\right). \tag{1.44}$$

Expanding the two fractions in powers of the ratio V/c and keeping powers of this ratio up to and including the second, we have

$$\Delta L = 2D\left[\left(1 + \frac{V^2}{c^2}\right) - \left(1 + \frac{V^2}{2c^2}\right)\right] \tag{1.45}$$

and, finally,

$$\Delta L = D\frac{V^2}{c^2}. \tag{1.46}$$

This difference will determine the position of the interference fringes in the optical field of the interferometer's telescope. It is of course impossible to determine the absolute position of the fringes with such accuracy as to be able to check if this is in agreement with the theoretical prediction. We can, however, measure with great sensitivity a shift of the fringes. If we rotate the whole interferometer by 90° (Fig. 1.19), the path difference will change sign.

The total change in the path difference for the two orientations will be

$$2\Delta L = 2D\frac{V^2}{c^2}. \tag{1.47}$$

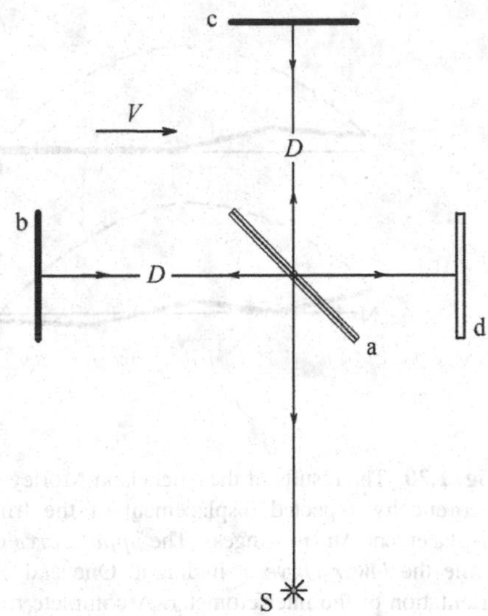

Fig. 1.19 The interferometer is rotated by 90° relative to its initial orientation (Fig. 1.16b). This will lead to a displacement of the interference fringes

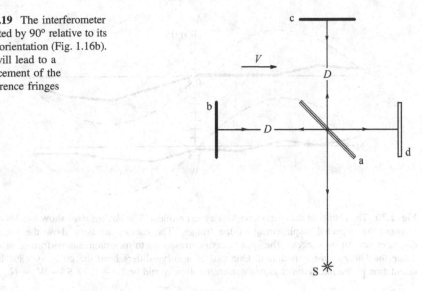

Given that a change in the path difference of the two beams by one wavelength λ of the light leads to a shift of the interference pattern by the distance between two adjacent bright fringes, which we call 'one fringe', for a change in the path difference by $2\Delta L$, the shift of the interference pattern in 'fringes' will be equal to

$$\Delta n = 2\frac{V^2}{c^2}\frac{D}{\lambda}. \tag{1.48}$$

The maximum shift between the two orientations of the interferometer will be observed in the case in which initially one of the arms was in the direction of the Earth's motion on its orbit and then it was rotated so that it is perpendicular to it. With the speed of the Earth on its orbit equal to $V \approx 10^{-4}c$, Eq. (1.48) gives for the maximum possible shift of the interference pattern, in fringes, the value of

$$\Delta n = 2 \times 10^{-8}\frac{D}{\lambda}. \tag{1.49}$$

In the initial series of measurements by Michelson, the length of the arms of the interference was such that $D = 2 \times 10^6 \lambda$ and, therefore, it was $\Delta n = 0.04$ of a fringe. This shift was difficult to detect, given the problems created by the vibrations of the apparatus.

In the second series of measurements, of Michelson and Morley, the length of the light's path was increased by a factor of 10 with multiple reflections and so it was $D = 2 \times 10^7 \lambda$. This would lead to a shift of the interference pattern by $\Delta n =$ 0.4 of a fringe as the interferometer was rotated. The rotation was now achieved with smaller vibrations because the block of granite on which the interferometer rested was floating in mercury. Michelson and Morley estimated that they would be

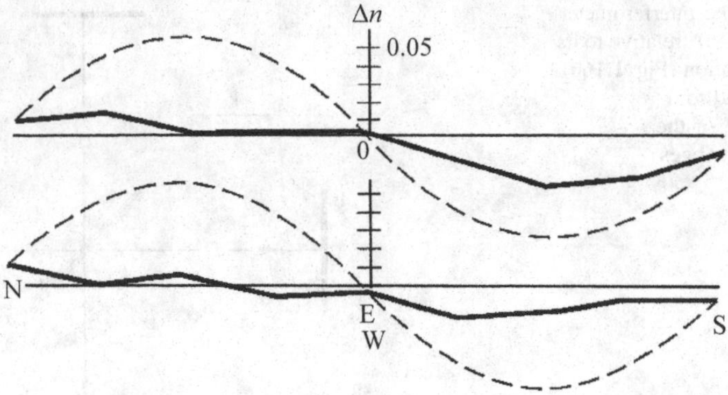

Fig. 1.20 The results of the Michelson-Morley experiment. The *dashed lines* show the 1/8 of the theoretically expected displacement of the fringes. The *continuous lines* show the measured displacement Δn (in fringes). The *upper curve* corresponds to measurements performed at noon, while the *lower curve* at midnight. One end of a curve differs from the other by 180° in the orientation of the interferometer. A complete rotation would be N → E → S → W → N

able to detect a shift as low as 1/100 of a fringe. Some of the results of the measurements of Michelson and Morley are shown in Fig. 1.20. The dashed lines show the fraction of 1/8 of the fringe displacement predicted theoretically. The continuous lines join the experimental points for the shifts observed. The upper curve corresponds to measurements performed at noon, while the lower corresponds to measurements made at midnight. One end of the curve differs from the other by 180° in the interferometer's orientation.

A complete rotation would correspond to N → E → S → W → N (N = North, E = East, S = South and W = West). It is obvious that the results are negative. They set an upper limit for any speed of the Earth relative to the aether, $\upsilon_{\text{Earth}-\text{aether}}$, equal to 8 km/s. The experiment has been repeated many times since, with light of various colors, light from the Sun and from the stars, at great heights in an air balloon, under the surface of the Earth, at different geographical regions and in different seasons. Some of the results of repetitions of the Michelson-Morley experiment by other observers are summarized in Table 1.1. All experiments gave negative results.[1] The conclusion is that the aether, which was considered necessary for the propagation of electromagnetic waves, does not exist. With it also disappears the absolute frame of reference used in Classical Physics.

[1]With the exception of one experiment! D.C. Miller reported [*Rev. Mod. Phys.* **5**, 203 (1933)] results of his experiments that lasted for a decade and showed a speed of the Earth relative to the aether equal to about 10 km/s. R.S. Shankland, S.W. McCuskey, F.C. Leone and G. Kuerty ["New analysis of the interferometric observations of Dayton C. Miller", *Rev. Mod. Phys.* **27**, 167 (1955)] showed that Miller's results were due to statistical fluctuations and variations in local temperature and that his results actually agree with those of the other researchers.

Table 1.1 Results of the repetitions of the Michelson-Morley experiment. [From the review article: Shankland et al., Rev. Mod. Phys., **27**, 167 (1955)]

Observers	l (m)	Δn_{theor}	Δn_{exp} (upper limit)	$\frac{\Delta n_{\text{theor}}}{\Delta n_{\text{exp}}}$	Upper limit $\upsilon_{\text{Earth-aether}}$ (km/s)
Michelson (1881)	1.2	0.04	0.02	2	15
Michelson and Morley (1887)	11.0	0.40	0.01	40	8
Morley and Miller (1902–4)	32.2	1.13	0.015	80	12
Miller (1921)	32.2	1.12	0.08	15	65
Miller (1923–4)	32.2	1.12	0.03	40	25
Miller (solar light, 1924)	32.2	1.12	0.014	80	11
Tomaschek (stellar light, 1924)	8.6	0.30	0.02	15	15
Miller (1925–6)	32.0	1.12	0.08	13	65
Kennedy (1926)	2.0	0.07	0.002	35	1.6
Illingworth (1927)	2.0	0.07	0.0004	175	0.3
Piccard and Stahel (1927)	2.8	0.13	0.006	20	5
Michelson et al. (1929)	25.9	0.90	0.01	90	8
Joos (1930)	21.0	0.75	0.002	375	1.6

The second column shows the total length of the light path, l, in the interferometer. As the interferometer moves with the Earth and rotates, the theoretically expected displacement of the interference pattern, in units of the width of a fringe, is Δn_{theor} and the upper limit set by the experiment is Δn_{exp}. Negative results are seen for values of $\Delta n_{\text{theor}}/\Delta n_{\text{exp}}$ up to 375.

1.11 The Lorentz-FitzGerald Contraction Hypothesis

In their effort to salvage the aether and interpret the negative results of the Michelson-Morley experiment, Lorentz [17] and FitzGerald [18] put forward an arbitrary hypothesis that bodies moving relative to the aether contract in the direction of their motion. If that arm of the interferometer of Michelson and Morley, which is orientated in the direction of the Earth's velocity $\mathbf{V}$ relative to the aether, contracts by a factor of $\sqrt{1 - V^2/c^2}$, the results of the experiment would be negative, as indeed they were. This factor is known as the *Lorentz contraction factor*.

As an explanation of length contraction, Lorentz put forward the idea that the electrostatic molecular forces that hold material bodies together might be affected by the motion of the body through the aether. The contraction hypothesis, although without any substantial physical foundation, saved the aether for an additional decade. Of course, it only interprets the results of experiments such as that of Michelson and Morley, in which the interferometer has arms perpendicular to each other and of approximately equal lengths. The contraction hypothesis would not, for example, interpret the results of the Kennedy and Thorndike experiment [19], which was performed much later, in 1932. In this experiment, the interferometer, similar to a Michelson interferometer, had arms which were not perpendicular to each other and had different lengths, as a result of which the light paths were of quite different lengths. The interferometer was stationary in the laboratory and changed orientation following the motions of the Earth. The interference fringes were photographed in order to record any shift they might suffer. Measurements were performed during different seasons, thanks to the stabilization of the temperature, which was kept constant to within 0.001°C. The results were negative, in agreement with the Michelson-Morley experiment. This, of course, would have put an end to the contraction hypothesis, had this survived for that long.

1.12 The Increase of the Mass of the Electron with Speed

With the discovery of radioactivity by Becquerel, in 1896, electrons became available for research, which had energies much higher than those that could be achieved at the time by accelerating them in electric fields. The use of these electrons led to the discovery of new phenomena in Physics. One very important such discovery was that of the observation by Kaufmann, in 1901 [20], of the variation of the inertial mass of the electron as a function of its speed. The experimental apparatus used by Kaufmann is shown in Fig. 1.21a. Between the poles, N and S, of a powerful electromagnet which produced a homogeneous magnetic field B_y, there were the plates of a charged capacitor, which created a homogeneous electric field, E_y. The two fields were parallel, in the y direction (say). A quantity of radium, which is radioactive, and was situated at a point P in the region of the fields, produced a thin, initially parallel beam of electrons moving in the z direction, which is normal to the plane of the figure and is directed outwards. The magnetic field caused a deflection of the electrons in the x direction, while the electric field caused them to deflect towards the positive plate of the capacitor, in the $\pm y$ directions, depending on the polarities of the plates. The electrons were recorded on a photographic plate, which was parallel to the plane of the figure and

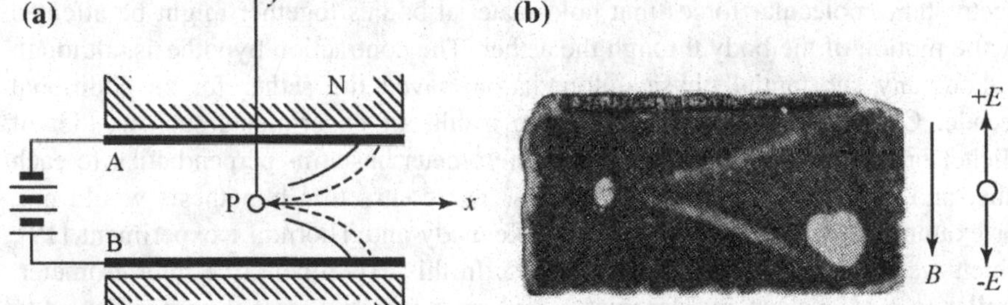

Fig. 1.21 **a** The experimental arrangement used for Kaufmann's measurements. **b** The photographic plate on which the dispersion of the beam's electrons was recorded. The energy spectrum of the electrons was continuous

in front of it. Figure 1.21b shows the image made by the electrons on the photographic plate, in Kaufmann's original experiment.

It can be easily shown that, an electron originally moving with speed v in the z direction, will suffer, after travelling a distance z, a deflection equal to

$$y = \frac{-e}{2m} E_y \frac{z^2}{v^2} \tag{1.50}$$

due to the electric field E_y and a deflection

$$x = \frac{e}{2m} B_y \frac{z^2}{v} \tag{1.51}$$

due to the magnetic field B_y.

If a beam consists of electrons with a continuous spectrum of speeds, these will suffer different deflections x and y each. The relationship between the corresponding x and y is found from Eqs. (1.50) and (1.51) by eliminating the velocity v between them. The relation

$$y = -\left(2 \frac{m}{e} \frac{E_y}{B_y^2 z^2}\right) x^2 \tag{1.52}$$

results, which shows that the beam of electrons with a continuous spectrum of speeds will be distributed on a parabola. It is known that the spectrum of the β particles from the radioactive source of radium used by Kaufmann is continuous and it was expected that the trace of the beam would be a parabolic curve. This, however, was found not to happen, as can be easily seen in Fig. 1.21b, by the fact that the two curves, obtained with opposite electrostatic fields, do not approach the point ($x = 0$, $y = 0$) tangentially to the x-axis and to each other (dashed lines in Fig. 1.21a, in contrast to the continuous lines which were observed). The point ($x = 0$, $y = 0$) is a very bright spot on the photographic plate, caused by the γ rays which are also emitted by the radioisotope and are not affected by the fields.

The curve obtained by Kaufmann is shown in Fig. 1.22 (continuous curve), together with parabolic curves from Eq. (1.52) corresponding to different values of the mass of the electron, in multiples of the mass m_0 of a low speed electron. The points of intersection of the parabola with the continuous, experimental, curve, give pairs of speed-mass for electrons of a certain speed. Also given in the figure is the axis of the kinetic energy of the electrons, based our knowledge of today for the properties of the electron. Kaufmann's results gave the first indication that the inertial mass of an electron increases with its speed.

Theoretical laws were proposed for the interpretation of these experimental results. In 1903, Abraham [21] assumed that the mass of the electron is purely of electromagnetic origin and that the rate of change of its momentum with time is equal to the total force applied on its own distribution of charge by the fields which

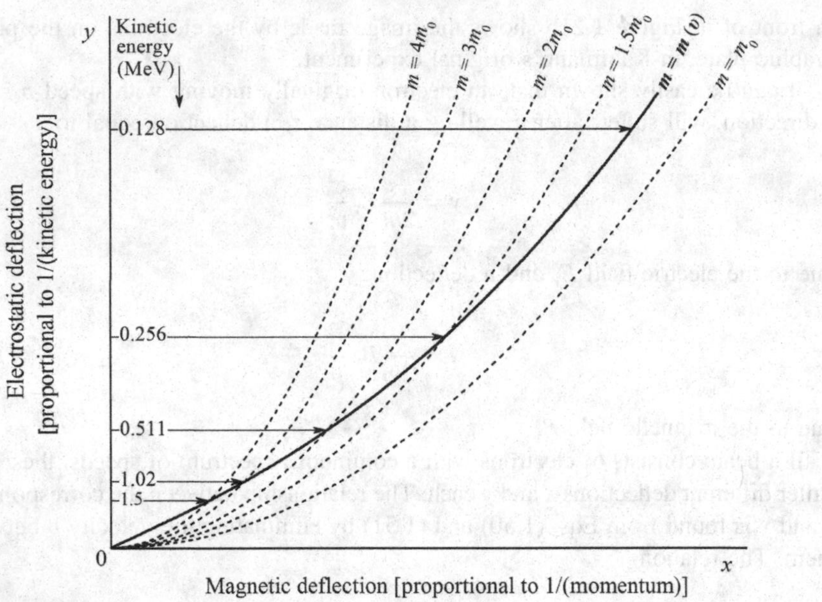

Fig. 1.22 Displacement y (due to the electric field) as a function of displacement x (due to the magnetic field), in Kaufmann's experiments (*continuous curve*). The parabolas in *dashed lines* show the expected displacements for various values of the electron mass, between m_0 and $4m_0$, where m_0 is the mass of a low speed electron. The vertical scale for the kinetic energies of the electrons is based on our present knowledge of the electron properties

the electron itself produces as it moves. Taking the electron to be completely spherical with its charge distributed uniformly on the surface of the sphere, he concluded that the mass of the particle depends on direction relative to its direction of motion. For the *transverse electromagnetic mass* of the electron, which is the one that is relevant in Kaufmann's measurements, he found the relation

$$m(v) = \frac{3m_0}{4\beta^2}\left[\frac{1+\beta^2}{2\beta}\ln\left(\frac{1+\beta}{1-\beta}\right) - 1\right], \tag{1.53}$$

where $m(v)$ is the mass at speed $v = \beta c$ and m_0 is the limit to which this mass tends to for low speeds.

In his theory of the electron, Lorentz [22] derived for the first time in 1904 the relation

$$m(v) = \frac{m_0}{\sqrt{1 - v^2/c^2}}. \tag{1.54}$$

For the derivation of this relation, he assumed that the electronic charge is uniformly distributed in a sphere which suffers contraction in its direction of motion, according to the Lorentz-FitzGerald contraction hypothesis. Lorentz compared the values of the ratio m/m_0 measured by Kaufmann with those predicted by Eq. (1.54), and found the agreement to be satisfactory, according to a publication of his in 1903.

A third theory, suggested in 1904 by Bucherer and Langevin (Alfred Heinrich Bucherer, 1863–1927, Paul Langevin, 1872–1946), assumes a spherical distribution of the charge of the electron, which contracts as in Lorentz's theory, but retains its volume constant. The result is

$$m(v) = \frac{m_0}{(1 - v^2/c^2)^{1/3}}. \tag{1.55}$$

The initial results of Kaufmann were not accurate enough for a decision to be made as to which of these theoretical predictions was the correct one. The issue was decided when more accurate measurements were performed by Kaufmann and others, later.

Bucherer, in 1909 [23], was the first to make very accurate measurements of the mass of the electron as a function of its speed. Again the source of electrons was a quantity of radium and the deflections occurred in mutually perpendicular electric and magnetic fields. Bucherer's method determined the variation of the ratio *e/m* as a function of the speed v. For theoretical reasons, the charge of the electron cannot vary with speed, something that has been verified experimentally with great accuracy later. The variation of the ratio *e/m* with the speed v is due, therefore, solely to the variation of the inertial mass of the electron. In Fig. 1.23 are also

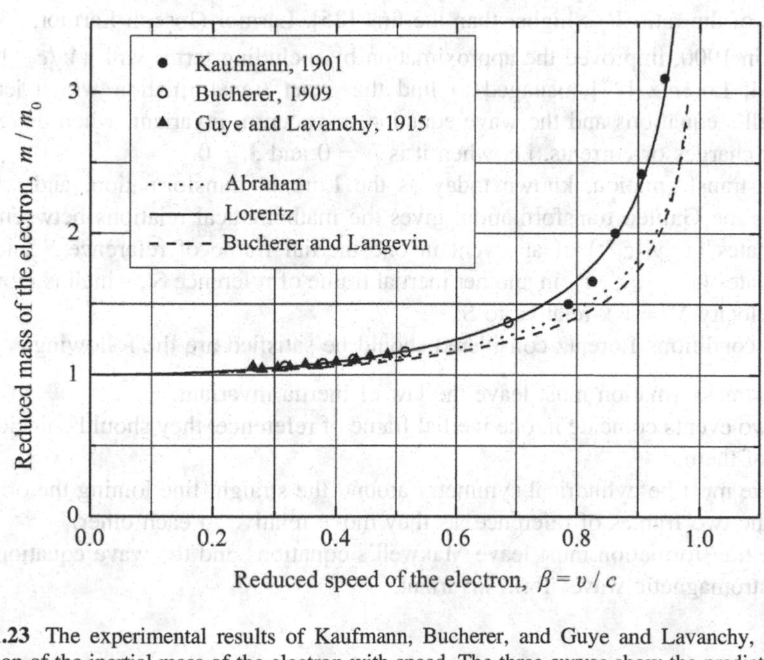

Fig. 1.23 The experimental results of Kaufmann, Bucherer, and Guye and Lavanchy, for the variation of the inertial mass of the electron with speed. The three curves show the predictions of the equations of Abraham, of Lorentz and of Bucherer and Langevin [Eqs. (1.53)–(1.55)]

presented, together with results of other experiments, the results of Kaufmann and Bucherer. Measurements by Bucherer for values $\beta = \upsilon/c > 0.7$ are not included, as their accuracy has been disputed.

Much more accurate measurements of the ratio m/m_0 were made in 1915 by Guye and Lavanchy [24]. They used a monoenergetic beam of electrons which were accelerated by a known potential difference. An electrostatic and a magnetic field were used for the determination of the ratio $m(\upsilon)/m(\upsilon_0)$ of the mass of the electrons $m(\upsilon)$ at speed υ to their mass $m(\upsilon_0)$ at a low reference speed υ_0. Some of the measurements of Guye and Lavanchy are presented in Fig. 1.23. With 2000 measurements of this ratio, for electron speeds between $\upsilon = 0.26\,c$ και $\upsilon = 0.48\,c$, the predictions of the Lorentz-Einstein equation, [Eq. (1.54)], were verified to an accuracy of 1 part in 2000 (or 0.05 %). The simpler equation finally proved to be the correct one!

1.13 The Invariance of Maxwell's Equations and the Lorentz Transformation

During the period of 1892–1904, Lorentz tried to find that transformation between two inertial frames of reference, one 'at rest' and one moving relative to the first with speed V, which would leave the laws of electromagnetism, as they are formulated in Maxwell's equations, invariant. In his initial work he ignored terms with powers of the ratio V/c higher than the first [25]. Larmor (Joseph Larmor, 1857–1942), in 1900, improved the approximation by including terms with $(V/c)^2$ [26]. In 1904, Lorentz [27] managed to find the exact transformation which leaves Maxwell's equations and the wave equation in vacuum, invariant, when there are no free charges or currents, i.e. when it is $\rho = 0$ and $\mathbf{J} = 0$.

This transformation, known today as the Lorentz transformation, and which replaces the Galileo transformation, gives the mathematical relations between the coordinates $(x,\ y,\ z,\ t)$ of an event in one inertial frame of reference S and its coordinates $(x',\ y',\ z',\ t')$ in another inertial frame of reference S′, which is moving with velocity $\mathbf{V} = V\hat{\mathbf{x}}$ relative to S.

The conditions Lorentz considered should be satisfied are the following:

1. The transformation must leave the law of inertia invariant.
2. If two events coincide in one inertial frame of reference, they should coincide in all of them.
3. There must be cylindrical symmetry around the straight line joining the origins of the two frames of reference, as they move relative to each other.
4. The transformation must leave Maxwell's equations and the wave equation for electromagnetic waves form invariant.

Naturally, in the classical limit the transformation must tend to that of Galileo. Conditions (1) and (2) were imposed for simplicity. From the first condition, there follows that the transformation is linear. The transformation has the form:

$$x' = \gamma\,(x - V\,t), \quad y' = y, \quad z' = z, \quad t' = \gamma\left(t - \frac{V}{c^2}x\right), \tag{1.56}$$

where $\gamma = 1\Big/\sqrt{1-(V/c)^2}$. In Appendix 4 it is proved that this transformation leaves Maxwell's equations invariant, with the assumption that the electric field **E**, the magnetic field **B**, the charge density ρ and the current density **J**, transform in a certain way. The transformation, therefore, also leaves the wave equation invariant. The reason Lorentz was not able to solve the complete problem and include in his solution the cases when space charges ρ and currents **J** are present, was the fact that he was unable to find how these magnitudes transform. This was achieved in 1905 by Poincaré [28], thus completing the transformations of coordinates and electromagnetic quantities that leave Maxwell's equation and the wave equation invariant. In the same work, Poincaré proves that the transformation has the properties of a group and uses, for the first time, the term, *the Lorentz transformation group*.

1.14 The Formulation of the Special Theory of Relativity

It must be clear to the reader by now that the approach of Physics to the Theory of Relativity stretches in time over a period of at least three centuries, ever since Galileo first stated the law of inertia and the principle of invariance of the laws of Physics for inertial frames of reference in relative motion. It is also obvious that a very great number of scientists contributed to this procedure. At the beginning of the twentieth century, the stage had been set and the ideas had matured for the formulation of the theory.

Already in 1899, Poincaré, commenting on the results of the Michelson-Morley experiment, expressed the opinion that 'it is very probable that the phenomena of Optics depend only on the *relative* velocities of the material bodies, the luminous sources and the experimental apparatus involved'. He also had the feeling that, not only by optical means, but with no physical method whatsoever is it possible to detect 'absolute' motion. In 1900 he disputed the existence of the aether as well as that it would ever be possible, using instruments of any sensitivity, to detect any but *relative* translational motions. In 1904, Poincaré formulated what he called *the principle of relativity* [29]:

> The laws of physical phenomena must be the same for a 'steady' observer as well for an observer in uniform translational motion.

He reached the logical conclusion that 'from all these results, a new kind of Dynamics will arise, that will be characterized by the fact that no speed will be greater than the speed of light in vacuum'. Poincaré came very close to formulating the Special Theory of Relativity. He did not, however, dare take the great conceptual step that was necessary.

At this point, the possibilities that presented themselves were the following:

1. Maxwell's equations are wrong. The correct electromagnetic theory, when discovered, would be invariant under the Galilean transformation.
2. Galileo's relativity applies to mechanical phenomena but not to electromagnetic ones, which required the existence of a privileged frame of reference, that of the aether.
3. There is a Theory of Relativity for all the physical phenomena, mechanical and electromagnetic, which is different than that of Galileo. In this case the laws of Mechanics must be modified accordingly.

It was then, in 1905, that Einstein's historical article was published [30]. Einstein adopted the third of the possibilities we mentioned. He rejected the concept of aether and he accepted a theory of relativity common to both mechanical and electromagnetic phenomena, which he based on two postulates. The postulates on which *The Special Theory of Relativity* is based are:

(a) *The laws of Physics are the same in all inertial frames of reference.*
(b) *The speed of light in vacuum is the same in all inertial frames of reference.*

The theory is called 'special' in the sense that it applies only to inertial frames of reference.

It appears that Einstein was not aware of the work of Lorentz and Poincaré on the subject. R. Shankland, in one of his publications in 1963, mentions that Einstein had told him during an interview he had with him in 1950, that, in 1905, he was not aware of the Michelson-Morley experiment [31]. Einstein based his arguments on the non-existence of the aether and, therefore, on the complete equivalence of all the inertial frames of reference and the experimental fact of the constancy of the speed of light in vacuum in all these frames of reference.

The consequences of these two postulates made by Einstein we will examine in the chapters to follow. We will very frequently refer to experiment for the foundation of the theory as well as the verification of its predictions. The subsequent development of the theory by the formulation of the General Theory of Relativity, will not concern us in this book.

References

1. E.T. Whittaker, *A History of the Theories of Aether and Electricity. Vol. 1: The Classical Theories* (2nd ed. 1951), *Vol. 2: The Modern Theories 1900–1926*, (Nelson, London, 1953). A. Pais, *'Subtle is the Lord...' The Science and the Life of Albert Einstein* (Clarendon Press, Oxford, 1982). O. Darrigol, *Electrodynamics from Ampère to Einstein* (Clarendon Press, Oxford, 2000)

2. G. Galilei, *Dialogue Concerning the Two Main World Systems—Ptolemaic and Copernican*, (1632). Passage quoted in V.A. Ugarov, *Special Theory of Relativity* (MIR Publishers, Moscow, 1979), p. 20
3. J. Bradley, Phil. Trans. Roy. Soc. **35**, 637 (1729). See also: A. Stewart, The Discovery of Stellar Aberration. Scientific American **210**(3), 100 (1964)
4. T. Tomson, *History of the Royal Society*, (London, 1812), p. 346
5. J. Mitchel, Phil. Trans. **74**, 35 (1784)
6. F. Arago, Compt. Rend. **8**, 326 (1839); **36**, 38 (1853)
7. J.G. Fox, Experimental evidence for the second postulate of special relativity. Am. J. Phys. **30**, 297 (1962)
8. V.C. Ballenegger, T.A. Weber, The Ewald-Oseen extinction theorem and extinction lengths. Am. J. Phys. **67**, 599 (1999)
9. A presentation of the main methods of measurement, up to about 1960, is given in: R.S. Longhurst, *Geometrical and Physical Optics* (Longmans, London, 1964), Chap. 24
10. A.J. Fresnel, Annls Chim. Phys. **9**, 57 (1818)
11. H. Fizeau, C.r. hebd. Seanc. Acad. Sci., Paris, **33**, 349 (1851)
12. H. Fizeau, Annln. phys. Chem. **3**, 457 (1853)
13. A.A. Michelson, E.W. Morley, Am. J. Sci. **31**, 377 (1886)
14. M. Hoek, Archs. Nèerl. Sci. **3**, 180 (1868)
15. A.A. Michelson, Am. J. Sci. **22**, 20 (1881)
16. A.A. Michelson, E.W. Morley, Am. J. Sci. **34**, 333 (1887)
17. H.A. Lorentz, Verh. K. Akad. Wet. **1**, 74 (1892)
18. O. Lodge, Phil. Trans. R. Soc. A **184**, 727 (1893)
19. R.J. Kennedy, E.M. Thorndike, Experimental establishment of the relativity of time. Phys. Rev. **42**, 400–418 (1932)
20. W. Kaufmann, Göttingen Nach. **2**, 143 (1901)
21. M. Abraham, Annln. Phys. **10**, 105 (1903)
22. H.A. Lorentz, Proc. Acad. Sci. Amsterdam **6**, 809 (1904). Reprinted in the book: A. Einstein, H.A. Lorentz, H. Weyl, H. Minkowski, *The Principle of Relativity* (Methuen and Co, London, 1923. Also, Dover Publications, New York, 1952)
23. A.H. Bucherer, Ann. d. Phys. **28**, 513 (1909)
24. C.E. Guye, C. Lavanchy, Compt. Rend. **161**, 52 (1915)
25. H.A. Lorentz, *Versuch einer Theorie der electrischen und optischen Erscheinungen in bewegten Korpern* (E.J. Brill, Leiden, 1895)
26. J. Larmor, *Aether and Matter* (Cambridge University Press, Cambridge, 1900)
27. H.A. Lorentz, Electromagnetic phenomena in a system moving with any velocity less than that of light. Proc. Acad. Sci., Amsterdam, **6**, 809 (1904). Also the collection of original publications: A. Einstein, H.A. Lorentz, H. Minkowski, H. Weyl, *The Principle of Relativity* (Methuen and Co, London, 1923. Also Dover Publications, New York, 1952)
28. H. Poincaré, Sur la dynamique de l'electron. C. R. Acad. Sci., Paris, **140**, 1504 (1905)
29. H. Poincaré, Bull. des Sc. Math. **28**, 302 (1904)
30. A. Einstein, *Ann. d. Phys.* **17**, 891 (1905). The English translation of the article "On the electrodynamics of moving bodies", can be found in the book: A. Einstein, H.A. Lorentz, H. Minkowski, H. Weyl, *The Principle of Relativity* (Methuen and Co, London, 1923. Also, Dover Publications, New York, 1952) which consists of a collection of original publications
31. For more on the subject see e.g. V.A. Ugarov, *Special Theory of Relativity* (MIR Publishers, Moscow, 1979), p. 345

Chapter 2
Prolegomena

2.1 Inertial Frames of Reference

The Special Theory of Relativity describes the physical laws that are valid in inertial frames of reference, i.e. frames of reference in which Newton's first law holds. Classical Mechanics deals mostly with inertial frames of reference. In these frames, a body on which the total external force applied is equal to zero is either stationary or moves with constant speed. There are, however, frames of reference in which, in order for Newton's first law to hold, it is necessary to include so called fictitious or d'Alembert forces, which have no apparent physical source of origin. To apply Newton's laws of motion on a rotating frame of reference, for example, we must assume the existence of fictitious forces, namely the centrifugal force and the Coriolis force [1] . These forces have no obvious physical source of origin, such as another body that exerts them, but depend entirely on the rotation of the frame of reference with respect to the rest of the matter in the universe.

Acceleration, in contrast to velocity, is an absolute magnitude. If we move with a uniform velocity, being acted on by zero forces, we are unable to sense this fact. Acceleration, on the other hand, is perceptible, due to the inertial properties of matter. If we try to accelerate a mass, we will feel a reaction to this effort of ours. Even in a closed room, we would be able to sense that the Earth rotates about its axis, either by the use of a Foucault pendulum or by studying the motion of a gyroscope. However, Galileo's principle of relativity, which we have mentioned in Sect. 1.2, considers the detection of absolute speed impossible.

If a body was the only body in the universe, its motion or its rotation would have no meaning, since these would not be able to be defined with reference to some other body. We are necessarily led to suspect that the inertial properties of matter are due to the other bodies present in the universe. These opinions were put forward by Mach, who disagreed radically with the concept of absolute space of Aristotle and Newton. We find ourselves in the apparently difficult position to need the rest of the matter in the universe for matter to have inertial properties, but have a body

© Springer International Publishing Switzerland 2016

C. Christodoulides, *The Special Theory of Relativity*,
Undergraduate Lecture Notes in Physics, DOI 10.1007/978-3-319-25274-2_2

on which no net external force is exerted, which would accelerate the body and make it a non-inertial frame of reference. We will assume that the matter in the universe is enough and isotropically distributed, so that the total force on a frame of reference we might choose is practically zero and the frame is inertial, but this matter imparts to the material bodies their inertial properties. It appears that mass in the universe is so sparsely distributed that we are able to regard a frame of reference as almost isolated if it is far enough from other bodies. We will see below to what degree this is possible, by examining various frames of reference which we might consider as even approximately inertial.

2.1.1 *The Earth, the Sun and the Galaxy as Inertial Frames of Reference*

The Earth is only approximately an inertial frame of reference, since it rotates about its axis and revolves around the Sun, a fact that leads to the appearance of fictitious forces when the Earth is assumed to be an inertial frame of reference. The centrifugal force is a fictitious force, which leads, in the case of the Earth, to a decrease of a body's weight as its geographical latitude is decreased. The Coriolis force is another fictitious force, to which cyclones and anticyclones are due in the motions of masses of air in the atmosphere. Depending on geographical latitude, the centrifugal acceleration at the surface of the Earth varies from zero at the poles up to a maximum value at the Equator. For a point on the Equator (Fig. 2.1), which is at a distance of $R_{\mathrm{E}} = 6.38 \times 10^6$ m from the axis of rotation of the Earth and has an angular velocity equal to that of the Earth, $\omega = 7.3 \times 10^{-5}$ rad/s, the centrifugal acceleration is $a_{\mathrm{I}} = 0.034$ m/s^2. The value of the centrifugal acceleration varies, therefore, from zero at the poles to this value at the Equator, which is equal to 0.4 % of the mean acceleration due to gravity, g, on the surface of the Earth, and is easily observable.

The Earth revolves about the Sun, at a mean distance of $D_{\mathrm{E}} = 1.50 \times 10^{11}$ m, with angular velocity equal to $\omega_{\mathrm{E}} = 2 \times 10^{-7}$ rad/s. This means that the centrifugal acceleration of the Earth relative to the Sun is equal to $a_{\mathrm{E}} = 6 \times 10^{-3}$ m/s^2.

The solar system revolves about the center of mass of the Galaxy, which is at a distance of $D_{\mathrm{S}} = 2.7 \times 10^{20}$ m, approximately. The speed of the solar system in its motion about the center of the Galaxy is about $\upsilon_{\mathrm{S}} = 220$ km/s. The angular velocity of the solar system about the center of the Galaxy is, therefore, $\omega_{\mathrm{S}} = 10^{-15}$ rad/s and its centrifugal acceleration, due to this motion, is of the order of $a_{\mathrm{S}} = 1.8 \times 10^{-10}$ m/s^2.

The Galaxy, in its turn, revolves about the center of mass of the group of galaxies to which it belongs. The centrifugal acceleration due to this motion is not known, but we may assume that it is much smaller than the last acceleration.

The conclusion we reach is that the solar system is a satisfactory inertial frame of reference for most applications. An even better inertial system is that defined by the

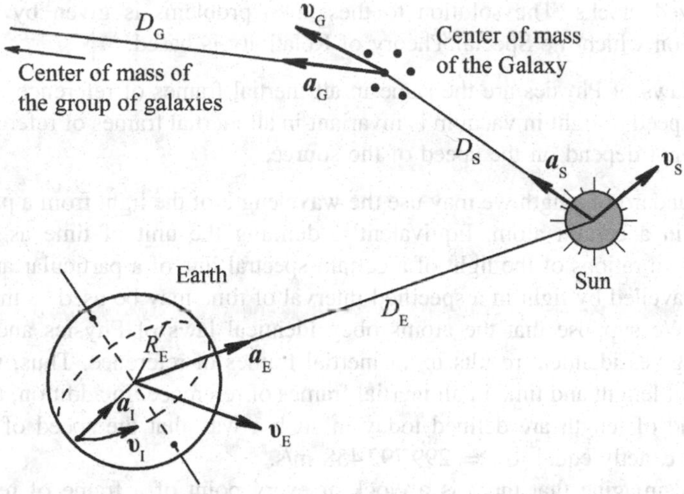

Fig. 2.1 The accelerations due to the rotation of the Earth around its axis, a_I, the revolution of the Earth about the Sun, a_E, the revolution of the Sun about the center of mass of our galaxy, a_S, and the revolution of our galaxy about the center of mass of the local group of galaxies to which it belongs, a_G

'distant fixed stars', i.e. the mean position of the distant galaxies. As we have already said, this is due to the fact that the distances between the galaxies are large enough and their distribution in space is isotropic enough.

In what will follow we will assume that it is possible for frames of reference to exist which are inertial to a satisfactory degree for most practical applications. Such would be frames of reference in interstellar space, at large distances from stars, and the infinite other frames of reference which move with constant velocities relative to these frames of reference.

2.2 The Calibration of a Frame of Reference and the Synchronization of Its Clocks

Something which occurs in a small enough volume and lasts for a short enough time we will call an *event*. The localization of an event in space and time requires giving its four coordinates (x, y, z, t). The Special Theory of Relativity deals with events in various frames of reference. It must be possible to measure lengths in every inertial frame of reference and for this we need a common length standard. In contrast to Classical Physics, time is not absolute and so we must also have the ability to measure time at all points of a frame of reference, with identical and synchronized clocks. The solution to these two problems is given by the two postulates on which the Special Theory of Relativity is based:

(a) The laws of Physics are the same in all inertial frames of reference, and
(b) The speed of light in vacuum is invariant in all inertial frames of reference and does not depend on the speed of the source.

As a standard of length we may use the wavelength of the light from a particular transition in a certain atom. Equivalently, defining the unit of time as a given number of vibrations of the light of a certain spectral line of a particular atom, the distance travelled by light in a specified interval of time may be used as the length standard. We suppose that the atoms obey identical laws of Physics and so our definition gives identical results in all inertial frames of reference. Thus, we have standards of length and time in all inertial frames of reference. In addition, the units of time and of length are defined today in such a way that the speed of light in vacuum is exactly equal to $c = 299\,792\,458$ m/s.

We will imagine that there is a clock at every point of a frame of reference, which shows the correct time of that particular frame of reference. This presupposes the *synchronization* of these clocks for every frame of reference. The method proposed by Einstein for the synchronization of the clocks of a system is the following: We define one of the clocks of the frame of reference as the reference clock, A (Fig. 2.2). From the point at which clock A is situated, a light pulse is emitted at the moment this clock's reading is t_1. The pulse propagates in vacuum to the point where clock B, which is to be synchronized with the reference clock A, is situated. The reading of clock B, t_B, at the moment of arrival of the pulse at B, is noted. The light is reflected, without delay, back towards A, where it arrives when clock A indicates time t_2. We suppose, in addition, that the speed of light from A to B is the same as the speed of light from B to A. Thus, we conclude that the time of arrival of the light at B is $t_1 + \frac{t_2 - t_1}{2} = \frac{t_1 + t_2}{2}$, when clock B had an indication t_B. The values of t_1 and t_2 may be sent to B for the necessary adjustment to be made in order that the two clocks are synchronized. The distance of clock B from clock A may be calculated from these measurements as being equal to $D_{AB} = (t_2 - t_1)c/2$. In the same way, all the clocks of the frame of reference may be synchronized and their positions found.

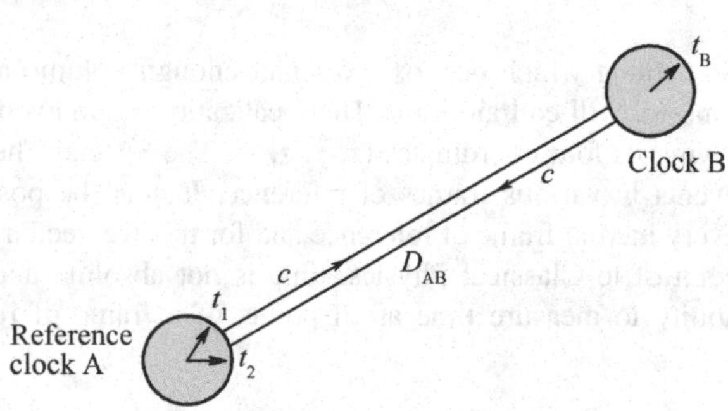

Fig. 2.2 The synchronization of the clocks and the calibration of distances in a frame of reference

The existence of these calibrated frames of reference is necessary in the Special Theory of Relativity. It will be found that the difference from Classical Physics is that every frame of reference has its own synchronized clocks which show the time of the frame of reference, and which may differ from those of other frames of reference.

2.3 The Relativity of Simultaneity

We will now examine one of the important consequences of the invariance of the speed of light: the collapse of the concept of simultaneity as this is known in Classical Physics.

Let us assume that a train is moving relative to an inertial observer O with constant speed V. We will assume that the train moves with a speed which is smaller than that of light in vacuum. Exactly at the middle of the train, and stationary relative to it, stands another observer O′. Two light sources, A and B, are situated at the two ends of the train. At the exact moment when O′ passes in front of O, two pulses of light arrive simultaneously at the two observers, having been emitted, respectively, one from source A and the other from source B (Fig. 2.3).

We examine first the train(!) of thought of observer O′, who is on the train and at its middle (Fig. 2.4):

'Points A and B are at equal distances $L_0/2$ from me, where L_0 is the length of the train. The speed of light is the same for both directions from A and B.

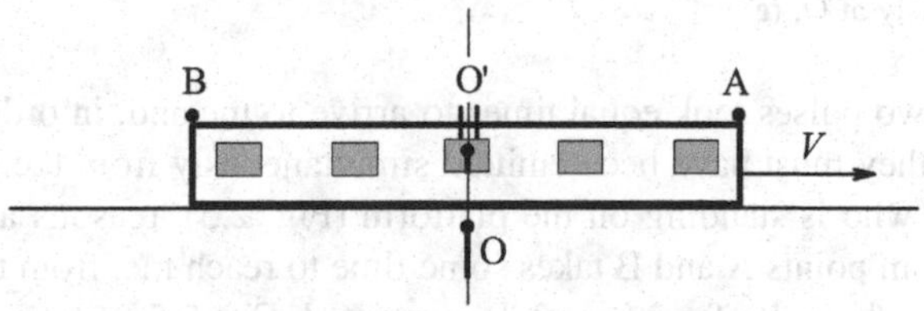

Fig. 2.3 A train moves with speed V relative to an observer O. At the middle of the train stands another observer, O′. When the two observers coincide, two pulses of light reach both of them, having been emitted, respectively, from the light sources A and B at the train's two ends

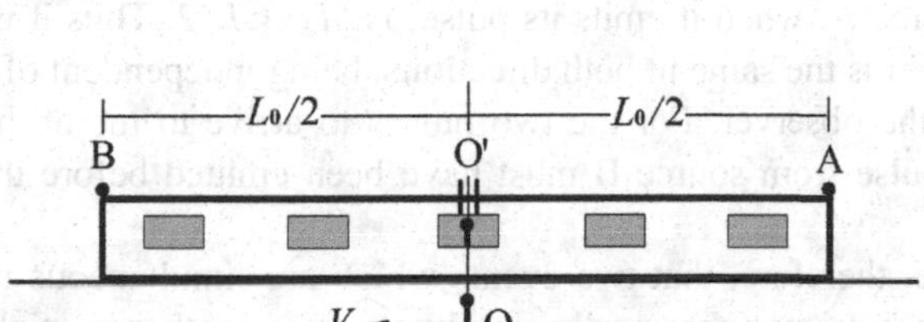

Fig. 2.4 Observer O′, on the train, concludes that the two pulses of light were emitted simultaneously from the two sources, A and B, which lie at equal distances from him

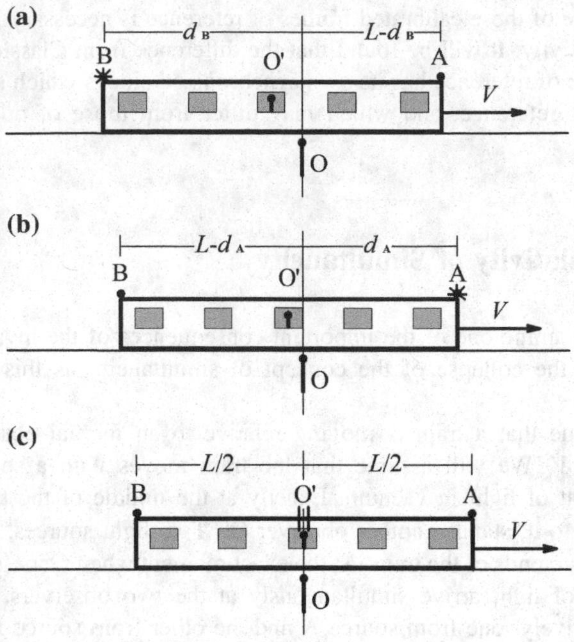

Fig. 2.5 Observer O, on the platform, concludes that the two light pulses were not emitted simultaneously from the two sources A and B. **a** When the pulse is emitted by source B, observer O′ has not yet reached observer O, and the distance of source B from O must be $|d_B| > L/2$, L being the length of the train as far as O is concerned. **b** When source A emits its pulse, observer O′ has still not reached observer O and source A must be at a distance from O which is $|d_A| < L/2$. It is, therefore, $|d_B| > |d_A|$. Thus, pulse B must have been emitted earlier than pulse A, since both arrive simultaneously at O, (**c**)

Therefore, the two pulses took equal times to arrive to me and, in order to arrive at the same time, they must have been emitted simultaneously from the two sources.'

Observer O, who is standing on the platform (Fig. 2.5), reasons as follows:

'The light from points A and B takes some time to reach me, from the moment it is emitted. When the pulse from source B is emitted, Fig. 2.5a, observer O′ has not yet reached me. Therefore, the distance from me of source B, when it emits its pulse, is $|d_B| > L/2$, where L is the length of the train. When the pulse from source A is emitted, Fig. 2.5b, observer O′ has still not reached me. Therefore, the distance from me of source A, when it emits its pulse, is $|d_A| < L/2$. Thus it is $|d_B| > |d_A|$. The speed of light is the same in both directions, being independent of the speeds of the sources or the observer. For the two pulses to arrive to me at the same time, Fig. 5.2c, the pulse from source B must have been emitted before the pulse from source A.'

We conclude, therefore, that two events which are simultaneous in one inertial frame of reference are not necessarily simultaneous in another inertial frame which is moving relative to the first. The invariance of the speed of light obliges us to abandon the concept of simultaneity which we take for granted in Classical Physics.

The reader must have noticed that, in the figures, the length of the train was taken as being different for the two observers: L_0 for the observer on the train and L for the observer on the platform. This is not accidental, as we will see in what follows. However, our ignorance, at present, of the exact lengths, does not influence the arguments developed above.

2.4 The Relativity of Time and Length

The invariance of the speed of light has very important consequences on time and length, as we will see in this section. We will begin by examining a quantity which remains invariant—any transverse dimension of a moving body. We will then examine the changes that have to be made to our understanding of time and length. We will return to these subjects in the next chapter, for a more general examination which will be based on the Lorentz transformation.

2.4.1 *The Invariance of the Dimensions of a Body Which Are Perpendicular to Its Velocity*

We examine two hollow cylinders, A and B (Fig. 2.6), which, when at rest, are identical, with radius r_0 and length L_0. Let the two cylinders be in relative motion with relative speed V, along their common axis. We will assume that to an observer a moving cylinder shrinks in directions perpendicular to its velocity, so that the radius of the moving cylinder is $r(V) < r_0$. The arguments that will follow would still hold had we assumed that the moving cylinder expands in directions perpendicular to its velocity. It will not affect our arguments, but we may assume that the

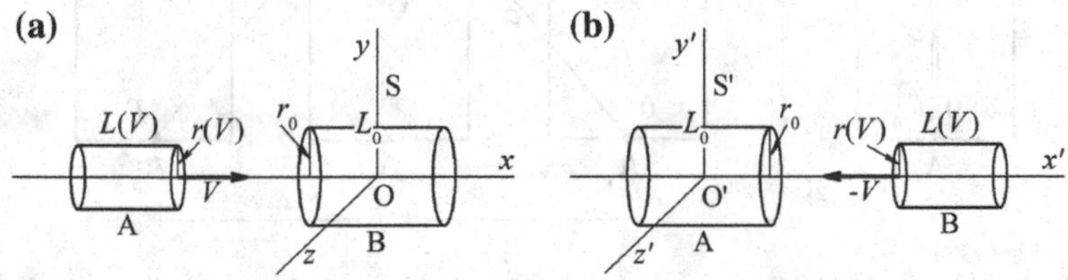

Fig. 2.6 Two identical *hollow cylinders*, A and B, in relative motion to each other, with their axes being coincident. The two cylinders, when at rest, have a radius equal to r_0. We will assume that when a cylinder is moving with speed V along its axis, its radius changes to $r(V)$ and its length to $L(V)$. Without loss of generality, we will take $r(V) < r_0$. In the reference frame of B, S, cylinder A moves with a velocity $\mathbf{V} = V\hat{\mathbf{x}}$ and its radius shrinks to $r(V)$. Cylinder A will, therefore, pass through cylinder B. In the reference frame of A, S′, cylinder B moves with a velocity $\mathbf{V} = -V\hat{\mathbf{x}}$ and its radius shrinks to $r(V)$. In the frame of reference S′, cylinder B will be expected to pass through cylinder A

length of the moving cylinder changes to $L(V)$. In the frame of reference S of cylinder B, cylinder A moves with a velocity $\mathbf{V} = V\hat{\mathbf{x}}$ and its radius is $r(V) < r_0$. Cylinder A will, therefore, pass through cylinder B (Fig. 2.6a).

In the frame of reference S′ of cylinder A, cylinder B moves with a velocity $\mathbf{V} = -V\hat{\mathbf{x}}$. Since space is isotropic, the radius of the moving cylinder will depend on its speed and so the radius of cylinder B will be $r(V) < r_0$. In this frame of reference, cylinder B is expected to pass through cylinder A (Fig. 2.6b).

Since it is impossible for both these predictions to be true, we must conclude that the transverse dimensions of a moving body do not change. This conclusion is purely a consequence of the non-existence of a privileged inertial frame of reference and the equivalence of all the inertial frames of reference.

2.4.2 *The Relativity of Time*

We will examine the rates of counting time by two clocks in relative motion to each other with a constant speed. The clocks we will consider work in the following way: Two mirrors, A and B in Fig. 2.7a, are at a distance D from each other. A very short light pulse (or a photon!) travels between the two mirrors. An observer, who is at rest relative to such a clock, will consider one unit of time, as measured by the clock, to be the time $T_0 = 2D/c$ needed for the pulse to travel from A to B and back to A, Fig. 2.7a.

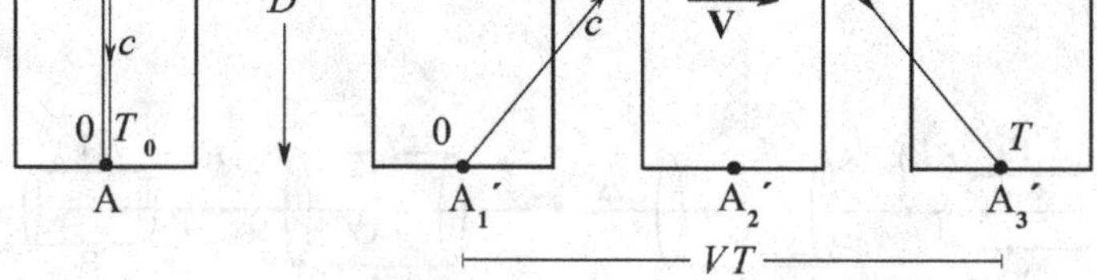

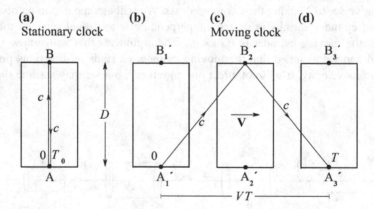

Fig. 2.7 *The dilation of time*. Two clocks, one at rest relative to an observer and one moving with a speed V relative to him. The operation of the clocks is based on a light pulse traveling between two mirrors, A and B (**a**). A unit of time (between two successive beats) is measured by the clock when the light pulse travels from A to B and then back to A. The moving clock moves in a direction perpendicular to the line joining the two mirrors of the clock at rest. The observer sees that the light in the moving clock has to cover a greater distance between successive beats than the light in the clock at rest relative to him (**b–d**). He therefore observes that more time passes between the successive beats of the moving clock as compared to the corresponding time in the clock at rest relative to him

Let now the observer who is at rest relative to a clock, observe another identical clock moving with speed V relative to himself in a direction perpendicular to the line AB, as shown in Figs. 2.7b, c, d. If for this observer the light needs a time T to travel from A to B and then back to A, the observer will see the moving clock shift by a distance $A_1'A_3' = VT$. The distance this observer will see as travelled by the light, always with a speed c, will be

$$cT = A_1'B_2'A_3' = 2\sqrt{D^2 + (A_1'A_2')^2} = 2\sqrt{D^2 + (VT/2)^2}, \tag{2.1}$$

where we have taken into account the fact we have already proved, that the distance D does not change due to the motion of the clock. But $D = cT_0/2$ and, hence, substituting, we have

$$c^2T^2 = c^2T_0^2 + V^2T^2, \tag{2.2}$$

from which we find that

$$T = \frac{T_0}{\sqrt{1 - V^2/c^2}}, \tag{2.3}$$

which is a time greater than T_0.

It follows that the observer sees the moving clock count time at a slower rate than that of the clock at rest relative to himself. A clock at rest registers that the time needed for light to travel the path ABA, from one of its mirrors to the other and back (*proper time*), is smaller by a factor of $\sqrt{1 - V^2/c^2}$ compared to the time that the clock at rest registers as needed for the light to travel from one mirror of the moving clock to the other and back. This phenomenon is known as *dilation of time*. We will return to it in the next chapter and in the one after, where we will also examine its experimental verification.

As a numerical example, consider two identical clocks, one at rest in a frame of reference S and one at rest in a frame of reference S′. S′ moves relative to S with a speed $V = (4/5)c$, for which it is $\sqrt{1 - V^2/c^2} = 3/5$. Let the clock in S′ count a time interval $T_0 = 1$ unit of time, as its light pulse moves from mirror A to mirror B and back. To an observer relative to which this clock is moving, the time taken for this process to be completed in the moving clock will be $T = T_0\big/\sqrt{1 - V^2/c^2} = 1/(3/5) = 5/3 = 1.67$ units of time. The observer in S sees that the moving clock needs 1.67 units of time to complete a procedure that corresponds to the counting of 1 unit of time. He, therefore, concludes that the moving clock is slow in counting time.

One may claim that the dilation of time was proved to apply only to a clock in which the light travels in a direction normal to the clock's velocity $\mathbf{V}$. We will now show that the same result holds for a clock at any orientation relative to $\mathbf{V}$. In Fig. 2.8 we have drawn two identical clocks, at rest in the same frame of reference,

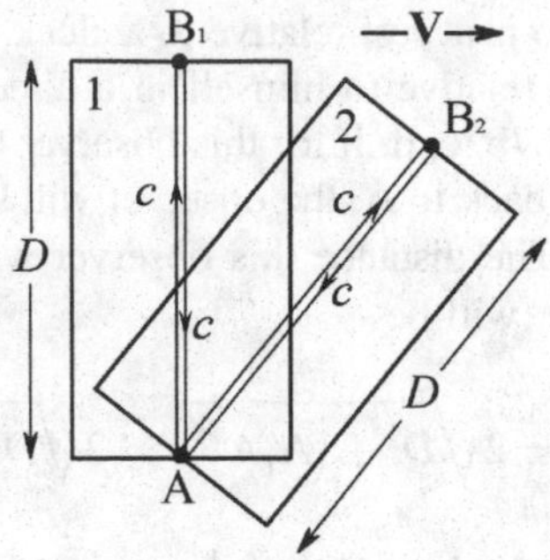

Fig. 2.8 Two identical clocks, similar to those of Fig. 2.7. The clocks are at rest in the same frame of reference and are oriented one (*1*) normal to the velocity **V** of an outside observer and the other (*2*) at some other angle. The two clocks emit their pulses of light at a common point, *A*

one (1) with its axis normal to **V** and the other (2) inclined at another angle. The two clocks have a common point of emission of their photons, A.

In the frame of reference of the clocks, if two pulses start simultaneously in the two clocks from point A, they will return to the same point simultaneously, because they have the same distance to travel, $2D$, and the speed of light is the same in all directions. In the frame of reference of the clocks, point A and the two pulses coincide initially and they will also coincide again when the two pulses return. It is obvious that the two pulses will also coincide at point A when viewed from any other frame of reference. If this did not happen, we might have a phenomenon, such as an interaction, in one frame of reference and not in the other. This would naturally be physically unacceptable. The two pulses will therefore coincide at point A in all other frames of reference as well. The rate at which the two clocks measure time will be the same and what is true for clock 1 will also be true for clock 2. The dilation of time is, therefore, the same, independent of the clock's orientation relative to the velocity **V**.

2.4.3 The Relativity of Length

We will now consider two clocks, similar to those of the previous section, one of which is at rest relative to some observer, and another that is moving with speed V, in a direction which is now parallel to the straight line AB between the two mirrors, as shown in Fig. 2.9. We must leave open the possibility that the length of the moving clock in the direction of its motion may differ from that of a stationary clock. If the distance measured by the observer between the two mirrors is L_0 for the clock at rest, let it be equal to L for the clock that moves relative to the observer (Fig. 2.9b, c, d).

In the clock which is at rest relative to the observer, light travels the distance ABA in time $T_0 = 2L_0/c$. We have already proved that the observer will measure

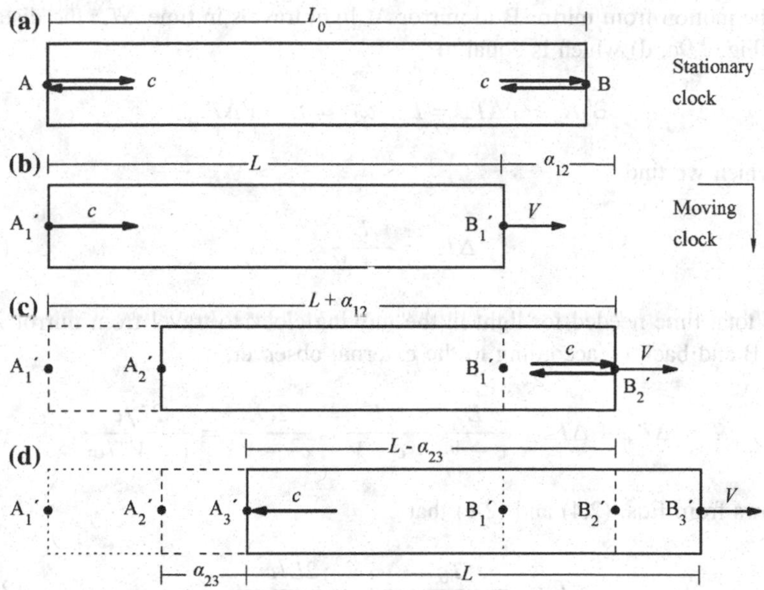

Fig. 2.9 *The contraction of length*. Two clocks, one at rest relative to an observer and one moving with a speed V relative to him. The operation of the clocks is based on a light pulse traveling between two mirrors, A and B. The moving clock moves in a direction parallel to the line AB. The distances travelled by the light in the one direction and in the other, as measured by the observer, are different in the moving clock

that for the moving clock the time needed for the light to travel from one mirror to the other and back is

$$T = \frac{T_0}{\sqrt{1 - V^2/c^2}}. \tag{2.4}$$

Alternatively, the observer sees light travel in the moving clock from mirror A to mirror B in time ΔT_+, covering a distance $A_1'B_2'$ (Fig. 2.9b, c) which, due to the motion of the clock with speed V, will be equal to

$$A_1'B_2' = c\Delta T_+ = L + \alpha_{12} = L + V\Delta T_+, \tag{2.5}$$

from which we find,

$$\Delta T_+ = \frac{L}{c - V}. \tag{2.6}$$

In the motion from mirror B to mirror A, light travels in time ΔT_- the distance $B_2'A_3'$ (Fig. 2.9c, d) which is equal to

$$B_2'A_3' = c\Delta T_- = L - \alpha_{23} = L - V\Delta T_-, \tag{2.7}$$

from which we find,

$$\Delta T_- = \frac{L}{c+V}. \tag{2.8}$$

The total time needed for light in the moving clock to travel from mirror A to mirror B and back is, according to the external observer,

$$T = \Delta T_+ + \Delta T_- = \frac{L}{c-V} + \frac{L}{c+V} = \frac{2cL}{c^2-V^2} = \frac{2L/c}{1-V^2/c^2}. \tag{2.9}$$

It follows from Eqs. (2.4) and (2.9) that

$$T = \frac{T_0}{\sqrt{1-V^2/c^2}} = \frac{2L/c}{1-V^2/c^2}. \tag{2.10}$$

However, it is $T_0 = 2L_0/c$, so that

$$\frac{2L_0/c}{\sqrt{1-V^2/c^2}} = \frac{2L/c}{1-V^2/c^2} \tag{2.11}$$

and, finally,

$$L = L_0\sqrt{1-V^2/c^2}. \tag{2.12}$$

This equation describes the phenomenon of the *contraction of length*, according to which an observer finds that the length of a body that is moving with a speed V relative to him is, in the direction of motion of the body, smaller by a factor $\sqrt{1-V^2/c^2}$ compared to the length of the same body in its own frame of reference, in which it is stationary (*rest length* or *proper length*). We will return to this effect in the next chapter.

2.5 The Inevitability of the Special Theory of Relativity

It is natural that someone learning about the Special Theory of Relativity for the first time will react to the proposition that he or she should accept the consequences of the theory. The conclusions of the theory are in direct contrast to our everyday experiences, given the limited means we have for acquiring them with our senses.

It must, however, be understood that the results produced by the Theory of Relativity are not a consequence of some choice we have made at some stage of the development of the theory. On the contrary, it is the way the natural world is and the only thing we do is to discover these physical laws, based on experiment and logical reasoning. So long as our knowledge was limited to the study of mechanical phenomena at low energies and low speeds, these laws remained unknown to us. When, however, we studied the laws of electromagnetism and of light, which is pre-eminently relativistic, the laws of relativity became inevitable. It is almost certain that, sooner or later, the theory will be modified to take into account new experimental results. At today's limits of energy, speed and phenomena observed, however, this improved theory must give the same results as today's theory does, for today's accuracy of observations. This is exactly what the Theory of Relativity does with respect to Classical Mechanics, with which it agrees at the limit of low energies and speeds. In the same way, Quantum Mechanics has Classical Mechanics as its limit macroscopically, and the General Theory of Relativity agrees, in the limit of weak gravitational fields, with the Newtonian theory of universal attraction.

To the question of whether things could have been different, the answer seems to be negative. Once it would have been unreasonable for an absolute frame of reference to exist in the universe, the Theory of Relativity is a necessity. It would have been impossible for Nature to describe which frame of reference should be privileged. It is, therefore impossible for such a privileged frame of reference to exist and all inertial frames of reference are equivalent. The Theory of Relativity necessarily follows.

With these facts in mind, perhaps it will be easier for the reader to accept the conclusions in what will follow, even if they radically disagree with our everyday experiences.

Reference

1. For a good introduction to the subject of fictitious forces and non-inertial frames of reference, see e.g. C. Kittel, W.D. Knight, M.A. Ruderman, A.C. Helmholz, B.J. Moyer, *Mechanics* (Berkeley Physics Course, vol. 1), (McGraw-Hill Book Company, 2nd ed., 1973), Chap. 4. For a more advanced analysis see e.g. K.R. Symon, *Mechanics* (Addison-Wesley, 3rd ed., 1971)

Chapter 3
Relativistic Kinematics

3.1 The Lorentz Transformation for the Coordinates of an Event

We will consider two inertial frames of reference, S and S′, in motion relative to each other. We choose the axes of the coordinate systems associated with the two frames of reference to coincide at some moment in time and define that moment to be, for the two frames, $t = t' = 0$. The velocity of the frame of reference S′ relative to S is constant and equal to $\mathbf{V} = V\hat{\mathbf{x}}$ (Fig. 3.1), where $\hat{\mathbf{x}}$ is the unit vector in the positive x direction.

In what will follow in this book, unless otherwise stated, the relationship between two inertial frames of reference, S and S′, will by that described above.

In the Special Theory of Relativity we will be dealing with events in four-dimensional space. Something which happens in a small enough space so that it can be characterized by a point and is localized in time so that it may be characterized by a single value of time we will call an *event*. An event will be represented by a point in four-dimensional space. Successions of points will be seen as successive events which may describe the development of a phenomenon in space and time.

A point P in four-dimensional space is characterized by its coordinates $(\mathbf{r}, t) = (x, y, z, t)$ in the frame of reference S, to which there also correspond its coordinates $(\mathbf{r}', t') = (x', y', z', t')$ in the frame of reference S′. If the point has a velocity $\boldsymbol{\upsilon}$ in S, the components of which are $(\upsilon_x, \upsilon_y, \upsilon_z)$, the point's velocity, $\boldsymbol{\upsilon}'$, in frame S′, will have components $(\upsilon'_x, \upsilon'_y, \upsilon'_z)$. Obviously, if (x, y, z) is the position of point P in the frame of reference S at time t, then

$$\upsilon_x = \frac{dx}{dt}, \quad \upsilon_y = \frac{dy}{dt}, \quad \upsilon_z = \frac{dz}{dt}, \tag{3.1}$$

© Springer International Publishing Switzerland 2016
C. Christodoulides, *The Special Theory of Relativity*,
Undergraduate Lecture Notes in Physics, DOI 10.1007/978-3-319-25274-2_3

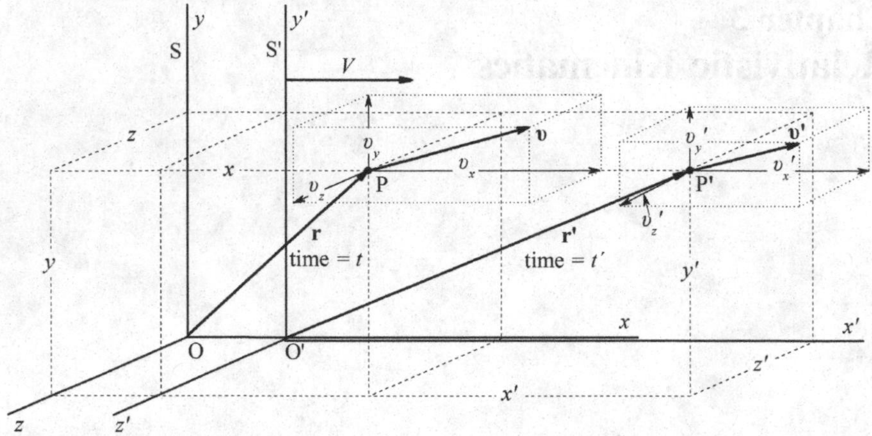

Fig. 3.1 Two inertial frames of reference, S and S′, in motion relative to each other. The velocity of frame S′ relative to frame S is constant and equal to $\mathbf{V} = V\hat{\mathbf{x}}$. At the moment $t = t' = 0$, the axes of the two frames of reference coincide. Shown in the figure, for a moving point P, are its position vectors, $\mathbf{r}$ and $\mathbf{r}'$, and its velocity vectors, $\boldsymbol{\upsilon}$ and $\boldsymbol{\upsilon}'$, in the two frames of reference. Also shown are the components of these. The figure was drawn so that all these components are positive

while in frame S′ it will be

$$\upsilon'_x = \frac{dx'}{dt'}, \quad \upsilon'_y = \frac{dy'}{dt'}, \quad \upsilon'_z = \frac{dz'}{dt'}. \tag{3.2}$$

Note: With reference to the mathematical notation, the components of speed in frame S′ should not be written as $(\upsilon'_x, \upsilon'_y, \upsilon'_z)$ but as $(\upsilon'_{x'}, \upsilon'_{y'}, \upsilon'_{z'})$, since in frame S′ the spatial axes are (x', y', z'). However, as we will be examining only transformations between two frames of reference S and S′ whose corresponding axes are parallel, we will avoid using primes in the indices x', y' and z', for simplicity.

For the two frames of reference, we assume a linear transformation $(x, y, z, t) \Leftrightarrow (x', y', z', t')$, of the form

$$x' = \alpha x + \varepsilon t, \quad y' = y, \quad z' = z, \quad t' = \delta x + \eta t, \tag{3.3}$$

where α, ε, δ and η are the coefficients of the transformation, which must be determined. Before we do that, however, we must justify two assumptions we have made:

(a) The transformation $(x, y, z, t) \Leftrightarrow (x', y', z', t')$ is linear, and
(b) For the choice of axes of S and S′ we have made, it must be $y' = y$ and $z' = z$.

The reasons that led us to these choices are the following:

(a) Had we chosen a non-linear transformation, there would appear in one frame of reference forces not present in the other. A motion with constant velocity in one frame of reference would appear accelerated in the other. Our choice also ensures that the transformation $(x, y, z, t) \Rightarrow (x', y', z', t')$ followed by the inverse transformation $(x', y', z', t') \Rightarrow (x, y, z, t)$, gives the initial coordinates.

(b) The result $y' = y$ and $z' = z$ follows from the proof we gave in Sect. 2.4.1 that the dimensions of a moving body which are perpendicular to its velocity remain unchanged.

We now proceed to the determination of the coefficients α, ε, δ and η of the transformation. We will base our arguments on the second postulate of the Special Theory of Relativity:

The speed of light in vacuum has the same value c in all inertial frames of reference.

From Eqs. (3.3), the infinitesimal variations in the two frames are related by the equations:

$$dx' = \alpha\,dx + \varepsilon\,dt, \quad dy' = dy, \quad dz' = dz, \quad dt' = \delta\,dx + \eta\,dt. \tag{3.4}$$

The components of velocity are, therefore,

$$\frac{dx'}{dt'} = \frac{\alpha\,dx + \varepsilon\,dt}{\delta\,dx + \eta\,dt}, \quad \frac{dy'}{dt'} = \frac{dy}{\delta\,dx + \eta\,dt}, \quad \frac{dz'}{dt'} = \frac{dz}{\delta\,dx + \eta\,dt}, \tag{3.5}$$

and, using Eqs. (3.1) and (3.2),

$$\upsilon'_x = \frac{\alpha\,\upsilon_x + \varepsilon}{\delta\,\upsilon_x + \eta}, \quad \upsilon'_y = \frac{\upsilon_y}{\delta\,\upsilon_x + \eta}, \quad \upsilon'_z = \frac{\upsilon_z}{\delta\,\upsilon_x + \eta}. \tag{3.6}$$

Now if a point P is at rest in frame S′, i.e. it is $\upsilon'_x = 0$, $\upsilon'_y = 0$, $\upsilon'_z = 0$, in frame S it will move with velocity $\upsilon_x \hat{\mathbf{x}} = \mathbf{V} = V\hat{\mathbf{x}}$. From the first of Eqs. (3.6), it follows that

$$0 = \frac{\alpha V + \varepsilon}{\delta V + \eta} \quad \text{or} \quad \frac{\varepsilon}{\alpha} = -V. \tag{3.7}$$

Similarly, if a point P is at rest in frame S, i.e. it is $\upsilon_x = 0$, $\upsilon_y = 0$, $\upsilon_z = 0$, in frame S′ it will move with velocity $\upsilon'_x \hat{\mathbf{x}} = -\mathbf{V} = -V\hat{\mathbf{x}}$. From the first of Eqs. (3.6), it now follows that

$$-V = \frac{\varepsilon}{\eta}. \tag{3.8}$$

Equations (3.7) and (3.8) are summarized in the relations

$$\frac{\varepsilon}{\alpha} = -V \quad \text{and} \quad \eta = \alpha. \tag{3.9}$$

The magnitudes of the velocity of point P in the two frames of reference,

$$\upsilon = \sqrt{\upsilon_x^2 + \upsilon_y^2 + \upsilon_z^2} \quad \text{and} \quad \upsilon' = \sqrt{\upsilon_x'^2 + \upsilon_y'^2 + \upsilon_z'^2},$$

are related through the equation

$$\upsilon'^2 = \frac{(\alpha\upsilon_x + \varepsilon)^2 + \upsilon_y^2 + \upsilon_z^2}{(\delta\upsilon_x + \eta)^2} \quad \text{or} \quad \upsilon'^2 = \frac{\upsilon^2 + [(\alpha^2 - 1)\upsilon_x^2 + 2\alpha\varepsilon\upsilon_x + \varepsilon^2]}{(\delta\upsilon_x + \eta)^2}. \quad (3.10)$$

Using Eqs. (3.9) and the second of Eqs. (3.10), we have

$$\upsilon'^2 = \frac{\upsilon^2 + [(\alpha^2 - 1)\upsilon_x^2 - 2\alpha^2 V\upsilon_x + \alpha^2 V^2]}{(\delta\upsilon_x + \alpha)^2}. \quad (3.11)$$

If the velocity of P in the one frame of reference has magnitude equal to c, it must also have magnitude equal to c in the other frame of reference as well. Equation (3.11) then becomes

$$c^2 = \frac{c^2 + [(\alpha^2 - 1)\upsilon_x^2 - 2\alpha^2 V\upsilon_x + \alpha^2 V^2]}{(\delta\upsilon_x + \alpha)^2}. \quad (3.12)$$

Expanding this relation and grouping the coefficients of the powers of υ_x, we have the identity

$$(\delta^2 c^2 - \alpha^2 + 1)\upsilon_x^2 + 2\alpha(\delta c^2 + \alpha V)\upsilon_x + (\alpha^2 c^2 - c^2 - \alpha^2 V^2) = 0, \quad (3.13)$$

which must be true for every value of υ_x. Equating to zero the coefficients of the powers of υ_x, we obtain the relations:

$$(\delta^2 c^2 - \alpha^2 + 1) = 0, \qquad 2\alpha(\delta c^2 + \alpha V) = 0, \qquad (\alpha^2 c^2 - c^2 - \alpha^2 V^2) = 0. \quad (3.14)$$

The third of these equations gives

$$\alpha = \pm\frac{1}{\sqrt{1 - V^2/c^2}}. \quad (3.15)$$

This, with the second of Eqs. (3.14) gives

$$\delta = \mp\frac{V/c^2}{\sqrt{1 - V^2/c^2}}. \quad (3.16)$$

Equations (3.15) and (3.16) also satisfy the first of Eqs. (3.14). The coefficients of the transformation are, therefore,

$$\alpha = \pm\frac{1}{\sqrt{1 - V^2/c^2}}, \quad \delta = \mp\frac{V/c^2}{\sqrt{1 - V^2/c^2}},$$
$$\varepsilon = \mp\frac{V}{\sqrt{1 - V^2/c^2}}, \quad \eta = \pm\frac{1}{\sqrt{1 - V^2/c^2}} \tag{3.17}$$

and the transformation of the position coordinates is:

$$x' = \pm\frac{x - Vt}{\sqrt{1 - V^2/c^2}}, \qquad y' = y, \qquad z' = z, \qquad t' = \pm\frac{t - (V/c^2)x}{\sqrt{1 - V^2/c^2}}. \tag{3.18}$$

Due to the fact that for $c \to \infty$ the transformation must be identical to the Galileo transformation and in the limit $V \to 0$ it must be $x' \to x$ and $t' \to t$, we must take the positive signs. We thus end up with *the special Lorentz transformation for the position coordinates of an event*, $(x, y, z, t) \Rightarrow (x', y', z', t')$:

$$x' = \frac{x - Vt}{\sqrt{1 - V^2/c^2}}, \qquad y' = y, \qquad z' = z, \qquad t' = \frac{t - (V/c^2)x}{\sqrt{1 - V^2/c^2}}. \tag{3.19}$$

The transformation is called *special* because it refers to the case when the axes of the two frames of reference are parallel and coincide at the moment $t = t' = 0$.

We denote the *reduced speed* of frame S′ relative to frame S by

$$\beta \equiv \frac{V}{c}. \tag{3.20}$$

It should be noted that the reduced speed of frame S relative to S′ is $-\beta$.

We also define the *Lorentz factor* corresponding to the speed V, as

$$\gamma \equiv \frac{1}{\sqrt{1 - V^2/c^2}}. \tag{3.21}$$

As already mentioned, with the term *Lorentz contraction factor* we refer to the factor $\sqrt{1 - V^2/c^2} = 1/\gamma$.

With these definitions, we have

$$x' = \gamma(x - \beta ct), \qquad y' = y, \qquad z' = z, \qquad t' = \gamma(t - (\beta/c)x). \tag{3.22}$$

The inverse transformation $(x', y', z', t') \Rightarrow (x, y, z, t)$ from frame S′ to frame S is found either by solving Eqs. (3.22) for x, y, z, t or by exchanging the primed variables with the unprimed ones and vice versa and putting $-V$ in place of V (or $-\beta$ in place of β). The transformation that results is:

$$x = \gamma(x' + \beta ct'), \qquad y = y', \qquad z = z', \qquad t = \gamma(t' + (\beta/c)x'). \tag{3.23}$$

Vectorially, for two frames of reference S and S′ which have their corresponding axes parallel, and with S′ moving with the general constant velocity **V** relative to S, we obtain the *general Lorentz transformation without rotation of the axes*,

$$\mathbf{r}' = \mathbf{r} + (\gamma - 1)\mathbf{V}\frac{\mathbf{V}\cdot\mathbf{r}}{V^2} - \gamma\mathbf{V}t, \qquad t' = \gamma\left(t - \frac{\mathbf{V}\cdot\mathbf{r}}{c^2}\right), \tag{3.24}$$

where $\mathbf{r} = (x, y, z)$ and $\mathbf{r}' = (x', y', z')$. The inverse transformation is

$$\mathbf{r} = \mathbf{r}' + (\gamma - 1)\mathbf{V}\frac{\mathbf{V}\cdot\mathbf{r}'}{V^2} + \gamma\mathbf{V}t', \qquad t = \gamma\left(t' + \frac{\mathbf{V}\cdot\mathbf{r}'}{c^2}\right). \tag{3.25}$$

It is worth noting that, as can be found by substitution from Eqs. (3.23), if in frame S it is

$$x^2 + y^2 + z^2 - c^2t^2 = 0, \tag{3.26}$$

which is the equation of a spherical wave front, emitted at $t = 0$, with center the origin of frame S and whose radius ct increases with a speed c, then in frame S′ it also is

$$x'^2 + y'^2 + z'^2 - c^2t'^2 = 0, \tag{3.27}$$

which is the equation of a spherical wave front, emitted at $t' = 0$, with center the origin of frame S′ and whose radius ct' increases with a speed c. In other words, a spherical wave front which is emitted from the (coinciding) origins of the axes of frames S and S′ at the moment $t = t' = 0$, will have in S as center the origin O of the axes and a radius which increases with a rate c, while in S′ it will have as center the origin O′ of the axes and a radius which increases with a rate c. It appears to us strange that this could happen, but it is as expected, because it constitutes the postulate of the invariance of the speed of light in vacuum, on which we based our derivation of the Lorentz transformation for the position coordinates of an event.

Example 3.1 An application of the Lorentz transformation

In the inertial frame of reference S, an event has coordinates ($x = 1$ m, $y = 2$ m, $z = 2$ m, $t = 1$ ns). Find the coordinates of the event in S′, which moves relative to S with constant velocity $\mathbf{V} = V\hat{\mathbf{x}}$, where $V = \frac{4}{5}c$. The axes of the two frames of reference coincided at $t = t' = 0$.

We apply the special Lorentz transformation, with $\beta = V/c = \dfrac{4}{5}$, $\gamma = 1/\sqrt{1 - V^2/c^2} = \dfrac{5}{3}$.

With these values and $x = 1$ m, $y = 2$ m, $z = 2$ m, $t = 1$ ns, Eqs. (3.22) give:

$$x' = \gamma(x - \beta ct) = \frac{5}{3}\left(1 - \frac{4}{5}(3 \times 10^8)(1 \times 10^{-9})\right) = \frac{5}{3}(1 - 0.24) = 1.27 \text{ m}$$

$$y' = y = 2 \text{ m}, \qquad z' = z = 2 \text{ m}$$

$$t' = \gamma(t - (\beta/c)x) = \frac{5}{3}\left(1 \times 10^{-9} - \frac{4}{5}\frac{1}{3 \times 10^8}\right) = \frac{5}{3}\left(1 - \frac{40}{15}\right) \times 10^{-9} \text{ s}$$
$$= -2.78 \text{ ns}.$$

It was found that the event which has coordinates ($x = 1$ m, $y = 2$ m, $z = 2$ m, $t = 1$ ns) in the inertial frame of reference S, has coordinates ($x = 1.27$ m, $y = 2$ m, $z = 2$ m, $t = -2.78$ ns) in S′. It is interesting to note that something which happened in frame S at the point ($x = 1$ m, $y = 2$ m, $z = 2$ m) at time $t = 1$ ns (i.e. 1 ns after the moment of coincidence of the axes of the two frames), in S′ it happened at the point ($x = 1.27$ m, $y = 2$ m, $z = 2$ m) at the time $t = -2.78$ ns (i.e. 2.78 ns before the moment of coincidence of the axes of the two frames).

Example 3.2 Finding the frame of reference in which two events are simultaneous

The coordinates of two events in an inertial frame of reference S are:

$$\text{Event 1}: x_1 = x_0, \quad y_1 = 0, \quad z_1 = 0, \quad t_1 = x_0/c.$$
$$\text{Event 2}: x_2 = 2x_0, \quad y_2 = 0, \quad z_2 = 0, \quad t_2 = x_0/2c.$$

There exists a frame of reference, S′, which moves at a constant velocity $\mathbf{V} = V\hat{\mathbf{x}}$ relative to S, in which these two events happen at the same moment. Find V. Assume that the axes of S and S′ coincide at $t = t' = 0$. At what time and at what points do these two events happen in S′?

Using the special Lorentz transformation, we find the coordinates of the two events in the frame of reference S′:

$$\text{Event 1:} \quad x_1' = \gamma(x_1 - Vt_1) = \gamma\left(x_0 - \frac{Vx_0}{c}\right) = x_0\gamma(1 - \beta) = x_0\sqrt{\frac{1-\beta}{1+\beta}}$$

$$t_1' = \gamma\left(t_1 - \frac{V}{c^2}x_1\right) = \gamma\left(\frac{x_0}{c} - \frac{Vx_0}{c^2}\right) = \frac{x_0}{c}\gamma(1-\beta) = \frac{x_0}{c}\sqrt{\frac{1-\beta}{1+\beta}}$$

$$\text{Event 2:} \quad x_2' = \gamma(x_2 - Vt_2) = \gamma\left(2x_0 - \frac{Vx_0}{2c}\right) = x_0\gamma\left(2 - \frac{1}{2}\beta\right)$$

$$t_2' = \gamma\left(t_2 - \frac{V}{c^2}x_2\right) = \gamma\left(\frac{x_0}{2c} - \frac{V}{c^2}2x_0\right) = \frac{x_0}{c}\gamma\left(\frac{1}{2} - 2\beta\right)$$

For $t'_1 = t'_2$ it must be $1-\beta = \frac{1}{2} - 2\beta$ or $\beta = -\frac{1}{2}$ and $\gamma = \frac{2\sqrt{3}}{3}$.

For $\beta = -\frac{1}{2}$, it is $t'_1 = t'_2 = \frac{x_0}{c}\sqrt{\frac{1-\beta}{1+\beta}} = \frac{x_0}{c}\sqrt{\frac{1+\frac{1}{2}}{1-\frac{1}{2}}}$ or $t'_1 = t'_2 = \sqrt{3}\frac{x_0}{c}$ is the time at which the two events happen in S′.

The positions at which the two events happen are

$$x'_1 = x_0\gamma(1-\beta) = \sqrt{3}x_0 \quad \text{and} \quad x'_2 = x_0\gamma\left(2 - \frac{1}{2}\beta\right) = x_0\frac{2+\frac{1}{4}}{\sqrt{1-\frac{1}{4}}} = \frac{3\sqrt{3}}{2}x_0.$$

It is $x_2 - x_1 = x_0$, while $x'_2 - x'_1 = \frac{3\sqrt{3}}{2}x_0 - \sqrt{3}x_0 = \frac{\sqrt{3}}{2}x_0 = \frac{x_0}{\gamma} = \frac{x_2 - x_1}{\gamma}$.

Example 3.3 Some relations concerning the Lorentz factor

Show that it is: (i) $\beta\gamma = \sqrt{\gamma^2 - 1}$, (ii) $d(\beta\gamma) = \gamma^3 d\beta$ and (iii) $c^2 d\gamma = \gamma^3 V dV = \gamma^3 \mathbf{V}\cdot d\mathbf{V}$

(i) From the definition $\gamma = \frac{1}{\sqrt{1-V^2/c^2}}$, it is $\gamma^2 = \frac{1}{1-V^2/c^2}$ and $\frac{V^2}{c^2} = 1 - \frac{1}{\gamma^2}$. Therefore, $\gamma^2 V^2 = c^2(\gamma^2 - 1)$ and, finally, $\gamma V = c\sqrt{\gamma^2 - 1}$ or $\beta\gamma = \sqrt{\gamma^2 - 1}$.

(ii) $d(\gamma V) = d\left(\frac{V}{\sqrt{1-V^2/c^2}}\right) = \left(\frac{1}{\sqrt{1-V^2/c^2}} + \frac{V^2/c^2}{(1-V^2/c^2)^{3/2}}\right)dV,$

and $d(\gamma V) = \left(\frac{1-V^2/c^2}{(1-V^2/c^2)^{3/2}} + \frac{V^2/c^2}{(1-V^2/c^2)^{3/2}}\right)dV = \frac{dV}{(1-V^2/c^2)^{3/2}} = \gamma^3 dV.$ Thus, $d(\beta\gamma) = \gamma^3 d\beta$.

(iii) $c^2 d\gamma = c^2 d\left(\frac{1}{\sqrt{1-V^2/c^2}}\right) = c^2\frac{V/c^2}{(1-V^2/c^2)^{3/2}}dV = \gamma^3 V dV$

Also, $c^2 d\gamma = \gamma^3 V dV = \frac{1}{2}\gamma^3 d(V^2) = \frac{1}{2}\gamma^3 d(\mathbf{V}\cdot\mathbf{V}) = \gamma^3 \mathbf{V}\cdot d\mathbf{V}$.

Problems

3.1 Show that for an event (x, y, z, t), the quantity $s^2 = c^2t^2 - x^2 - y^2 - z^2$ is invariant under the special Lorentz transformation. Also show that s^2 is not invariant under the Galilean transformation.

3.2 Show that if in an inertial frame of reference S it is $x = ct$ (light pulse), then in every other inertial frame of reference S′ it will be $x' = ct'$.

3.3 In an inertial frame of reference S, two events are separated by a distance of $\Delta x = 600$ m and an interval of time $\Delta t = 0.8$ μs. There exists a frame of reference S′, which moves with a constant velocity $\mathbf{V} = V\hat{\mathbf{x}}$ with respect to S, in which these two events happen at the same moment. Find the value of V. What is the spatial distance of the two events in S′? Ans.: $V = 0.4c$, $\Delta x' = 550$ m

3.4 Two events happen at the same point in an inertial frame of reference S. Show that the temporal sequence in which the two events happen is the same in all inertial frames of reference. Also show that the time interval between them is minimum in frame S.

3.5 Define the parameter ϕ, usually called *rapidity*, according to the relation $\tanh\phi = \beta$. Show that

$$\sinh\phi = \beta\gamma, \qquad \cosh\phi = \gamma, \qquad e^{\phi} = \sqrt{\frac{1+\beta}{1-\beta}}$$

and that the special Lorentz transformation for x and t may be written as $x' = x\cosh\phi - ct\sinh\phi, \qquad ct' = -x\sinh\phi + ct\cosh\phi$.

Also show that $ct' + x' = e^{-\phi}(ct + x)$ and $ct' - x' = e^{\phi}(ct - x)$.

3.6 Show that, as the speed υ approaches that of light in vacuum, c, the Lorentz factor corresponding to υ can be written, in S.I. units, as $\gamma \approx 12\,243/\sqrt{c-\upsilon}$, approximately.

3.7 Show that in the frames of reference S and S′, in relative motion with respect to each other, there exists a plane on which the clocks of the two frames show the same time. Find the positions of the plane in the two frames.

Ans.: In S: $x = \frac{c}{\beta}\left(1 - \frac{1}{\gamma}\right)t$. In S′: $x' = -\frac{c}{\beta}\left(1 - \frac{1}{\gamma}\right)t'$

3.8 A rectilinear rod is parallel to the x'-axis in the inertial frame of reference S′, which moves with a velocity $\mathbf{V} = V\hat{\mathbf{x}}$ relative to another frame of reference, S. In S′, the rod moves in the direction of positive y''s, with constant speed υ. The rest length of the rod is L_0 and its center remains on the y'-axis. Find the position of the rod in frame S, as a function of time.

Ans.: Position of rod's center: $x_{\mathrm{A}}(0) = Vt$, $y_{\mathrm{A}}(0) = \dfrac{\upsilon t}{\gamma}$.

Angle formed by rod with x-axis: $\theta = -\arctan\left(\dfrac{\upsilon V}{c^2}\gamma\right)$

3.1.1 The Contraction of Length

How can we measure the length of a rod in a frame of reference in which the rod is in motion? The measurement will have a meaning if performed 'instantaneously'. In other words, we will note the positions of the ends of the moving rod at the same moment and later we will measure, in this frame, the distance between these two points. We will assume that the rod is at rest in the inertial frame of reference S, is parallel to the x-axis and that in this frame it has length L_0. Length L_0 is known as the *rest length or proper length* of the rod (Fig. 3.2). We want to find the length of the rod in another frame of reference, S′, which is moving relative to S with velocity $\mathbf{V} = V\hat{\mathbf{x}}$.

In the frame of reference S, let the two ends of the rod, A and B, be, respectively, at the points x_{A} and x_{B}, where $L_0 = x_{\mathrm{B}} - x_{\mathrm{A}}$. In the frame of reference S′, at the moment t' the two ends of the rod are, respectively, at the points x'_{A} and x'_{B}. The length of the rod in frame S′ is equal to $L' = x'_{\mathrm{B}} - x'_{\mathrm{A}}$. The Lorentz transformation

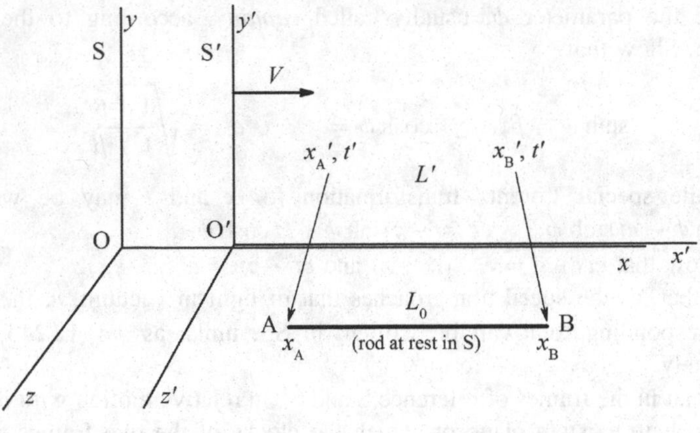

Fig. 3.2 The measurement of the length of a rod in two inertial frames of reference, S and S′, in relative motion to each other. In its own frame of reference, S, the rod has length L_0. In the frame of reference S′, which is moving relative to the rod, its length is $L' = L_0/\gamma$

gives, for the moment t', the relations $x_A = \gamma(x_A' + Vt')$ και $x_B = \gamma(x_B' + Vt')$. Subtracting the first of these equations from the second, we find

$$x_B - x_A = L_0 = \gamma(x_B' - x_A') = \gamma L'. \tag{3.28}$$

We notice that

$$L' = L_0/\gamma = L_0\sqrt{1 - V^2/c^2}. \tag{3.29}$$

This relation expresses the phenomenon of the *contraction of length*. The measurements give for the length of the moving rod a value which is smaller than its proper or rest length by a factor equal to the Lorentz contraction factor, $1/\gamma$. The rod has its maximum possible length, L_0, in that frame of reference in which it is at rest.

It must be understood that nothing happens to the rod due to its motion which causes its contraction. There is no change in its atomic structure, for example. The difference in the results of the measurements of the length in the two frames of reference may be understood if we examine the moments in time at which the measurements were made, as these are observed in the two frames of reference. In frame S′, the measurements at the points x_A' and x_B' were both performed at time t'. The times at which the measurements were seen to be performed in frame S, are, respectively,

$$t_A = \gamma\left(t' + \frac{V}{c^2}x_A'\right) \quad \text{and} \quad t_B = \gamma\left(t' + \frac{V}{c^2}x_B'\right). \tag{3.30}$$

We see that there is a time difference between the two measurements equal to

$$t_B - t_A = \frac{\gamma V}{c^2}(x'_B - x'_A) \quad \text{or} \quad t_B - t_A = \frac{V}{c^2} L_0. \tag{3.31}$$

If the observer in S is informed that the observer in S′ finds a length for the rod which is smaller than the rest length, he will not be surprised, since he observes that the measurement in S′ at the position x'_A of the leading end of the rod A, was performed earlier than the measurement at the position x'_B of the other end of the rod, B. In this time interval the rod moves towards the negative x''s in S′. It is reasonable, therefore, that a smaller length should be found for the rod. Length contraction, as well as other effects of Special Relativity, appear to stem from the disagreement between observers regarding the simultaneity of events.

Problems

3.9 An airplane has rest length equal to $L_0 = 10$ m. It moves, relative to an observer on the ground, with a speed equal to that of sound in air, $V = 343$ m/s. Calculate the contraction in the length of the airplane that the observer will measure.
Ans.: $\Delta L = 6.5 \times 10^{-12}$ m

3.10 At what speed does the length of a body contract to 99 % of its rest length?
Ans.: $\upsilon = 0.141c \approx c/7$

3.11 An observer S is stationary at the middle of a straight line AB, which has a length that the observer measures to be equal to $2L$. Another observer, S′, moves along the line with a speed of $V = \frac{3}{5}c$ relative to S. Both observers are at the origin of their respective frame of reference and coincide when their clocks show $t = t' = 0$.

(a) What is the length of AB as measured by S′? Ans.: $8L/5$

(b) In frame S, at the moment $t = 0$ two pulses are emitted simultaneously from the points A and B. Find, in the frame of reference of S′, the positions at which the pulses are emitted as well as the time at which each pulse is emitted.
Ans.: $x'_A = -(5/4)L$, $t'_A = (3/4)(L/c)$ and $x'_B = (5/4)L$, $t'_B = -(3/4)(L/c)$

(c) At which values of time, T_A and T_B, will the pulses arrive at S, and at which (T'_A and T'_B) will they arrive at S′? Ans.: $T_A = L/c$, $T_B = L/c$ and $T'_A = 2L/c$, $T'_B = L/2c$

3.12 A rod, having a rest length of L_0, lies along the x'-axis of the inertial frame of reference S′ and moves with a constant velocity $\mathbf{V} = V\hat{\mathbf{x}}$ with respect to another frame of reference, S. An observer in frame S notes the times t_A and t_B at which the two ends of the rod pass in front of him and estimates the length of the rod as $(t_B - t_A)V$. Show that the phenomenon of length contraction is exhibited in his result.

3.13 *The transformation of an angle*. A straight line lies in the xy plane and forms an angle θ with the x-axis, in the inertial frame of reference S. What is the angle θ' formed by the line with the x'-axis, in the frame of reference S′ which moves with a constant velocity $\mathbf{V} = V\hat{\mathbf{x}}$ relative to the frame S′? Ans.: $\theta' = \arctan(\gamma \tan\theta)$

3.14 A straight rod moves in an inertial frame of reference S with a constant velocity $\boldsymbol{\upsilon} = \upsilon\hat{\mathbf{x}}$, and has a slope λ with respect to the x-axis. Another frame of reference, S′, moves with a constant velocity $\mathbf{V} = V\hat{\mathbf{x}}$ in S. Show, by two methods, that, in frame S′, the slope of the rod with respect to the x'-axis is $\lambda\gamma(1 - V\upsilon/c^2)$.

Example 3.4 Where we decide to settle the question of length contraction once and for all by grabbing a moving rod at a certain moment, immobilizing it and measuring its length!

Assume that a rod is moving with a constant velocity in our frame of reference. We decide to measure the rod's length in our frame of reference by applying the following method: Along the line of motion of the rod, we place a large number of (very strong) observers, who have been instructed to grab and immobilize the point of the rod passing in front of them, at a certain moment in time. We then proceed to measure the length of the immobilized rod. What will actually happen?

Let a rod of proper length $2L_0$ move in our frame of reference, S, with a constant velocity $\mathbf{V} = V\hat{\mathbf{x}}$. According to theory, the rod will have a length $2L_0/\gamma$ in S, where $\gamma = \frac{1}{\sqrt{1-V^2/c^2}}$. We place a large number of observers along the x-axis, on which the rod is moving, at distances equal to $L_0/\gamma N$ between them, where N is an integer. This is shown in (a) of the figure that follows, where the observers are represented by short vertical lines.

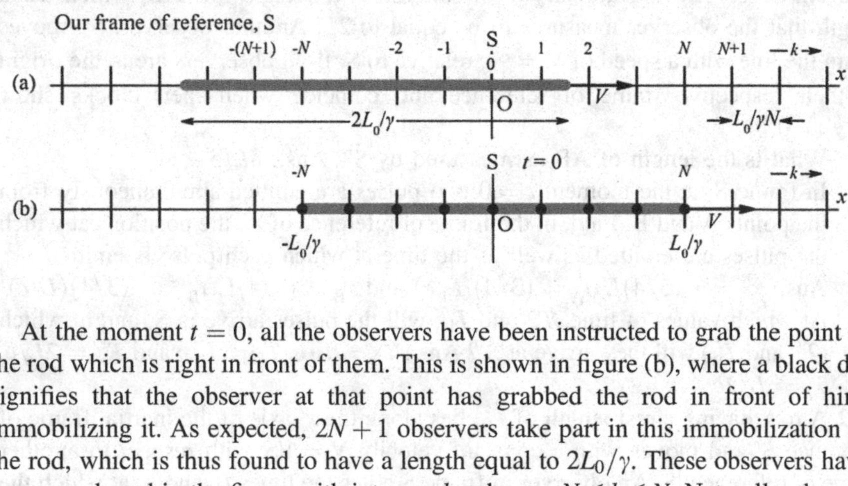

At the moment $t = 0$, all the observers have been instructed to grab the point of the rod which is right in front of them. This is shown in figure (b), where a black dot signifies that the observer at that point has grabbed the rod in front of him, immobilizing it. As expected, $2N + 1$ observers take part in this immobilization of the rod, which is thus found to have a length equal to $2L_0/\gamma$. These observers have been numbered in the figure with integers k, where $-N \leq k \leq N$. Naturally, the rod will in fact be in an extreme state of compression along its axis, something which no material would be able to withstand, for any substantial value of γ. We will ignore this difficulty!

To understand what really happened at $t = 0$, we will observe the whole procedure from another frame of reference, S′, in which the rod was initially at rest (see figure below). In this frame of reference, all the observers of frame S move with a velocity $-\mathbf{V} = -V\hat{\mathbf{x}}$. The distance between them is contracted by a factor of γ due to this motion in S′, becoming $L_0/\gamma^2 N$. The figure was drawn for $\gamma = 2$.

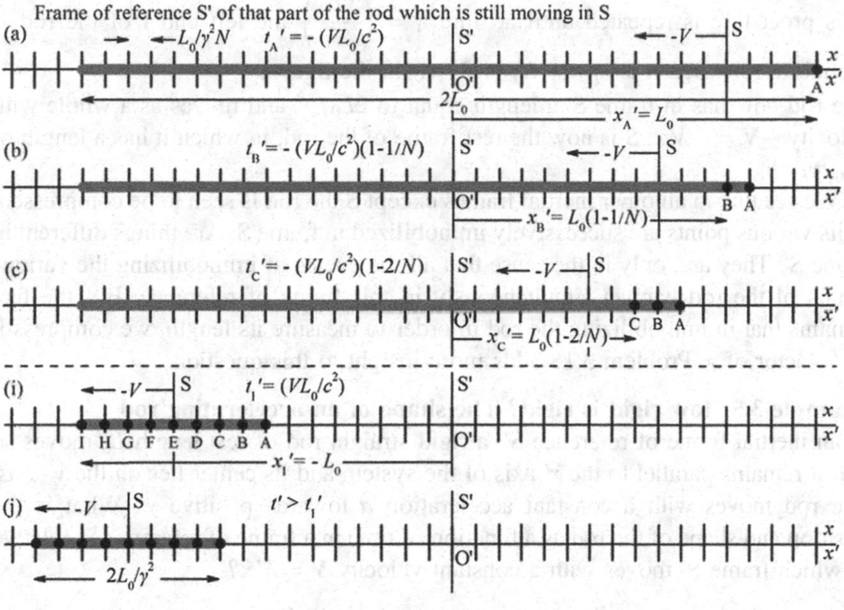

In frame S, we have $2N+1$ events happening at $t=0$, namely the immobilization of the bar by grabbing it at $2N+1$ equally spaced points. When and where do these events occur in S′? We use the Lorentz transformation to answer this question.
Frame S: $2N+1$ events happen at $t_k=0$, at the points $x_k = k\frac{L_0}{\gamma N}$, for $-N \le k \le N$.
Frame S′: these events happen at $t_k' = \gamma\left(t_k - \frac{V}{c^2}x_k\right) = -\gamma\frac{V}{c^2}x_k = -k\frac{V}{c^2}\frac{L_0}{N}$, at the points $x_k' = \gamma(x_k - Vt_k) = \gamma x_k = k\frac{L_0}{N}$.

We see that these events happen in frame S′ not simultaneously but with a time difference between successive events equal to $\Delta t' = t_{k+1}' - t_k' = -\frac{V}{c^2}\frac{L_0}{N}$ and at positions which differ from each other, successively, by a distance $\Delta x' = x_{k+1}' - x_k' = \frac{L_0}{N}$.

Step by step, what happened is this:

At time $t_A' = -\left(\dfrac{VL_0}{c^2}\right)$, observer A immobilizes the point of the rod in front of him, at $x_A' = L_0$. The rod is compressed at its right end as observer A moves [(a) in the figure]. At time $t_B' = -\left(\dfrac{VL_0}{c^2}\right)\left(1-\dfrac{1}{N}\right)$, observer B immobilizes the point of the rod in front of him, when it is $x_B' = L_0\left(1-\dfrac{1}{N}\right)$ in S′ [(b) in the figure].

The rod is compressed at its right end as observers A and B move. The part of the rod between A and B is now immobilized in frame S and it moves with velocity $-\mathbf{V} = -V\hat{\mathbf{x}}$ relative to frame S′, which is the frame of reference of the rest of the rod.

This procedure is repeated until at time $t'_I = \left(\frac{VL_0}{c^2}\right)$ the left end I of the rod is immobilized, when it is $x'_I = -L_0$ in S′ [(i) in the figure].

The rod now has in frame S′ a length equal to $2L_0/\gamma^2$ and moves as a whole with velocity $-\mathbf{V} = -V\hat{\mathbf{x}}$. S is now the rest frame of the rod, in which it has a length of $2L_0/\gamma$.

We see that in all other inertial frames except S the rod is seen to be compressed, as its various points are successively immobilized in frame S. Are things different in frame S? They are, only in the sense that all the events of immobilizing the various points of the rod happen simultaneously in this frame of reference. But the fact remains that in immobilizing the rod in order to measure its length, we compress it by a factor of γ. Problem 3.18 adds more insight to this question.

Example 3.5 How rigid is rigid? The shape of an accelerating rod
In an inertial frame of reference S′, a rigid straight rod of rest length L_0 moves so that it remains parallel to the x'-axis of the system and its center lies on the y'-axis. The rod moves with a constant acceleration a towards positive y'. What is the position and shape of the rod as a function of time in a frame of reference S, relative to which frame S′ moves with a constant velocity $\mathbf{V} = V\hat{\mathbf{x}}$?

In frame S′, the rod extends from end A at $x'_A = -L/2$ to end B at $x'_B = L/2$. Its center is at $x'_O = 0$. If we assume that at $t' = 0$ the rod coincided with the x'-axis, then for all the points of the rod it is $y' = \frac{1}{2}at'^2$. In the first figure we have drawn the position of the rod in frame S′ for six values of time t' between 0 and 5 ns. We have drawn the figure using $\frac{1}{2}a\gamma^2 = 1$ m/(ns)^2, with $\gamma = 5/3$.

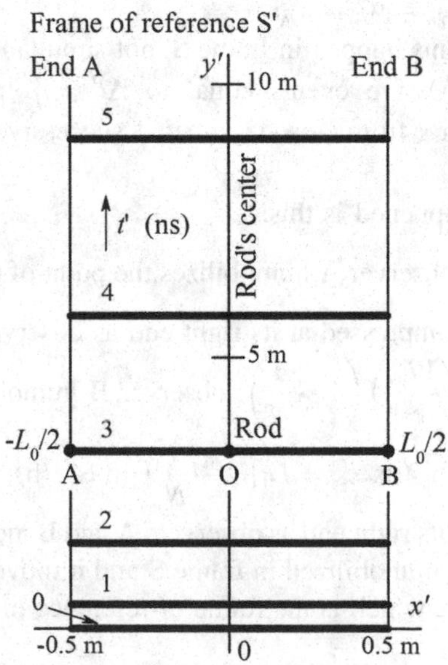

In frame S, the center of the rod will be at $x_O(t)$ and the rod's two ends at $x_A(t)$ and $x_B(0)$. Transforming for y' into S, we have $y = y' = \frac{1}{2}at'^2 = \frac{1}{2}a\gamma^2\left(t - \frac{V}{c^2}x\right)^2$. This function, $y(x)$, gives, for a given value of t and $x_A \leq x \leq x_B$, the shape of the rod in frame S. We notice that in S the rod lies on a parabola. This parabola retains its shape constant and its peak lies on the x-axis, along which it shifts with a speed c^2/V. Its axis is parallel to the y-axis.

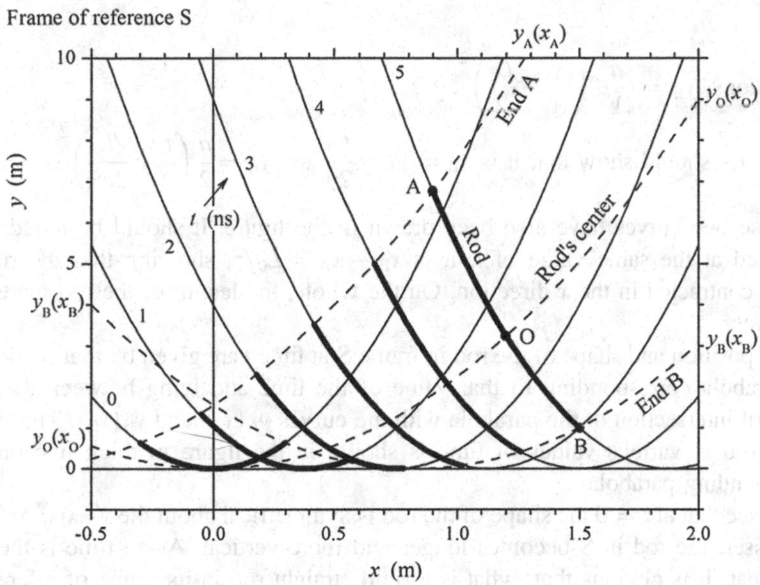

The second figure (above) shows the xy plane of frame S and has been drawn for $\beta = 4/5$, $\gamma = 5/3$, $V = 0.24$ m/ns, $L_O = 1$ m and $\frac{1}{2}a\gamma^2 = 1$ m/(ns)^2. Six parabolas have been drawn, for values of the time t between 0 and 5 ns. We wish to find the position of the rod on each one of these parabolas. To do this, we will need the loci $y_A(x_A)$ and $y_B(x_B)$ of the two ends of the rod, as well as $y_O(x_O)$, which is the curve on which what is the center of the rod in S′ moves in S.

Rod's center

The center of the rod has coordinates, in frame S′, $x'_O = 0$, $y'_O = \frac{1}{2}at'^2$.
In S, it is $y_O = y'_O = \frac{1}{2}at'^2$ and $x_O = \gamma\left(x'_O + Vt'\right) = \gamma Vt' = \gamma V\sqrt{2y_O/a}$. This is solved to give $y_O = \dfrac{a}{2\gamma^2 V^2}x_O^2$. This is a parabola in the xy plane, on which point O lies. The curve $y_O(x_O)$ has been drawn in the figure in dashed line.
The reader should show that, in terms of the time, it is $x_O = Vt$ and $y_O = \dfrac{a}{2\gamma^2}t^2$.

End A

The end A of the rod has coordinates, in frame S′, $x'_A = -\frac{1}{2}L_O$, $y'_A = \frac{1}{2}at'^2$.

In S, $x_A = \gamma\left(-\frac{L_0}{2} + Vt'\right)$ and $y_A = y'_A = \frac{1}{2}at'^2$, from which we find $t' = \sqrt{\frac{2y_A}{a}}$.

We substitute in x_A and solve to get $y_A = \frac{a}{2V^2}\left(\frac{x_A}{\gamma} + \frac{L_0}{2}\right)^2$.

The reader should show that it is $x_A = Vt - \frac{L_0}{2\gamma}$ and $y_A = \frac{a}{2}\left(\frac{t}{\gamma} + \frac{VL_0}{2c^2}\right)^2$.

End B

By analogy, $y_B = \frac{a}{2V^2}\left(\frac{x_B}{\gamma} - \frac{L_0}{2}\right)^2$.

The reader should show that it is $x_B = Vt + \frac{L_0}{2\gamma}$ and $y_B = \frac{a}{2}\left(\frac{t}{\gamma} - \frac{VL_0}{2c^2}\right)^2$.

These two curves have also been drawn in the figure. It should be noted that, estimated at the same value of t, it is $x_B - x_A = L_0/\gamma$, showing that the rod is simply contracted in the x direction. On the whole, the length of the rod increases with time.

The position and shape of the rod in frame S at time t are given by that section of the parabola corresponding to that value of the time and lying between the two points of intersection of the parabola with the curves $y_A(x_A)$ and $y_B(x_B)$. The shape of the rod at various values of time is shown in the figure in thick line on the corresponding parabola.

We see that at $t = 0$ the shape of the rod is symmetrical about the y-axis. As time progresses, the rod in S becomes longer and more vertical. At no time is the rod rectilinear. It is obvious that, what is a rigid straight rod in its frame of reference, appears as neither rigid nor straight in another frame of reference in relative motion with the frame in which the rod is in accelerated motion as described. This does not mean that the rod is not rigid. However, what appears as the rod in another frame of reference S is formed with geometrical points taken at different time moments in frame S′. These events are simultaneous in frame S but not in frame S′, leading to the deformation of the rod observed. Finally, it is interesting to notice that, in the example drawn, part of the rod near and including end B of the rod are actually moving in the negative y direction for the first 2 ns!

3.1.2 The Dilation of Time

We will compare the time intervals between two events measured by two observers in relative motion to each other. Let a clock be situated at the point x of the inertial frame of reference S. As two events we will consider the moments at which the clock shows time t_1 and t_2, respectively, Fig. 3.3. In the frame of reference S, the time interval between the two events is, therefore,

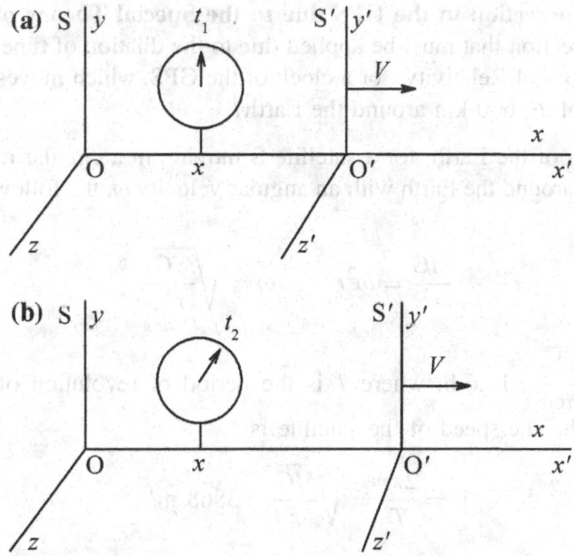

Fig. 3.3 The measurement of the time interval between two events in two inertial frames of reference, S and S′, in relative motion to each other. A clock is at rest at x in frame S. The first event is considered to be the moment the clock shows time t_1, (a). The second event is considered to be the moment the clock shows time t_2, (b). In the rest frame of the clock, S, the time interval measured is $\tau = t_2 - t_1$. In the other frame of reference, S′, which moves with constant velocity $\mathbf{V} = V\hat{\mathbf{x}}$ relative to S, the time interval measured is $\gamma\tau$

$$\tau = t_2 - t_1. \tag{3.32}$$

As the clock is at rest in the frame S, this time is called *rest time* or *proper time* between these two events.

For an observer in another frame, S′, which is moving with a velocity of $\mathbf{V} = V\hat{\mathbf{x}}$ relative to S and the clock, the times at which the two events will be observed will be, respectively,

$$t_1' = \gamma\left(t_1 - \frac{V}{c^2}x\right) \quad \text{and} \quad t_2' = \gamma\left(t_2 - \frac{V}{c^2}x\right). \tag{3.33}$$

Subtracting, we get for the interval of time between the two events, T', as this is measured by the observer in the frame of reference S′,

$$T' = t_2' - t_1' = \gamma(t_2 - t_1) \quad \text{or} \quad T' = \gamma\tau. \tag{3.34}$$

This relation expresses the phenomenon of the *dilation of time*. The smaller time, τ, between two events, is measured by a clock which is at rest in that frame of reference in which the two events happen at the same point. In other words: An observer sees the clocks which move relative to him to count time at a rate which is slower than that of the clocks which are at rest in his frame of reference.

Example 3.6 Correction in the GPS due to the Special Theory of Relativity What is the correction that must be applied due to the dilation of time predicted by the Special Theory of Relativity, for a clock of the GPS, which moves in a circular orbit of radius of 26 600 km around the Earth?

If M is the mass of the Earth, for a satellite S moving in a circular orbit of radius $r = 26\ 600$ km around the Earth with an angular velocity ω, the following relations will hold:

$$\frac{MG}{r^2} = \omega^2 r \qquad \omega = \sqrt{\frac{MG}{r^3}}$$

and $T = 2\pi\sqrt{\frac{r^3}{MG}} = 11.6$ h, where T is the period of revolution of the satellite around the Earth. The speed of the satellite is

$$V = \frac{2\pi r}{T} = \sqrt{\frac{MG}{r}} = 3868 \text{ m/s},$$

to which there correspond the values of

$$\beta = 0.000\ 129 \quad \text{and} \quad \gamma = 1.000\ 000\ 000\ 083.$$

Compared to a clock on the surface of the Earth, the clock in the satellite will fall behind, over a period of time equal to t, by a time interval equal to

$$\Delta t = t(1 - 1/\gamma),$$

according to the Special Theory of Relativity. In one day, the delay will be

$$\Delta t = 24 \times 60 \times 60 \times 8.3 \times 10^{-11} = 7.2 \times 10^{-6} \text{ s} = 7.2\,\mu\text{s}.$$

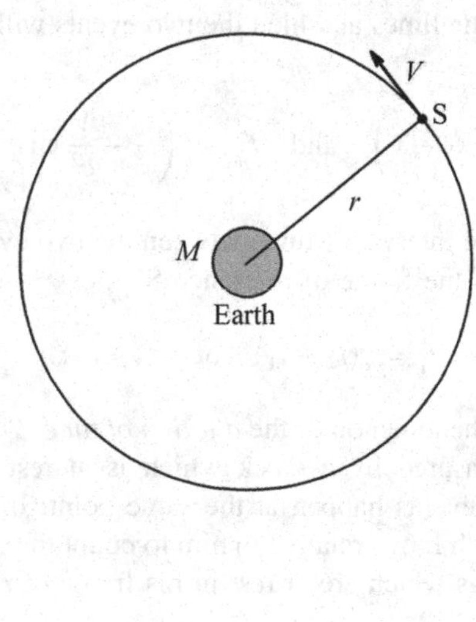

In this time interval light travels a distance equal to $\Delta s = c\Delta t = 2151\ \text{m} \approx 2\ \text{km}$ and this will be the error that would be added for every day without correction, to the error in the position of a point on the Earth, as this is determined by the GPS.

Example 3.7 A moving array of clocks
Synchronized clocks are situated on the x'-axis of an inertial frame of reference S′ at constant distances s between them. Frame S′ moves with a constant velocity $\mathbf{V} = V\hat{\mathbf{x}}$ relative to another frame of reference S. What will someone in frame S observe about the behavior of the clocks?

In the frame of reference S′, the clocks are situated at the positions $x'_n = sn'$, where n' is an integer $-\infty < n' < \infty$. These clocks are synchronized in their own frame of reference, S′, as they are shown at time $t' = 0$ in the first figure. Apart from the clocks' hands, which indicate the time, the clocks of the figures also have digital time displays and indicators of the position of each one of them in frame S′.

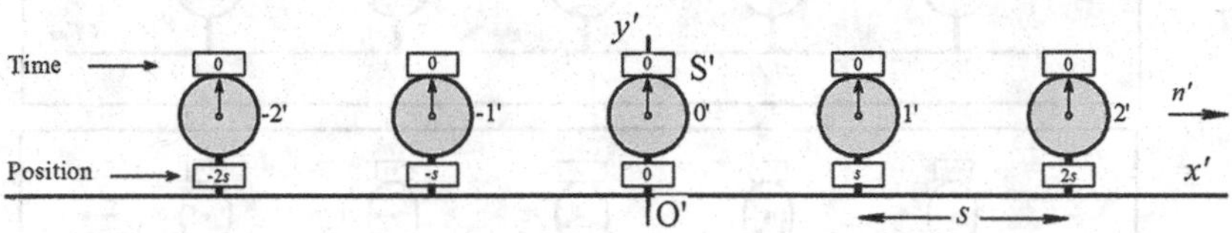

Using the Lorentz transformation, $x = \gamma(x' + \beta ct')$ and $t = \gamma[t' + (\beta/c)x']$, we find, in frame S, for the n'-th clock: $x_{n'} = \gamma\left(sn' + \beta ct'_{n'}\right)$, $t_{n'} = \gamma\left[t'_{n'} + (\beta/c)sn'\right]$. For a certain moment of time $t_{n'} = t$, it is $x_{n'} = \gamma\left(sn' + \beta ct'_{n'}\right)$ and $t = \gamma\left[t'_{n'} + (\beta/c)sn'\right]$. Solving, we find $x_{n'} = Vt + \dfrac{s}{\gamma}n'$ and $t'_{n'} = \dfrac{t}{\gamma} - \dfrac{\tau}{\gamma}n'$, where we have defined the time constant $\tau \equiv \dfrac{\gamma}{c^2}Vs$.

Given in the figures that follow, for three values of time t in frame S ($t = 0,\ \ t = \tau,\ \ t = 2\tau$), are the positions of the clocks as observed by observers in frame S, as well as the times these clocks appear to indicate. The clocks of frame S′ are characterized by their increasing number n'. For comparison, at the positions of these clocks, are also drawn the corresponding clocks of frame S. It is of paramount importance to realize that (here) the upper parts of the figures show the pictures seen by observers in the frame of reference S at the moment t and not the situation in frame S′ at any particular moment in time. For this reason the clocks of frame S′ are drawn contracted, in the direction of their motion, as they should appear to observers in frame S.

[In order to simplify the figures, we have chosen $V\tau = s/\lambda$, which means that $\beta = \sqrt{(\sqrt{5}-1)/2} = 0.786$ and $\gamma = \sqrt{(3+\sqrt{5})/2} = 1.62$.]

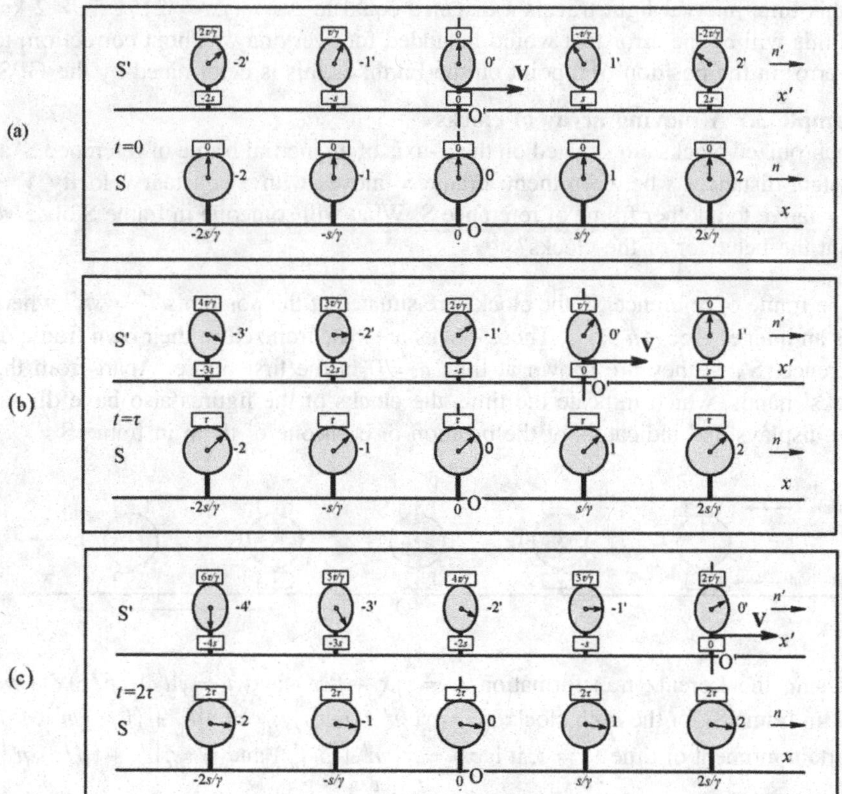

We notice that in frame S the clocks which coincide with those of S′, are at distances s/γ between them (contraction of length). Also, when in frame S time τ passes, it is observed from frame S that for all the clocks of frame S′ the time that passes is equal to τ/γ (dilation of time).

Something that is very important to mention is the fact that, to an observer in frame S, the clocks of frame S′ at no moment appear to be all synchronized with each other.

Example 3.8 A thin luminescent rod

A thin rod with rest length equal to L_0, moves with a constant velocity $\mathbf{V} = V\hat{\mathbf{x}}$ on the x-axis of an inertial frame of reference S. Let S′ be the frame of reference of the rod. Simultaneously in S′, at the moment $t' = 0$, a large number of points on the rod emit a photon each. These photons escape without being scattered anywhere. Explain what the observers at rest in frame S will see.

In frame S′, the rod is at rest and stretches from point $x_2' = -L_0/2$ (back end) to point $x_1' = L_0/2$ (leading end). The midpoint of the rod is at point $x_0' = 0$. At the moment $t' = 0$ all the photons are emitted simultaneously from the rod.

If an observer in frame S could see the whole train at the moment $t = 0$, he would see the midpoint of the rod at the point $x_0 = 0$, its back end at the point $x_2 = -L_0/2\gamma$ and its front end at the point $x_1 = L_0/2\gamma$. He would see the total length of the rod to be equal to $L = L_0/\gamma$, due to the contraction of length.

The observer in frame S will see, however, every point of the rod when the photon emitted by it will reach him. We assume that the observer in frame S makes the necessary corrections for the time needed for the light from each point to reach him, and so he may calculate the time of emission of each photon in the frame S.

The emission of a photon from the point x' of S′ at the moment $t' = 0$ will happen in frame S at point x at the moment t. Putting $t' = 0$ in the Lorentz transformation, we find

$$x = \gamma x', \qquad t = \gamma(\beta/c)x'.$$

For example, in frame S,

the back end of the rod, $x_2' = -L_0/2$, will emit its photon at point $x_2 = -\gamma L_0/2$ at the moment $t_2 = -\gamma(\beta/c)L_0/2$,

the midpoint of the rod, $x_0' = 0$, will emit its photon at point $x_0 = 0$ at the moment $t_0 = 0$,

the front end of the rod, $x_1' = L_0/2$, will emit its photon at the point $x_1 = \gamma L_0/2$ at the moment $t_1 = \gamma(\beta/c)L_0/2$.

In the frame of reference S, the photon emission happens first at the back end of the rod, then at its midpoint and, finally, at the leading end of the rod.

The observer in frame S will calculate that, at every moment, photons are emitted from the points of a plane of constant x, normal to $\mathbf{V}$, which appears to move from the back end of the rod to the front, with constant speed

$$V_\phi = \frac{x}{t} = \frac{\gamma x'}{\gamma(\beta/c)x'} = \frac{c}{\beta}.$$

This speed is always greater than c. It must be stressed, of course, that we are dealing with a kind of 'phase velocity', since neither mass nor information is transmitted by this 'wave'.

If the observer in frame S 'photographs' the rod from a large distance, leaving the camera's shutter open until all the incoming photons are registered, reaching the camera at different times, the image of the rod will appear to correspond to a rod of length

$$L_\phi = x_1 - x_2 = \gamma L_0,$$

i.e. greater than the rest length L_0 of the rod, by a factor γ. This does not violate the prediction for length contraction, because the various points of the train become visible at different times in frame S.

If the observer in S filmed what was happening, from a large distance so that the photons from different points on the rod need the same time to reach the camera, he would see, at every moment, photons being emitted from a plane normal to the velocity of the rod, which will appear to move from the back end of the rod to the front, with a constant speed V_ϕ. This *tsunami* of photon emission will appear to the observer in S to need, in order to 'scan' the whole length of the rod, a total time equal to

$$T_\phi = \frac{L_\phi}{V_\phi} = \frac{\gamma L_0}{c/\beta} = \beta\gamma\frac{L_0}{c} = \frac{\beta}{\sqrt{1-\beta^2}}\frac{L_0}{c}.$$

This time varies from 0 to ∞ as β varies between 0 and 1.

Problems

3.15 With what constant speed must a spaceship be moving, relative to the Earth, if it is to travel across the Galaxy in 40 years, as this time is measured inside the spaceship? In the Earth's frame of reference, which is considered to be inertial, the diameter of the Galaxy is $D = 10^5$ light years. What will the diameter of the Galaxy appear to be to an observer inside the spaceship? How much time, t_a, would be required, in the Earth's frame of reference, for the spaceship to reach this speed, if it could move at a constant acceleration of $g = 9.81\ \text{m/s}^2$?

Ans.: $V/c = 1 - 8\times10^{-8}$, $D' = 40$ light years, $t_a \approx 1$ year

3.16 Two events happen at the same point in the inertial frame of reference S′, which is moving with a constant velocity $\mathbf{V} = V\hat{\mathbf{x}}$ with respect to another frame of reference S. An observer at rest in frame S notes the positions x_A and x_B at which these two events happen in his own frame of reference and estimates the time interval between the events as $(x_B - x_A)/V$. Show that the phenomenon of time dilation is exhibited in the observer's result.

3.17 An observer A remains on Earth and watches his twin sister B move away from him with a speed of 5000 km/s for 6 months, and then return to him with the same speed. How much younger than A will B be when they meet again?

Ans.: $\Delta t = 1.22$ hour

3.18 Frame of reference S′ moves with velocity $\mathbf{V} = V\hat{\mathbf{x}}$ relative to frame S. In S′, $2N+1$ events occur (N being a positive integer). The events are equally separated in space and in time. Event k $(-N \le k \le N)$ occurs at the point $x'_k = ka$ when $t'_k = k\tau$, with a and τ constants. In frame S′, the events are spread over a time interval $T' = t'_N - t'_{-N} = 2N\tau$ and a length $L' = x'_N - x'_{-N} = 2Na$. What are the corresponding T and L in frame S? What is the condition for all the events to be simultaneous in frame S? What is then the value L_0 of L?

Ans.: $L = 2\gamma N(a + V\tau)$, $T = 2\gamma N(\tau + Va/c^2)$, $V = -\dfrac{\tau}{a}c^2$, $L_0 = L'/\gamma$

3.2 The Transformation of Velocity

3.2.1 The Transformation of the Components of Velocity

The transformation of the components of the velocity of point P, $(\upsilon_x, \upsilon_y, \upsilon_z) \Rightarrow (\upsilon'_x, \upsilon'_y, \upsilon'_z)$, follows from Eqs. (3.6) and (3.17) (with the upper signs) to be

$$\upsilon'_x = \frac{\upsilon_x - V}{1 - \upsilon_x V/c^2}, \quad \upsilon'_y = \frac{\upsilon_y}{\gamma(1 - \upsilon_x V/c^2)}, \quad \upsilon'_z = \frac{\upsilon_z}{\gamma(1 - \upsilon_x V/c^2)}. \tag{3.35}$$

The inverse transformation is found, by replacing the primed quantities with unprimed and vice versa and putting $V \to -V$, to be

$$\upsilon_x = \frac{\upsilon'_x + V}{1 + \upsilon'_x V/c^2}, \quad \upsilon_y = \frac{\upsilon'_y}{\gamma(1 + \upsilon'_x V/c^2)}, \quad \upsilon_z = \frac{\upsilon'_z}{\gamma(1 + \upsilon'_x V/c^2)}. \tag{3.36}$$

We note that a constant velocity in one inertial frame of reference is transformed to a constant velocity in all other inertial frames of reference, as expected.

Vectorially, for two frames of reference S and S′, with S′ moving with a general constant velocity $\mathbf{V}$ relative to S, the transformation of the velocity $\boldsymbol{\upsilon}$ is

$$\boldsymbol{\upsilon}' = \frac{\boldsymbol{\upsilon} + (\gamma - 1)\dfrac{\boldsymbol{\upsilon} \cdot \mathbf{V}}{V^2}\mathbf{V} - \gamma \mathbf{V}}{\gamma\left(1 - \dfrac{\boldsymbol{\upsilon} \cdot \mathbf{V}}{c^2}\right)} \tag{3.37}$$

and, inversely,

$$\boldsymbol{\upsilon} = \frac{\boldsymbol{\upsilon}' + (\gamma - 1)\dfrac{\boldsymbol{\upsilon}' \cdot \mathbf{V}}{V^2}\mathbf{V} + \gamma \mathbf{V}}{\gamma\left(1 + \dfrac{\boldsymbol{\upsilon}' \cdot \mathbf{V}}{c^2}\right)}. \tag{3.38}$$

Example 3.9 A Lorentz transformation, followed by its inverse
At time t, particle P is situated at position $\mathbf{r}$ and is moving with velocity $\boldsymbol{\upsilon}$ in the inertial frame of reference S. Show that transforming to another inertial frame of reference S′ and then back, using the Lorentz transformation, will result in the same initial values of $\mathbf{r}$ and $\boldsymbol{\upsilon}$.

Let the position and velocity of particle P at time t be given, in frame of reference S, by

$$x = x_1, \quad y = y_1, \quad z = z_1 \quad \text{and} \quad \upsilon_x = \upsilon_{1x}, \quad \upsilon_y = \upsilon_{1y}, \quad \upsilon_z = \upsilon_{1z}.$$

Transforming to frame of reference S′, which is moving with constant velocity $\mathbf{V} = V\hat{\mathbf{x}}$ relative to frame S, we have

$$x_1' = \gamma(x_1 - \beta c t_1), \quad y_1' = y_1, \quad z_1' = z_1, \quad t_1' = \gamma(t_1 - (\beta/c)x_1),$$

where $\gamma = \frac{1}{\sqrt{1-V^2/c^2}}$. For the components of the particle's velocity, we have

$$\upsilon_{1x}' = \frac{\upsilon_{1x} - V}{1 - \upsilon_{1x}V/c^2}, \quad \upsilon_{1y}' = \frac{\upsilon_{1y}}{\gamma(1 - \upsilon_{1x}V/c^2)}, \quad \upsilon_{1z}' = \frac{\upsilon_{1z}}{\gamma(1 - \upsilon_{1x}V/c^2)}$$

If we now apply the inverse Lorentz transformation to these values, we get

$$x = \gamma(x' + \beta c t') = \gamma^2\left[x_1 - \beta c t_1 + \beta c\left(t_1 - \frac{\beta}{c}x_1\right)\right] = \gamma^2 x_1(1-\beta^2) = x_1, \quad y = y', \quad z = z',$$

$$t = \gamma\left(t' + \frac{\beta}{c}x'\right) = \gamma^2\left[t_1 - \frac{\beta}{c}x_1 + \frac{\beta}{c}(x_1 - \beta c t_1)\right] = \gamma^2 t_1(1-\beta^2) = t_1.$$

For the velocity components, the inverse Lorentz transformation gives

$$\upsilon_x = \frac{\upsilon_{1x}' + V}{1 + \frac{\upsilon_{1x}'V}{c^2}} = \frac{\frac{\upsilon_{1x} - V}{1 - \upsilon_{1x}V/c^2} + V}{1 + \frac{V}{c^2}\left(\frac{\upsilon_{1x} - V}{1 - \upsilon_{1x}V/c^2}\right)} = \frac{\upsilon_{1x} - V + V - \frac{\upsilon_{1x}V^2}{c^2}}{1 - \frac{\upsilon_{1x}V}{c^2} + \frac{\upsilon_{1x}V}{c^2} - \frac{V^2}{c^2}} = \upsilon_{1x}$$

$$\upsilon_y = \frac{\upsilon_{1y}'}{\gamma\left(1 + \frac{\upsilon_{1x}'V}{c^2}\right)} = \frac{\frac{\upsilon_{1y}}{\gamma(1 - \upsilon_{1x}'V/c^2)}}{\gamma\left[1 + \frac{V}{c^2}\left(\frac{\upsilon_{1x} - V}{1 - \upsilon_{1x}V/c^2}\right)\right]} = \frac{\frac{\upsilon_{1y}}{\gamma^2}}{1 - \frac{\upsilon_{1x}V}{c^2} + \frac{\upsilon_{1x}V}{c^2} - \frac{V^2}{c^2}} = \upsilon_{1y}$$

$$\upsilon_z = \frac{\upsilon_{1z}'}{\gamma\left(1 + \frac{\upsilon_{1x}'V}{c^2}\right)} = \frac{\frac{\upsilon_{1z}}{\gamma(1 - \upsilon_{1x}'V/c^2)}}{\gamma\left[1 + \frac{V}{c^2}\left(\frac{\upsilon_{1x} - V}{1 - \upsilon_{1x}V/c^2}\right)\right]} = \frac{\frac{\upsilon_{1z}}{\gamma^2}}{1 - \frac{\upsilon_{1x}V}{c^2} + \frac{\upsilon_{1x}V}{c^2} - \frac{V^2}{c^2}} = \upsilon_{1z} \qquad \text{Q.E.D.}$$

3.2.2 *The Transformation of the Magnitude of Velocity*

As shown in Fig. 3.4, let the velocity of a particle be $\mathbf{\upsilon}$ in the frame of reference S and $\mathbf{\upsilon}'$ in the frame of reference S′.

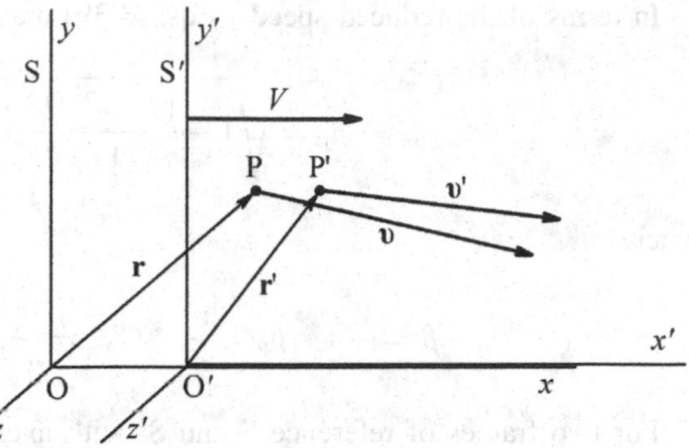

Fig. 3.4 The velocity of point P in two inertial frames of reference, S and S′, in relative motion to each other. In frame S the velocity of P is $\boldsymbol{\upsilon}$. In frame S′ it is $\boldsymbol{\upsilon}'$

The transformations of the components of the velocity are given by Eqs. (3.35). The magnitude of the velocity, υ, is found from the relation $\upsilon^2 = \upsilon_x^2 + \upsilon_y^2 + \upsilon_z^2$, while υ' from $\upsilon'^2 = \upsilon_x'^2 + \upsilon_y'^2 + \upsilon_z'^2$. Substituting in this last equation, we have:

$$\upsilon'^2 = \upsilon_x'^2 + \upsilon_y'^2 + \upsilon_z'^2 = \frac{1}{\left(1 - \frac{V\upsilon_x}{c^2}\right)^2}\left[(\upsilon_x - V)^2 + \upsilon_y^2\left(1 - \frac{V^2}{c^2}\right) + \upsilon_z^2\left(1 - \frac{V^2}{c^2}\right)\right] =$$

$$= \frac{1}{\left(1 - \frac{V\upsilon_x}{c^2}\right)^2}\left[\upsilon_x^2 - 2V\upsilon_x + V^2 + \upsilon^2\left(1 - \frac{V^2}{c^2}\right) - \upsilon_x^2\left(1 - \frac{V^2}{c^2}\right)\right] =$$

$$= \frac{1}{\left(1 - \frac{V\upsilon_x}{c^2}\right)^2}\left[\frac{V^2\upsilon_x^2}{c^2} - 2V\upsilon_x + V^2 + \upsilon^2\left(1 - \frac{V^2}{c^2}\right)\right] =$$

$$= \frac{c^2}{\left(1 - \frac{V\upsilon_x}{c^2}\right)^2}\left[\frac{V^2\upsilon_x^2}{c^4} - 2\frac{V\upsilon_x}{c^2} + 1 - 1 + \frac{V^2}{c^2} + \frac{\upsilon^2}{c^2}\left(1 - \frac{V^2}{c^2}\right)\right] =$$

$$\upsilon'^2 = \frac{c^2}{\left(1 - \frac{V\upsilon_x}{c^2}\right)^2}\left[\left(1 - \frac{V\upsilon_x}{c^2}\right)^2 - \left(1 - \frac{\upsilon^2}{c^2}\right)\left(1 - \frac{V^2}{c^2}\right)\right].$$

This may be written as

$$\upsilon'^2 = c^2\left[1 - \frac{\left(1 - \frac{\upsilon^2}{c^2}\right)\left(1 - \frac{V^2}{c^2}\right)}{\left(1 - \frac{\upsilon_x V}{c^2}\right)^2}\right] \quad \text{or} \quad \upsilon' = c\sqrt{1 - \frac{\left(1 - \frac{\upsilon^2}{c^2}\right)\left(1 - \frac{V^2}{c^2}\right)}{\left(1 - \frac{\upsilon_x V}{c^2}\right)^2}} \qquad (3.39)$$

or, also, as

$$\upsilon' = \frac{\sqrt{\left(1 - \frac{V^2}{c^2}\right)\left(\upsilon^2 - \upsilon_x^2\right) + (V - \upsilon_x)^2}}{\left|1 - \frac{\upsilon_x V}{c^2}\right|}. \qquad (3.40)$$

In terms of the reduced speeds, Eqs. (3.39) may be written as

$$\beta'_{\mathrm{P}} = \sqrt{1 - \frac{(1-\beta^2)(1-\beta_{\mathrm{P}}^2)}{(1-\beta\beta_{\mathrm{P}x})^2}}, \tag{3.41}$$

where

$$\beta = \frac{V}{c}, \quad \beta_{\mathrm{P}} = \frac{\upsilon}{c}, \quad \beta_{\mathrm{P}x} = \frac{\upsilon_x}{c}, \quad \beta'_{\mathrm{P}} = \frac{\upsilon'}{c}. \tag{3.42}$$

For two frames of reference S and S′ with axes that are mutually parallel and with S′ moving with the general velocity **V** with respect to S, the transformation of the magnitude of $\boldsymbol{\upsilon}$ is found by taking the inner product of υ' with itself, using Eqs. (3.37). After some algebraic manipulation, we find

$$\upsilon'^2 = c^2\left[1 - \frac{\left(1-\dfrac{\upsilon^2}{c^2}\right)\left(1-\dfrac{V^2}{c^2}\right)}{\left(1-\dfrac{\boldsymbol{\upsilon}\cdot\mathbf{V}}{c^2}\right)^2}\right] \tag{3.43}$$

and, inversly,

$$\upsilon^2 = c^2\left[1 - \frac{\left(1-\dfrac{\upsilon'^2}{c^2}\right)\left(1-\dfrac{V^2}{c^2}\right)}{\left(1+\dfrac{\boldsymbol{\upsilon}'\cdot\mathbf{V}}{c^2}\right)^2}\right]. \tag{3.44}$$

The combination of Eqs. (3.43) and (3.44), gives the interesting relation

$$(1-\boldsymbol{\upsilon}\cdot\mathbf{V}/c^2)(1+\boldsymbol{\upsilon}'\cdot\mathbf{V}/c^2) = 1 - V^2/c^2. \tag{3.45}$$

3.2.3 The Transformation of the Lorentz Factor, γ

Let the point P move in the frame of reference S with velocity $\boldsymbol{\upsilon}_{\mathrm{P}}$. To this velocity there corresponds a Lorentz factor $\gamma_{\mathrm{P}} = 1\Big/\sqrt{1-\upsilon_{\mathrm{P}}^2/c^2} = 1\Big/\sqrt{1-\beta_{\mathrm{P}}^2}$. In frame S′, the velocity of P is $\boldsymbol{\upsilon}'_{\mathrm{P}}$, to which there corresponds a Lorentz factor $\gamma'_{\mathrm{P}} = 1\Big/\sqrt{1-\upsilon'^2_{\mathrm{P}}/c^2} = 1\Big/\sqrt{1-{\beta'_{\mathrm{P}}}^2}$.

If to the relative speed V of the two frames there corresponds a Lorentz factor $\gamma = 1\Big/\sqrt{1-V^2/c^2} = 1\Big/\sqrt{1-\beta^2}$, which is the relationship between γ'_{P} and γ_{P}? From Eqs. (3.39) it follows that

$$\frac{1}{\gamma'^2_{\rm P}} = \frac{1}{\gamma^2\gamma^2_{\rm P}\left(1 - \frac{\upsilon_{\rm Px}V}{c^2}\right)^2} \quad \text{and, finally,} \quad \gamma'_{\rm P} = \gamma\gamma_{\rm P}\left(1 - \frac{\upsilon_{\rm Px}V}{c^2}\right), \tag{3.46}$$

where we took the positive root of the square in parentheses, because the γ's are positive and, as we will prove below, it is $\upsilon_{\rm Px} \leq c$ and $V < c$. The inverse transformation is:

$$\gamma_{\rm P} = \gamma\gamma'_{\rm P}\left(1 + \frac{\upsilon'_{\rm Px}V}{c^2}\right). \tag{3.47}$$

Vectorially, for two frames of reference S and S′, with S′ moving with a general constant velocity $\mathbf{V}$ relative to S, the transformation is

$$\gamma'_{\rm P} = \gamma\gamma_{\rm P}\left(1 - \frac{\boldsymbol{\upsilon}_{\rm P}\cdot\mathbf{V}}{c^2}\right) \tag{3.48}$$

and, inversely,

$$\gamma_{\rm P} = \gamma\gamma'_{\rm P}\left(1 + \frac{\boldsymbol{\upsilon}'_{\rm P}\cdot\mathbf{V}}{c^2}\right). \tag{3.49}$$

The transformation of the Lorentz factor will be needed below, when we will be examining the mass, the momentum and the energy of a particle in different frames of reference.

3.2.4 Speed c as an Upper Limit for Speeds

According to the first of Eqs. (3.36), if a body moves with a velocity $\boldsymbol{\upsilon}' = \upsilon'_x\hat{\mathbf{x}}$ relative to an observer S′ and observer S′ moves with velocity $\mathbf{V} = V\hat{\mathbf{x}}$ relative to observer S (in the same direction as υ'), then, the speed of the body relative to the observer S will not be $\upsilon_{\rm nr} = V + \upsilon'_x$, as predicted by non-relativistic theory, but

$$\upsilon_x = \frac{V + \upsilon'_x}{1 + V\upsilon'_x/c^2}. \tag{3.50}$$

This relativistic composition of the velocities $\mathbf{V} = V\hat{\mathbf{x}}$ and $\boldsymbol{\upsilon}' = \upsilon'_x\hat{\mathbf{x}}$ is shown in Fig. 3.5. For $|V| < c$ and $|\upsilon'_x| < c$, it is seen that the resultant speed is always $|\upsilon_x| < c$.

Equation (3.50) may also be written as

$$c - \upsilon_x = \frac{1}{c}\frac{(c - V)(c - \upsilon'_x)}{1 + V\upsilon'_x/c^2}. \tag{3.51}$$

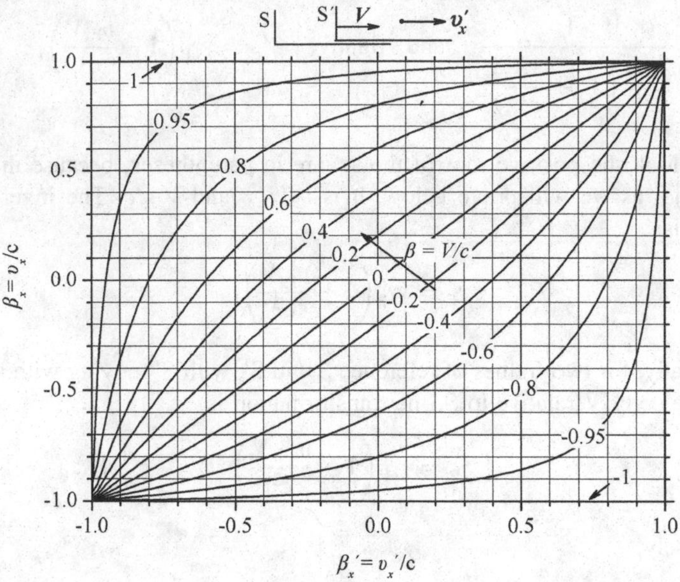

Fig. 3.5 The relativistic composition of the velocities $\mathbf{V} = V\hat{\mathbf{x}}$ and $\boldsymbol{\upsilon}' = \upsilon'_x\hat{\mathbf{x}}$. It is seen that for $|V| < c$ and $|\upsilon'_x| < c$, the resultant speed is always $|\upsilon_x| < c$

Since for $|V| < c$ and $|\upsilon'_x| < c$ the fraction on the right is positive, it follows that the composition of any two speeds which are smaller than the speed of light in vacuum leads to a speed that is also smaller than the speed of light in vacuum. Obviously, if one of these two speeds is equal to c, then the resultant υ_x of the two speeds is equal to c. It therefore appears that adding velocities smaller than c it is impossible to exceed c. We will also find out below that, for a body to reach the speed of c, an infinite time is needed, as this is measured by an observer in an inertial frame of reference, whether the body moves with a constant acceleration as measured in its own frame of reference (Sect. 4.8) or it moves under the influence of a constant force (Sect. 6.13). In addition, from energy considerations, we will find that it is impossible for a body to be accelerated to a speed c (Chap. 6).

From the first of Eqs. (3.39) for the magnitudes of velocities, $\upsilon' = c\sqrt{1 - \frac{(1-\upsilon^2/c^2)(1-V^2/c^2)}{(1-\upsilon_x V/c^2)^2}}$, we note the following:

If it is $|V| < c$ and $\upsilon < c$, then the fraction in the equation is positive and, therefore, it also is $\upsilon' < c$.

For a photon, for which it is $\upsilon = c$ it will also be $\upsilon' = c$.

For a frame of reference moving with speed $|V| = c$, it is $\upsilon'_x = -c$, $\upsilon'_y = 0$ and $\upsilon'_z = 0$, and therefore also $\upsilon' = c$. The observer S′ (a photon?) will see everything moving with speed $-c$.

For a *tachyon*, if it exists, it is $\upsilon > c$ and then, for every $|V| < c$, it will also be $\upsilon' > c$. A tachyon remains, therefore a tachyon in all inertial frames of reference, provided it is $|V|<c$. How would a tachyon see things? Let the tachyon move relative to us with a speed V, where $|V|>c$. If the tachyon sees a particle move with speed $\upsilon'<c$, the fraction in the last equation must be positive and, therefore, it must be $\upsilon>c$. The particles appearing to the tachyon as not tachyons will appear to us as tachyons. The inverse is also true, i.e. the particles that a tachyon would see as tachyons would appear to us as non-tachyons.

Example 3.10 The relative velocity of two particles

Two particles are projected simultaneously from the same point in an inertial frame of reference, with equal speeds υ, in mutually perpendicular directions. What is the velocity of each particle relative to the other?

We take particle A to move with speed υ along the y-axis of a frame of reference S. We also assume that particle B rests at the origin of a frame of reference S′ and we take the axes x and x' to coincide with the direction of motion of particle B. Frame S′, therefore, moves with velocity $\boldsymbol{\upsilon}=\upsilon\,\hat{\mathbf{x}}$ relative to frame S [figure (a)].

In frame S, particle A has velocity components $\upsilon_{Ax}=0,\quad \upsilon_{Ay}=\upsilon,\quad \upsilon_{Az}=0.$

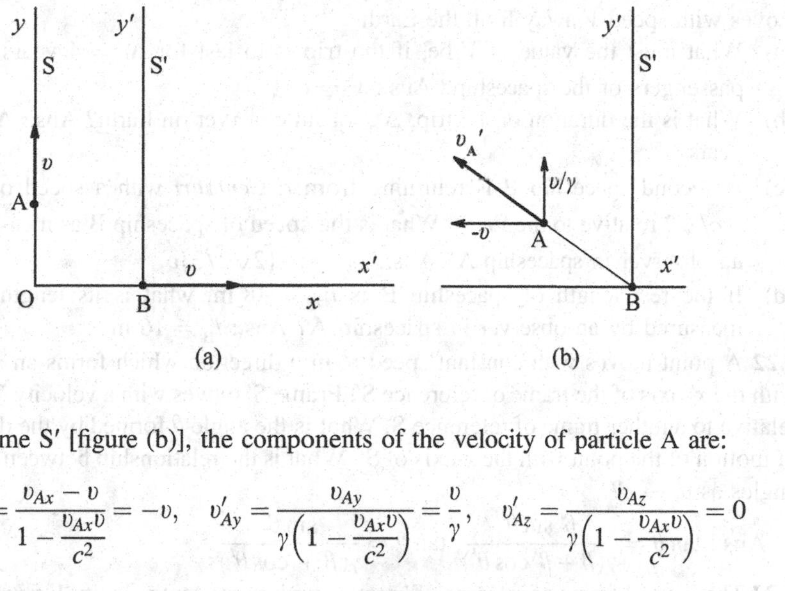

In frame S′ [figure (b)], the components of the velocity of particle A are:

$$\upsilon'_{Ax}=\frac{\upsilon_{Ax}-\upsilon}{1-\dfrac{\upsilon_{Ax}\upsilon}{c^2}}=-\upsilon,\quad \upsilon'_{Ay}=\frac{\upsilon_{Ay}}{\gamma\left(1-\dfrac{\upsilon_{Ax}\upsilon}{c^2}\right)}=\frac{\upsilon}{\gamma},\quad \upsilon'_{Az}=\frac{\upsilon_{Az}}{\gamma\left(1-\dfrac{\upsilon_{Ax}\upsilon}{c^2}\right)}=0$$

The magnitude of the velocity of particle A in the frame of reference S′ is

$$\upsilon'_{A}=\sqrt{\upsilon'^2_{Ax}+\upsilon'^2_{Ay}+\upsilon'^2_{Az}}=\sqrt{\upsilon^2+\upsilon^2/\gamma^2}=\upsilon\sqrt{1+(1-\upsilon^2/c^2)}$$

or

$$\upsilon'_{A}=\upsilon\sqrt{2-\upsilon^2/c^2},$$

which is also the magnitude of the velocity of particle A relative to B, and vice versa. We note that for $\upsilon \ll c$ the result agrees with the result of Classical Mechanics, $\upsilon'_A = \sqrt{2}\upsilon$.

Problems

3.19 Three galaxies, A, B and C, lie on a straight line. Relative to A, which lies between B and C, the other two galaxies are moving away with speeds $0.7\,c$. The rate at which the distance between B and C is changing is $1.4\,c$, as measured by A. Does this violate the conclusion derived in the Special Theory of Relativity that nothing can travel with a speed higher than c? What is the speed of B as measured by C? Ans.: $\upsilon'_x = 0.94c$

3.20 Relative to the Earth, spaceship A moves in one direction with speed $0.8c$ and another spaceship, B, moves in the opposite direction with speed $0.6c$. What is the speed of spaceship A as measured by B? Ans.: $\upsilon'_x = 0.946c$

3.21 A spaceship A departs from Earth for a trip to *α Centauri*, at constant speed. The distance of the star from the Earth is $D = 4$ light years, approximately. Consider the Earth as frame of reference S and the spaceship as frame of reference S′, which moves with speed V away from the Earth.

(a) What must the value of V be, if the trip is to last for $\Delta t' = 4$ years for the passengers of the spaceship? Ans.: $V = c/\sqrt{2}$

(b) What is the duration of the trip, Δt, for an observer on Earth? Ans.: $\Delta t = 5.7$ years

(c) A second spaceship B is returning from *α Centauri* with a speed of $\upsilon_{\mathrm{B}x} = -c/\sqrt{2}$ relative to the Earth. What is the speed of spaceship B as measured by an observer in spaceship A? Ans.: $\upsilon'_{\mathrm{B}x} = -(2\sqrt{2}/3)c$

(d) If the rest length of spaceship B is $l_{\mathrm{B}0} = 48$ m, what is its length, l'_{B}, as measured by an observer in spaceship A? Ans.: $l'_{\mathrm{B}} = 16$ m

3.22 A point moves with constant speed υ' in a direction which forms an angle θ' with the x'-axis of the frame of reference S′. Frame S′ moves with a velocity $\mathbf{V} = V\hat{\mathbf{x}}$ relative to another frame of reference S. What is the angle θ formed by the direction of motion of the point with the x-axis of S? What is the relationship between the two angles as $\upsilon' \to c$?

$$\text{Ans.: } \tan\theta = \frac{\beta' \sin\theta'}{\gamma(\beta + \beta' \cos\theta')},\quad \tan\theta = \frac{\sin\theta'}{\gamma(\beta + \cos\theta')}$$

3.23 The inertial frame of reference S′ moves with a constant velocity $V\,\hat{\mathbf{x}}$ relative to another frame of reference S. In S′, a photon has velocity components $\upsilon'_x = c\ \cos\theta'$ and $\upsilon'_y = c\ \sin\theta'$. Find the values of the velocity components in S and show that the speed of the photon in S is c. What angle is formed with the x-axis by the direction of motion of the photon in S?

$$\text{Ans.: } \upsilon_x = c\frac{\beta + \cos\theta'}{1 + \beta\cos\theta'},\quad \upsilon_y = c\frac{\sin\theta'}{\gamma(1 + \beta\cos\theta')},\quad \theta = \arctan\left[\frac{\sin\theta'}{\gamma(\beta + \cos\theta')}\right]$$

3.3 The Transformation of Acceleration

The components of the velocity transform in the way shown by Eqs. (3.35):

$$\upsilon_x' = \frac{\upsilon_x - V}{1 - \dfrac{\upsilon_x V}{c^2}} \quad \upsilon_y' = \frac{\upsilon_y}{\gamma\left(1 - \dfrac{\upsilon_x V}{c^2}\right)} \quad \upsilon_z' = \frac{\upsilon_z}{\gamma\left(1 - \dfrac{\upsilon_x V}{c^2}\right)}. \tag{3.52}$$

From these equations, by differentiating with respect of time, we may derive the transformation of the components of acceleration. These components are in frame S

$$a_x = \frac{d\upsilon_x}{dt}, \quad a_y = \frac{d\upsilon_y}{dt}, \quad a_z = \frac{d\upsilon_z}{dt}, \tag{3.53}$$

and in frame S′

$$a_x' = \frac{d\upsilon'_x}{dt'}, \quad a'_y = \frac{d\upsilon'_y}{dt'}, \quad a'_z = \frac{d\upsilon'_z}{dt'}. \tag{3.54}$$

For the differentiations with respect to time, we will also need the relation $t' = \gamma\left(t - \frac{V}{c^2}x\right)$, which gives:

$$dt' = \gamma\left(dt - \frac{V}{c^2}dx\right) = \gamma\left(1 - \frac{\upsilon_x V}{c^2}\right)dt. \tag{3.55}$$

For the x component of acceleration we have

$$a_x' = \frac{d\upsilon_x'}{dt'} = \frac{d}{dt}\left(\frac{\upsilon_x - V}{1 - \upsilon_x V/c^2}\right)\frac{dt}{dt'} = \frac{d}{d\upsilon_x}\left(\frac{\upsilon_x - V}{1 - \upsilon_x V/c^2}\right)\frac{d\upsilon_x}{dt}\frac{dt}{dt'} =$$
$$= \left[\frac{1}{1 - \upsilon_x V/c^2} + \frac{(\upsilon_x - V)V/c^2}{(1 - \upsilon_x V/c^2)^2}\right]\frac{d\upsilon_x}{dt}\left[\gamma\left(1 - \frac{\upsilon_x V}{c^2}\right)\right]^{-1} = \frac{a_x}{\gamma^3(1 - \upsilon_x V/c^2)^3}. \tag{3.56}$$

In a similar manner, we find the other two components. The final results are:

$$a_x' = \frac{a_x}{\gamma^3(1 - V\upsilon_x/c^2)^3}, \quad a_y' = \frac{a_y}{\gamma^2(1 - V\upsilon_x/c^2)^2} + \frac{V\upsilon_y/c^2}{\gamma^2(1 - V\upsilon_x/c^2)^3}a_x, \tag{3.57}$$

$$a_z' = \frac{a_z}{\gamma^2(1 - V\upsilon_x/c^2)^2} + \frac{V\upsilon_z/c^2}{\gamma^2(1 - V\upsilon_x/c^2)^3}a_x. \tag{3.58}$$

The inverse transformation is found, by replacing the primed quantities with unprimed and vice versa, and putting $-V$ instead of V, as

$$a_x = \frac{a'_x}{\gamma^3\left(1+\upsilon'_x V/c^2\right)^3}, \quad a_y = \frac{a'_y}{\gamma^2\left(1+\upsilon'_x V/c^2\right)^2} - \frac{\upsilon'_y V/c^2}{\gamma^2\left(1+\upsilon'_x V/c^2\right)^3} a'_x,$$

$$a_z = \frac{a'_z}{\gamma^2\left(1+\upsilon'_x V/c^2\right)^2} - \frac{\upsilon'_z V/c^2}{\gamma^2\left(1+\upsilon'_x V/c^2\right)^3} a'_x. \qquad (3.59)$$

We notice that a constant acceleration in one inertial frame of reference S, does not necessarily imply that the acceleration will be constant in any other inertial frame of reference S′. This is due to the dependence of the components of acceleration not only on the acceleration of the point we are considering in frame S, but also on its velocity in this frame of reference.

3.3.1 Proper Acceleration

Speed is a relative magnitude and, therefore, an observer is not able to determine whether he is at rest in an inertial frame of reference or he is moving with a constant velocity. He can only determine by measurements his speed relative to another frame of reference. By contrast, the observer is able to feel an acceleration, even in his own frame of reference, in which the observer is at rest. According to the Principle of Equivalence of the General Theory of Relativity, this acceleration is equivalent to a gravitational field in the opposite direction.

Let us assume that a spaceship starts from Earth, with zero initial speed at $t = 0$, and that it moves along the x-axis with a constant acceleration α as this is sensed by the passengers of the spaceship (*proper acceleration*). We assume that in the frame of reference S of the Earth, which we will take to be inertial, at time t, the spaceship is at a distance $x(t)$ from the Earth and that it moves with velocity $\mathbf{V} = V\hat{\mathbf{x}}$ (Fig. 3.6, dotted line), with $V(t)$ a function of time. At time t, we consider an inertial frame of reference S′ moving with a constant velocity $\mathbf{V} = V(t)\hat{\mathbf{x}}$ relative to frame S, and in which the spaceship is momentarily at rest. The phrase 'the spaceship moves with a constant acceleration α, as this is sensed by its passengers', means that the acceleration of the spaceship in frame S′ is equal to α.

If time $\Delta\tau$ passes in the frame S′, and thus also for the spaceship, the spaceship will acquire a speed equal to $\alpha\Delta\tau$ in frame S′. The corresponding time interval in the frame of the Earth is

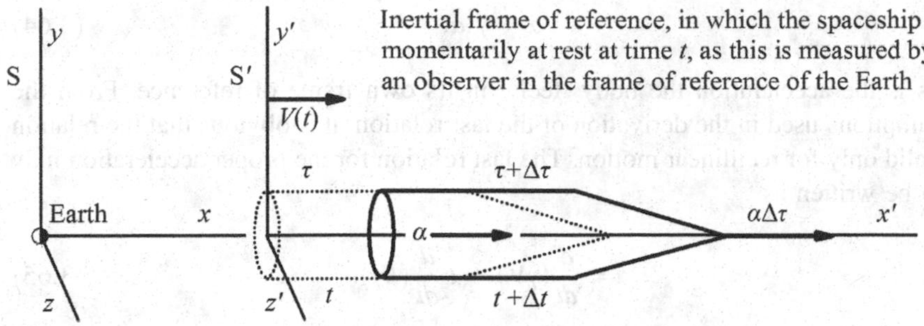

Fig. 3.6 In the frame of reference of the Earth, at time t, the spaceship moves with velocity $\mathbf{V} = V(t)\,\hat{\mathbf{x}}$. At that moment, the spaceship is momentarily at rest in a frame of reference S′, which is moving with a constant velocity $\mathbf{V} = V\hat{\mathbf{x}}$ relative to the Earth, V being the value of $V(t)$ at time t. The acceleration of the spaceship in this frame of reference is constant and equal to α (its proper acceleration). The change of the speed of the spaceship during the time interval $\Delta\tau$, in the inertial frame of reference S′, is $\alpha\Delta\tau$

$$\Delta t = \gamma\,\Delta\tau, \quad \text{where} \quad \gamma = 1\Big/\sqrt{1 - V^2/c^2} \tag{3.60}$$

and the new speed of the spaceship in the frame of reference of the Earth is (using the transformation of velocities)

$$V(t+\Delta t) = \frac{V(t) + \alpha\,\Delta\tau}{1 + \dfrac{V(t)\,\alpha}{c^2}\Delta\tau} = \frac{V(t) + \dfrac{\alpha}{\gamma}\Delta t}{1 + \dfrac{V(t)\,\alpha}{c^2\gamma}\Delta t}. \tag{3.61}$$

From this relation we find that

$$\frac{V(t+\Delta t) - V(t)}{\Delta t} = \frac{\dfrac{\alpha}{\gamma} - \dfrac{V^2(t)\,\alpha}{c^2\gamma}}{1 + \dfrac{V(t)\,\alpha}{c^2\gamma}\Delta t}. \tag{3.62}$$

In the limit, as $\Delta t \to 0$, this relation gives

$$\frac{dV}{dt} = \frac{\alpha}{\gamma}\left(1 - \frac{V^2}{c^2}\right) = \alpha\left(1 - \frac{V^2}{c^2}\right)^{3/2} = \frac{\alpha}{\gamma^3}, \tag{3.63}$$

i.e. that the acceleration of the spaceship in the frame S of the Earth is γ^3 times smaller than its proper acceleration, where $\gamma = 1\Big/\sqrt{1 - V^2/c^2}$ is the Lorentz factor corresponding to the instantaneous speed V of the spaceship in the frame of reference S. Inversely, if the acceleration of a body in an inertial frame of reference S is $a = dV/dt$, the body's proper acceleration is

$$\alpha = \gamma^3 a. \tag{3.64}$$

This is the acceleration the body 'feels' in its own frame of reference. From the assumptions used in the derivation of the last relation, it is obvious that the relation is valid only for rectilinear motion. The last relation for the proper acceleration may also be written as

$$\alpha = \frac{d}{dt}(\gamma V) = c\frac{d}{dt}(\beta\gamma). \tag{3.65}$$

The general equation for the proper acceleration of a body moving with a general velocity is derived in Sect. 8.2.3.1.

Example 3.11 Proper acceleration in the motion on a circle with constant speed

A particle is moving with constant speed V on a circle of radius R. What is its proper acceleration?

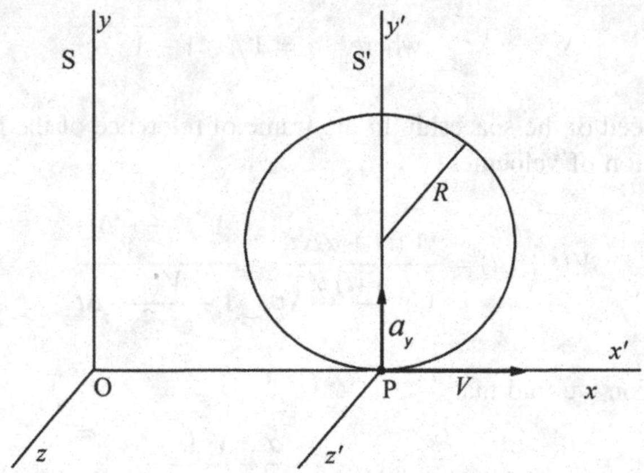

Let the particle (P in the figure) move on a circle of radius R lying on the xy plane of the inertial frame of reference S. At one particular moment in time, the velocity of the particle in frame S is $\mathbf{V} = V\hat{\mathbf{x}}$. At that moment, the acceleration of the particle has, in S, components

$$a_x = 0, \quad a_y = V^2/R, \quad a_z = 0.$$

We define an inertial frame of reference S′ in which the particle is momentarily at rest. The acceleration of the particle in frame S′ is its proper acceleration. The frame of reference S′ moves with velocity $\mathbf{V} = V\hat{\mathbf{x}}$ relative to S. The transformation of the components of the acceleration gives, for $\upsilon_x = V$, $\upsilon_y = 0$ and $\upsilon_z = 0$,

$$a'_x = \frac{a_x}{\gamma^3(1 - Vv_x/c^2)^3} = 0,$$

$$a'_y = \frac{a_y}{\gamma^2(1 - Vv_x/c^2)^2} + \frac{Vv_y/c^2}{\gamma^2(1 - Vv_x/c^2)^3}a_x = \frac{V^2/R}{\gamma^2(1 - V^2/c^2)^2} = \gamma^2\left(\frac{V^2}{R}\right),$$

$$a'_z = \frac{a_z}{\gamma^2(1 - Vv_x/c^2)^2} + \frac{Vv_z/c^2}{\gamma^2(1 - Vv_x/c^2)^3}a_x = 0.$$

The proper acceleration of the particle is, therefore, $\alpha = a'_y = \gamma^2\left(\frac{V^2}{R}\right)$.

This relation must not be considered as disagreeing with Eqs. (3.64), which is valid only for rectilinear motion.

As an application, we will evaluate the proper acceleration of the particles in an experiment which was performed CERN at the end of the decade of 1970, to test the dilation of time, as predicted by the Special Theory of Relativity (a brief description of the experiment will be given in Sect. 4.1.1). In this experiment, muons were stored in a ring of radius $R = 7\,\text{m}$, where they circulated with a speed that corresponded to $\beta = 0.9994$ and $\gamma = 29.3$. The acceleration of the particles in the frame of reference of the laboratory was, therefore, equal to the centripetal acceleration $a = V^2/R = 1.3 \times 10^{16}\,\text{m/s}^2$. The proper acceleration of the particles is γ^2 times this value, i.e. it is equal to $\alpha = \gamma^2(V^2/R) = 1.1 \times 10^{19}\ \text{m/s}^2$. It is very interesting that the predictions of the Special Theory of Relativity hold even at such enormous values of the acceleration!

Chapter 4
Applications of Relativistic Kinematics

4.1 The 'Meson' Paradox

There are in space charged particles, ions and photons of X and γ rays, with energies that may sometimes reach very large values. Collectively they are known as *cosmic radiation*. The main component of cosmic radiation consists of protons, with energies of the order of GeV. These protons react in the upper layers of the Earth's atmosphere with the nucleons of the nuclei of nitrogen and oxygen, producing the *pions* π^0, π^+ and its antiparticle, π^-. These are the particles whose existence, as carriers of the nuclear forces, was predicted by Yukawa in 1935. They were originally called *mesons*, because they had masses with values between those of the *baryons*, such as the proton and the neutron, which are *hadrons*, and the *leptons*, such as the electron. The pions are unstable and they decay according to the law of radioactivity,

$$N(t) = N_0 \mathrm{e}^{-t/\tau}, \tag{4.1}$$

where N_0 is the number of particles at time $t = 0$, $N(t)$ is the number of particles surviving at time t, and τ is a constant, characteristic of the particle, known as its *mean lifetime*. The ways pions decay are

$$\pi^0 \to 2\gamma \quad (0.83 \times 10^{-16}\ \mathrm{s}), \qquad \pi^- \to \mu^- + \bar{\nu}_\mu \quad (2.6 \times 10^{-8}\ \mathrm{s}),$$
$$\pi^+ \to \mu^+ + \nu_\mu \quad (2.6 \times 10^{-8}\ \mathrm{s}),$$

where, in parentheses is given the value of τ for every mode of decay. In the equations, symbolized by ν are the various kinds of *neutrinos* and with μ the leptons known as *muons*. The pions are created at heights of 6–10 km in the atmosphere (Fig. 4.1). Even at speeds approaching the speed of light they will travel only a few meters, on the average, before decaying.

© Springer International Publishing Switzerland 2016

C. Christodoulides, *The Special Theory of Relativity*,
Undergraduate Lecture Notes in Physics, DOI 10.1007/978-3-319-25274-2_4

Fig. 4.1 The production of π and μ particles in the atmosphere

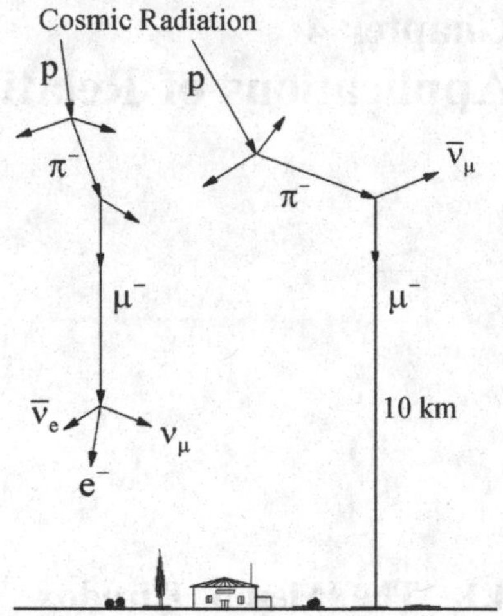

It must be stressed that the values of the mean lifetimes given in tables for radioactive nuclei and unstable particles, refer to the rest frame of the particular nucleus or particle.

Muons, in their turn, are unstable and decay in the following ways

$$\mu^- \to e^- + \bar{\nu}_e + \nu_\mu \quad \mu^+ \to e^+ + \nu_e + \bar{\nu}_\mu \quad (\tau = 2.2 \times 10^{-6}\ \text{s})$$

where now the (common) mean lifetime is of the order of μs. With speeds very close to the speed of light in vacuum, they travel before they decay 660 m on average, according to non-relativistic Physics. A very small proportion of them is expected to survive travelling the 6–10 km in the atmosphere and be observed at sea level. This was actually found to happen. The paradox was, however, that their number was much higher than expected.

The muon was discovered by C. Anderson in 1938, before the discovery of the pion (in cosmic radiation!) by C.F. Powel and G.P.S. Occhialini in 1947, and was originally thought that the μ was the meson predicted by Yukawa. The paradox of the unexpectedly large number of muons created by cosmic radiation that survive down to sea level became known as the *meson paradox*, although we know today that muons are leptons and not mesons and that Yukawa's particles are actually the pions.

Rossi [1] and his co-workers studied, in the beginning of the decade of the 1940s, the decrease with decreasing height of the number of muons (then known as *mesotrons*), by performing measurements at two positions, at Denver and at lake Echo on mount Evans (Fig. 4.2), which had an elevation difference of $3240 - 1616 = 1624$ m. For the muons which they studied and which had an energy of 520 MeV (with speeds above $0.99c$), they found a decrease in their flux by a factor of 1.43 between the two altitudes.

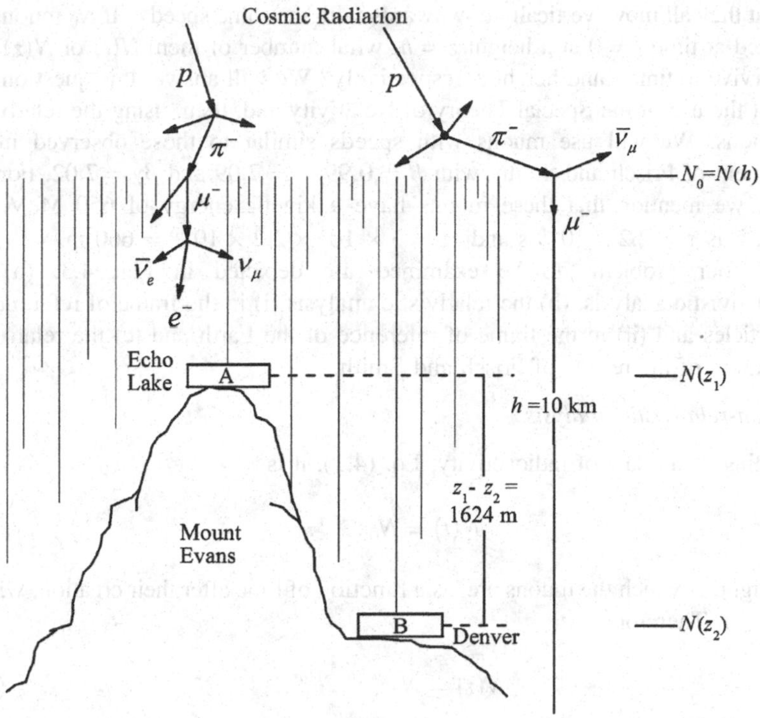

Fig. 4.2 The production of π and μ particles at high altitudes in the atmosphere. The π particles travel only a few meters before they disintegrate. The μ particles are also unstable, so their number decreases with decreasing altitude. Rossi and his co-workers measured a decrease by a factor of 1.43 in the number of muons between two points with a difference in altitude equal to 1624 m. μ particles are about 100 times more long-lived than π particles and a small proportion of them survives and reaches the surface of the sea. However, the number observed is much greater than the non-relativistic predictions. This is called *the meson paradox*

A more recent and much more precise experiment was carried out by Frisch and Smith [2] in 1963. Their measurements were performed at Mount Washington and at Cambridge, Massachusetts, with a difference in elevation of 1907 m. At Mount Washington they counted approximately 563 muons per hour with mean speeds of between $0.995c$ and $0.9954c$. At Cambridge they counted 412 muons per hour, instead of the expected 27 according to non-relativistic theory. This is a decrease in the particle flux by a factor of only 1.37, instead of the non-relativistically expected factor of approximately 21, based on a mean lifetime of $\tau = 2.2 \times 10^{-6}$ s for the muons.

An experiment which can be performed in an undergraduate laboratory in order to demonstrate the phenomenon of the dilation of time using muons, has been described by Coan et al. [3].

To interpret the meson paradox, we will make certain assumptions, in order to simplify things. We will assume that all muons are created at the same altitude h

and that they all move vertically downwards with the same speed υ. If N_0 muons are produced at time $t = 0$ at a height $z = h$, what number of them $N(t)$ or $N(z)$ will still survive at time t and height z, respectively? We will answer this question first without the use of the Special Theory of Relativity and, then, using the relativistic predictions. We will use muons with speeds similar to those observed in the experiment of Frisch and Smith, with $\beta = 0.99$, $\gamma = 7.09$ and $\beta\gamma = 7.02$. For reference, we mention that these muons have a kinetic energy of 650 MeV. For muons, it is $\tau = 2.2 \times 10^{-6}$ s and $c\tau = 3 \times 10^8 \times 2.2 \times 10^{-6} = 660$ m.

The four problems to be examined are depicted in Fig. 4.3: (a) the non-relativistic analysis, (b) the relativistic analysis (i) in the frame of reference of the particles and (ii) in the frame of reference of the Earth and (c) the relativistic explanation of the results of Frisch and Smith.

(a) *Non-relativistic analysis*

According to the law of radioactivity, Eq. (4.1), it is

$$N(t) = N_0 e^{-t/\tau}.$$

The height, at which the muons are, as a function of time after their creation, will be $z = h - \upsilon t$. Therefore,

$$N(z) = N_0 e^{-(h-z)/\upsilon\tau}. \tag{4.2}$$

On travelling a distance equal to $l = \upsilon\tau$, the number of the particles decreases by a factor of $e = 2.718$. For $\beta = 0.99$ it is $l = \beta c\tau = 0.99 \times 660 = 653$ m.

At sea level, with $z = 0$, the number of surviving muons is

$$N(0) = N_0 e^{-h/\upsilon\tau} = N_0 e^{-h/l}. \tag{4.3}$$

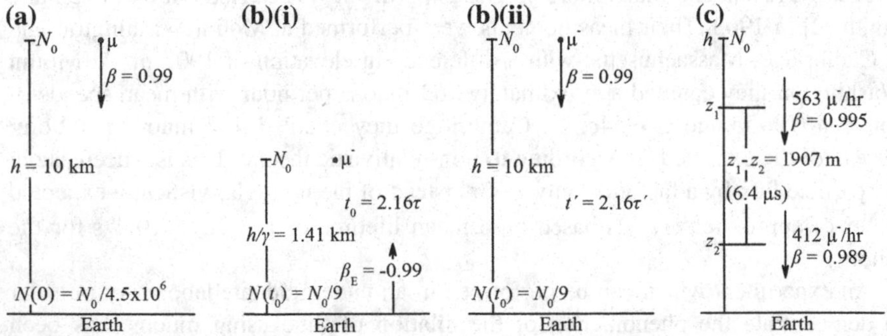

Fig. 4.3 The decay of muons in the Earth's atmosphere, as seen **a** non-relativistically and **b** relativistically, (i) in the frame of reference of the muons and (ii) in the frame of reference of the Earth. The results of the Frisch-Smith experiment are presented in figure (**c**)

For muons produced at a height of $h = 10$ km, it is

$$N(0) = N_0 e^{-h/l} = N_0 e^{-10000/653} = N_0 e^{-15.3} = 2.23 \times 10^{-7} N_0 = N_0/4.5 \times 10^6. \quad (4.4)$$

In other words, only one muon in 4.5 million would reach sea level.

(b) *Relativistic analysis*

According to the Special Theory of Relativity, the error made in the analysis above is that the value of τ measured for the muons at rest is used in another frame of reference (that of the Earth), in which the muons are moving. Time dilation is thus ignored. The passage of time equal to τ for the particles, corresponds to the passage of time $\gamma\tau$ in the frame of reference of the Earth. The mean lifetime of the particles in the frame of reference of the Earth will, therefore, be equal to $\gamma\tau$.

We may solve the problem in two ways: (i) using the frame of reference of the particles and the value of τ given in the tables or (ii) using the frame of reference of the Earth and the value of $\gamma\tau$ for the mean lifetime of the particles in that frame.

(i) *In the frame of reference of the particles*

We have $N(t) = N_0 e^{-t/\tau}$.

If the trip of the particles in the atmosphere lasts for a proper time t_0 (for them), then a proportion of $N(t_0)/N_0 = e^{-t_0/\tau}$ of them will reach sea level. The particles see the Earth moving towards them with a speed υ, to which there correspond $\beta = 0.99$ and $\gamma = 7.09$. The thickness of the atmosphere will be, due to length contraction, equal to h/γ. In the particles' frame of reference, the trip of the atmosphere will last for a time $t_0 = h/\upsilon\gamma = h/c\beta\gamma$.

At sea level, therefore, the proportion of particles surviving will be $N(t_0)/N_0 = e^{-h/c\beta\gamma\tau}$.

We compare this with the non-relativistic result $N(0)/N_0 = e^{-h/c\beta\tau}$.

For $h = 10$ km, the thickness of the atmosphere will appear to the particles to be $h/\gamma = 10000/7.09 = 1410$ m and their trip in the atmosphere to last for time $t_0 = h/c\beta\gamma = 10^4/(3 \times 10^8 \times 0.99 \times 7.09) = 4.75 \times 10^{-6}$ s or $t_0 = (h/c\beta\gamma\tau)\tau = 2.16\tau$.

The proportion of muons that survive for long enough to reach sea level is

$$N(t_0)/N_0 = e^{-h/c\beta\gamma\tau} = e^{-2.16} = 0.115 = 1/8.7 \approx 1/9 \quad (4.5)$$

i.e. on average, one muon in about 9 manages to reach sea level.

(ii) *In the frame of reference of the Earth*

If in the frame of reference of the Earth the trip of the muons in the atmosphere lasts for a time t', then a proportion $N(t')/N_0 = e^{-t'/\tau'}$ of them will reach sea level, where τ' is the mean lifetime of the particles in the frame of reference of the Earth.

An observer on the Earth sees the particles move in the atmosphere with speed υ. The trip in the atmosphere will last, in the Earth's frame of reference, for time $t' = h/\upsilon = h/c\beta$. Due to time dilation, the particles will live, on average, for time $\tau' = \gamma\tau$ (instead of τ). The proportion of particles reaching sea level will, therefore, be equal to $N(t')/N_0 = \mathrm{e}^{-t'/\tau'} = \mathrm{e}^{-h/c\beta\gamma\tau}$.

Here, it is $\beta = 0.99$, $\gamma = 7.09$, $\tau = 2.2 \times 10^{-6}$ s and $h = 10$ km. The mean lifetime of these particles in the frame of reference of the Earth is $\tau' = \gamma\tau = 1.56 \times 10^{-5}$ s. For an observer on the Earth, the particles' trip in the atmosphere lasts for a time $t' = h/c\beta = 10^4/(3 \times 10^8 \times 0.99) = 3.37 \times 10^{-5}\ \mathrm{s} = 2.16\tau'$.

The proportion of muons surviving for long enough to reach sea level is:

$$N(t')/N_0 = \mathrm{e}^{-t'/\tau'} = \mathrm{e}^{-2.16} = 0.115 = 1/8.7 \approx 1/9 \tag{4.6}$$

i.e. we find again that, on average, one muon in 9 manages to reach sea level.

We note that, according to the Special Theory of Relativity, for muons with speed $0.99c$, the proportion of particles reaching sea level is $4.5 \times 10^6/8.7 =$ 520 000 times higher than that predicted by non-relativistic Physics. The observations are thus explained by theory.

(c) *Interpretation of the results of the Frisch and Smith experiment*

The results of the Frisch and Smith experiment can be analyzed and compared with theory. To the values of β between 0.995 and 0.9954 there corresponds an average Lorentz factor of $\gamma = 10.2$ and an average value of $\beta\gamma = 10.18$. The time needed for the particles to travel the 1907 m of the difference in elevations is about 6.4 μs. Owing to the interaction of the muons with the atoms of the atmosphere, by the time they reached the elevation of Cambridge, Massachusetts, their speeds were reduced to values between $0.9881c$ and $0.9897c$, corresponding to a mean Lorentz factor of 6.7 and an average $\beta\gamma = 6.65$. The average value of $\beta\gamma$ for the whole trip of the muons is, therefore, 8.41. Using $N(z_1)/N(z_2) = \exp[(z_1 - z_2)/c\beta\gamma\tau]$ and taking $\tau = 2.2 \times 10^{-6}$ s for the muons, we find $N(z_1)/N(z_2) = \mathrm{e}^{0.34} = 1.41$. This factor is in very good agreement with the measured factor of 1.37.

4.1.1 Experimental Verification of the Dilation of Time with Muon Experiments at CERN

The experimental test of the phenomenon of dilation of time may be achieved in the laboratory with very great accuracy. One such experiment was performed at CERN at the end of the decade of 1970 by Bailey and his co-workers [4]. Muons $\mu^{\pm}$ from the CERN muon beam were stored in a ring with a diameter of 14 m, where they circulated at a speed corresponding to $\beta = 0.9994$ and $\gamma = 29.3$. The decays of the muons were recorded by detecting the electrons and positrons emitted (Fig. 4.4). The mean lifetime of these fast muons was thus determined. For the μ^+ particles,

Fig. 4.4 The experiment performed at CERN in order to test the dilation of time using μ particles

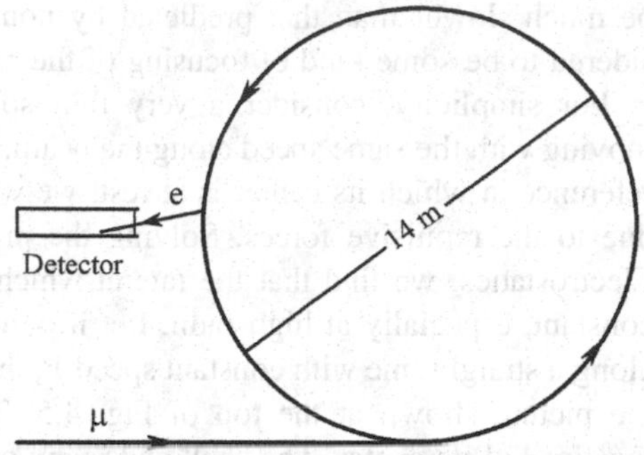

a mean lifetime of $\tau_L^+ = 64.419 \pm 0.058\ \mu s$ was measured and for the μ^- particles, $\tau_L^- = 64.368 \pm 0.029\ \mu s$, where the index L shows that these measurements refer to the laboratory frame of reference. From these measurements the mean lifetime of the particles in their own frame of reference, $\tau_L^{\pm}/\gamma$, can be predicted, according to the theory.

These values could be compared with the proper mean lifetimes $\tau^{\pm}$ of the particles, performed with particles moving at very much smaller speeds. The comparison gave, for the μ^+ particles

$$\frac{\tau^+ - \tau_L^+/\gamma}{\tau^+} = (2 \pm 9) \times 10^{-4}, \tag{4.7}$$

with similar results for the μ^- particles. The dilation of the mean lifetime of muons by the expected factor of $\gamma = 29.3$ was thus verified to an accuracy of a few parts in 10 000. It is worth mentioning that the centripetal acceleration of the muons, which was of the order of $10^{16}\ \text{m/s}^2$, did not affect the agreement between the experiment and the Special Theory of Relativity.

4.2 The Apparent Focusing of Fast Charged Particle Beams Due to the Dilation of Time

Time dilation is demonstrated in a simple way by the behavior of beams consisting of fast charged particles. If we assume that we have a narrow beam of very energetic charged particles moving with the same speed, we expect that the repulsion between the charges will cause the diameter of the beam to increase as the particles move. This is indeed the case but the rate of increase of the beam's diameter is observed to be much slower than that predicted by non-relativistic theory. This may be considered to be some kind of focusing of the beam. The explanation is very simple.

For simplicity, consider a very thin spherical shell of charged particles all moving with the same speed along the beam. If we examine the shell in the frame of reference in which its center is at rest, we will observe the increase with its radius due to the repulsive forces. Solving the problem using classical Mechanics and Electrostatics, we find that the rate at which the shell's radius increases is almost constant, especially at high radii. Let it be equal to υ. If now the shell is moving along a straight line with constant speed V, the non-relativistic treatment would give the picture shown at the top of Fig. 4.5. The shell was drawn at 9 equidistant locations of its center. The shell and many other similar ones would define a beam of particles that diverges as it moves at a certain rate. In time Δt the center of the shell moves a distance $\Delta x = V\Delta t$ and its radius increases by $\Delta R = \upsilon\Delta t$. The half-angle defined by the conical beam will, therefore, be $\theta(1) = \arctan(\upsilon/V)$.

Examining the problem relativistically, we must take into account the dilation of time. For the case of particle shells whose centers move with a speed V to which there corresponds a Lorentz factor $\gamma = 2$, as shown in the figure, when time $\tau = 1$ unit has elapsed in the particles' frame of reference, the time that passed in the laboratory frame of reference will be $t = \gamma\tau = 2$ units. The center of the shell will therefore be at the position $x = 2$ units. Due to length contraction, the spherical shell will now be an ellipsoid by revolution but its radius normal to its direction of motion will not be affected and in this case it will be R_1. The half-angle of the

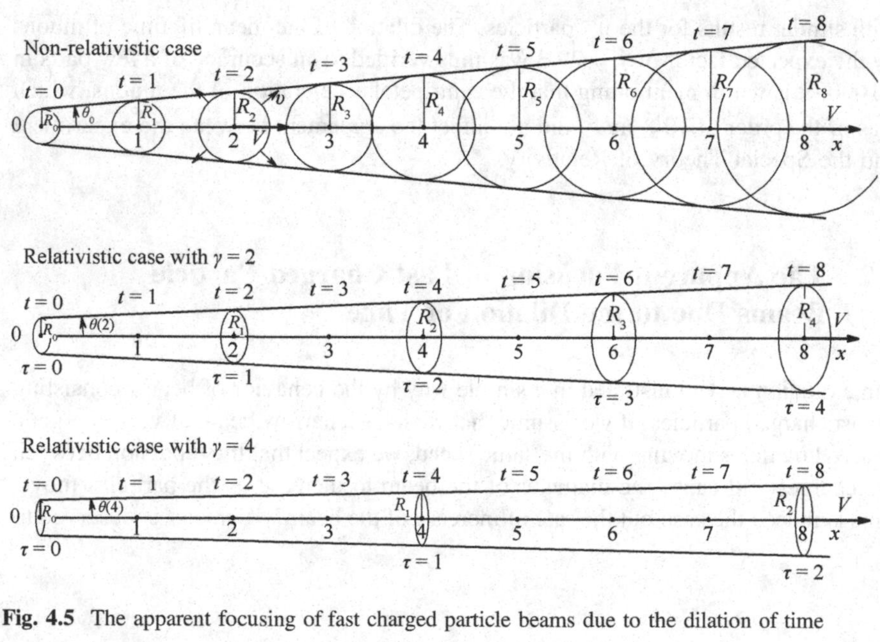

Fig. 4.5 The apparent focusing of fast charged particle beams due to the dilation of time

conical beam will now be smaller and the beam will appear as if it has been focused. For $\gamma = 4$, also shown in the figure, the effect is more pronounced.

In general, for a speed V to which there corresponds a Lorentz factor γ, when the center of the shell moves by a distance $\Delta x = V\Delta t$, its radius increases by $\Delta R = \upsilon(\Delta t/\gamma)$. The half-angle defined by the conical beam will be $\theta(\gamma) = \arctan(\upsilon/\gamma V)$. This effect is observed in high-energy particle beams, produced by particle accelerators. For high γ values the beam appears to be completely collimated.

An alternative way of looking at the effect is to consider the magnetic field produced by the moving charges of the beam and evaluate the magnetic forces exerted by it on the charged particles of the beam, pushing them towards the beam's axis. These oppose the repulsive electrostatic forces also acting. As $\gamma \to \infty$, the two forces cancel each other out completely. It is interesting that the problem can be solved by two methods which are so different from each other. We will see later that this is due to the close connection between magnetism and relativity.

Problems

4.1 At a certain point, particles are produced, which move at a speed of $\upsilon = \frac{4}{5}c$ in the laboratory frame of reference. The particles are unstable, with a mean lifetime of $\tau = 10^{-8}$ s (in their own frame of reference). After how much time, as measured in the laboratory, will the number of particles be reduced by a factor e = 2.71828...? What is the distance the particles will travel in this time? What is this distance in the frame of reference of the particles? Ans.: $\Delta t = 1.67 \times 10^{-8}$ s, $l = 4$ m, $l' = 2.4$ m

4.2 The mean lifetime of muons at rest is $\tau = 2.2 \times 10^{-6}$ s. The muons in a beam are observed in the laboratory to have a mean lifetime of $\tau_L = 1.5 \times 10^{-5}$ s. What is the speed of the muons in the laboratory frame of reference? Ans.: $V = 0.989c$

4.3 The particles in a beam move in the laboratory with a common speed of $V = \frac{4}{5}c$. It is observed that the number of the particles is reduced by a factor of e = 2.71828... after they travel a distance of 300 m in the laboratory frame of reference. What is the mean lifetime τ of the particles in their own frame of reference?
Ans.: $\tau = 0.75$ ns

4.4 The mean lifetime of π^+ mesons, in their own frame of reference, is $\tau_0 = 2.8 \times 10^{-8}$ s. A pulse of 10^4 π^+ mesons travels a distance of 59.4 m in the laboratory, with a speed of $\upsilon = 0.99c$

(a) Approximately how many mesons will survive till the end of the trip? Ans.: $N \approx 3700\pi^+$

(b) How many mesons would have survived after the same interval of time, had they been at rest? Ans.: $N \approx 9\pi^+$

4.5 A beam of μ particles is produced at some height in the atmosphere. The particles move with a speed of $\upsilon_\mu = 0.99\,c$ vertically downwards. μ particles disintegrate into electrons and neutrinos ($\mu^- \to e^- + \bar{\nu}_e + \nu_\mu$) with a mean lifetime of $\tau_\mu = 20.2$ μs in their own frame of reference.

(a) Find the height at which the particles are produced, if a fraction of 1 % of them survives for long enough to reach the surface of the Earth. Ans.: 21 km
(b) What is this distance as seen by the particles? Ans.: 3 km

4.6 A spaceship, which will serve as frame of reference S′, has a rest length $L_0 = 10$ m. It moves with a speed $\frac{4}{5}c$ with respect to another frame of reference, S. Inside the spaceship and at the point $x = x' = 0$, $y = y' = 0$, $z = z' = 0$, there is a quantity of radioactive material, which at the moment $t = t' = 0$ consists of N_0 nuclei. The half-life of the nuclei is $\tau_{1/2} = 2$ μs. When and at what position will the number of the surviving nuclei be $N_0/2$? Ans.: At $t_2 = 3.33$ μs, at the point $x_2 = 800$ m

4.3 The Sagnac Effect

An observer shut in a room without the possibility of receiving messages from outside, has no way, mechanical or optical, to determine whether he is moving uniformly relative to any external inertial frame of reference. Uniform motion is a relative magnitude. Acceleration, however, is not. The observer will sense any acceleration of himself relative to an inertial frame of reference, as a mechanical force. In other words, d'Alembert forces appear, such as a centrifugal or a Coriolis force for a rotating system. Linear acceleration will appear to be a gravitational field. Experiments and instruments have been devised in order to test whether a frame of reference is not inertial. Foucault's pendulum demonstrates the rotation of the Earth in a spectacular manner. Gyroscopes are very sensitive instruments that point towards a constant direction in space, relative to which the rotation of a frame of reference may be detected.

Apart from the mechanical ones, an optical method may be used for the same purpose. Sagnac [5], in 1913, was the first to demonstrate that it is possible, by optical means, to detect and measure the rotation of a system relative to the inertial frames of reference. The apparatus he used is shown, in simplified form, by Fig. 4.6. A parallel beam of monochromatic light from a source S, after passing through a narrow vertical slit, is split into two beams by a half-silvered mirror D. One beam, reflected on mirrors A, B and C, traverses the closed path DABCD and is then reflected by D towards a photographic plate P. The other beam of light, moves in the opposite sense, DCBAD, and also falls on the photographic plate.

If the apparatus is at rest in an inertial frame of reference, the two beams will need the same time in order to travel from S to P and, on the photographic plate, an interference pattern, with fringes, will be formed and registered. The two beams will reach the central points of the photographic plate in phase, which means that the central fringe of the interference pattern will be a bright one.

Fig. 4.6 The experimental arrangement used in the observation of the Sagnac effect

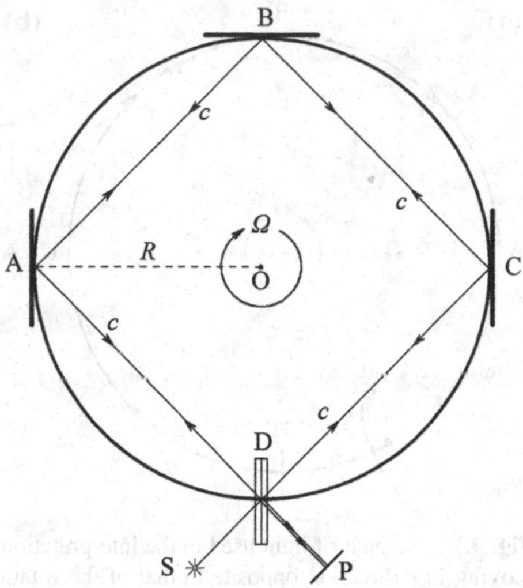

If, now, the apparatus is rotated at a constant angular speed, Ω, we will show below that the times needed by the two beams to travel to the photographic plate are different. The result will be that the interference fringes will be displaced relative to their positions in the non-rotating apparatus. In this way, the rotation of the apparatus relative to the inertial frames of reference may be detected, and quantitative observations be made concerning the rotation.

For mathematical simplicity, we will assume that, with a great number of reflections, the light paths are circular to a satisfactory approximation (Fig. 4.7). Examined in the figure are the cases in which (a) the direction of motion of the beam of light is opposite to that of the rotation of the apparatus and (b) the direction of motion of the beam of light is the same as that of the rotation of the apparatus.

With reference to Fig. 4.7a, we see that in the time τ_1 that is needed for the beam of light, moving in the opposite direction to that of the rotation of the apparatus, starting from D, to meet again the half-silvered mirror, the mirror will have shifted by a certain angle, from point D to point D′. The angular displacement of D will be equal to $\Delta\phi = \Omega\tau_1$. The beam of light will therefore have to travel an angle of $2\pi - \Omega\tau_1$ and it will need a time τ_1 for this. Thus,

$$(2\pi - \Omega\tau_1)R = c\tau_1, \tag{4.8}$$

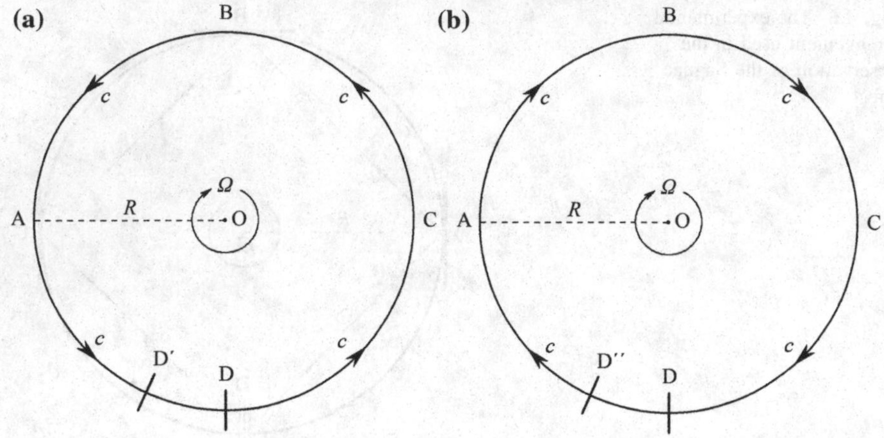

Fig. 4.7 The path of light used in the interpretation of the Sagnac effect. **a** With the beam of light moving in a direction opposite to that of the rotation of the apparatus. **b** With the beam of light moving in the same direction as that of the rotation of the apparatus

from which it follows that

$$\tau_1 = \frac{2\pi R}{c + R\Omega}. \tag{4.9}$$

The other beam will need time τ_2 to travel along a complete circle plus the angle $\Delta\phi = \Omega\tau_2$ in order to meet again the half-silvered mirror at point D″. The time needed is, therefore,

$$\tau_2 = \frac{2\pi R}{c - R\Omega}. \tag{4.10}$$

The difference of the two values of time is

$$\Delta\tau = \tau_2 - \tau_1 = \frac{2\pi R}{c - R\Omega} - \frac{2\pi R}{c + R\Omega} = \frac{4\pi R^2\Omega}{c^2 - R^2\Omega^2} = \gamma^2 \frac{4\pi R^2\Omega}{c^2}. \tag{4.11}$$

where $\gamma = 1 \Big/ \sqrt{1 - (R\Omega/c)^2}$. This is the time as measured in the laboratory inertial frame. Given that the mirrors and the photographic plate rotate with a speed equal to $R\Omega$, the time difference for them is $\Delta\tau_0 = \Delta\tau/\gamma$. Therefore,

$$\Delta\tau_0 = \gamma \frac{4\pi R^2\Omega}{c^2}. \tag{4.12}$$

This time interval is equivalent to $\Delta n = \Delta\tau_0/T = c\Delta\tau_0/\lambda_0$ periods of the light and will lead to a displacement of the interference pattern by Δn fringes. Here, λ_0 is the wavelength of the light used in vacuum and in the inertial frame of reference in which we examine the phenomenon. The displacement of the interference pattern, in fringes, is, therefore,

$$\Delta n = \gamma \frac{4\pi R^2 \Omega}{c\lambda_0}. \tag{4.13}$$

If we use $S = \pi R^2$ for the area enclosed by the beams of light with their paths, this relation may be written as

$$\Delta n = \gamma \frac{4S\Omega}{c\lambda_0}. \tag{4.14}$$

It is proved that the equation

$$\Delta n = \frac{4\gamma}{c\lambda_0} \mathbf{S} \cdot \mathbf{\Omega} \tag{4.15}$$

gives the displacement of the fringes independently of the shape of the area enclosed by the paths of the two beams, which is denoted by the vector **S**, and independently of the position of the center of rotation of the apparatus or even the presence of an optical medium in which the beams of light propagate.

The displacement of the fringes is easier to detect and measure if these are photographed when the apparatus is rotating in one direction and then again in the opposite direction. A shift equal to $2\Delta n$ is then measured.

Sagnac used light with $\lambda_0 = 436$ nm, an apparatus with $S = 866\ \text{cm}^2$ and rotation rate of 2 revolutions per second and he measured a displacement of the diffraction pattern by $2\Delta n = 0.07$ of a fringe, which was easily measurable. The experiment has since been repeated with greater sensitivity. More on the subject can be found in Sect. 10.7.

4.4 Clocks Moving Around the Earth

If a clock is moved very slowly around the Earth (or another rotating body) will relativistic phenomena be observed? The temptation is to state that measurable effects will only exist if the clock moves with a great speed. We will see, however, that differences are observed between clocks moving even at negligible speeds on the surface of the Earth and those that are stationary on the Earth. The differences are of the order of a few ns, easy to measure with today's atomic clocks.

Let us examine what happens when a clock P is moved, with a small speed υ relative to the Earth, towards east, along the Equator (Fig. 4.8). Another clock, P_0,

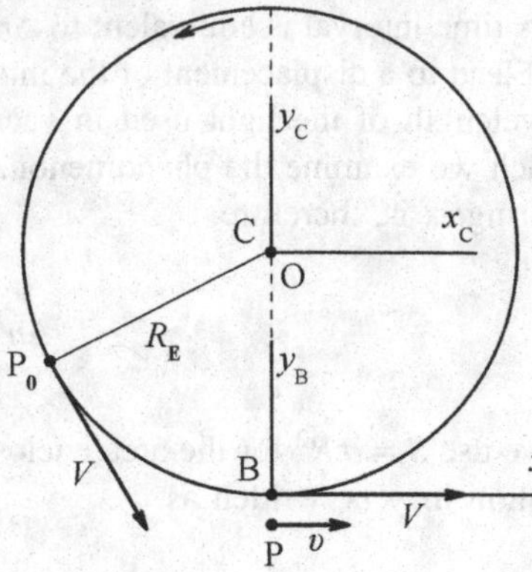

Fig. 4.8 Two clocks, P and P_0, in relative motion to each other on the rotating Earth. The inertial frame of reference C is at rest at the center of the Earth

remains at rest on the Earth, at the point from which the moving clock starts its journey and where it will return after a complete revolution around the Earth. For an inertial observer C, say at the center of the Earth, clock P_0 moves with the velocity V of the points on the Equator, as the Earth rotates about its axis. For the same observer, clock P moves with a speed υ_C, which we will evaluate. At some moment, let clock P coincide with a point B at rest on the Earth, relative to which the clock moves with speed υ in the direction of positive x_C. The speed of clock P relative to the observer C is

$$\upsilon_C = \frac{\upsilon + V}{1 + \frac{\upsilon V}{c^2}},$$

to which there corresponds a Lorentz factor of

$$\gamma_C = \left[1 - \frac{(\upsilon + V)^2}{c^2}\left(1 + \frac{\upsilon V}{c^2}\right)^{-2}\right]^{-1/2}. \tag{4.16}$$

Due to the phenomenon of the dilation of time, if time Δt_C passes for C, the time that will pass for clock P_0 will be

$$\Delta t_0 = \Delta t_C \sqrt{1 - \frac{V^2}{c^2}} \tag{4.17}$$

and for clock P

$$\Delta t = \Delta t_C \sqrt{1 - \frac{(\upsilon + V)^2}{c^2}\left(1 + \frac{\upsilon V}{c^2}\right)^{-2}}. \tag{4.18}$$

The ratio of the two time intervals is

$$\frac{\Delta t}{\Delta t_0} = \sqrt{1 - \frac{(\upsilon + V)^2}{c^2}\left(1 + \frac{\upsilon V}{c^2}\right)^{-2}} \Bigg/ \sqrt{1 - \frac{V^2}{c^2}}. \tag{4.19}$$

If the moving clock P revolves once around the Earth, along the Equator, in time T_0 as this is measured by clock P_0, and time T as measured by the clock P itself, it will be

$$\frac{T}{T_0} = \left[1 - \frac{(\upsilon + V)^2}{c^2}\left(1 + \frac{\upsilon V}{c^2}\right)^{-2}\right]^{1/2}\left(1 - \frac{V^2}{c^2}\right)^{-1/2}. \tag{4.20}$$

Expanding, for $V \ll c$, and keeping terms with powers V^2/c^2, υ^2/c^2 and $V\upsilon/c^2$ only, we have

$$\frac{T}{T_0} \approx \left(1 - \frac{1}{2}\frac{(\upsilon + V)^2}{c^2}\right)\left(1 + \frac{1}{2}\frac{V^2}{c^2}\right) \approx 1 + \frac{1}{2}\frac{V^2}{c^2} - \frac{1}{2}\frac{\upsilon^2}{c^2} - \frac{\upsilon V}{c^2} - \frac{1}{2}\frac{V^2}{c^2} \tag{4.21}$$

and, finally,

$$\frac{T}{T_0} \approx 1 - \frac{\upsilon^2}{2c^2} - \frac{\upsilon V}{c^2} = 1 - \frac{\upsilon V}{c^2}\left(1 + \frac{\upsilon}{2V}\right). \tag{4.22}$$

We note that, with the approximations made, the relativistic addition of speeds used in the evaluation of υ_C, was not actually necessary.

The difference in the indication of the clock that travelled around the Earth from that of the 'stationary' clock will be

$$\Delta T \equiv T - T_0 \approx -\frac{V\upsilon}{c^2} T_0 \left(1 + \frac{\upsilon}{2V}\right) \tag{4.23}$$

with υ taken positive for motion towards east. If it is also $\upsilon \ll V$, then

$$\Delta T \equiv T - T_0 \approx -\frac{V\upsilon}{c^2} T_0. \tag{4.24}$$

However, it is $\upsilon T_0 = 2\pi R_E$, where R_E is the equatorial radius of the Earth. Also, $V = 2\pi R_E/T_E$, where $T_E = 1$ day is the period of rotation of the Earth about its axis. Therefore,

$$\Delta T = -\frac{(2\pi R_E)^2}{c^2 T_E} = -207 \text{ ns}, \tag{4.25}$$

which is independent of the magnitude of υ, to the approximation used. For westward motion, the sign of υ in Eq. (4.24) changes, and the equation gives for the time difference the value of $\Delta T = +207$ ns.

We note that, for very small speeds, $\upsilon \ll V$, the difference in the clocks' readings is a constant, characteristic of the Earth. This is expected, because with low speeds the differences in the rates of the clocks are indeed smaller, but the journey lasts longer. The effect of time dilation is measurable, in this case, thanks to the sensitivity of atomic clocks and also due to the appearance of the product υV of a speed υ, which may be very small, with a relatively large velocity V. Whatever the value of υ, its integration with respect to time will always give the circumference of the Earth, so it does not affect the final result, to a first approximation.

If the two clocks are not situated on the Equator but on a circle of constant geographic latitude λ, then, instead of the equatorial radius of the Earth, R_E, the radius of the orbit should be taken as $R_E \cos\lambda$, and Eq. (4.25) is modified to

$$\Delta T = -\frac{(2\pi R_E \cos\lambda)^2}{c^2 T_E} = -207\cos^2\lambda \text{ ns}. \tag{4.26}$$

It is proved that, in general, for a displacement P_0P such as the one shown in Fig. 4.9, with any speed which is much smaller than the speed of the Earth's surface, the moving clock P will be left behind relative to the stationary clock P_0, solely as a result of the Sagnac effect, by a time which is proportional to the shaded area of the figure. The conversion coefficient is 1.62 ns per 10^6 km^2 of area. The area is considered to be positive for westward motion and negative for eastward motion. This effect is easily observable with moving atomic clocks. The proof of the theorem will be given in Example 4.1.

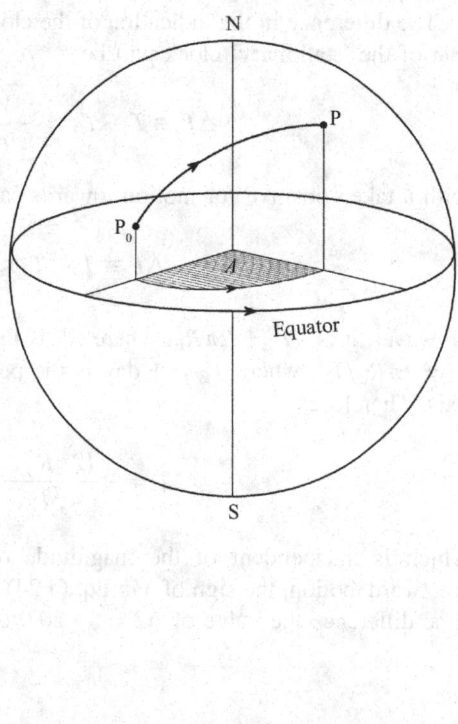

Fig. 4.9 A clock P, moving on the surface of the Earth with a component of its velocity in the eastward direction, lags behind a clock at rest at the point P_0, solely due to the Sagnac effect, by an amount that is proportional to the *shaded area, A*, defined by the projection of the position of the moving clock on the plane of the Equator

Phileas Fogg, in going around the Earth in 80 days, moving eastwards, did not only forget to subtract 1 day from the date when he crossed the International Date Line at the geographic latitude of ±180°, but he also forgot to add about one fifth of a millionth of a second to the indication of his watch, when he returned to his starting point, to correct for the consequences of the Sagnac effect!

For speeds υ which are not negligible compared to V, Eq. (4.23) is true, which may also be written as

$$\Delta T = T - T_0 = \pm 207\left(1 + \frac{\upsilon}{2V}\right)\text{ ns} = \pm 207\left(1 \mp \frac{T_\text{E}}{2T_0}\right)\text{ ns} \qquad (4.27)$$

(upper signs and negative υ for motions towards the West and lower signs and positive υ for motions towards the East).

With $R_\text{E} = 6378$ km, the equatorial speed of the Earth is $V = 1670$ km/h $= 464$ m/s. For an airplane moving with the speed of sound in air, 340 m/s, the terms in parentheses in Eq. (4.27) have, for $\upsilon = +340$ m/s, the value of $\left(1 + \frac{\upsilon}{2V}\right) = 1.37$, in which case it is $\Delta T \equiv T - T_0 \approx -283$ ns for motion towards the East and, for $\upsilon = -340$ m/s the value of $\left(1 + \frac{\upsilon}{2V}\right) = 0.634$, in which case it is $\Delta T \equiv T - T_0 \approx 131$ ns for motion towards the West.

Example 4.1 The Sagnac effect for the motion of a clock on the surface of the Earth

The theorem of Fig. 4.9 will be proved.

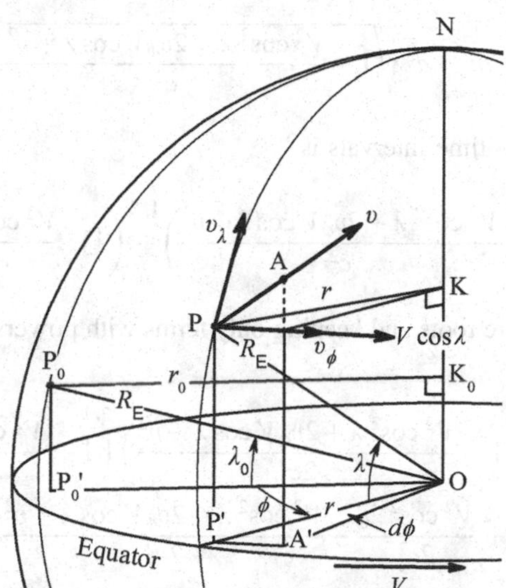

Let a clock P move on the surface of the Earth following a certain path. At the starting point of the clock, there is, at rest relative to the Earth, another clock P_0. The geographic latitude of P_0 is λ_0, while clock P is, at a certain moment, at geographic latitude λ. Clocks P_0 and P are at distances from the Earth's axis equal to r_0 and r, respectively. If the angular velocity of the Earth due to its rotation about its axis is ω and the speed of the points on the Equator is V, the speeds of the clocks, due to the Earth's rotation alone, are $\omega r_0 = V\cos\lambda_0$ and $\omega r = V\cos\lambda$, respectively. The speed of clock P relative to the surface of the Earth is $\boldsymbol{\upsilon}$, with components υ_ϕ eastwards and υ_λ towards the North (see figure). As we have seen above, given the approximations we will make, for the evaluation of the total speed of P it is not necessary to use the relativistic law for the addition of velocities. Relative to an inertial observer situated, for example, at the center of the Earth, the speed of clock P_0 is $V\cos\lambda_0$, while that of P has components $\upsilon_\phi + V\cos\lambda$ towards East and υ_λ towards North. The total speed of P is, therefore, equal to $\sqrt{\upsilon_\lambda^2 + (\upsilon_\phi + V\cos\lambda)^2}$.

Let time $d\tau$ pass for an inertial observer, say at the center of the Earth. Clock P_0 registers time

$$dt_0 = d\tau\sqrt{1 - \frac{V^2\cos^2\lambda_0}{c^2}}$$

while clock P registers time

$$dt = d\tau\sqrt{1 - \frac{\upsilon_\lambda^2 + (\upsilon_\phi + V\cos\lambda)^2}{c^2}}$$

or

$$dt = d\tau\sqrt{1 - \frac{V^2\cos^2\lambda + 2\upsilon_\phi V\cos\lambda + \upsilon^2}{c^2}}.$$

The ratio of the two time intervals is

$$\frac{dt}{dt_0} = \left(1 - \frac{V^2\cos^2\lambda + 2\upsilon_\phi V\cos\lambda + \upsilon^2}{c^2}\right)^{1/2}\left(1 - \frac{V^2\cos^2\lambda_0}{c^2}\right)^{-1/2}.$$

Expanding the square roots and keeping only terms with powers of the speeds up to the second, we have

$$\begin{aligned}\frac{dt}{dt_0} &\approx \left(1 - \frac{V^2\cos^2\lambda + 2\upsilon_\phi V\cos\lambda + \upsilon^2}{2c^2}\right)\left(1 + \frac{V^2\cos^2\lambda_0}{2c^2}\right)\\ &\approx 1 + \frac{V^2\cos^2\lambda_0}{2c^2} - \frac{V^2\cos^2\lambda + 2\upsilon_\phi V\cos\lambda + \upsilon^2}{2c^2}\end{aligned}$$

If the difference in the readings of the two clocks is defined as $\Delta t = t - t_0$, the change in this difference is $d(\Delta t) = dt - dt_0$

$$d(\Delta t) = \frac{dt_0}{2c^2}\left(V^2\cos^2\lambda_0 - V^2\cos^2\lambda - 2\upsilon_\phi V\cos\lambda - \upsilon^2\right).$$

The first two terms give the difference which results purely from the dilation of time due to the rotation of the Earth. The last two terms are due to the motion of clock P relative to the surface of the Earth, i.e. to the Sagnac effect. The difference in the indications of the two clocks which is due purely to the Sagnac effect is:

$$d(\Delta t)_{\mathrm{S}} = -\frac{dt_0}{2c^2}\left(2\upsilon_\phi V\cos\lambda + \upsilon^2\right).$$

If it is $\lambda = 0$ and $\upsilon_\lambda = 0$, the last equation reduces to Eq. (4.23).

For υ low enough that it is $V\cos\lambda \gg \upsilon^2/2\upsilon_\phi$, the second term may be omitted. Then, it is

$$d(\Delta t)_{\mathrm{S}} \approx -\frac{dt_0}{c^2}\upsilon_\phi V\cos\lambda.$$

Putting $\upsilon_\phi dt_0 = (\mathrm{P'A'})$ and $V\cos\lambda = \omega r$, we have

$$d(\Delta t)_{\mathrm{S}} \approx -\frac{1}{c^2}(\mathrm{P'A'})\omega r = -\frac{2\omega}{c^2}\left[\frac{1}{2}(\mathrm{P'A'})r\right] = -\frac{2\omega}{c^2}dS,$$

where dS is the area of the triangle P′A′O. This area is positive if it is described in a clockwise sense, as seen from the North pole.

For a given path on the surface of the Earth, with relatively low speeds, the difference in the indications of the two clocks due to the Sagnac effect (indication of clock P minus the indication of clock $\mathrm{P_0}$) is

$$(\Delta t)_{\mathrm{S}} = -\frac{2\omega}{c^2}S,$$

where S is the area defined on the equatorial plane by the projections of the path described by the moving clock and the two meridians passing through the initial and the final positions of the clock (see Fig. 4.9). The difference is negative for motion with an eastward component and positive for motion with a westward component. We have, of course, assumed that the moving clock does not change direction in such a way that the surface S folds back on itself. If this happens, however, the result still holds, as the negative and positive areas, corresponding to loss and gain of the moving clock, cancel each other. The coefficient of proportionality is $2\omega/c^2 = 1.62 \times 10^{-21}$ s/m^2 or 1.62 ns per 10^6 km^2.

4.5 The Experiment of Hafele and Keating

The accuracy with which atomic clocks can measure time, makes it possible to test the predictions made by the Special Theory of Relativity for moving clocks. Such a test is the experiment of Hafele and Keating performed in 1971 [6]. These two researchers put 4 cesium beam atomic clocks in a commercial airplane which circumnavigated the Earth, first in 41 h flying eastwards and then in 49 h flying westwards and, in each case, compared their indications with atomic clocks that had remained stationary on the surface of the Earth, at the point from which the flying clocks started their journey and where they finally returned. The prediction of the Special Theory of Relativity is that the moving clocks will differ from the stationary ones by the time interval we evaluated in the previous section.

Apart from this difference, a difference must also be taken into account which is predicted by the General Theory of Relativity, due to the fact that the clocks on the surface of the Earth are at a different gravitational potential than those in the airplane. According to the General Theory of Relativity, a clock at a height h in a homogeneous gravitational field of intensity (acceleration of gravity) g, will be in advance compared to a clock at height h_0 by an amount of time equal to

$$d(\Delta T)_{\mathrm{G}} = \frac{g}{c^2}(h - h_0)dt_0 \tag{4.28}$$

when the clock at the lower potential measures the passage of a time interval equal to dt_0. For the same time interval dt_0, as found in Example 4.1, according to the Special Theory of Relativity, the flying clock will be in advance by an amount of time equal to

$$d(\Delta t)_{\mathrm{S}} = \frac{dt_0}{2c^2}\left(V^2\cos^2\lambda_0 - V^2\cos^2\lambda - 2\upsilon_\phi V\cos\lambda - \upsilon^2\right) \tag{4.29}$$

where V is the equatorial speed of the Earth, λ_0 is the geographic latitude of the clock which is at rest on the surface of the Earth, λ is that of the flying clock, which has a total speed υ and an eastward component of its velocity equal to υ_ϕ. The total interval of time by which the flying clock will be in advance, after a time dt_0, will be

$$d(\Delta T) = \frac{dt_0}{2c^2}\left[2g(h - h_0) + V^2\cos^2\lambda_0 - V^2\cos^2\lambda - 2\upsilon_\phi V\cos\lambda - \upsilon^2\right] \tag{4.30}$$

with υ_ϕ being positive for motion towards the East.

Of course, neither the speed υ, nor the latitude λ and, naturally, the height h were constant during the flight. With a detailed recording of these magnitudes as functions of time, during the flight, the total time difference in the indications of the clocks is estimated to be equal to

$$\Delta T = \frac{g}{c^2}\int_0^{T_0}(h-h_0)dt_0 + \frac{V_2}{2c^2}\int_0^{T_0}(\cos^2\lambda_0 - \cos^2\lambda)dt_0$$
$$-\frac{1}{2c^2}\int_0^{T_0}(2v_\phi V\cos\lambda + v^2)dt_0, \tag{4.31}$$

where T_0 is the total time of the experiment. The first term is due to the difference in the gravitational potential with height, the second is due to the dilation of time caused by the rotation of the Earth and the third is due entirely to the Sagnac effect. Using the data of the flights, the partial predictions were calculated for the two journeys, as well as the total expected differences in the readings of the clocks. The theoretical predictions and the results of the experiment are summarized in Table 4.1. The differences between the theoretical values from those measured are within the values of the experimental errors, especially for the westward motion. The agreement of theory with the differences observed is very satisfactory.

After the experiment of Hafele and Keating, a similar experiment was performed in 1975, in which 3 cesium beam atomic clocks and 3 rubidium ones were loaded on an airplane which travelled a total of 15 h with a speed of 240 knots (445 km/h) at a height of about 10 km. The airplane followed paths in the shape of the figure of 8 at an almost constant geographical latitude, to minimize the consequences of the Sagnac effect due to the rotation of the Earth. The low speed of the airplane also limited the consequences of the dilation of time. The effect of the gravitational field on the rates of the clocks was measured by comparison with a similar group of 6 atomic clocks on the surface of the Earth. In a typical flight, the flying clocks were expected to be leading by 52.8 ns due to gravity and lag behind by 5.7 ns due to speed. The expected clear difference was 47.1 ns, which agreed very well with the measured difference of 47 ns which had an uncertainty of ±1.5 ns.

The ability of modern technology to detect the differences predicted by theory among clocks in relative motion with each other, made it possible to verify these predictions experimentally but also made it necessary for corrections to be made in

Table 4.1 The results of the Hafele and Keating experiment (1971)

		Time ΔT by which the moving clocks were in advance compared to the clocks which remained stationary on the surface of the Earth (ns)	
		Flight towards west (duration: 49 h)	Flight towards east (duration: 41 h)
Theoretical predictions:	Due to motion (special theory of relativity)	96 ± 10	-184 ± 18
	Due to gravity (general theory of relativity)	179 ± 18	144 ± 14
Total theoretically predicted time difference		275 ± 21	-40 ± 23
Measured time difference		273 ± 7	-59 ± 10
| Difference of theory–experiment |		2 ± 22	19 ± 25

the measurement of time at the limits of today's possibilities. Jones [7] cites the following example which demonstrates the kind of corrections that must be made in such cases: An atomic clock is moved, for comparison, from the German Office of Measures (Physikalisch-Technische Bundesanstalt, PTB) in Braunschweig, to the United States Naval Observatory, in Washington. If the journey from Berlin to Washington, a distance of 6700 km, is made with an airplane flying at a constant speed of 800 km/h at a height of 10 km, when the atomic clock arrives at the USNO, it will have lost 8.3 ns due to the dilation of time, have gained 32.9 ns due to the difference in gravitational potential at the height of 10 km and have gained 18.8 ns due to the Sagnac effect. In total, the clock will have gained 43.4 ns, a time interval that must certainly be taken into account. In the return journey, the sign of the difference due to the Sagnac effect will change and the clock will gain a total of only 5.8 ns. As Jones very aptly remarks, the jetlag effect is always worse when one travels towards the West!

4.6 Einstein's Train

In Sect. 2.3 we examined the following problem: A train moves relative to an inertial observer O with speed V. At exactly the midpoint of the train, and at rest relative to it, stands an observer O′. At the two ends of the train there are two sources of light, A and B. At the exact moment the observers O′ and O pass near each other, two pulses of light reach them simultaneously, having been emitted from the sources A and B, respectively (Fig. 4.10). We consider that the train moves with a speed smaller than that of light in vacuum. We associate with observers O and O′ the inertial frames of reference S and S′, respectively. The axes of the two frames of reference coincide at the moment $t = t' = 0$. We will examine the events associated with the pulses and their observation first in the frame of reference S′ of observer O′ on the train (Fig. 4.11):

If the rest length of the train is L_0 and the two sources are at equal distances $L_0/2$ from O′, the light from the two sources needed a time equal to $L_0/2c$ to reach O′. Since the two pulses arrive at the point $x' = 0$ simultaneously at the moment $t' = 0$,

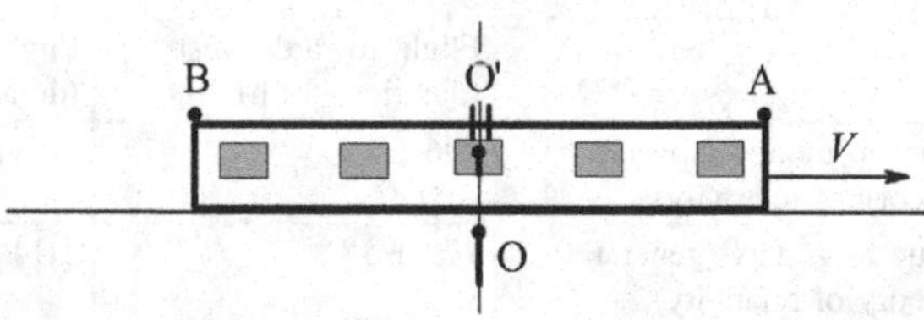

Fig. 4.10 A train moves with speed V relative to an observer O. At the *middle* of the train stands another observer, O′. When the two observers coincide, two pulses of light reach both of them, having been emitted, respectively, from the light sources A and B at the train's two ends

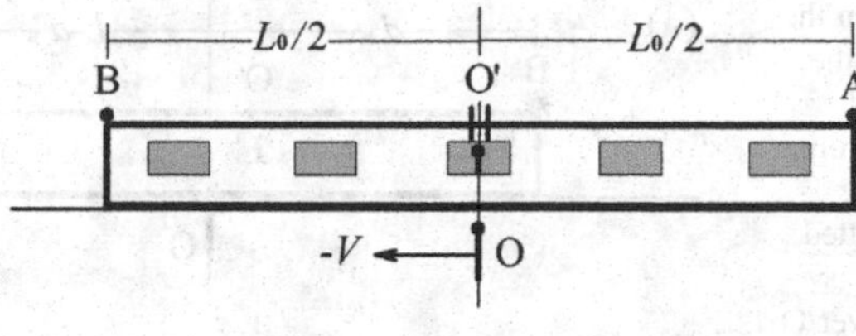

Fig. 4.11 Observer O′, on the train, concludes that the two pulses of light were emitted simultaneously from the two sources, A and B, which lie at equal distances from him

the two events of the emission of the pulses from sources, A and B, have coordinates:

$$\text{Pulse A}: \qquad x'_{\mathrm{A}} = \frac{L_0}{2}, \ t'_{\mathrm{A}} = -\frac{L_0}{2c}. \tag{4.32}$$

$$\text{Pulse B}: \qquad x'_{\mathrm{B}} = -\frac{L_0}{2}, \ t'_{\mathrm{B}} = -\frac{L_0}{2c}. \tag{4.33}$$

In the frame of reference S of the observer O, on the platform, the coordinates of the two events are found using the Lorentz transformation from frame S′ to frame S. We find:

Pulse A:

$$x_{\mathrm{A}} = \gamma(x'_{\mathrm{A}} + \beta c t'_{\mathrm{A}}) = \gamma\left(\frac{L_0}{2} - \beta\frac{L_0}{2}\right) = \frac{L_0}{2}\sqrt{\frac{1-\beta}{1+\beta}}, \tag{4.34}$$

$$t_{\mathrm{A}} = \gamma(t'_{\mathrm{A}} + (\beta/c)x'_{\mathrm{A}}) = \gamma\left(-\frac{L_0}{2c} + \frac{\beta}{c}\frac{L_0}{2}\right) = -\frac{L_0}{2c}\sqrt{\frac{1-\beta}{1+\beta}}. \tag{4.35}$$

Pulse B:

$$x_{\mathrm{B}} = \gamma(x'_{\mathrm{B}} + \beta c t'_{\mathrm{B}}) = \gamma\left(-\frac{L_0}{2} - \beta\frac{L_0}{2}\right) = -\frac{L_0}{2}\sqrt{\frac{1+\beta}{1-\beta}}, \tag{4.36}$$

$$t_{\mathrm{B}} = \gamma(t'_{\mathrm{B}} + (\beta/c)x'_{\mathrm{B}}) = \gamma\left(-\frac{L_0}{2c} - \frac{\beta}{c}\frac{L_0}{2}\right) = -\frac{L_0}{2c}\sqrt{\frac{1+\beta}{1-\beta}}. \tag{4.37}$$

We notice that it is $x_{\mathrm{A}}/t_{\mathrm{A}} = -c$ and $x_{\mathrm{B}}/t_{\mathrm{B}} = c$, as expected.

Figure 4.12 shows the positions of the train at the moments of emission of the two pulses. Obviously, due to the contraction of length, the length of the train will be, for observer O, equal to $L = L_0/\gamma$. When source B emits its pulse (Fig. 4.12a) this source is at a distance from O equal to $d_B = |x_{\mathrm{B}}| = \frac{L_0}{2}\sqrt{\frac{1+\beta}{1-\beta}}$, which is greater than $L_0/2$. When source A emits its pulse (Fig. 4.12b) this source is at a distance

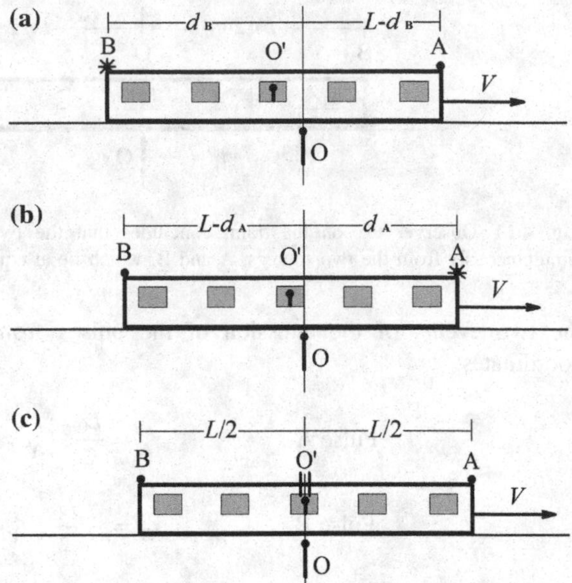

Fig. 4.12 Observer O, on the platform, concludes that the two light pulses were not emitted simultaneously from the two sources A and B. **a** When the pulse is emitted by source B, that source is further away from observer O than source A, when it emits its pulse (**b**), i.e. it is $|d_B| > |d_A|$. Pulse B, therefore, must have been emitted earlier than pulse A, since both arrive simultaneously at O (**c**)

from O equal to $d_{\mathrm{A}} = x_{\mathrm{A}} = \dfrac{L_0}{2}\sqrt{\dfrac{1-\beta}{1+\beta}}$, which is smaller than $L_0/2$. We see that it is $d_B > d_A$, as expected.

The time difference in the emission of the pulses is

$$\Delta t = t_{\mathrm{B}} - t_{\mathrm{A}} = -\frac{L_0}{2c}\sqrt{\frac{1+\beta}{1-\beta}} + \frac{L_0}{2c}\sqrt{\frac{1-\beta}{1+\beta}} = -\frac{\beta L_0/c}{\sqrt{1-\beta^2}} = -\gamma\frac{VL_0}{c^2}, \qquad (4.38)$$

with pulse B being emitted earlier than pulse A in the frame of reference of observer O.

4.7 The Twin Paradox

Certainly the best known 'paradox' of the Special Theory of Relativity is the *twin paradox*. It is also known as the *clock paradox* or *Langevin's paradox*, as Langevin [8] was the first to state it. Let us examine how two observers see each other, when they are in relative motion with a constant speed between them (Fig. 4.13). Observer O, in the inertial frame of reference S, sees observer O′, in the frame of reference S′, move relative to himself with a constant speed V (Fig. 4.13a). The clocks of O′, including the biological ones, appear to O to count time more slowly than his own. According to O, therefore, O′ grows older at a rate which is slower than his own rate of growing old. Now, the picture from the frame of reference of

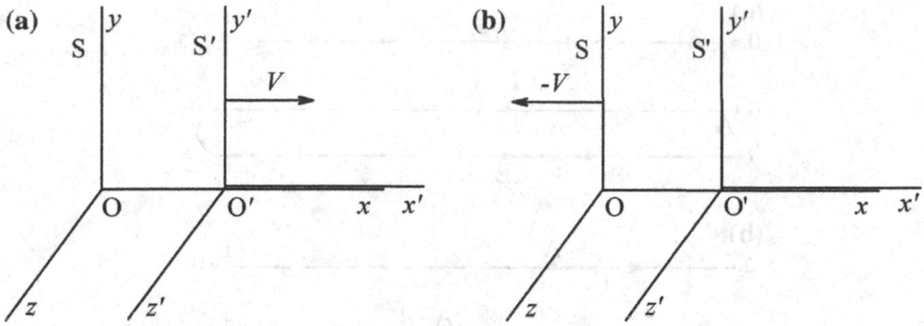

Fig. 4.13 *The twin paradox.* **a** Observer O, in the frame of reference S, sees observer O′, in frame S′, moving with speed *V*. All the clocks of O′, including the biological, appear to O to count time more slowly than his own. According to O, therefore, O′ ages more slowly than himself. **b** Observer O′, in the frame of reference S′, sees observer O, in frame S, moving with speed $-V$. All the clocks of O, including the biological, appear to O to count time more slowly than his own. According to O′, therefore, O ages more slowly than himself

O′ is absolutely symmetrical to this. Observer O′, in frame S′, sees the observer O, in frame S, move relative to himself with a speed of $-V$ (Fig. 4.13b). O's clocks, including the biological, appear to O′ to be slower than his own. According to O′, therefore, O grows old at a slower rate than he does. Each one of the two observers, therefore, sees the other grow older at a slower rate than himself. Who is right?

The answer is that, according to the Special Theory of Relativity, both observers are right. If the clocks of O and O′ were synchronized when the two frames coincided, O will observe the clock of O′ to lag behind his own clocks in frame S, and O′ will observe the clock of O lag behind his own clocks in frame S′.[1]

The paradox is supposed to arise if one of the observers reverses direction of motion and returns to the position of the other observer, so that their clocks may be compared (Fig. 4.14). Which one's clock will be found to lag behind that of the other? The effect is dramatized by assuming that the two observers, O and O′, are twins and O′ departs for a trip, moving with a constant speed *V* relative to his brother O. If after some time O′ returns to the position where O is situated, who will be younger? Due to the 'symmetry' of relative motion, each one of them will expect his brother to have remained younger than himself!

However, is the relative motion symmetrical? A more careful examination shows that it is not. For O′ to return to O's position, he must reverse his velocity relative to O, from *V* to $-V$, or some other negative value. In order to achieve this, he must

[1]The reader may have noticed that we referred to the (one) clock of O′ when this is observed by O, but to the (many) clocks in his own frame of reference. The same holds in the symmetrical case when O′ is making the observations. The reason is a simple one: for each of the observers, all the clocks of his own frame of reference remain synchronized, while the clocks at various points of the frame that is moving relative to him appear to him not to be synchronized, even when they are synchronized in their own frame of reference (Example 3.7).

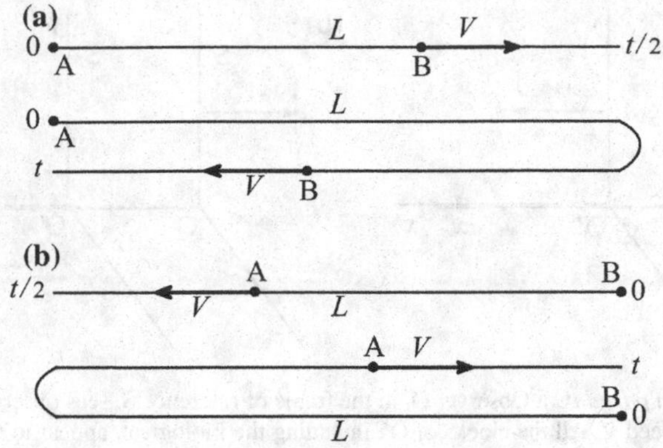

Fig. 4.14 *The twin paradox.* **a** B moves away from A with speed V for a time interval $t/2$, as measured by B, and then returns, having travelled for a total time t. A will find that for himself the time elapsed is longer than that for B, by an amount $t(\gamma - 1)$. **b** For B, however, it is A that moves and A's clocks should lag behind his

decelerate for some period of time. This is equivalent to transitions to successive different inertial frames of reference. This procedure destroys the symmetry in the motions of the twins. O′ will suffer a deceleration while O will not. Acceleration, in contrast to velocity, is an absolute magnitude and not a relative one. Accelerometers may be used to measure changes in the value of velocity and tell us which one of the two observers actually reverses his direction of motion.

Does this process of deceleration leave O″s clock unaffected? The answer is given by the General Theory of Relativity. The acceleration in the direction of O is equivalent to a gravitational field in the opposite direction, in the direction of the original velocity V. It is known that a clock in a more intense gravitational field than another, counts time at a slower rate than that clock. This retardation will explain the fact that the twin who changes direction of motion and returns to the other, will be the younger of the two when they meet.

Does the General Theory of Relativity give the correct answer quantitatively? A complete analysis using the General Theory of Relativity does indeed give the correct answer [9]. An approximate proof up to powers $(V/c)^2$ of the speed may be easily given. We will assume that O′ moves with speed V relative to O and at some moment passes very close to him. At that moment, their cocks show the same time, $t = t' = 0$. O′ then moves for time $\Delta t/2$ with speed V relative to O and covers a distance $l = V\Delta t/2$. Then, O′ suffers a constant acceleration $-a$ for a time interval T. For a change of $-2V$ to be achieved in the speed of O′, it must be $T = 2V/a$. If we make the acceleration time T negligible compared to the total duration Δt of the trip, as measured by O, and $\Delta t'$ is the total duration of the trip, as measured by O′, it will be $\Delta t = \gamma \Delta t'$, according to the Special Theory of Relativity.

We will now evaluate the time as measured by O′. We calculate first the effect that the acceleration will have on O's clock, according to the General Theory of Relativity. If two clocks are in a gravitational field of intensity a, with a difference of l in their heights, and time T passes for the clock at the highest point, then the clock at the lower point will lose, relative to the other, time alT/c^2. Because, here, it is $l = V\Delta t/2$ and $T = 2V/a$, this delay is equal to $(V/c)^2\Delta t = \beta^2\Delta t$.

O finds, according to the Special Theory of Relativity, $\Delta t = \gamma\Delta t' \approx \Delta t'(1 + \frac{1}{2}\beta^2)$. If O′ ignores the effect of acceleration, he will find $\Delta t' = \gamma\Delta t \approx \Delta t(1 + \frac{1}{2}\beta^2)$. If he also subtracts the delay of the clock due to acceleration, he will find $\Delta t' \approx \Delta t(1 + \frac{1}{2}\beta^2) - \Delta t\beta^2 = \Delta t(1 - \frac{1}{2}\beta^2)$. Since, keeping only terms up to the square of β, it is $(1 - \frac{1}{2}\beta^2) \approx (1 + \frac{1}{2}\beta^2)^{-1}$, we see that O′ agrees with O that it is $\Delta t = \Delta t'(1 + \frac{1}{2}\beta^2)$. Both agree that the clock of the observer who will be accelerated in order to return to the other observer will lag behind.

These arguments may have persuaded the reader that there is no paradox in this case. The fact is, however, that these conclusions differ radically from our everyday experiences. It is not, therefore, surprising that the discussion for the clock or the twin paradox has been revived in scientific journals many times, ever since it was first described.

4.8 Motion with a Constant Proper Acceleration. Hyperbolic Motion

We will now examine the motion of a body moving with constant proper acceleration. We will not be concerned with the way by which this is achieved or the variation of the body's mass that this motion implies. We will simply examine the motion as a problem in kinematics.

Let us assume that a spaceship starts from the Earth, with zero initial speed at $t = 0$, and that it moves along the x-axis with a constant acceleration g as this is sensed by the passengers of the spaceship (proper acceleration). The value of the acceleration may be chosen so that the passengers of the spaceship feel an artificial gravity equal or similar to that on the surface of the Earth. We will evaluate the distance of the spaceship from the Earth, $x(t)$, as a function of the time t measured by someone on the Earth and the corresponding time τ which passes for the passengers of the spaceship.

We have found in Sect. 3.3.1 that, for a body moving with a constant proper acceleration, g, it is [Eq. (3.63)]

$$\frac{dV}{dt} = g\left(1 - \frac{V^2}{c^2}\right)^{3/2}, \tag{4.39}$$

where V is the body's (spaceship's) speed relative to an inertial frame of reference, which, here, we will take to be that of the Earth. This relation may be integrated with respect to t to give us the speed $V(t)$. Thus,

$$\int_0^{V(t)} \frac{dV}{(1 - V^2/c^2)^{3/2}} = \int_0^t g dt \tag{4.40}$$

with the result $\left[\frac{V}{\sqrt{1 - V^2/c^2}}\right]_0^{V(t)} = [gt]_0^t$ or $\frac{V/c}{\sqrt{1 - V^2/c^2}} = \frac{g}{c} t.$ (4.41)

This may be solved to give, with $\tau_0 = c/g$,

$$V = \frac{c}{\sqrt{1 + \frac{c^2}{g^2 t^2}}} = \frac{c}{\sqrt{1 + \frac{\tau_0^2}{t^2}}}, \tag{4.42}$$

which is the speed of the spaceship as a function of time, as these quantities are measured in the frame S of the Earth. From Eq. (4.42), it is seen that, in order to reach the speed of c, a body moving with a constant proper acceleration needs an infinite amount of time (even if the energy required did not prevent such a thing happening). Figure 4.15 shows the speed acquired by the spaceship as a function of Earth time.

For the value of the speed found, the corresponding Lorentz factor, $\gamma = 1 \Big/ \sqrt{1 - V^2/c^2}$, is

$$\gamma = \sqrt{1 + \frac{g^2}{c^2} t^2} \quad \text{or} \quad \gamma = \sqrt{1 + \frac{t^2}{\tau_0^2}}. \tag{4.43}$$

Which is the distance of the spaceship from the Earth after the passage of time t in the frame of the Earth? From the relation $V = dx/dt$ we have

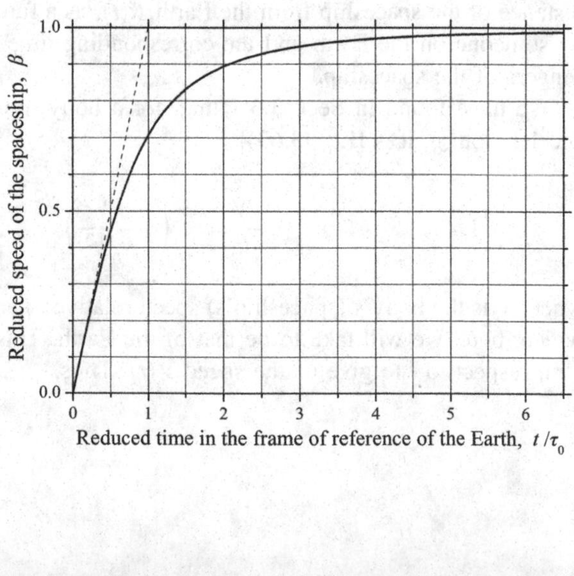

Fig. 4.15 Motion with constant proper acceleration. The variation of the speed of the spaceship, as a function of time, as this is measured on the Earth

$$dx = \frac{c\,dt}{\sqrt{1+\dfrac{\tau_0^2}{t^2}}} = c\frac{t\,dt}{\sqrt{\tau_0^2+t^2}}. \qquad (4.44)$$

Integrating, we find

$$\int_0^{x(t)} dx = c\int_0^t \frac{t\,dt}{\sqrt{\tau_0^2+t^2}} \quad \text{or} \quad x(t) = c\left[\sqrt{\tau_0^2+t^2}\right]_0^t \qquad (4.45)$$

and finally

$$x(t) = c\left(\sqrt{t^2+\tau_0^2}-\tau_0\right). \qquad (4.46)$$

This relationship may also be expressed as $(x+c\tau_0)^2 - c^2t^2 = c^2\tau_0^2$, which is the equation of a hyperbola when x is plotted as a function of t. It is due to this relationship that this kind of motion is called *hyperbolic motion*. The displacement, in units of $c\tau_0$, has been plotted as a function of time, in units of τ_0, in Fig. 4.16.

To find the time required, in the frame of the Earth, for the spaceship to travel to a distance x, we solve Eq. (4.46) for t. This gives:

$$t = \sqrt{\frac{x}{c}\left(\frac{x}{c}+2\tau_0\right)}. \qquad (4.47)$$

We will now find the time which passes for the passengers of the spaceship, τ, as a function of the time t which passes in the frame of the Earth. The relation between the infinitesimal time interval on Earth, dt, and the corresponding time interval on the spaceship, $d\tau$, is $dt = \gamma d\tau$. Therefore, when a total time t passes on the Earth, the corresponding total time for the spaceship is

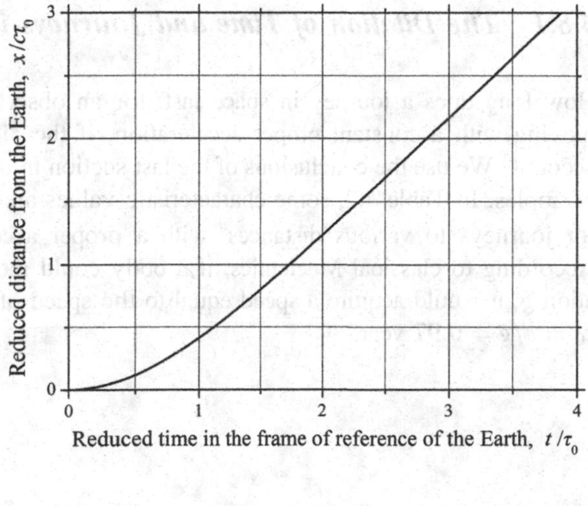

Fig. 4.16 Hyperbolic motion. The displacement x, in units of $c\tau_0$, is plotted as a function of time t, as measured by an observer on the Earth, expressed in units of τ_0

$$\tau = \int_0^t dt\sqrt{1-\frac{V^2}{c^2}} = \int_0^t dt\sqrt{1-\frac{1}{1+\tau_0^2/t^2}} = \int_0^t \frac{dt}{\sqrt{1+t^2/\tau_0^2}}$$
$$= \tau_0\left[\ln\left(t+\sqrt{\tau_0^2+t^2}\right)\right]_0^t \tag{4.48}$$

and finally
$$\tau = \tau_0 \ln\left(\frac{t}{\tau_0}+\sqrt{1+\frac{t^2}{\tau_0^2}}\right). \tag{4.49}$$

We may solve for t, the time measured by an observer on the Earth, as a function of the time for the spaceship, τ:

$$t = \tau_0\frac{e^{\tau/\tau_0}-e^{-\tau/\tau_0}}{2} = \tau_0 \sinh\left(\frac{\tau}{\tau_0}\right). \tag{4.50}$$

If we substitute t as a function of τ in $x(t)$ of Eq. (4.46), we find the distance $x(\tau)$ as a function of τ:

$$x(\tau) = c\tau_0\left[\cosh\left(\frac{\tau}{\tau_0}\right)-1\right]. \tag{4.51}$$

Substituting for the time t from Eq. (4.47) into Eq. (4.49), we find that the time needed in the frame of the spaceship for it to travel to a distance x, as this is measured in the frame of the Earth, is

$$\tau = \tau_0 \ln\left[1+\frac{x}{c\tau_0}+\sqrt{\frac{x}{c\tau_0}\left(\frac{x}{c\tau_0}+2\right)}\right]. \tag{4.52}$$

4.8.1 The Dilation of Time and Journeys in Space

How long does a journey in space last, for an observer in a spaceship which is moving with a constant proper acceleration, if the dilation of time is taken into account? We use the conclusions of the last section in order to give a few numerical examples. In Table 4.2, some characteristic values are given of the times required for journeys to various distances, with a proper acceleration of $g = 9.81\ \mathrm{m/s^2}$. According to classical Mechanics, if a body could move with a constant acceleration g, it would acquire a speed equal to the speed of light in vacuum, c, in time $\tau_0 = c/g = 0.97$ year.

Table 4.2 Data for a spaceship moving with constant proper acceleration equal to $g = 9.81$ m/s^2

t (y)	τ (y)	β	γ	x (l.y.)	Object Reached
0.00554[a]	0.00554	0.006 32	1.000 02	1.585×10^{-5}	Radius of the solar system
1	0.88	0.719 5	1.44	0.43	
5.26	2.32	0.983 7	5.52	4.4	Proxima Centauri
9.5	2.89	0.994 8	9.86	8.6	Sirius
10	2.94	0.995 33	10.4	9.1	
100	5.17	0.999 952	102	98	
10^3	7.40	0.999 999 53	1028	997	
10^4	9.64	0.999 999 995 3	10 309	10^4	
10^5	11.9	$1 - 0.47 \times 10^{-10}$	103 092	10^5	Diameter of the Galaxy
10^6	14.1	$1 - 0.47 \times 10^{-12}$	1.03×10^6	10^6	
2.5×10^6	15.0	$1 - 0.75 \times 10^{-13}$	2.57×10^6	2.5×10^6	Andromeda galaxy, M31
10^7	16.3	$1 - 0.47 \times 10^{-14}$	1.03×10^7	10^7	
5.4×10^7	18.0	$1 - 1.61 \times 10^{-16}$	5.56×10^7	5.4×10^7	Virgo cluster
10^8	18.6	$1 - 0.47 \times 10^{-16}$	1.03×10^8	10^8	
10^9	20.8	$1 - 0.47 \times 10^{-18}$	1.03×10^9	10^9	
10^{10}	23.0	$1 - 0.47 \times 10^{-20}$	1.03×10^{10}	10^{10}	
1.32×10^{10}	23.3	$1 - 0.27 \times 10^{-20}$	1.36×10^{10}	1.32×10^{10}	Galaxy Abell 1835 IR1916
9.3×10^{10}	25.2	$1 - 0.54 \times 10^{-22}$	9.59×10^{10}	9.3×10^{10}	Diameter of the visible universe

[a]2 days

Proxima Centauri is the star nearest to us. The Andromeda galaxy (M31) is the galaxy nearest to us. The Virgo cluster belongs, with our Galaxy and another 2000 galaxies, to the Virgo super cluster. Galaxy Abell 1835 IR1916 is the most distant object whose distance has been measured, as of 2012

In the table:

$t =$ time as measured by an observer on the Earth, in years (y)
$\tau =$ time as measured by an observer in the spaceship, in years (y)
$\beta =$ reduced speed of the spaceship relative to the Earth
$\gamma =$ Lorentz factor of the spaceship relative to the Earth at time t
$x =$ distance of the spaceship from the Earth, as measured by an observer on the Earth, in light years (l.y.)

The data of the table are presented in Fig. 4.17.

The times calculated above do not include stops of the spaceship at the various destinations. If the spaceship is to stop at its destination, the most 'comfortable' method would be, after the first half of the distance has been travelled, to apply a deceleration $-g$ for the remaining journey. In this case the time for the whole trip up to a distance x, will be equal to 2 times the time needed for half the distance to the

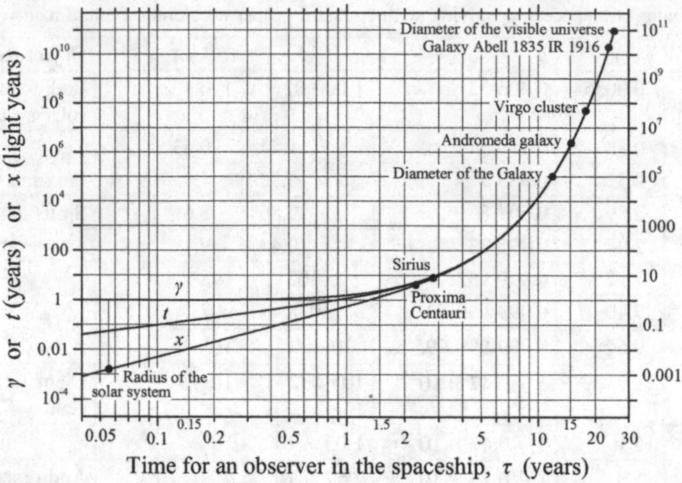

Fig. 4.17 Given for a spaceship moving with a constant proper acceleration $g = 9.81$ m/s^2, as functions of the time τ measured by an observer in the spaceship, in years, are: the final Lorentz factor, γ, achieved by the spaceship, its distance x from the Earth, in light years, and the time t, in years, that has elapsed for an observer on the Earth

final destination to be covered. For a trip to some destination at a distance L, with return, in which the successive distances of $L/2$, $L/2$, $-L/2$ and $-L/2$ are travelled with proper accelerations, g, $-g$, $-g$ and g, respectively, the total time is

in the frame of reference of the Earth
$$t_{\text{tot}} = 4\sqrt{\frac{L}{2c}\left(\frac{L}{2c} + 2\tau_0\right)} \tag{4.53}$$

and in the frame of reference of the spaceship

$$\tau_{\text{tot}} = 4\tau_0 \ln\left[1 + \frac{L}{2c\tau_0} + \sqrt{\frac{L}{2c\tau_0}\left(\frac{L}{2c\tau_0} + 2\right)}\right]. \tag{4.54}$$

The use of these relations is demonstrated in the problem that follows.

Problem 4.7

A spaceship will travel to *Proxima Centauri*, at a distance of $L = 4.4$ light years from the Earth, and return. The spaceship will cover the first half of the distance accelerating at a constant proper acceleration of $g = 9.81$ m/s^2 and the other half at a constant proper deceleration of $g = -9.81$ m/s^2, and following the reverse procedure on the return trip. How much time will be needed for the whole trip, in the frame of reference of the Earth and in that of the spaceship? Ans.: $t_{\text{tot}} = 12$ y and $\tau_{\text{tot}} = 7.2$ y

The numerical results seem to suggest that long trips in space are made possible by the dilation of time predicted by Relativity. Things are seen to be very different, however, if certain very important difficulties are considered:

The energy needed for the acceleration of a spaceship to such speeds is enormous. From Relativistic Dynamics (Chap. 6) it is known that when a body of rest mass m_0 is accelerated to a speed corresponding to a Lorentz factor γ, it behaves as if it had a mass equal to $m = \gamma m_0$, its energy is $E = \gamma m_0 c^2$ and its kinetic energy $K = (\gamma - 1) m_0 c^2$. A body of rest mass equal to 1 kg moving with $\gamma = 1000$, will have a mass which is equal to a tonne and a kinetic energy of $K = 9 \times 10^{19}$ J. For such an amount of energy to be produced, a mass equal to 999 times 1 kg must undergo fission and be converted completely to energy. Nuclear reactions have an efficiency of about 3 % and would not provide but a small fraction of the energy required even to accelerate the fissionable material itself.

In addition, although moving with speeds close to the speed of light in vacuum we make the journeys short for the travelers, the time passes at a very much faster rate on Earth. The return to Earth would be rather pointless for long trips.

These comments show that journeys into space, apart perhaps at very small distances, present considerable difficulties, at least for the time being.

4.9 Two Successive Lorentz Transformations. The Wigner Rotation

We will study the transformation that results from two successive applications of the Lorentz transformation. This will lead us to the presentation of the phenomenon of the *Wigner rotation* [10] and to the special case of the *Thomas precession*. For the purposes of this section, we will follow the methods of Fisher [11] and of Ben-Menahem [12]. The interested reader may find other approaches to the subject in the literature [13]. A detailed analysis of the effect using matrices is given in Appendix 6.

Consider three inertial frames of reference, S_1, S_2 and S_3. Frame S_2 moves relative to S_1 with velocity $\mathfrak{v}_{21}$ and frame S_3 moves relative to S_2 with velocity $\mathfrak{v}_{32}$, Fig. 4.18a, b.

From the transformation of velocity, Eqs. (3.38), we find that:

the velocity of S_3 relative to S_1 is

$$\mathfrak{v}_{31} = \frac{\mathfrak{v}_{32} + \{[(\gamma_{21} - 1)/\upsilon_{21}^2](\mathfrak{v}_{21} \cdot \mathfrak{v}_{32}) + \gamma_{21}\}\mathfrak{v}_{21}}{\gamma_{21}(1 + \mathfrak{v}_{21} \cdot \mathfrak{v}_{32}/c^2)} \tag{4.55}$$

and the velocity of S_1 relative to S_3 is

$$\mathfrak{v}_{13} = -\frac{\mathfrak{v}_{21} + \{[(\gamma_{32} - 1)/\upsilon_{32}^2](\mathfrak{v}_{21} \cdot \mathfrak{v}_{32}) + \gamma_{32}\}\mathfrak{v}_{32}}{\gamma_{32}(1 + \mathfrak{v}_{21} \cdot \mathfrak{v}_{32}/c^2)}. \tag{4.56}$$

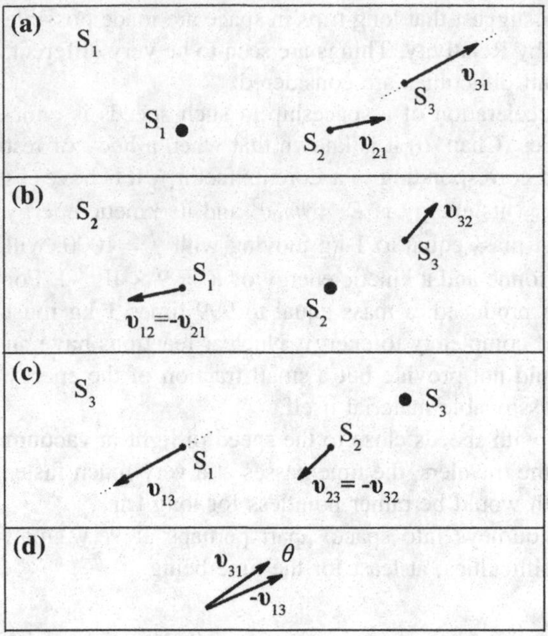

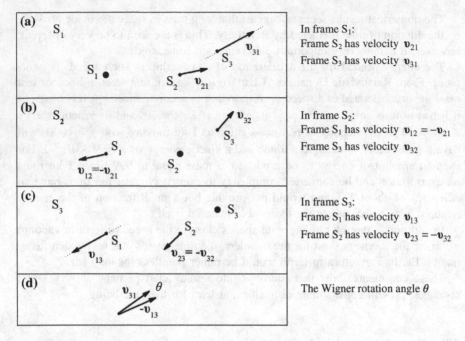

Fig. 4.18 Two successive Lorentz transformations. The Wigner rotation

If we evaluate the magnitudes of these two velocities, we find that they are both equal to

$$|\boldsymbol{v}_{13}| = |\boldsymbol{v}_{31}| = \sqrt{1 - \frac{\left(1 - v_{21}^2/c^2\right)\left(1 - v_{32}^2/c^2\right)}{\left(1 + \boldsymbol{v}_{21} \cdot \boldsymbol{v}_{32}/c^2\right)^2}} \tag{4.57}$$

and that it is

$$\gamma_{13} = \gamma_{31} = \gamma_{21}\gamma_{32}\left(1 + \frac{\boldsymbol{v}_{21} \cdot \boldsymbol{v}_{32}}{c^2}\right). \tag{4.58}$$

We may examine whether the velocities $-\boldsymbol{v}_{13}$ and $\boldsymbol{v}_{31}$ are parallel, calculating the angle between them, θ, from the relation

$$\sin\theta\,\hat{\boldsymbol{\theta}} = \frac{(-\boldsymbol{v}_{13}) \times \boldsymbol{v}_{31}}{|\boldsymbol{v}_{13}|\,|\boldsymbol{v}_{31}|}. \tag{4.59}$$

It follows that

$$\sin\theta\,\hat{\boldsymbol{\theta}} = \frac{\gamma_{21}\gamma_{32}}{\gamma_{13}^2 - 1}\left\{1 - \left[\frac{\gamma_{21}-1}{\upsilon_{21}^2}(\boldsymbol{\upsilon}_{21}\cdot\boldsymbol{\upsilon}_{32}) + \gamma_{21}\right]\left[\frac{\gamma_{32}-1}{\upsilon_{32}^2}(\boldsymbol{\upsilon}_{21}\cdot\boldsymbol{\upsilon}_{32}) + \gamma_{32}\right]\right\}\frac{\boldsymbol{\upsilon}_{21}\times\boldsymbol{\upsilon}_{32}}{c^2}. \tag{4.60}$$

This equation expresses the phenomenon of the Wigner rotation. It states that two successive Lorentz transformations for the velocity, without rotation of the axes, are equivalent to a transformation without rotation, followed by a rotation of the velocity vector, relative to its initial direction, by an angle θ.

We may simplify Eq. (4.60), using the notation

$$\boldsymbol{\upsilon}_{21} = \boldsymbol{\upsilon}, \quad \boldsymbol{\upsilon}_{32} = \boldsymbol{\upsilon}', \quad \gamma_{21} = \gamma, \quad \gamma_{32} = \gamma', \quad \gamma_{13} = \gamma'', \tag{4.61}$$

when it is

$$\sin\theta\,\hat{\boldsymbol{\theta}} = \frac{\gamma\gamma'}{\gamma''^2 - 1}\left\{1 - \left[\frac{\gamma-1}{\upsilon^2}\boldsymbol{\upsilon}\cdot\boldsymbol{\upsilon}' + \gamma\right]\left[\frac{\gamma'-1}{\upsilon'^2}\boldsymbol{\upsilon}\cdot\boldsymbol{\upsilon}' + \gamma'\right]\right\}\frac{\boldsymbol{\upsilon}\times\boldsymbol{\upsilon}'}{c^2}. \tag{4.62}$$

We should not forget that the velocity $\boldsymbol{\upsilon}_{21}$ (or $\boldsymbol{\upsilon}$) of frame S_2 is given in the frame of reference S_1, while the velocity $\boldsymbol{\upsilon}_{32}$ (or $\boldsymbol{\upsilon}'$) of frame S_3 is given in the frame of reference S_2. More convenient, for cases of physical applications, is to give both velocities in the same frame of reference. This is done in Fig. 4.19, where the three inertial frames of reference, S_1, S_2 and S_3, are shown. S_2 moves relative to S_1 with velocity $\boldsymbol{\upsilon}_{21}$ and S_3 moves relative to S_1 with velocity $\boldsymbol{\upsilon}_{31}$. From the transformation of velocities, we find that:

the velocity of S_3 relative to S_2 is

$$\boldsymbol{\upsilon}_{32} = \frac{\boldsymbol{\upsilon}_{31} + \{[(\gamma_{21}-1)/\upsilon_{21}^2](\boldsymbol{\upsilon}_{21}\cdot\boldsymbol{\upsilon}_{31}) - \gamma_{21}\}\boldsymbol{\upsilon}_{21}}{\gamma_{21}(1 - \boldsymbol{\upsilon}_{21}\cdot\boldsymbol{\upsilon}_{31}/c^2)} \tag{4.63}$$

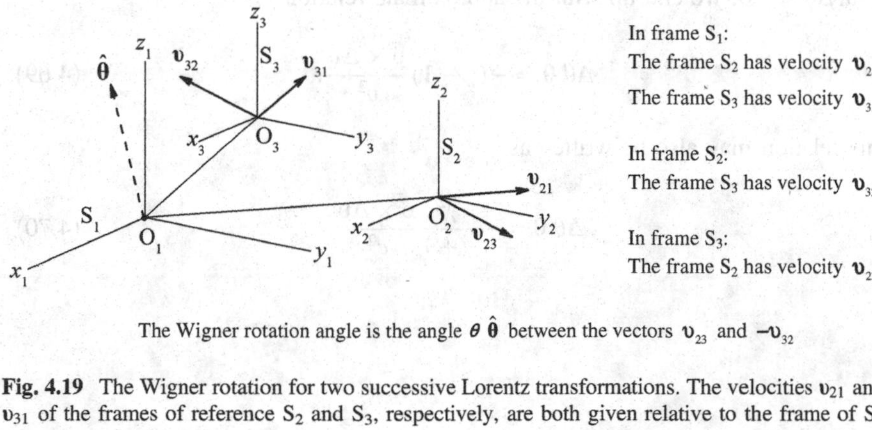

Fig. 4.19 The Wigner rotation for two successive Lorentz transformations. The velocities $\boldsymbol{\upsilon}_{21}$ and $\boldsymbol{\upsilon}_{31}$ of the frames of reference S_2 and S_3, respectively, are both given relative to the frame of S_1

and the velocity of S_2 relative to S_3 is

$$\mathbf{v}_{23} = \frac{\mathbf{v}_{21} + \{[(\gamma_{31} - 1)/v_{31}^2](\mathbf{v}_{21} \cdot \mathbf{v}_{31}) - \gamma_{31}\}\mathbf{v}_{31}}{\gamma_{31}(1 - \mathbf{v}_{21} \cdot \mathbf{v}_{31}/c^2)}. \tag{4.64}$$

The magnitudes of these two velocities are both equal to

$$|\mathbf{v}_{13}| = |\mathbf{v}_{31}| = \sqrt{1 - \frac{(1 - v_{21}^2/c^2)(1 - v_{32}^2/c^2)}{(1 + \mathbf{v}_{21} \cdot \mathbf{v}_{32}/c^2)^2}} \tag{4.65}$$

and, also,

$$\gamma_{23} = \gamma_{32} = \gamma_{21}\gamma_{31}\left(1 - \frac{\mathbf{v}_{21} \cdot \mathbf{v}_{31}}{c^2}\right). \tag{4.66}$$

The Wigner angle, θ, of the relative rotation of the axes of frame S_3 with respect to the axes of frame S_2 is evaluated using the relation

$$\sin\theta\,\hat{\boldsymbol{\theta}} = \frac{(-\mathbf{v}_{32}) \times \mathbf{v}_{23}}{|\mathbf{v}_{32}|\,|\mathbf{v}_{23}|} \tag{4.67}$$

where the unit vector $\hat{\boldsymbol{\theta}}$ gives the direction of the axis, around which the rotation happens. It turns out that it is

$$\sin\theta\,\hat{\boldsymbol{\theta}} = \frac{\gamma_{21}\gamma_{31}}{\gamma_{23}^2 - 1}\left\{1 - \left[\frac{\gamma_{21} - 1}{v_{21}^2}(\mathbf{v}_{21} \cdot \mathbf{v}_{31}) - \gamma_{21}\right]\left[\frac{\gamma_{31} - 1}{v_{31}^2}(\mathbf{v}_{21} \cdot \mathbf{v}_{31}) - \gamma_{31}\right]\right\}\frac{\mathbf{v}_{21} \times \mathbf{v}_{31}}{c^2}. \tag{4.68}$$

We will now assume that $\mathbf{v}_{21} = \mathbf{v}$ and $\gamma = \gamma_{21} = 1/\sqrt{1 - (v/c)^2}$, while the velocity $\mathbf{v}_{31}$ differs from $\mathbf{v}$ only by a very small quantity, $\Delta\mathbf{v}$, and $\mathbf{v}_{31} = \mathbf{v} + \Delta\mathbf{v}$. Expanding the quantities containing $\mathbf{v}_{31}$, $\gamma_{31} = 1/\sqrt{1 - (v_{31}/c)^2}$ and γ_{23}, up to the third power of $\Delta\mathbf{v}$, and since the angle of rotation $\Delta\theta$ is very small and, therefore, it is $\sin\Delta\theta \approx \Delta\theta$, we end up with the approximate relation

$$\Delta\theta\,\hat{\boldsymbol{\theta}} = -(\gamma - 1)\frac{\mathbf{v} \times \Delta\mathbf{v}}{v^2}. \tag{4.69}$$

This relation may also be written as

$$\Delta\theta\,\hat{\boldsymbol{\theta}} = -\frac{\gamma^2}{\gamma + 1}\frac{\mathbf{v} \times \Delta\mathbf{v}}{c^2}. \tag{4.70}$$

If the variation happens in a time interval Δt, Eq. (4.70) gives the angular velocity

$$\boldsymbol{\omega} = \frac{\Delta\theta}{\Delta t}\hat{\boldsymbol{\theta}} = -\frac{\gamma^2}{\gamma+1}\frac{\boldsymbol{\upsilon}\times\boldsymbol{a}}{c^2}, \tag{4.71}$$

where $\boldsymbol{a} = \Delta\boldsymbol{\upsilon}/\Delta t$ is the acceleration of frame S_3 relative to frame S_1. If the speed υ is very small compared to c, then $\gamma \to 1$ and the angular velocity of Eq. (4.71) becomes

$$\boldsymbol{\omega}_T = -\frac{\boldsymbol{\upsilon}\times\boldsymbol{a}}{2c^2}. \tag{4.72}$$

The result states that the frame of reference of a body which moves with velocity $\boldsymbol{\upsilon}$ and acceleration $\boldsymbol{a}$ in an inertial frame of reference, rotates in this frame of reference with angular velocity $\boldsymbol{\omega}_T$ about an axis which is normal to the plane defined by $\boldsymbol{\upsilon}$ and $\boldsymbol{a}$. The angular velocity $\boldsymbol{\omega}_T$ is called angular velocity of the Thomas precession.

4.9.1 *The Thomas Precession*

In the classical picture of the atom, the motion of an electron about a nucleus has the result that a magnetic field appears at the point where the electron is situated. If we examine the motion in the frame of reference of the electron, then a positive charge Ze, the charge of the nucleus, revolves around the electron with a frequency of $1/T$, where T is the period of revolution of the electron on its orbit. This is equivalent to Z/T charges e passing a point on the path of the nucleus per unit time, which is equivalent to a current Ze/T. The orbital angular momentum of the electron is $l = m_0\upsilon r$, where m_0 is the mass of the electron. We thus find that $1/T = \upsilon/(2\pi r) = l/(2\pi m_0 r^2)$ and the current to which the electron revolving around the nucleus is equivalent to is

$$I = \frac{Zel}{2\pi m_0 r^2}. \tag{4.73}$$

The Biot-Savart law gives us the magnetic field at the position of the electron as having the value of

$$\mathbf{B}_l = \frac{\mu_0}{4\pi}\frac{Ze}{m_0 r^3}\boldsymbol{l} \tag{4.74}$$

with $\boldsymbol{l}$ standing for the vector orbital angular momentum of the electron.

The intrinsic angular momentum (spin) of the electron, executes a (Larmor) precession about the direction of the vector of orbital angular momentum $\boldsymbol{l}$ with an angular frequency of

$$\boldsymbol{\omega}_{\mathrm{L}} = \frac{e}{m_0}\mathbf{B}_l = \frac{\mu_0}{4\pi}\frac{Ze^2}{m_0^2 r^3}\boldsymbol{l}. \tag{4.75}$$

According to what we have proved above, an electron, as it moves in a circular orbit around the nucleus, executes a precession as predicted by the Special Theory of Relativity based purely on kinematic arguments. The frame of reference of the electron, as it moves with velocity $\boldsymbol{\upsilon}$ and has a centrifugal acceleration $\boldsymbol{a}$ in the laboratory frame of reference, executes a precession with angular frequency that of the *Thomas precession* (Thomas 1903–92) [14]. The electron has an axis of symmetry which is defined by the vector of its spin. This vector executes, therefore, a precession, with angular frequency $\boldsymbol{\omega}_{\mathrm{T}}$. This is given by Eq. (4.72) and is the angular velocity of the precession of the spin vector of the electron about the direction of the vector of its orbital angular momentum, $\boldsymbol{l}$, as the electron moves around the nucleus. These are shown in Fig. 4.20.

The Bohr model for the atom gives for the speed of an electron in its orbit in the ground state ($n = 1$) of a hydrogen-like atom the value of

$$\upsilon = \sqrt{\frac{Ze^2}{4\pi\varepsilon_0 m_0 r}} \tag{4.76}$$

while its centrifugal acceleration is $a = \upsilon^2/r$. Its orbital angular momentum has magnitude $l = m_0 \upsilon r$.
Therefore,

$$\boldsymbol{\omega}_{\mathrm{T}} = -\frac{\boldsymbol{\upsilon}\times\boldsymbol{a}}{2c^2} = -\frac{\mu_0}{8\pi}\frac{Ze^2}{m_0^2 r^3}\boldsymbol{l}. \tag{4.77}$$

We note that it is

$$\boldsymbol{\omega}_{\mathrm{T}} = -\frac{\boldsymbol{\omega}_{\mathrm{L}}}{2}. \tag{4.78}$$

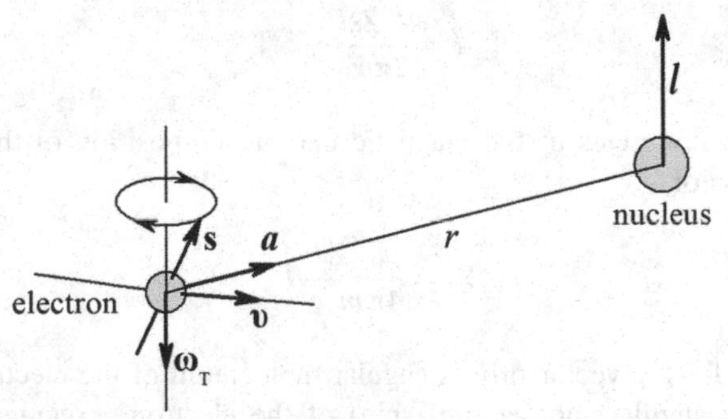

Fig. 4.20 *Thomas precession*. An electron revolves around a nucleus in an orbit of radius r with velocity $\boldsymbol{\upsilon}$ and centripetal acceleration $\boldsymbol{a}$. The orbital momentum of the electron is $\boldsymbol{l}$ and its spin is **s**. The spin vector precesses about the direction of $\boldsymbol{l}$ with an angular velocity $\boldsymbol{\omega}_{\mathrm{T}}$, as shown in the figure

The electron will, therefore, execute a precession with a resultant angular frequency of

$$\omega = \omega_{\mathrm{L}} + \omega_{\mathrm{T}} = \frac{\omega_{\mathrm{L}}}{2} = \frac{\mu_0}{8\pi}\frac{Ze^2}{m_0^2 r^3}\mathbf{l}. \tag{4.79}$$

Due to this, the Thomas precession affects by a factor of 2 the energies of interaction between the spin of the electron and its orbital angular momentum. Given that these energies are observable in the fine structure of atomic spectra, the utilization of the Thomas precession was of vital importance in the interpretation of this fine structure. Contrary to the usual relativistic effects which give small corrections, in the best cases of the order of the ratio υ/c, the phenomenon of the Thomas precession makes a dynamic appearance with a correction by a factor of 2!

4.9.2 The Thomas Precession in the Cases of a Planetary or a Stellar System

The reader may wonder whether the Thomas precession could be observed in a planetary system or a binary star system. Using the gravitational force and the mass M of the Sun, the relation corresponding to Eq. (4.77) gives for the period T_{T} of the Thomas precession of the planet's axis of rotation the value of

$$T_{\mathrm{T}} = \left(\frac{\sqrt{2}}{\pi}\frac{c^3}{GM}\right)^{2/3} T^{5/3}, \tag{4.80}$$

where T is the period of revolution of the planet about the Sun. For the case of a satellite, M is the mass of the planet and T the period of revolution of the satellite about the planet. For a binary star system, M is the reduced mass of the system.

Substituting in Eq. (4.80), we find for the Earth $T_{\mathrm{T}} = 2.03 \times 10^{12}$ years and for the Moon $T_{\mathrm{T}} = 1.29 \times 10^{14}$ years. Given that the age of the Earth's crust is approximately 4.5×10^9 y, this means that since the time the Earth cooled down to a solid planet, its axis has suffered a Thomas precession of only 0.8 of a degree! Binary star systems, in which one member (or both in at least one case) is a pulsar, have been discovered, in which the orbital periods are small, in some cases of the order of an hour. For the double pulsar PSR 1908+00 in the cluster NGC 6760, for which it is $T = 3.4$ hours only, it turns out that it is $T_{\mathrm{T}} = 6.10 \times 10^7$ years. Even in the case of the binary star SDSS J0926+3624, which has the smallest known today orbital period of 28 minutes (!), it is $T_{\mathrm{T}} = 1.7 \times 10^5$ years. The Thomas precession does not appear, therefore, to play any role at all in planetary or stellar systems and is not observable, at least at present.

Thomas precession is predicted by the Special Theory of Relativity for flat space-time. For the gravitational field of a spherical distribution of mass, around which a spinning body revolves, the General Theory of Relativity predicts two precessions:

1. The *geodetic precession* or *de Sitter precession*, which is due to the presence of the central mass and has an angular velocity equal to

$$\boldsymbol{\omega}_{\mathrm{G}} = \frac{3}{2}\frac{GM}{r^3}\mathbf{r} \times \boldsymbol{\upsilon}, \tag{4.81}$$

where M is the mass of the central body and $\boldsymbol{\upsilon}$ and $\mathbf{r}$ are the velocity and the position vector of the revolving body relative to the center of mass M. The functional dependence of $\boldsymbol{\omega}_{\mathrm{G}}$ on r, $\mathbf{r}$ and $\boldsymbol{\upsilon}$ is the same as in the case of the angular frequency $\boldsymbol{\omega}_{\mathrm{T}}$ for the Thomas precession. For a body moving in a circular orbit of radius r, it is

$$\omega_{\mathrm{G}} = \frac{3}{2}\frac{GM}{c^2 r^2}\sqrt{\frac{GM}{r}}. \tag{4.82}$$

2. The *Lense-Thirring precession*, which is due to the rotation of the central mass and has an angular velocity

$$\boldsymbol{\omega}_{\mathrm{LT}} = \frac{G}{r^3}\left[\frac{3\mathbf{r}}{r^2}(\mathbf{S}_{\mathrm{E}} \cdot \mathbf{r}) - \mathbf{S}_{\mathrm{E}}\right], \tag{4.83}$$

where $\mathbf{S}_{\mathrm{E}}$ is the spin of the central body.

Gravity Probe B was an experiment designed by Stanford University, with C.W.F. Everit as the chief investigator, which was executed by NASA and the university, and had as its aim to measure the two precessions in spherical gyroscopes which revolved around the Earth in an artificial satellite moving in a circular polar orbit of radius equal to 7025 km. The experiment lasted for 6.5 years and was concluded in December 2010. Four gyroscopes were used, of a spherical shape of such perfection that any deviation from the spherical shape was no higher than 40 atoms, in a sphere the size of a ping-pong ball. The gyroscopes were kept at a temperature of 2 K, so that the material of which they were made was rendered superconducting. The orientation of the axis of rotation of each gyroscope was determined using a SQUID sensor. A telescope aimed at a certain star provided the reference direction in space.

The geodetic precession of the gyroscope G with its axis in the plane of its orbit around the Earth (Fig. 4.21), had an angular velocity which was predicted by theory to be $\omega_{\mathrm{G}} = -6.606''$ per year. For the Lense-Thirring precession of a gyroscope L-T with its axis perpendicular to the plane of its orbit, the predicted angular velocity was $\omega_{\mathrm{LT}} = -0.0392''$ per year. The angular velocities measured by the

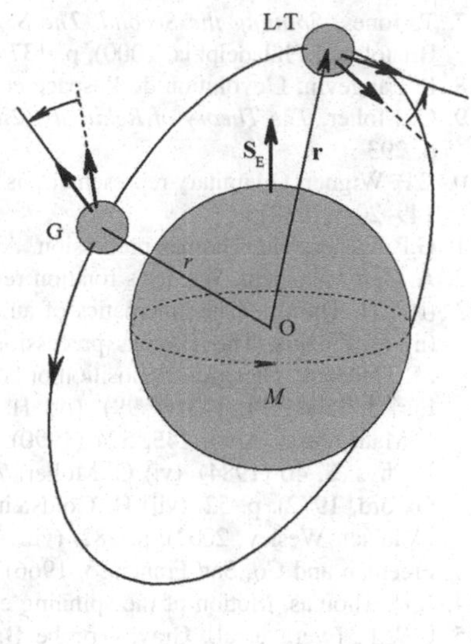

Fig. 4.21 The Gravity Probe B experiment

experiment [15] for the two precessions were $\omega_G = -6.602'' \pm 0.018''$ per year and $\omega_{LT} = -0.0372'' \pm 0.0072''$ per year, respectively. The precision with which the General Theory of Relativity was verified, appears to be very good, especially for the geodetic precession.

References

1. B. Rossi, D.B. Hall, Variation of the rate of decay of mesotrons with momentum. Phys. Rev. **59**, 223 (1941). B. Rossi, K. Greisen, J.C. Stearns, D.K. Froman, P.G. Koontz, Further measurements of the mesotron lifetime. Phys. Rev. **61**, 675 (1942)
2. D.H. Frisch, J.H. Smith, Measurement of the relativistic time dilation using μ-mesons. Am. J. Phy. **31**(5), 342–355 (1963)
3. T. Coan, T. Liu, J. Ye, A compact apparatus for muon lifetime measurement and time dilation demonstration in the undergraduate laboratory. Am. J. Phys. **74**, 161–164 (2006)
4. J. Bailey, K. Borer, F. Combley, H. Drumm, F. Krienen, F. Lange, E. Picasso, W. von Ruden, F.J.M. Farley, J.H. Field, W. Flegel, P.M. Hattersley, Measurements of relativistic time dilatation for positive and negative muons in a circular orbit. Nature **268**, 301–5 (1977). J. Bailey, K. Borer, F. Combley, H. Drumm, C. Eck, F.J.M. Farley, J.H. Field, W. Flegel, P.M. Hattersley, F. Krienen, F. Lange, G. Lebée, E. McMillan, G. Petrucci, E. Picasso, O. Rúnolfsson, W. von Rüden, R.W. Williams, S. Wojcicki, Final report on the CERN muon storage ring including the anomalous magnetic moment and the electric dipole moment of the muon, and a direct test of relativistic time dilation. Nuclear Physics B **150**, 1–75 (1979)
5. G. Sagnac, L'éther lumineux démontré par l'effet du vent relatif d'éther dans un interféromètre en rotation uniforme. C. R. Acad. Sci. (Paris), **157**, 708–10 (1913). Sur la preuve de la réalité de l'éther lumineux par l'expérience de l'interférographe tournant. C. R. Acad. Sci. (Paris), **157**, 1410–13 (1913). Effet tourbillonnaire optique. La circulation de l'éther lumineux dans un interférographe tournant. J. Phys. Radium Ser.5, **4**, 177–195 (1914)
6. J.C. Hafele, R.E. Keating, Science **177**, 166 (1972), and Science **177**, 168 (1972). J.C. Hafele, Relativistic time for terrestrial circumnavigations. Am. J. Phys. **40**, 81–85 (1971)

7. T. Jones, *Splitting the Second. The Story of Atomic Time* (Institute of Physics Publishing, Bristol and Philadelphia, 2000), p. 137
8. P. Langevin, L'évolution de l'espace et du temps. Scientia **10**, 31–54 (1911)
9. C. Møller, *The Theory of Relativity*, 2nd edn. (Clarendon Press, Oxford, 1972), Sect. 8.17, p. 293
10. E.P. Wigner, On unitary representations of the inhomogeneous Lorentz group. Ann. Math. **40**, 149–204 (1939)
11. G.P. Fisher, The Thomas precession. Am. J. Phys. **40**, 1772 (1972)
12. A. Ben-Menahem, Wigner's rotation revisited. Am. J. Phys. **53**, 62 (1985)
13. (i) L.H. Thomas, The kinematics of an electron with an axis. Phil. Mag. Ser. 7, **3**, 1 (1927). (ii) H. Zatzkis, The Thomas precession. J. Frankl. Inst. **269**, 268 (1960). (iii) R. Ferraro, M. Thibeault, Generic composition of boosts: an elementary derivation of the Wigner rotation. Eur. J. Phys. **20**, 143 (1999). (iv) H. Gelman, Sequences of co-moving Lorentz frames. J. Math. Anal. Appl. **145**, 524 (1990). (v) E.G. Peter Rowe, The Thomas precession. Eur. J. Phys. **5**, 40 (1984). (vi) C. Møller, *The Theory of Relativity,* 2nd edn. (Clarendon Press, Oxford, 1972), p. 52. (vii) H. Goldstein, C. Poole, J. Safko, *Classical Mechanics,* 3rd edn. (Addison Wesley, 2002), p. 282. (viii) E.F. Taylor, J.A. Wheeler, *Spacetime Physics* (W.H. Freeman and Co, San Francisco, 1966), p. 169
14. L.H. Thomas, Motion of the spinning electron. Nature **117**, 514 (1926)
15. C.W.F. Everit et al., Gravity probe B: final results of a space experiment to test general relativity. Phys. Rev. Lett. **106** (22), 221101 (2011)

Chapter 5
Optical Phenomena

In this chapter we will examine some phenomena related to light, from the point of view of the Special Theory of Relativity.

5.1 The Aberration of Light

We will discuss the interpretation of the phenomenon of the aberration of light both by non-relativistic and by relativistic Mechanics.

(*a*) *Non-relativistic analysis*

According to Bradley, if the velocity **c** of the light from a star is normal to the velocity **V** of the Earth on its orbit (Fig. 5.1), then the vector addition of the velocities **c** and −**V** gives a velocity of the light relative to the Earth which has magnitude $c' = \sqrt{c^2 + V^2} > c$ and direction making an angle

$$\alpha = \arctan\frac{V}{c} = \arctan\beta \tag{5.1}$$

with the direction of the straight line from the star to the Earth.

(*b*) *Relativistic analysis*

The relativistic explanation of the aberration of light is based on the transformation of the velocity of the light emitted by a star, from the star's frame of reference to that of the observer (Fig. 5.2).

Let the star emit light in the negative direction of the y-axis in frame S. The components of the velocity of this light in the frame of reference S are

$$\upsilon_x = 0, \quad \upsilon_y = -c, \quad \upsilon_z = 0. \tag{5.2}$$

Transforming these components to the frame of reference S′ of the observer, we have:

$$\upsilon'_x = -V, \quad \upsilon'_y = -c/\gamma, \quad \upsilon'_z = 0. \tag{5.3}$$

© Springer International Publishing Switzerland 2016

C. Christodoulides, *The Special Theory of Relativity*,
Undergraduate Lecture Notes in Physics, DOI 10.1007/978-3-319-25274-2_5

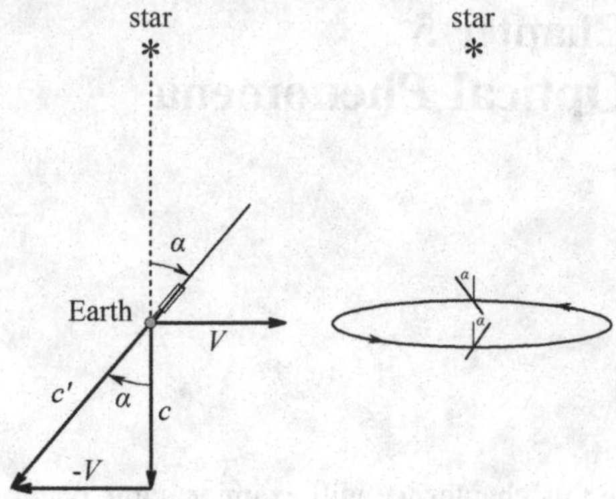

Fig. 5.1 The non-relativistic interpretation of the phenomenon of the aberration of light

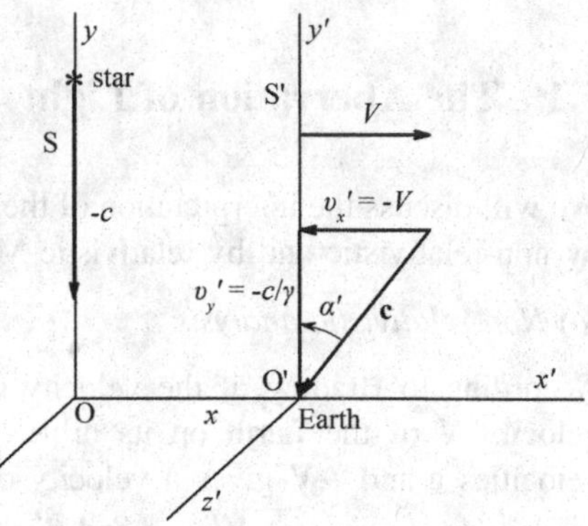

Fig. 5.2 The relativistic interpretation of the phenomenon of the aberration of light

We note that, as expected, the magnitude of the velocity in frame S′ is equal to c. The angle formed by the direction of propagation of the light with the y'-axis of the frame S′ is

$$\alpha' = \arctan\frac{v_x'}{v_y'} = \arctan(\beta\gamma). \tag{5.4}$$

The factor γ is the relativistic difference in the tangent of the angle. Because it is $\tan\alpha' = \beta\gamma = \dfrac{\beta}{\sqrt{1-\beta^2}}$, it follows that $\sin\alpha' = \beta$ or

$$\alpha' = \arcsin\beta. \tag{5.5}$$

The difference between the classical and the relativistic value

The classical relation (5.1) gives $\tan\alpha_C = \beta$ and, therefore,

$$\alpha_C = \arctan\beta = \beta - \frac{\beta^3}{3} + \frac{\beta^5}{5} - \ldots \tag{5.6}$$

The relativistic relation (5.5) gives $\sin\alpha_R = \beta$ and

$$\alpha_R = \arcsin\beta = \beta + \frac{\beta^3}{2\cdot 3} + \frac{1\cdot 3\beta^5}{2\cdot 4\cdot 5} + \ldots \tag{5.7}$$

The difference of the two values is, approximately,

$$\alpha_R - \alpha_C \approx \frac{1}{2}\beta^3. \tag{5.8}$$

For the speed of the Earth on its orbit, $V = 30$ km/s, it is $\beta = 10^{-4}$ and

$$\alpha_R - \alpha_C \approx \frac{1}{2}\beta^3 = 5\times 10^{-13}\ \text{rad} = (10^{-7})''. \tag{5.9}$$

The fractional difference is

$$\frac{\alpha_R - \alpha_C}{\alpha_R} \approx \frac{1}{2}\beta^2 = 5\times 10^{-9}. \tag{5.10}$$

It is not possible to measure such a small difference, even with today's instruments.

5.2 Fizeau's Experiment. Fresnel's Aether Dragging Theory

In 1851, Fizeau measured the speed of light in a medium (water) which was in motion in the frame of reference of the laboratory (see Sect. 1.8.1). We will show how the Special Theory of Relativity explains the results of Fizeau's experiments, as well as the equation derived by Fresnel for the drag coefficient of the aether by a moving optical medium. We will consider a beam of light which propagates in a column of water, which is moving with a speed υ relative to the laboratory. The index of refraction of water is n and, therefore, the speed of light relative to the water is c/n, where c is the speed of light in vacuum. Let the laboratory frame of reference be S, in which the water is moving with a speed υ. We will associate with the moving water a frame of reference S′, which, therefore, moves with a velocity $\boldsymbol{\upsilon} = \upsilon\,\hat{\mathbf{x}}$ relative to frame S (Fig. 5.3).

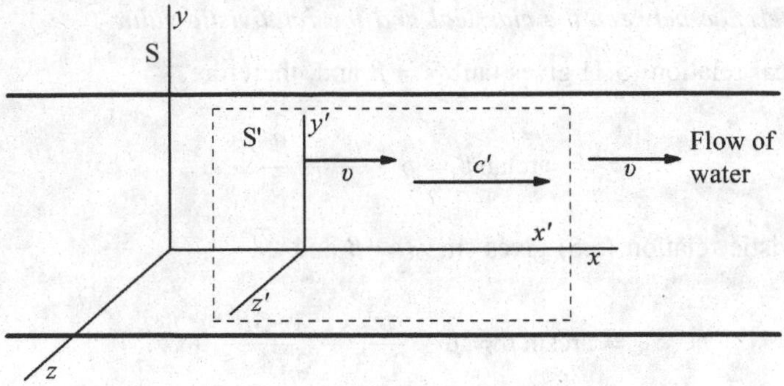

Fig. 5.3 The two frames of reference involved in the experiments of Fizeau for the detection of aether dragging by a moving optical medium

The speed of light in water and, therefore, also relative to frame S′, is $\upsilon'_x = c' = c/n$. Transforming, we find for the speed of light relative to the laboratory frame of reference

$$u = \upsilon_x = \frac{\upsilon'_x + \upsilon}{1+\upsilon'_x\upsilon/c^2} = \frac{c'+\upsilon}{1+c'\upsilon/c^2} = \frac{c/n+\upsilon}{1+\upsilon/nc} \quad \text{or} \quad u = c\left(\frac{c+n\upsilon}{\upsilon+nc}\right). \tag{5.11}$$

The last equation may also be written as $u = \frac{c}{n}\left(\frac{1+n\upsilon/c}{1+\upsilon/nc}\right)$, which gives, for $\upsilon/nc \ll 1$,

$$u \approx \frac{c}{n}\left(1+\frac{n\upsilon}{c}\right)\left(1-\frac{\upsilon}{nc}\right) = \frac{c}{n}\left(1+\frac{n\upsilon}{c}-\frac{\upsilon}{nc}\right) = \frac{c}{n}+\upsilon\left(1-\frac{1}{n^2}\right). \tag{5.12}$$

This is the relation derived by Fresnel in order to interpret the negative results of Arago's observations (Sect. 1.6) and which also explains the results of Fizeau's experiments,

$$u = \frac{c}{n} + \upsilon f, \quad \text{where} \quad f = 1 - \frac{1}{n^2} \tag{5.13}$$

is the drag coefficient. We may, therefore, say that the experiments of Fizeau verify the prediction of the Special Theory of Relativity, at least up to terms with the first power of the ratio υ/c.

5.3 The Doppler Effect

We will first examine the case of the longitudinal Doppler effect, which is observed when the relative motion of the source and the observer takes place along the line joining the two. The general effect will be examined immediately after.

5.3.1 The Longitudinal Doppler Effect

A source of light emits pulses with a period of T_0 in its own frame of reference. The source is situated at the origin O′ of the axes of a frame of reference S′, which moves with a speed of $\mathbf{V} = V\hat{\mathbf{x}}$ relative to another frame, S. An observer is at rest at the origin O of the axes of S. When the points O and O′ coincide (Fig. 5.4a) the source emits a pulse.

In frame S, the first pulse is emitted at the point $x_1 = 0$ at time $t_1 = 0$.

In frame S′, the first pulse is emitted at the point $x_1' = 0$ at time $t_1' = 0$.

After a time T_0, as measured in the frame of reference S′, the source is at a distance x_2 from the observer in frame S and emits a second pulse (Fig. 5.4b).

In frame S′, the second pulse is emitted at the point $x_2' = 0$ at time $t_2' = T_0$.

For the observer, O, the emission of the second pulse happened at time $t_2 = \gamma T_0$.

At this moment, the source is at the point $x_2 = Vt_2 = \gamma V T_0$.

The light will need to travel for a time equal to $\Delta t = x_2/c$ to reach O. The time interval between the observation of the two pulses, as *measured* by O, is

$$T = t_2 + \Delta t = \gamma T_0 + \frac{\gamma V T_0}{c} = T_0 \gamma (1+\beta) \quad \text{or} \quad T = T_0 \sqrt{\frac{1+\beta}{1-\beta}}. \tag{5.14}$$

β is taken positive when the source is moving away from the observer.

As the frequency is $f = c/T$, the relationship between the frequencies is

$$f = f_0 \sqrt{\frac{1-\beta}{1+\beta}}. \tag{5.15}$$

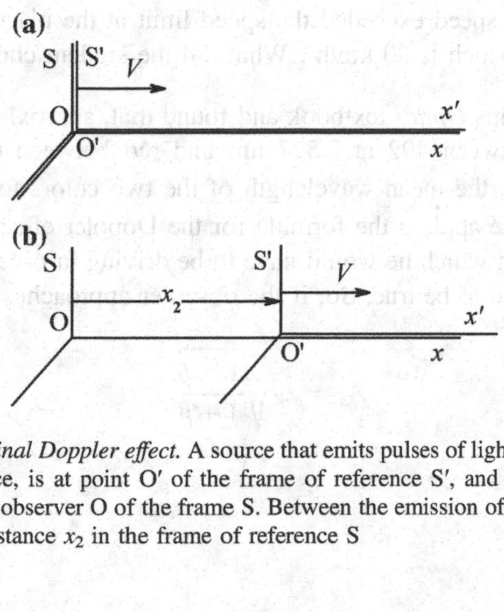

Fig. 5.4 *The longitudinal Doppler effect.* A source that emits pulses of light with a period T_0 in its own frame of reference, is at point O′ of the frame of reference S′, and moves with a velocity $\mathbf{V} = V\hat{\mathbf{x}}$ relative to an observer O of the frame S. Between the emission of two successive pulses, the source moves a distance x_2 in the frame of reference S

For the wavelengths, from $\lambda = cT$ it follows that

$$\lambda = \lambda_0 \sqrt{\frac{1+\beta}{1-\beta}}. \tag{5.16}$$

We may mention the classical relations for the longitudinal Doppler effect,

$$T = T_0 \frac{c + \upsilon_S}{c + \upsilon_O} \quad \text{and} \quad \lambda = \lambda_0 \frac{c + \upsilon_S}{c + \upsilon_O}. \tag{5.17}$$

These equations are very different compared to the relativistic ones, mainly because there appear in them two speeds apart from c, that of the source, υ_S, and that of the observer, υ_O, all these speeds taken relative to a medium which is the absolute frame of reference (the aether) and in which the wave propagates. In the relativistic equation, only the relative speed of source and observer, υ, appears:

$$\lambda = \lambda_0 \sqrt{\frac{1+\beta}{1-\beta}} = \lambda_0 \sqrt{\frac{c+\upsilon}{c-\upsilon}}. \tag{5.18}$$

Example 5.1 The student of Physics and the Doppler effect
A student of Physics was caught violating the red light when driving. In his defense he said: 'Due to the Doppler effect, as I was approaching the traffic lights the red color appeared to me as green, so I drove through'. As his bad luck would have it, his case was being tried by Judge Aristeides Just, known to his friends as 'Einstein', due to the fact that his hobby is Physics. The judge told the student: 'Unless I am mistaken, the effect you refer to gives us the possibility of calculating your speed at the time. I will postpone my decision by one hour, to give you enough time to decide what you choose to do. Either to pay a fine of 1000 euros or pay one cent for each km/h that your speed exceeded the speed limit at the place where you committed the offence, which is 50 km/h'. What did the student choose to do?

The student opened his *Optics* textbook and found that, approximately, green light has wavelengths between 492 and 577 nm and red between 622 and 780 nm. Taking, respectively, the mean wavelength of the two colors to be $\bar{\lambda}_G = 535$ nm and $\bar{\lambda}_R = 700$ nm, he applied the formula for the Doppler effect to determine the velocity $V = \beta c$ with which he would have to be driving in order for the fairy tale he told in his defense to be true. So, if the observer approaches the source, it is

$$\bar{\lambda}_G = \bar{\lambda}_R \sqrt{\frac{1-\beta}{1+\beta}}.$$

Defining the parameter

$$\kappa = \left(\frac{\bar{\lambda}_G}{\bar{\lambda}_R}\right)^2 = \left(\frac{535\ \text{nm}}{700\ \text{nm}}\right)^2 = 0.584,$$

the first equation gives

$$\kappa = \frac{1-\beta}{1+\beta} \quad \Rightarrow \quad \beta = \frac{1-\kappa}{1+\kappa}.$$

Substituting, the student found

$$\beta = \frac{1-0.584}{1+0.584} = 0.2626$$

and from this result he calculated his 'speed' as equal to

$$V = 0.2626c = 7.88 \times 10^4\ \text{km/s} \quad \text{or} \quad V = 2.84 \times 10^8\ \text{km/h}.$$

This meant that he would have to pay 284 million euros. Obviously the student chose to pay the fine of 1000 euros.

Example 5.2 The Doppler effect. A spaceship with two sources of light

A very long spaceship moves with speed V towards an observer O. At the two ends of the spaceship, A and B, there are two sources of light which emit light of wavelengths, in the spaceship's frame of reference, $\lambda_{A0} = 400$ nm and $\lambda_{B0} = 700$ nm, respectively. These correspond to the limits of the visible spectrum. At some moment, the observer is situated between the ends A and B (and very near the line AB) with A receding from him and B approaching him. For which value, $V_0 = c\beta_0$, of the spaceship's speed, will the observer see the light from the two ends as having the same wavelength, λ_0, and what is the value of λ_0?

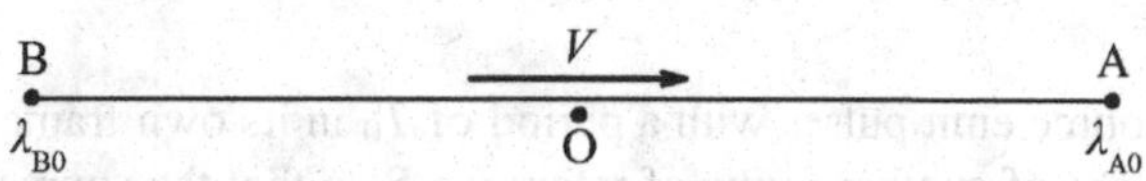

Since source A is moving away from O, the wavelength of the light from source A, as seen by the observer O, will be $\lambda_A = \lambda_{A0}\sqrt{\dfrac{1+\beta}{1-\beta}}$, where $\beta = V/c$. Source B is approaching O and the wavelength of the light from it, as seen by O, will be $\lambda_B = \lambda_{B0}\sqrt{\dfrac{1-\beta}{1+\beta}}$.

For $\lambda_A = \lambda_B$, it must be: $\lambda_{A0}\sqrt{\dfrac{1+\beta_0}{1-\beta_0}} = \lambda_{B0}\sqrt{\dfrac{1-\beta_0}{1+\beta_0}}$.

Therefore, $\dfrac{1+\beta_0}{1-\beta_0} = \dfrac{\lambda_{B0}}{\lambda_{A0}}$ and $\beta_0 = \dfrac{\lambda_{B0}-\lambda_{A0}}{\lambda_{B0}+\lambda_{A0}}$.

For this value of β, it is

$$\lambda_0 = \lambda_A = \lambda_B = \lambda_{A0}\sqrt{\frac{1+\dfrac{\lambda_{B0}-\lambda_{A0}}{\lambda_{B0}+\lambda_{A0}}}{1-\dfrac{\lambda_{B0}-\lambda_{A0}}{\lambda_{B0}+\lambda_{A0}}}} = \lambda_{A0}\sqrt{\frac{\lambda_{B0}}{\lambda_{A0}}} = \sqrt{\lambda_{A0}\lambda_{B0}}.$$

We have found that $\beta_0 = \dfrac{\lambda_{B0}-\lambda_{A0}}{\lambda_{B0}+\lambda_{A0}}$ and $\lambda_0 = \sqrt{\lambda_{A0}\lambda_{B0}}$.
Substituting in these the values of $\lambda_{A0} = 400$ nm and $\lambda_{B0} = 700$ nm, we find

$$\beta_0 = \frac{700-400}{700+400} = \frac{3}{11} = 0.273 \quad \text{and} \quad \lambda_0 = \lambda_A = \lambda_B = \sqrt{400 \times 700} = 529\,\text{nm}.$$

Problems

5.1 A certain line in the spectrum of the light from a nebula has a wavelength of 656 nm instead of the 434 nm measured in the laboratory. If the nebula is moving radially, what is its speed relative to the Earth? Ans.: It is moving away from the Earth with a speed of $V = 0.391c$.

5.2 Show that the light from a source which is moving away from us with a speed of $0.6c$ has twice the wavelength it has in the frame of reference of the source.

5.3 A spaceship moves with speed V relative to an observer O, on a straight line which passes very near the observer. A source on the spaceship emits light of wavelength $\lambda_0 = 500$ nm in the frame of reference of the spaceship. For what range of the spaceship's speed will the light be visible to the observer? The wavelengths of visible light stretches from $\lambda_A = 400$ nm to $\lambda_B = 700$ nm, approximately.
Ans.: $-0.22 < \beta < 0.33$

5.3.2 *The General Doppler Effect*

Let a light source emit pulses with a period of T_0 in its own frame of reference. We choose the axes of another frame of reference S so that the source moves, in the xy plane, with velocity $\mathbf{V} = V\hat{\mathbf{x}}$ in this frame. At some moment, the source is at a point in a direction that forms an angle θ with the x-axis of frame S, in which the observer is at rest (Fig. 5.5). At that moment the source emits a pulse. Obviously, the position of the source will have changed by the time the signal arrives at the observer. Angle θ corresponds, therefore, to the position of the source when it emitted the pulse. It defines the direction from which the observer will see the pulse moving towards him.

In frame S, at the moment t_1 the source is at position x_1 when it emits a pulse. After a time interval equal to T_0 has passed in the frame of the source, it emits a second pulse, from position x_2 in S. Due to the dilation of time, it is $t_2 - t_1 = \gamma T_0$, where $\gamma = 1/\sqrt{1-V^2/c^2}$. The pulses have to travel, respectively, distances equal

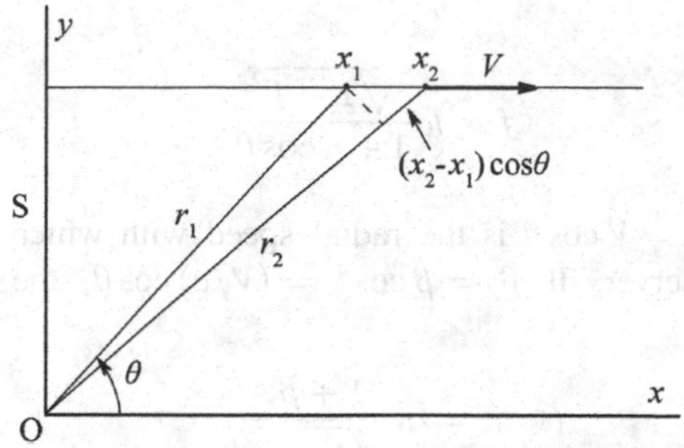

Fig. 5.5 *The general Doppler effect.* A source that emits pulses with a period T_0 in its own frame of reference, moves with a velocity $\mathbf{V} = V\hat{\mathbf{x}}$ with respect to the frame of reference S of observer O. At the emission of a pulse, the source is at a distance x_1 from the y-axis, in a direction forming an angle θ with the x-axis in frame S. The next pulse is emitted when the source is at a distance x_2 from the y-axis. The positions of the source shown in the figure are those at the moments of emission of the pulses and not those at the arrival of the pulses at the observer O

to r_1 and r_2 in order to reach the observer O. The difference in the times of arrival of the two pulses at O will, therefore, be

$$T = \left(t_2 + \frac{r_2}{c}\right) - \left(t_1 + \frac{r_1}{c}\right) = (t_2 - t_1) + \frac{r_2 - r_1}{c} = \gamma T_0 + \frac{r_2 - r_1}{c}. \qquad (5.19)$$

If the distance travelled by the source between the emission of the two successive pulses is small compared to the distances r_1 and r_2, it will be

$$r_2 - r_1 \approx (x_2 - x_1)\cos\theta = \gamma T_0 V \cos\theta. \qquad (5.20)$$

Equation (5.19) may now be written as

$$T = \gamma T_0 + \gamma T_0 (V/c)\cos\theta = \gamma T_0[1 + (V/c)\cos\theta] \qquad (5.21)$$

or

$$T = T_0 \frac{1 + \beta\cos\theta}{\sqrt{1 - \beta^2}}, \quad \text{where} \quad \beta = \frac{V}{c}. \qquad (5.22)$$

The relation for the wavelengths is

$$\lambda = \lambda_0 \frac{1 + \beta\cos\theta}{\sqrt{1 - \beta^2}}. \qquad (5.23)$$

For the frequencies,

$$f = f_0 \frac{\sqrt{1-\beta^2}}{1+\beta\cos\theta}. \tag{5.24}$$

We notice that $\upsilon_r = V\cos\theta$ is the radial speed with which the source moves away from the observer. If $\beta_r = \beta\cos\theta = (V/c)\cos\theta$, the equation for the wavelengths is

$$\lambda = \lambda_0 \frac{1+\beta_r}{\sqrt{1-\beta^2}}. \tag{5.25}$$

The factor $\sqrt{1-\beta^2}$ is a purely relativistic factor, due to the dilation of time.
For the frequencies we have

$$f = f_0 \frac{\sqrt{1-\beta^2}}{1+\beta_r}. \tag{5.26}$$

A brief examination of Eq. (5.24) will show some of its main characteristics (Fig. 5.6):
For $\theta = 0$, $\cos\theta = 1$, and

$$\frac{f}{f_0} = \sqrt{\frac{1-\beta}{1+\beta}} \tag{5.27}$$

and we have the purely *longitudinal Doppler effect*, with the source moving away from the observer.
For $\theta = 90^\circ$, $\cos\theta = 0$, and

$$\frac{f}{f_0} = \sqrt{1-\beta^2}, \tag{5.28}$$

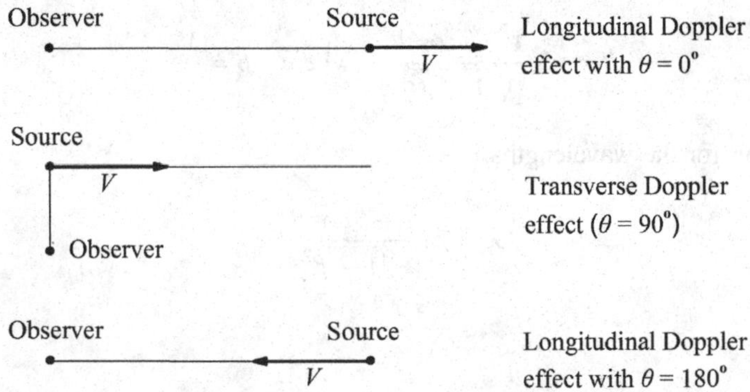

Fig. 5.6 Special cases of the Doppler effect

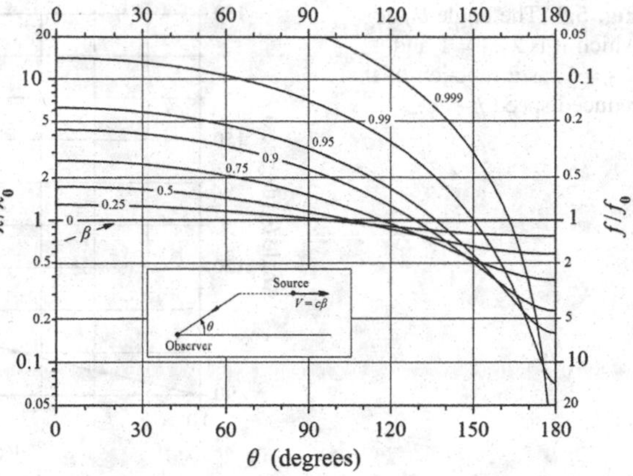

Fig. 5.7 The Doppler effect. The ratios λ/λ_0 and f/f_0, as functions of the angle θ, for various values of β

which describes the purely *transverse Doppler effect*.

For $\theta = 180°$, $\cos\theta = -1$, and

$$\frac{f}{f_0} = \sqrt{\frac{1+\beta}{1-\beta}} \tag{5.29}$$

which describes the purely *longitudinal Doppler effect*, with the source approaching the observer.

Using Eqs. (5.23) and (5.24), we plot the variations of the ratios λ/λ_0 and f/f_0 as functions of the angle θ for various values of β, which are shown in Fig. 5.7.

An interesting feature that follows from Eq. (5.24) is that there exists a value of the angle θ, say θ_0, for which it is $f = f_0$ (Fig. 5.8). This happens when $1+\beta\cos\theta_0 = \sqrt{1-\beta^2}$ or

$$\cos\theta_0 = -\frac{1}{\beta}\left(1-\sqrt{1-\beta^2}\right) \qquad (90° < \theta_0 < 180°). \tag{5.30}$$

For $\beta \ll 1$, it is $\sqrt{1-\beta^2} \approx 1 - \frac{1}{2}\beta^2$ and $\cos\theta_0 \approx -\frac{1}{2}\beta$. The explanation of this effect is the following:

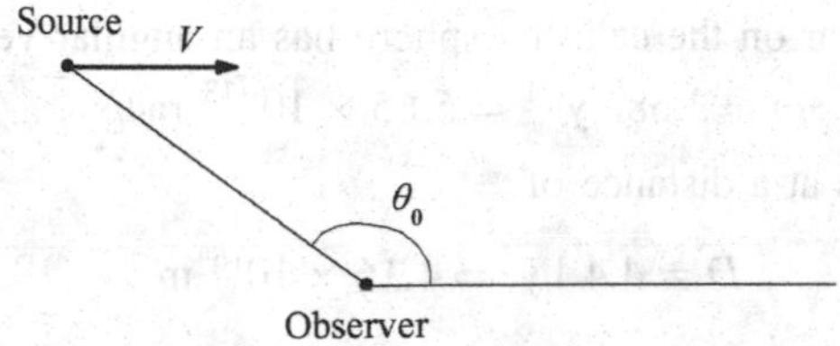

Fig. 5.8 The angle θ_0 at which it is $f = f_0$. The source is shown at its position when it emits the pulse and not that at which it is found when the signal reaches the observer

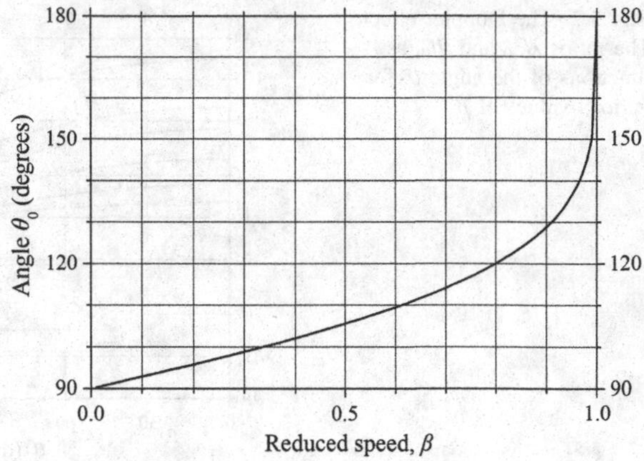

Fig. 5.9 The angle θ_0 at which it is $\lambda/\lambda_0 = 1$ and $f/f_0 = 1$, as a function of the reduced speed β

We have two competing effects:

1. The relativistic decrease of the frequency, due to the dilation of time (transverse Doppler effect), and
2. The increase of the frequency, due to the radial speed of approach of the source to the observer (longitudinal Doppler effect).

For $\theta = \theta_0$ the two effects cancel each other completely. There is a value of θ_0 between 90° and 180° for each value of $0 \leq \beta \leq 1$:

β	0	0.1	0.5	0.9	0.99	0.999	1
θ_0	90°	92.9°	105.5°	128.8°	150.2°	163°	180°

In Fig. 5.7, the values of θ_0 may be read at the points where the curves for the various values of β intersect the straight line $\lambda/\lambda_0 = 1$. Figure 5.9 shows the angle θ_0 as a function of β.

Example 5.3 The Doppler effect. Stellar motion
The star α Centauri is at a distance of 4.4 light years from the Sun and has a *proper motion* (angular displacement on the celestial sphere per unit time) of 3.68″ y^{-1}. The spectral line of calcium, which in the laboratory has a wavelength $\lambda_0 = 396.820$ nm, appears in the spectrum of the star to be displaced by $\Delta\lambda = \lambda - \lambda_0 = -0.029$ nm in the frame of reference of the Sun. Find the velocity (magnitude and direction) of the star relative to the Sun.

The motion of the star on the celestial sphere has an angular velocity of

$$\omega = 3.68'' \text{ y}^{-1} = 5.65 \times 10^{-13} \text{ rad/s}.$$

Given that the star is at a distance of

$$D = 4.4 \text{ l.y.} = 4.16 \times 10^{16} \text{ m},$$

its transverse speed is $\upsilon_\theta = \omega D = 5.65 \times 10^{-13}$ rad/s $\times\, 4.16 \times 10^{16}$ m $= 23.5$ km/s and, therefore, $\beta_\theta \equiv \upsilon_\theta/c = 7.83 \times 10^{-5}$.

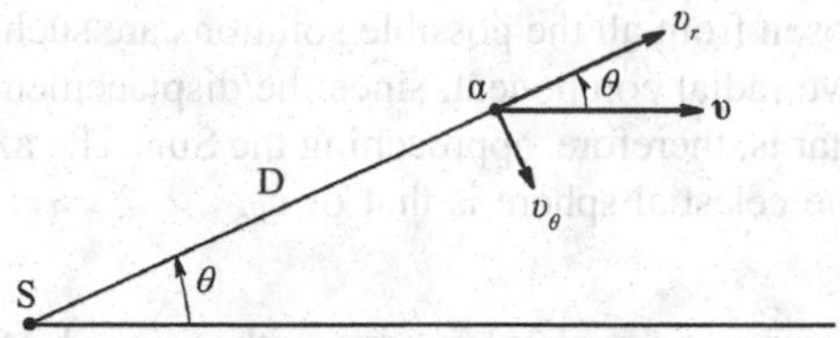

The equation for the wavelengths is $\lambda = \lambda_0 \dfrac{1+\beta_r}{\sqrt{1-\beta^2}}$,

where $\beta_r = \beta\cos\theta = v_r/c$ is the reduced radial velocity of the star.

It is $v^2 = v_r^2 + v_\theta^2$ or $\beta^2 = \beta_r^2 + \beta_\theta^2$ and, therefore, $\lambda = \lambda_0 \dfrac{1+\beta_r}{\sqrt{1-\beta_r^2-\beta_\theta^2}}$ and

$1-\beta_r^2-\beta_\theta^2 = \left(\dfrac{\lambda_0}{\lambda}\right)^2 (1+\beta_r)^2$, which gives

$$\left[1+\left(\frac{\lambda_0}{\lambda}\right)^2\right]\beta_r^2 + 2\left(\frac{\lambda_0}{\lambda}\right)^2\beta_r + \beta_\theta^2 - 1 + \left(\frac{\lambda_0}{\lambda}\right)^2 = 0$$

Substituting $\beta_\theta = 7.83\times 10^{-5}$ and $\lambda_0/\lambda = 1.000\,073$, we have

$$2.000\,146\,\beta_r^2 + 2.000\,292\beta_r + 0.000\,146 = 0$$
$$\beta_r^2 + 1.000\,073\,\beta_r + 0.000\,073 = 0,$$

whose roots are $\beta_{r1} = -1$ and $\beta_{r2} = -7.3\times 10^{-5}$. The root $\beta_{r1} = -1$ is rejected, as it would lead to $|\beta| > 1$. For the root $\beta_{r2} = -7.3\times 10^{-5}$, it is $v_r = -21.9$ km/s.

The magnitude of the star's velocity is found from $|\beta| = \sqrt{\beta_r^2+\beta_\theta^2} = 1.07\times 10^{-4}$ to be equal to $|v| = 32.1$ km/s.

The angle formed by the velocity of the star with the line joining it with the Sun is

$$\theta = \arctan\frac{v_\theta}{v_r} = \arctan\frac{\pm 23.5}{-21.9} = \arctan(\pm 1.073) = \pm 133°,$$

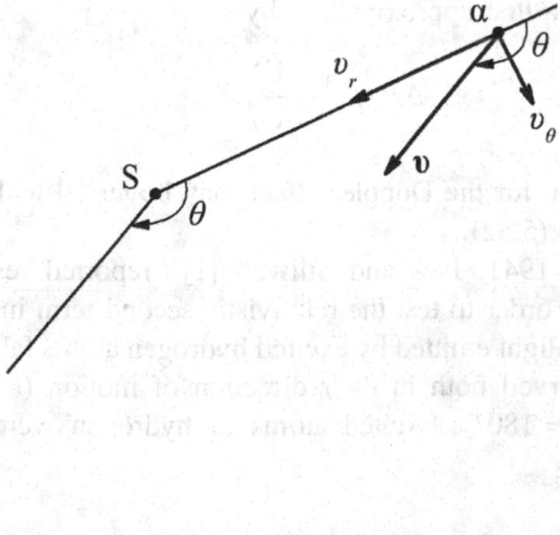

where the angles chosen from all the possible solutions are such that the velocity of the star has a negative radial component, since the displacement of the wavelength is negative and the star is, therefore, approaching the Sun. The azimuthal orientation of the velocity on the celestial sphere is that of υ_θ.

Problems

5.4 A light source moves in a circular orbit with a speed of $0.5c$. What is the displacement, due to the Doppler effect, of the sodium yellow line, as observed at the center of the circle? The line has a wavelength of 589 nm in the laboratory.

Ans.: $\Delta\lambda = 91$ nm

5.5 The Sun has a radius of 7.0×10^8 m, approximately, and a period of rotation about its axis equal to 24.7 days. What is the Doppler shift of a spectral line with laboratory wavelength 500 nm, in the light emitted (a) from the center of the Sun and (b) from the edges of the Sun's disk at its equator?

Ans.: (a) $\Delta\lambda = 1.8 \times 10^{-8}$ nm, (b) $\Delta\lambda = \pm 0.00344$ nm

5.3.3 *The Experimental Verification of the Relativistic Terms of the Doppler Effect*

The equation describing the Doppler effect for wavelengths may be written as

$$\lambda = \lambda_0 \frac{1+\beta\cos\theta}{\sqrt{1-\beta^2}} \approx \lambda_0\left(1+\beta\cos\theta+\frac{1}{2}\beta^2\right) \tag{5.31}$$

if the denominator is expanded up to the term containing the square of β. The Doppler displacement of the wavelength is, therefore,

$$\Delta\lambda = \lambda - \lambda_0 \approx \lambda_0\left(\frac{V}{c}\right)\cos\theta + \frac{\lambda_0}{2}\left(\frac{V}{c}\right)^2. \tag{5.32}$$

The first term is the classical one, while the second is due to the relativistic dilation of time. For the transverse Doppler effect, it is $\theta = 90^\circ$, $\cos\theta = 0$, and the displacement is given, to a first approximation by

$$\Delta\lambda \approx \frac{\lambda_0}{2}\left(\frac{V}{c}\right)^2. \tag{5.33}$$

The relativistic equation for the Doppler effect may be tested with accuracy up to the second term of Eq. (5.32).

Between 1938 and 1941, Ives and Stilwell [1] reported results of the first measurements made in order to test the relativistic second term in the equation for the Doppler effect. The light emitted by excited hydrogen atoms falling back to their ground state was observed both in their direction of motion ($\theta = 0$) and in the opposite direction ($\theta = 180^\circ$). Excited atoms of hydrogen were produced with collisions of ionized hydrogen molecules, H_2^+ and H_3^+, which had been accelerated in a van de Graaf generator, with atoms of neon gas at low pressure (of the order of 10^{-3} Torr). By the reactions

$$H_2^+ + \mathrm{Ne} \to H_f^* + [H^+ + \mathrm{Ne}] \quad \text{and} \quad H_3^+ + \mathrm{Ne} \to H_f^* + [H_2^+ + \mathrm{Ne}]$$

fast and excited hydrogen atoms, H_f^*, were produced. In this way, beams of fast hydrogen atoms with values of V/c in the region of 0.005 were available.

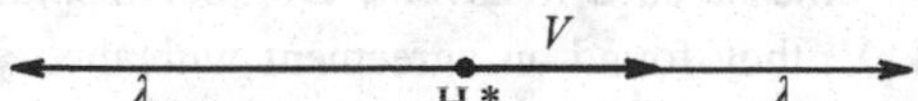

For the light emitted by the atoms on returning to their ground state in the directions with $\theta = 0$ and $\theta = 180^\circ$, we have, respectively,

$$\Delta\lambda_0 \approx \lambda_0\left(\frac{V}{c}\right) + \frac{\lambda_0}{2}\left(\frac{V}{c}\right)^2 \quad \text{and} \quad \Delta\lambda_{180} \approx -\lambda_0\left(\frac{V}{c}\right) + \frac{\lambda_0}{2}\left(\frac{V}{c}\right)^2. \tag{5.34}$$

If the two displacements are measured, their half sum gives

$$\Delta\lambda = \frac{\Delta\lambda_0 + \Delta\lambda_{180}}{2} \approx \frac{\lambda_0}{2}\left(\frac{V}{c}\right)^2, \tag{5.35}$$

the purely relativistic term. In the 1938 experiment, for the H_α spectral line of hydrogen, with a wavelength of 656.3 nm, values of the term in Eq. (5.35) were measured, which, depending on the value of V/c, ranged from 0.0011 nm (the theoretically expected value being 0.00109 nm) up to 0.0047 nm (theoretical prediction 0.00469 nm). The results of the 1941 experiment of Ives and Stilwell are shown in Fig. 5.10. The agreement between experiment and theory is very good, verifying the relativistic predictions.

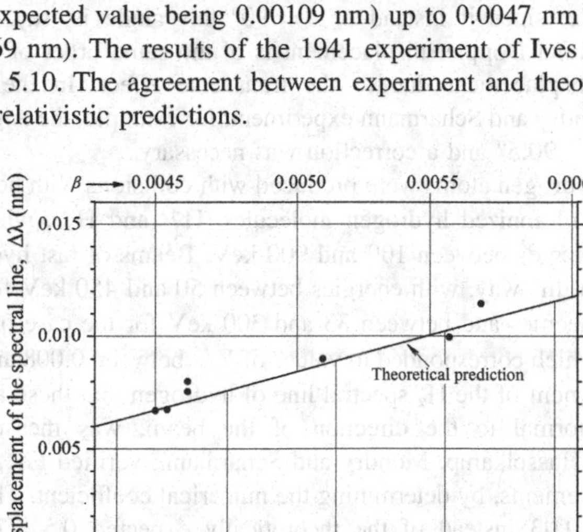

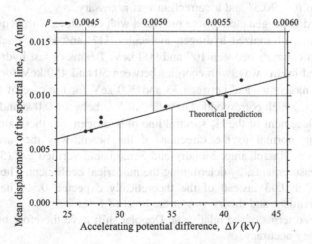

Fig. 5.10 The results of the Ives and Stilwell experiment of 1941. The points are the results of the measurements, $\Delta\lambda_{\text{exp}}$. The continuous line shows the theoretical relativistic prediction, $\Delta\lambda_{\text{th}}$. The values of β for the atoms are shown at the top of the figure

Similar experiments by Mandelberg and Witten [2] gave, instead of the theoretically expected value of 1/2 for the numerical coefficient in Eq. (5.35), the experimental value of 0.498 ± 0.025.

Olin, Alexander, Häuser and McDonald [3] performed similar measurements, not using visible light but the high energy photons, with energies of 8.64 MeV, which are emitted during the de-excitation of the ^{20}Ne nucleus. A narrow beam of excited ^{20}Ne ions was produced using the reactions $^{16}_{8}\mathrm{O}(^{4}_{2}\mathrm{He},\gamma)^{20}_{10}\mathrm{Ne}$ and $^{4}_{2}\mathrm{He}(^{16}_{8}\mathrm{O},\gamma)^{20}_{10}\mathrm{Ne}$. Having measured a transverse Doppler displacement equal, in energy, to 10.1 ± 0.4 keV, they found an agreement with theory, at the level of 3.5 %, for speeds corresponding to $\beta = 0.05$. The numerical coefficient in Eq. (5.35), which is equal to ½, was found to be 0.491 ± 0.017.

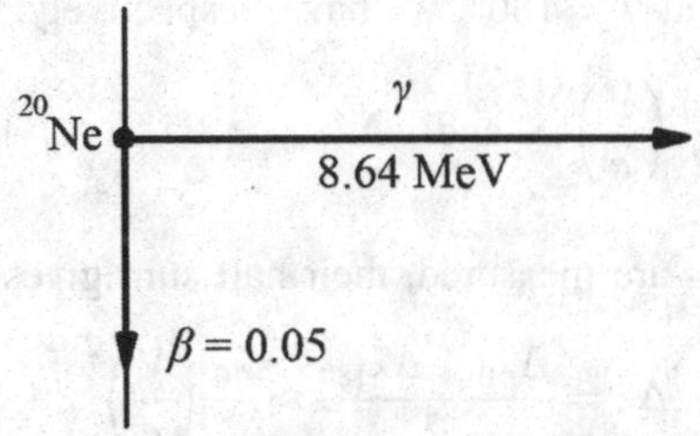

In 1979, Hasselkamp, Mondry and Scharmann [4] reported the results for the Doppler shift in light emitted by hydrogen atoms in a direction normal to the direction of their motion. In this case it is $\theta = 90^{\circ}$, $\cos\theta = 0$, the first term in Eq. (5.32) vanishes and only the second is observed. However, in practice, things are not so simple, since if instead of $\theta = 90^{\circ}$ this angle is, say, $\theta = 91^{\circ}$, the classical longitudinal Doppler displacement is of the same order of magnitude as the relativistic displacement due to the transverse effect. In the case of the Hasselkamp, Mondry and Scharmann experiment, for example, the angle was found to be equal to $\theta = 90.5^{\circ}$ and a correction was necessary.

The excited hydrogen atoms were produced with collisions with neon gas atoms at low pressure, of ionized hydrogen molecules, H_2^+ and H_3^+, which had been accelerated to energies between 100 and 900 keV. Beams of fast hydrogen atoms were produced in this way, with energies between 50 and 450 keV for the case of incident H_2^+ molecules and between 33 and 300 keV for the case of the incident H_3^+ molecules, which corresponded to values of V/c between 0.008 and 0.031. The Doppler displacement of the H_α spectral line of hydrogen that these atoms emitted in a direction normal to the direction of the beam, was measured using a monochromator. Hasselkamp, Mondry and Scharmann, verified Eq. (5.35) for $\Delta\lambda$ with their measurements, by determining the numerical coefficient. They found the value of 0.52 ± 0.03 instead of the theoretically expected 0.5. The agreement between experiment and theory was at the level of 6 %.

The first two terms of the relativistic Doppler shift have therefore been verified to a satisfactory accuracy.

5.4 Relativistic Beaming or the Headlight Effect

Let a point source, which is at rest in the frame of reference S′, emit photons isotropically in that frame. Frame S′ moves with velocity $\mathbf{V} = c\beta\hat{\mathbf{x}}$ relative to another frame of reference S. We will define the angle $\theta_{1/2}$ in frame S and $\theta'_{1/2}$ in frame S′, such that half the photons in each frame of reference are emitted in directions forming with the corresponding positive x-axes angles smaller than $\theta_{1/2}$ and $\theta'_{1/2}$. Obviously, due to symmetry, in frame S′, in which the source is at rest, it is $\theta'_{1/2} = 90^\circ$. We want to find the corresponding value of this angle in frame S, $\theta_{1/2}$.

We will examine a photon which in frame S′ has a direction of propagation forming an angle θ' with the x'-axis, as shown in Fig. 5.11. In this frame of reference, the photon has components of velocity equal to $\upsilon'_x = c\cos\theta'$ and $\upsilon'_y = c\sin\theta'$.

Transforming to frame S,

$$\upsilon_x = \frac{\upsilon'_x + V}{1 + \dfrac{\upsilon'_x V}{c^2}} = c\frac{\cos\theta' + \beta}{1 + \beta\cos\theta'} \tag{5.36}$$

$$\upsilon_y = \frac{\upsilon'_y}{\gamma\left(1 + \dfrac{\upsilon'_x V}{c^2}\right)} = c\frac{\sin\theta'}{\gamma(1 + \beta\cos\theta')}. \tag{5.37}$$

The angle θ formed by the vector of the photon's velocity with the x-axis of frame S is given by the relation:

$$\cos\theta = \frac{\upsilon_x}{c} = \frac{\cos\theta' + \beta}{1 + \beta\cos\theta'}. \tag{5.38}$$

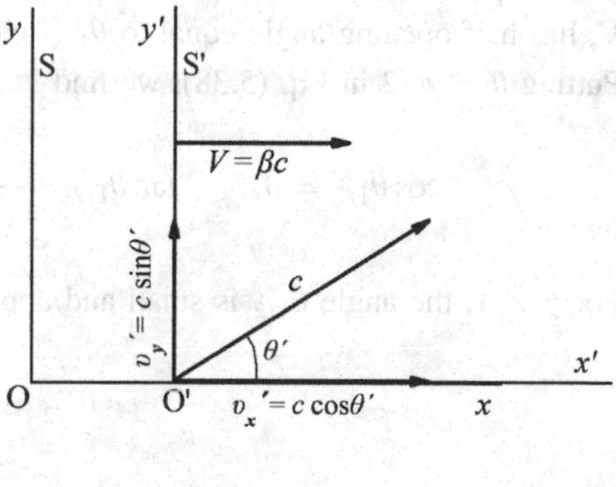

Fig. 5.11 A photon moves in the frame of reference S′ in a direction at an angle θ' with respect to the x'-axis

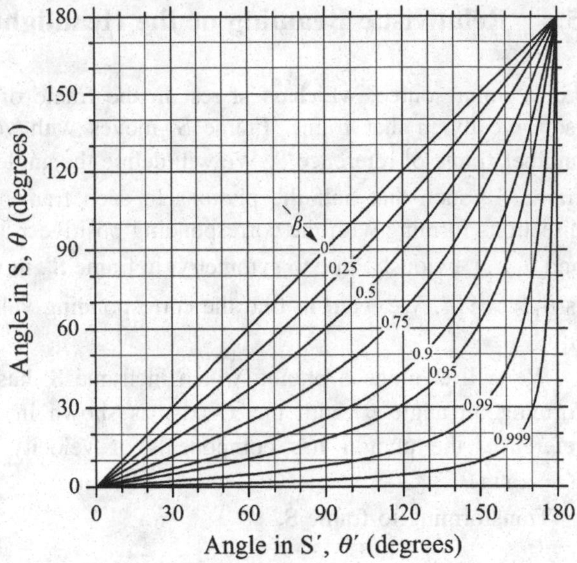

Fig. 5.12 A photon moves in the inertial frame of reference S′ in a direction at an angle θ' with respect to the positive x' axis. The frame of reference S′ moves with constant velocity $\mathbf{V} = c\beta\hat{\mathbf{x}}$ relative to another frame of reference, S. In frame S, the photon moves in a direction at an angle θ with respect to the positive x axis. Angle θ is given in the figure as a function of θ', for various values of β

Inversely,

$$\cos\theta' = \frac{\cos\theta - \beta}{1 - \beta\cos\theta}. \tag{5.39}$$

The angle θ is plotted as a function of the angle θ', for various values of β in Fig. 5.12. The angle $\theta_{1/2}$ may be found, for a given β with the aid of Fig. 5.12 from the point of intersection of the straight line with $\theta' = 90^\circ$ and the curve corresponding to the value of β.

We know that in the frame of reference S′ half the photons are emitted at angles smaller than $\theta'_{1/2}$, where $\theta'_{1/2} = 90^\circ$. In the frame of reference S, the cone in which half the photons are emitted, with a component of their velocity in the direction of **V**, has half opening angle equal to $\theta_{1/2}$ which corresponds to the angle $\theta' = \pi/2$. Putting $\theta' = \pi/2$ in Eq. (5.38), we find

$$\cos\theta_{1/2} = \beta, \qquad \tan\theta_{1/2} = \frac{1}{\sqrt{\gamma^2 - 1}}, \qquad \sin\theta_{1/2} = \frac{1}{\gamma}. \tag{5.40}$$

For $\gamma \gg 1$, the angle $\theta_{1/2}$ is small and approximately equal to

$$\theta_{1/2} \approx \frac{1}{\gamma}. \tag{5.41}$$

The angle $\theta_{1/2}$ is plotted as a function of β in Fig. 5.13.

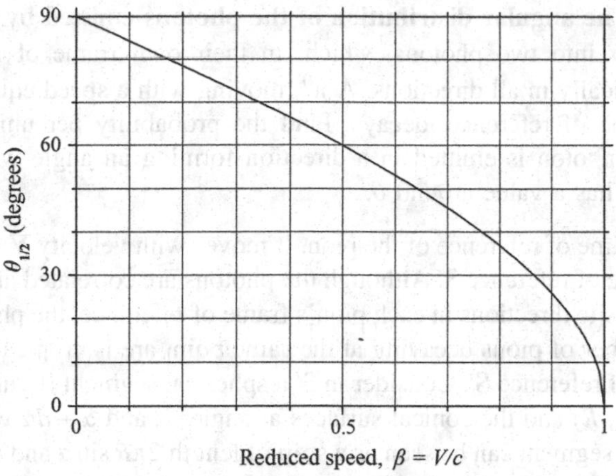

Fig. 5.13 The half opening angle $\theta_{1/2}$ of the cone in which half the photons are emitted from an isotropic source moving in the frame of reference S with speed $V = c\beta$, plotted as a function of β. The photons are emitted with a component of their velocity in the direction of motion of the source

In Fig. 5.14 are shown, for various values of the speed of a point source, a number of photons moving on a plane from the center outwards. In the frame of reference in which the source is at rest ($\beta = 0$), the photons are emitted isotropically, with the same flux per unit solid angle in all directions. As the speed of the source increases, the cone in which half the photons are emitted, with a component of their speeds in the direction of motion of the source, has a smaller half opening angle $\theta_{1/2}$. This angle tends to zero as β tends to unity. This effect is known as *relativistic beaming* or the *headlight effect*. As the speed of the source increases, the light it emits is concentrated in directions forming smaller and smaller angles with the vector of its velocity.

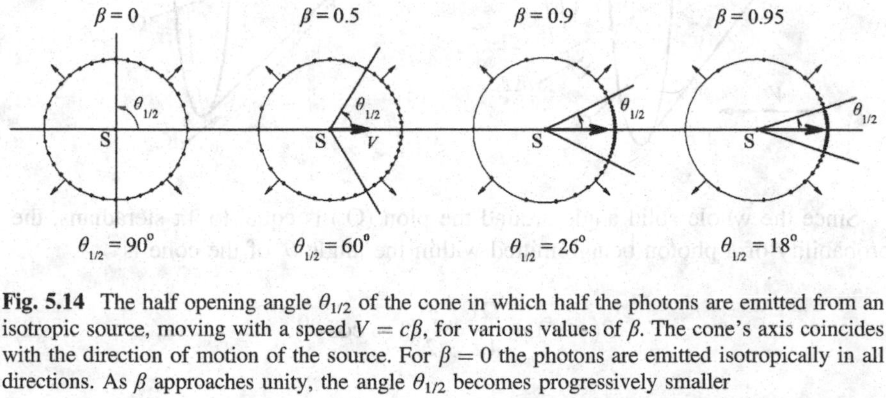

Fig. 5.14 The half opening angle $\theta_{1/2}$ of the cone in which half the photons are emitted from an isotropic source, moving with a speed $V = c\beta$, for various values of β. The cone's axis coincides with the direction of motion of the source. For $\beta = 0$ the photons are emitted isotropically in all directions. As β approaches unity, the angle $\theta_{1/2}$ becomes progressively smaller

Example 5.4 The angular distribution of the photons emitted by π^0 particles
Pions π^0 decay into two photons, which, in their own frame of reference, are emitted isotropically in all directions. A π^0, moving with a speed equal to V in the laboratory frame of reference, decays. Find the probability per unit solid angle, $dP/d\Omega$, that a photon is emitted in a direction forming an angle with the pion's velocity which has a value around θ.

Let S′ be the frame of reference of the pion. It moves with velocity $\mathbf{V} = V\hat{\mathbf{x}}$ relative to another frame of reference S. Although the photons are correlated and are emitted in pairs in opposite directions in each pion's frame of reference, the photons emitted by a large number of pions decaying at the same point are isotropically distributed in their frame of reference S′. Consider in S′ a spherical segment [figure (a)] defined by a sphere (O′, R) and the conical surfaces at angles α and $\alpha + d\alpha$ with respect to the x'-axis. The segment can be seen as a strip of length $2\pi R \sin\alpha$ and width $R d\alpha$. Its spherical area is, therefore, equal to $dS = 2\pi R^2 \sin\alpha\, d\alpha$. This area subtends at the center of the sphere a solid angle equal to $d\Omega' = dS/R^2 = 2\pi \sin\alpha\, d\alpha$. The cone with half opening angle θ' defines a solid angle equal to

$$\Omega' = \int_0^{\theta'} 2\pi \sin\alpha\, d\alpha = 2\pi(1 - \cos\theta'). \tag{1}$$

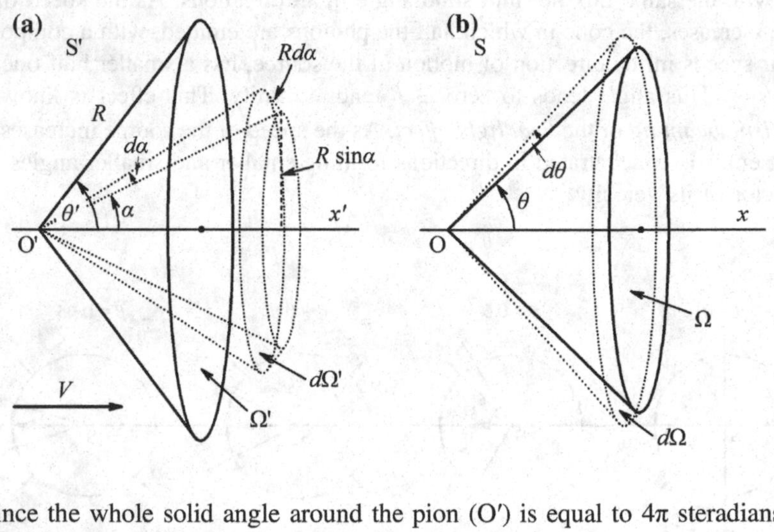

Since the whole solid angle around the pion (O′) is equal to 4π steradians, the probability of a photon being emitted within the angle θ' of the cone is

$$P(<\theta') = \frac{\Omega'}{4\pi} = \frac{1}{2}(1 - \cos\theta'). \tag{2}$$

This is the same as the probability $P(<\theta)$ that the photon is emitted within the solid angle Ω defined by a cone of half opening angle θ in the laboratory frame of reference S [figure (b)]:

$$P(<\theta) = \frac{1}{2}(1 - \cos\theta'). \tag{3}$$

The relation between the angles θ' and θ is given by Eq. (5.39): $\cos\theta' = \frac{\cos\theta - \beta}{1 - \beta\cos\theta}$. Substituting in Eq. (3), we get

$$P(<\theta) = \frac{1+\beta}{2}\frac{1-\cos\theta}{1-\beta\cos\theta}. \tag{4}$$

It should be noted that for $\beta = 1$ it is $P(<\theta) = 1$, which is independent of θ, implying that all the photons are emitted with $\theta = 0$

Differentiating with respect to θ,

$$\frac{dP}{d\theta} = \frac{1-\beta^2}{2}\frac{\sin\theta}{(1-\beta\cos\theta)^2}. \tag{5}$$

dP is the probability that a photon is emitted at an angle to the x-axis between θ and $\theta + d\theta$. To find the probability per unit solid angle, we substitute in Eq. (5) $d\theta = d\Omega/2\pi\sin\theta$. We obtain

$$\frac{dP}{d\Omega} = \frac{1}{4\pi}\frac{1-\beta^2}{(1-\beta\cos\theta)^2} \tag{6}$$

which is the probability per unit solid angle that a photon will be emitted, in the laboratory frame of reference, at an angle infinitesimally near the angle θ with respect to the positive x-axis.

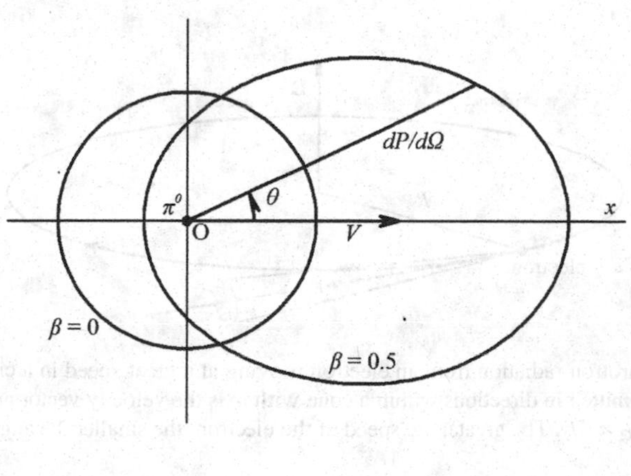

The function $dP/d\Omega$ is plotted as a function of θ, in polar form, in the figure, for β equal to 0 and to 0.5. The relativistic beaming effect for the photons emitted by moving decaying π^0 particles is clearly visible.

5.4.1 *Synchrotron Radiation or Magnetic Bremsstrahlung*

The headlight effect is dramatically demonstrated by the emission of *synchrotron radiation* or *magnetic bremsstrahlung*, as charged particles move in a circular orbit with high speeds in a magnetic field, for example in a synchrotron. The centripetal acceleration of the particles due to their motion in a circular orbit, leads to the emission of electromagnetic radiation, i.e. photons, with energies between a fraction of an eV (infrared radiation) up to visible light and up to energies of the order of magnitude of keV (X rays). It is observed that these photons move in directions forming very small angles with the velocity of the charges at every point of the accelerator, i.e. they are emitted in directions near the plane of the orbit and, approximately, tangentially to the circular orbits, as shown in Fig. 5.15.

As an electron is forced by a magnetic field **B** to move in a circular orbit of radius R with a speed equal to V, the photons it emits have a probability of 1/2 to move in a cone with axis the vector of the velocity of the electron and a half opening angle equal to $\theta_{1/2} \approx 1/\gamma$.

It is proved that the energy loss per revolution of the electron is equal to

$$w = \frac{e^2}{3\varepsilon_0}\frac{\beta^3\gamma^4}{R} = \frac{e^2}{3\varepsilon_0 R}\left(\frac{E}{m_0c^2}\right)^4\beta^3 \tag{5.42}$$

where m_0 is the mass and E the energy of the electron. As it is $\beta \approx 1$, this magnitude is proportional to E^4 and increases very rapidly with the energy. For $R = 1$ m and $E = 100$ MeV, it is 5.8 eV per revolution and per electron. For a proton, which has a mass 1836 times that of the electron, the loss w is negligible, for the same energy.

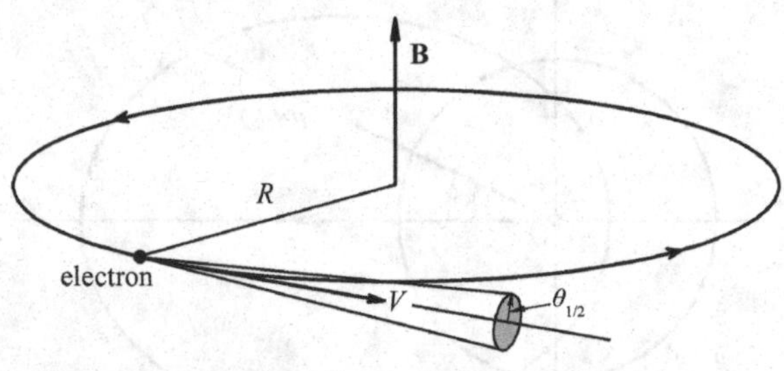

Fig. 5.15 Synchrotron radiation from an electron moving at a great speed in a circular orbit. Half the photons are emitted in directions within a cone with axis the velocity vector and a half opening angle equal to $\theta_{1/2} \approx 1/\gamma$. The greater the speed of the electron, the smaller the angle $\theta_{1/2}$, due to the headlight effect

The spectrum of the synchrotron radiation emitted is continuous and has a maximum at the frequency of

$$f_{\max} = \frac{3}{4\pi}\omega_0\gamma^3 = \frac{3}{4\pi}\omega_0\left(\frac{E}{m_0c^2}\right)^3 \tag{5.43}$$

where ω_0 is the cyclotron angular frequency of the electron in a particular magnetic field, $\omega_0 = qB/\gamma m_0$ (see Sect. 6.14). For electrons of energy $E = 100$ MeV and $R = 1$ m, the maximum is in the region of visible light and the synchrotron radiation can be seen by the naked eye.

In Astrophysics, the emission of synchrotron radiation is an important mechanism by which the cosmic ray electrons lose energy. In a characteristic galactic magnetic field of 10^{-10} T, electrons with energies between 1.5 and 1500 GeV emit synchrotron radiation with wavelengths between 30 m and 30 μm, respectively. An electron with an energy equal to 10 TeV, which emits synchrotron radiation in the region of visible light, will emit half its energy in 600 000 years if it moves in a space where the magnetic field is equal to 10^{-10} T. This effect has as a consequence the modification of the expected energy spectrum of cosmic ray electrons.

5.5 The Forces Exerted by Light

Consider a photon which falls on a perfectly reflecting surface at an incidence angle θ (Fig. 5.16). With reference to Fig. 5.16b, the initial momentum of the photon is seen to be $\mathbf{p}_\mathrm{p} = (E/c)\hat{\mathbf{x}}$. The reflection does not change the energy of the photon, as the mirror is assumed to suffer negligible recoil. After reflection, the photon has momentum equal to

$$\mathbf{p}'_\mathrm{p} = \frac{E}{c}\left[\cos 2\left(\frac{\pi}{2} - \theta\right)\hat{\mathbf{x}} - \sin 2\left(\frac{\pi}{2} - \theta\right)\hat{\mathbf{y}}\right] = -\frac{E}{c}(\cos 2\theta\,\hat{\mathbf{x}} + \sin 2\theta\,\hat{\mathbf{y}}). \tag{5.44}$$

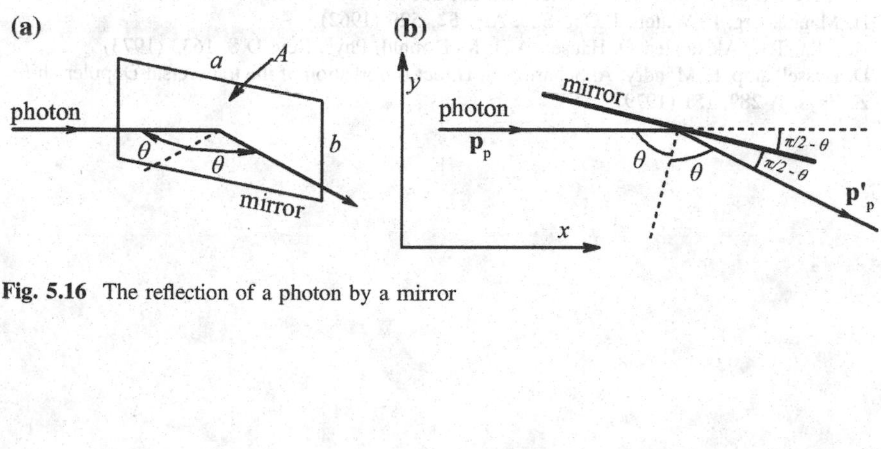

Fig. 5.16 The reflection of a photon by a mirror

The momentum given to the mirror by each photon is

$$\Delta \mathbf{p}_{\mathrm{m}} = \mathbf{p}_{\mathrm{p}} - \mathbf{p}'_{\mathrm{p}} = \frac{E}{c}(1 + \cos 2\theta)\,\hat{\mathbf{x}} + \frac{E}{c}\sin 2\theta\,\hat{\mathbf{y}} = 2\frac{E}{c}\cos\theta\,(\cos\theta\,\hat{\mathbf{x}} + \sin\theta\,\hat{\mathbf{y}}). \tag{5.45}$$

We now consider a mirror of area A exposed to photons of energy E at an energy flux density of w (J/m^2 s). The angle of incidence of the photons on the mirror is θ. The number of photons being reflected by the mirror per unit time is $N = (wA/E)\cos\theta$. Each photon imparts to the mirror a momentum equal to $\Delta \mathbf{p}_{\mathrm{m}}$. The rate of change of the mirror's momentum per unit time, i.e. the force exerted on it by the photons, is

$$\mathbf{F} = N\Delta \mathbf{p}_{\mathrm{m}} = 2\frac{wA}{c}\cos^2\theta\,(\cos\theta\,\hat{\mathbf{x}} + \sin\theta\,\hat{\mathbf{y}}). \tag{5.46}$$

This force is normal to the surface of the mirror. The magnitude of the force is $F = 2\frac{wA}{c}\cos^2\theta$. The pressure exerted by the photons on the mirror is

$$P = 2\frac{w}{c}\cos^2\theta. \tag{5.47}$$

Notice that this does not depend on the energy of the photons but only on w.

If the surface is perfectly absorbing, the momentum imparted to the mirror by each photon is $\Delta \mathbf{p}_{\mathrm{m}} = \mathbf{p}_{\mathrm{p}}$ and the total force on the mirror becomes

$$\mathbf{F} = N\Delta \mathbf{p}_{\mathrm{p}} = 2\frac{wA}{c}\cos\theta\,\hat{\mathbf{x}}. \tag{5.48}$$

References

1. H.E. Ives, G.R. Stilwell, J. Opt. Soc. Am. **28**, 215 (1938) and **31**, 369 (1941)
2. H. Mandelberg, L. Witten, J. Opt. Soc. Am. **52**, 529 (1962)
3. A. Olin, T.K. Alexander, O. Häuser, A.B. McDonald, Phys. Rev. D **8**, 1633 (1973)
4. D. Hasselkamp, E. Mondry, A. Scharmann, Direct observation of the transversal Doppler-shift. Z. Phys. A **289**, 151 (1979)

Chapter 6
Relativistic Dynamics

6.1 The Definition of Relativistic Momentum. Relativistic Mass

We suspect that, given our revision of the concepts of length, time and velocity, we will have to re-examine the magnitude of momentum and its conservation. Indeed, if we examine the collision of two masses in one inertial frame of reference and enforce the conservation of momentum as the letter is defined in Classical Mechanics, we will find that viewed from another inertial frame of reference, the law of conservation of momentum is violated, if we transform the velocities involved from the first frame of reference to the other using the Lorentz transformation.

We wish to preserve the validity of the law of conservation of momentum, whose usefulness is indisputable. We will find that this is possible, if we define momentum in a different manner. We will examine a simple problem in order to find the new expression for the relativistic mass which allows us to retain the laws of conservation of mass and of momentum. In Appendix 3 the problem is re-examined in its generality. Finally, we will find that the new definition of momentum has consequences on the definition of energy as well.

We will assume that it is possible to define the relativistic mass as such a function of speed, $m = m(\upsilon)$, which preserves in the Special Theory of Relativity the validity of

the law of conservation of relativistic mass $m(\upsilon)$ and
the law of conservation of momentum, which is still defined as $\mathbf{p} = m\boldsymbol{\upsilon}$.

We may find the function $m(\upsilon)$ by examining the following special case: Let a body of mass M_0 be at rest in an inertial frame of reference S (Fig. 6.1). The body decays into two identical bodies, 1 and 2, each with relativistic mass m_S and moving with equal and opposite velocities, $\pm V\hat{\mathbf{x}}$. We assume that, in frame S,

© Springer International Publishing Switzerland 2016

C. Christodoulides, *The Special Theory of Relativity*,
Undergraduate Lecture Notes in Physics, DOI 10.1007/978-3-319-25274-2_6

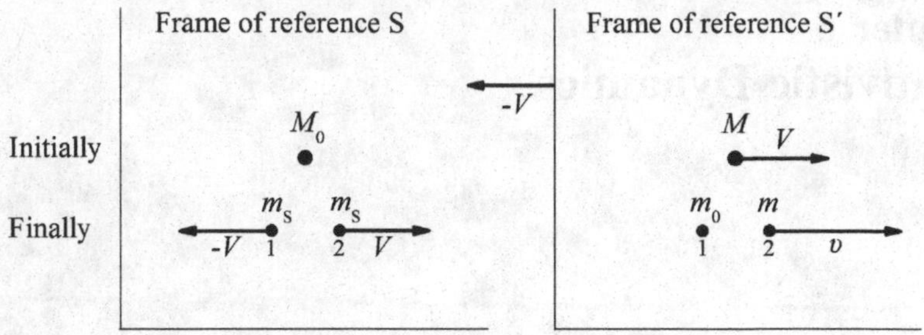

Fig. 6.1 The decay of a body at rest in the frame of reference S into two identical bodies

mass and momentum are conserved during this process. We will examine the decay from a frame of reference S′, which is moving with a constant velocity $-V\hat{\mathbf{x}}$ relative to frame S. In frame S′ the initial body has relativistic mass M and is moving with velocity $V\hat{\mathbf{x}}$. It decays into two bodies, 1 and 2, with relativistic masses m_0 and m, respectively. M_0 and m_0 are the masses of the initial body and of one of the two identical bodies produced, when they are at rest. The transformation of velocities gives for the velocity of body 1 in frame S′ the value of zero and for body 2,

$$\upsilon = \frac{V-(-V)}{1-V(-V)/c^2} = \frac{2V}{1+V^2/c^2}. \tag{6.1}$$

The laws of conservation give:

conservation of relativistic mass $\quad M = m_0 + m \quad$ (6.2)

conservation of momentum $\quad MV = m\upsilon. \quad$ (6.3)

These equations give $\quad \dfrac{m_0}{m} = \dfrac{\upsilon}{V} - 1. \quad$ (6.4)

From Eq. (6.1) we have

$$1+\frac{V^2}{c^2} = 2\frac{V}{\upsilon} \quad \Rightarrow \quad \frac{\upsilon^2}{V^2} + \frac{\upsilon^2}{c^2} = 2\frac{\upsilon}{V} \quad \Rightarrow \quad \frac{\upsilon^2}{V^2} - 2\frac{\upsilon}{V} + 1 = 1 - \frac{\upsilon^2}{c^2} \tag{6.5}$$

and, finally,

$$\frac{\upsilon}{V} - 1 = \pm\sqrt{1-\frac{\upsilon^2}{c^2}}. \tag{6.6}$$

Substituting in Eq. (6.4) we get

$$\frac{m_0}{m} = \sqrt{1 - \frac{\upsilon^2}{c^2}}, \tag{6.7}$$

where we have taken the positive sign as we want mass to be positive.
It follows that

$$m(\upsilon) = \frac{m_0}{\sqrt{1 - \upsilon^2/c^2}}. \tag{6.8}$$

Given that the two bodies are identical and that m_0 is the mass of one of them when at rest, it follows that the mass of either of them when moving with speed υ appears to be given by Eq. (6.8).

Mass $m(\upsilon)$ is what we called *relativistic mass* of the body. It is seen that, as expected, $m(0) = m_0$. For this reason, m_0 is called the *rest mass* of the body. It has been shown that

$$m(\upsilon) = m_0\gamma. \tag{6.9}$$

With this definition of the relativistic inertial mass and with the relativistic momentum being defined now as $\mathbf{p} = m(\upsilon)\boldsymbol{\upsilon}$, both relativistic mass and momentum are conserved, at least in the process examined. Momentum is now defined in terms of the relativistic mass as $\mathbf{p} = m(\upsilon)\boldsymbol{\upsilon}$. It may now be verified that, defining mass and momentum in this way, if relativistic mass and momentum are conserved in one inertial frame of reference during a decay of a body for the creation of two bodies, they will also be conserved in any other inertial frame of reference, if velocities are transformed according to the Lorentz transformation. Later we will prove that this leads to the conservation laws also in cases of a body breaking up into any number of bodies as well as in the opposite process of two or more bodies collide to produce one body.

Attention: the Lorentz factor $\gamma = \frac{1}{\sqrt{1-\upsilon^2/c^2}}$ corresponds to the speed υ of the body in the frame of reference in which it is observed. This is usually stated by using the notation $\gamma(\upsilon)$. Whenever necessary, we will denote by γ the Lorentz factor for the transformation between frames S and S′, and by γ_P the Lorentz factor of a particle.

The variation with speed υ, of the relativistic mass $m(\upsilon)$ of a body which has a rest mass m_0, is shown in Fig. 6.2. The mass increases significantly for speeds approaching the speed of light in vacuum, c, becoming infinite for $\upsilon = c$.

It must be understood that the body does not suffer some change due to which its mass varies. For this reason there are some objections in the use of terms such as *relativistic mass* and *variation of mass with speed*, which, however, we will use because of their usefulness in the description of phenomena. The body is

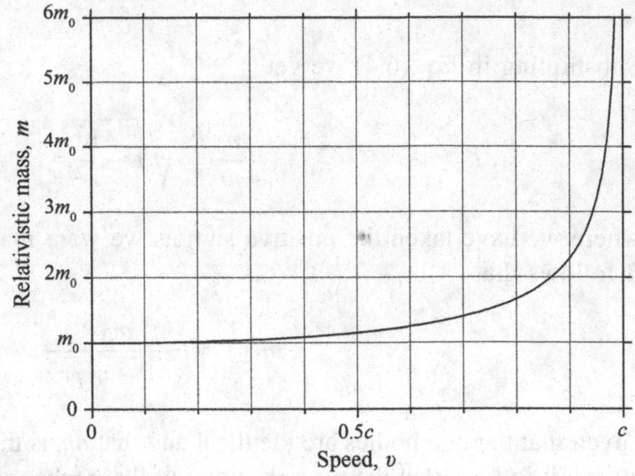

Fig. 6.2 The variation of the relativistic mass of a body, $m(\upsilon)$, in units of its rest mass m_0, as a function of its speed υ, in units of c

characterized by one mass and this is its rest mass, m_0. The inertial properties of the body are such that a more accurate definition of momentum is necessary if the conservation of momentum is to hold. The relativistic mass is just a marhematical convenience as it enables us to retain the definition of momentum as the product of a mass and its velocity.

With the relativistic mass of the body being given by the relation

$$m(\upsilon) = \frac{m_0}{\sqrt{1 - \upsilon^2/c^2}} = \gamma m_0, \tag{6.10}$$

where m_0 is the rest mass of the body, momentum is defined as

$$\mathbf{p} = m\boldsymbol{\upsilon} = \frac{m_0 \boldsymbol{\upsilon}}{\sqrt{1 - \upsilon^2/c^2}} = \gamma m_0 \boldsymbol{\upsilon}. \tag{6.11}$$

The magnitude of the momentum may be written as

$$p = m_0 c \beta \gamma. \tag{6.12}$$

Figure 6.3 shows the variation of the relativistic momentum of a body as a function of its speed. Also shown in the figure, for the purposes of comparison, is the variation of classical momentum, p_{cl}, with speed.

6.2 Relativistic Energy

We wish to preserve Newton's second law of motion in the form

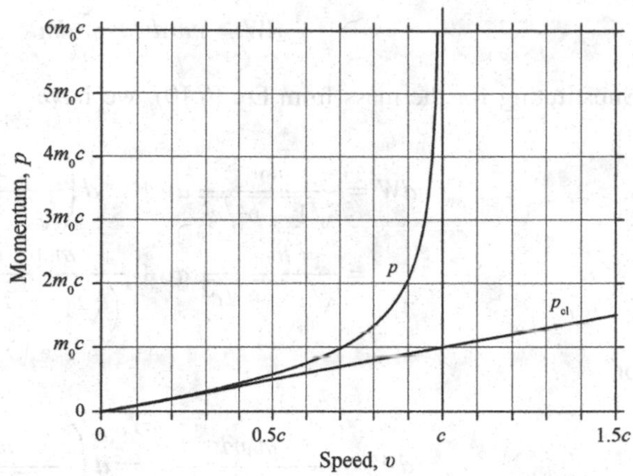

Fig. 6.3 The variation of the relativistic momentum p of a body, whose rest mass is m_0, as a function of its speed. Also shown is the classical momentum p_{cl}

$$\mathbf{F} \equiv \frac{d\mathbf{p}}{dt} \tag{6.13}$$

it has in classical Mechanics, but with the relativistic definition of momentum $\mathbf{p}$, given by Eq. (6.11):

$$\mathbf{F} = \frac{d}{dt}\left(\frac{m_0 \boldsymbol{\upsilon}}{\sqrt{1-\upsilon^2/c^2}}\right). \tag{6.14}$$

Naturally, the values of the quantities t, $\boldsymbol{\upsilon}$, $\mathbf{p}$ and $\mathbf{F}$ are all taken in the same frame of reference. We keep m_0 inside the brackets, in order to cover the possibility that m_0 may vary with time, such as it is the case, for example, when we have the transfer of heat, which, as we will see below, is equivalent to a change in the inertia of a body.

We will also keep the definition of the work dW produced by a force $\mathbf{F}$ when its point of application moves by a distance $d\mathbf{r}$, in its present form,

$$dW \equiv \mathbf{F} \cdot d\mathbf{r}. \tag{6.15}$$

We assume that a force $\mathbf{F}$ is applied on a free body which has a rest mass of m_0 and it causes it to accelerate. Expressing the force as

$$\mathbf{F} = \frac{d}{dt}(m\boldsymbol{\upsilon}) = m\frac{d\boldsymbol{\upsilon}}{dt} + \frac{dm}{dt}\boldsymbol{\upsilon} \tag{6.16}$$

and substituting in Eq. (6.15), we find, for a displacement $d\mathbf{r}$,

$$dW = m\frac{d\boldsymbol{\upsilon}}{dt}\cdot d\mathbf{r} + \frac{dm}{dt}\boldsymbol{\upsilon}\cdot d\mathbf{r} = md\boldsymbol{\upsilon}\cdot\frac{d\mathbf{r}}{dt} + dm\boldsymbol{\upsilon}\cdot\frac{d\mathbf{r}}{dt} = m\boldsymbol{\upsilon}\cdot d\boldsymbol{\upsilon} + \upsilon^2 dm. \tag{6.17}$$

However, it is $\boldsymbol{\upsilon}\cdot d\boldsymbol{\upsilon} = \frac{1}{2}d(\boldsymbol{\upsilon}\cdot\boldsymbol{\upsilon}) = \frac{1}{2}d(\upsilon^2) = \upsilon\, d\upsilon$ and so,

$$dW = m\upsilon\, d\upsilon + \upsilon^2\, dm. \tag{6.18}$$

Substituting for the mass from Eq. (6.10), we have

$$\begin{aligned} dW &= \frac{m_0\upsilon}{\sqrt{1-\upsilon^2/c^2}}\, d\upsilon + \upsilon^2\, d\left(\frac{m_0}{\sqrt{1-\upsilon^2/c^2}}\right) \\ &= \frac{m_0\upsilon}{\sqrt{1-\upsilon^2/c^2}}\, d\upsilon + \frac{m_0\upsilon^3/c^2}{\left(\sqrt{1-\upsilon^2/c^2}\right)^3}\, d\upsilon \end{aligned} \tag{6.19}$$

or

$$dW = \frac{m_0\upsilon d\upsilon}{\left(\sqrt{1-\upsilon^2/c^2}\right)^3} = d\left(\frac{m_0}{\sqrt{1-\upsilon^2/c^2}}\right). \tag{6.20}$$

Integrating Eq. (6.20), we have

$$W = \frac{m_0c^2}{\sqrt{1-\upsilon^2/c^2}} + a \tag{6.21}$$

where a is an integration constant. If we assume that the body was initially at rest when the force started accelerating it, it will be $W = 0$ when $\upsilon = 0$. Therefore,

$$0 = m_0c^2 + a \quad \text{and} \quad a = -m_0c^2. \tag{6.22}$$

Finally,

$$W = \frac{m_0c^2}{\sqrt{1-\upsilon^2/c^2}} - m_0c^2 = m_0c^2(\gamma - 1). \tag{6.23}$$

The production of work W by the force gives the body a speed υ. We consider that this work, produced on a free body, is the *kinetic energy* K of the body:

$$K = \frac{m_0c^2}{\sqrt{1-\upsilon^2/c^2}} - m_0c^2 = m_0c^2(\gamma - 1). \tag{6.24}$$

The quantity
$$E \equiv \frac{m_0c^2}{\sqrt{1-\upsilon^2/c^2}} = K + m_0c^2 \tag{6.25}$$

has the dimensions of energy and consists of two terms:

1. The term m_0c^2, which is the value of E for $\upsilon = 0$: $E(0) \equiv E_0 = m_0c^2$, and
2. The relativistic kinetic energy K of the body.

We call the quantity *E relativistic energy* or, simply, *energy* of the body. Its value for $\upsilon = 0$, $E(0) = E_0 = m_0c^2$, is called *rest energy* of the body. For this reason, E is also called *total relativistic energy* or, simply, *total energy* of the body, in order to stress that it consists of the sum of the rest energy of the body and its kinetic energy

$$E(\upsilon) = E_0 + K = m_0c^2 + K. \tag{6.26}$$

It should be emphasized that, in contrast to classical Mechanics, where by the term total energy we always imply (kinetic energy) + (potential energy), in relativistic Mechanics the term total energy means (rest energy) + (relativistic kinetic energy), for a free body. In those cases when the body is not free but it moves in a conservative field, its potential energy is also added to the quantity E of Eq. (6.26), to give the total energy of the body. The rest energy may be considered as a kind of internal potential energy, characteristic of the body.

The relativistic energy is also written as

$$E = mc^2 = \gamma m_0c^2. \tag{6.27}$$

Figure 6.4 shows the variation of the relativistic K and E as functions of the speed υ. Also shown is the classical kinetic energy $K_{\text{cl}} = \frac{1}{2}m_0\upsilon^2$.

The fact that the relativistic mass and the relativistic energy of a free body are proportional to each other, means that the conservation of the one also implies the conservation of the other. For this reason, in the Special Theory of Relativity, we refer to the conservation of both quantities as *the law of conservation of mass-energy*.

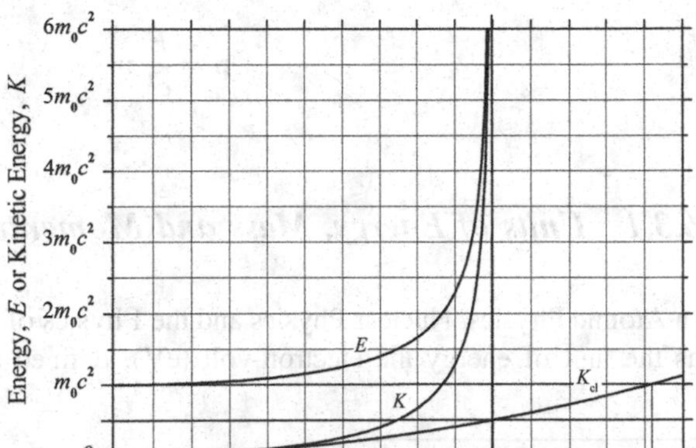

Fig. 6.4 The variation, as functions of the speed υ of a body, of its relativistic energy E, its relativistic kinetic energy K and its classical kinetic energy $K_{\text{cl}} = \frac{1}{2}m_0\upsilon^2$, expressed in units of its rest energy $m_0\,c^2$, where m_0 is the rest mass of the body

6.3 The Relationship Between Momentum and Energy

From the relation $\gamma = \frac{1}{\sqrt{1-\beta^2}}$, we have $\gamma^2 - \beta^2\gamma^2 = 1$. Multiplying throughout by $m_0^2c^4$, it follows that

$$\gamma^2 m_0^2 c^4 - \beta^2\gamma^2 m_0^2 c^4 = m_0^2 c^4. \tag{6.28}$$

Because it is $E = \gamma m_0 c^2$ and $p = \beta\gamma m_0 c$, the last equation may also be written as

$$E^2 - p^2c^2 = m_0^2 c^4. \tag{6.29}$$

The magnitude $m_0^2c^4$ is an invariant, because, evaluated in any inertial frame of reference, takes the same value. It follows that the quantity $E^2 - p^2c^2$ is also invariant.

The relation

$$E^2 = m_0^2c^4 + p^2c^2 \tag{6.30}$$

connects the magnitudes of the rest mass, the momentum and the energy of a body. Geometrically, it may be considered as an application of the Pythagorean theorem on a four-dimensional[1] 'right-angled triangle' with sides $E_0 = m_0c^2$, p_xc, p_yc and p_zc. The triangle of Fig. 6.5 may serve as a memory aid for relation (6.30).

The energy E may also be expressed as

$$E = \sqrt{m_0^2c^4 + p^2c^2}. \tag{6.31}$$

In addition, since it is $E = mc^2$ and $\mathbf{p} = m\boldsymbol{\upsilon}$, it follows that

$$\mathbf{p} = \frac{E}{c^2}\,\boldsymbol{\upsilon}. \tag{6.32}$$

6.3.1 *Units of Energy, Mass and Momentum*

In Atomic Physics, Nuclear Physics and the Physics of Elementary Particles, we use as the unit of energy the electron-volt (eV), defined as the change in the kinetic

[1]In fact, the quantity m_0c^2 is the modulus of the four-vector of energy-momentum, which has as components the quantities E, p_xc, p_yc and p_zc. The norm of this four-vector is, according to the definition, equal to $m_0^2c^4 = E^2 - p_x^2c^2 - p_y^2c^2 - p_z^2c^2 = E^2 - p^2c^2$ and is invariant. The difference between the magnitude of this and the magnitude of a vector in a four-dimensional Euclidean space is in the negative signs which appear. For more on four-vectors, see Chap. 8.

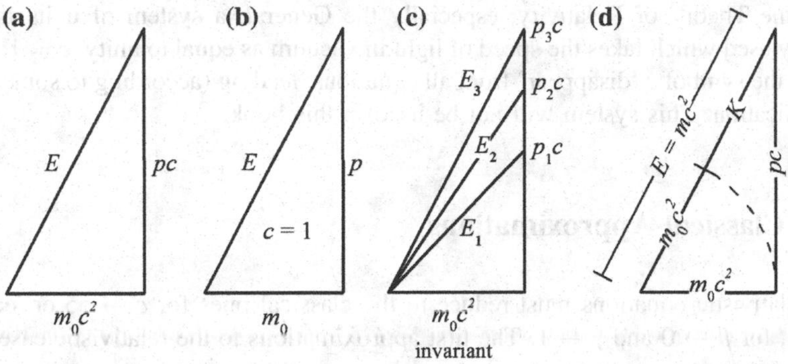

Fig. 6.5 Geometrical memory aid for the relationship $E^2 = m_0^2c^4 + p^2c^2$. Pythagoras' theorem, applied to the right-angle triangle of figure (**a**), with sides equal to m_0c^2 and pc and the total energy E as its hypotenuse, gives the relationship. In figure (**b**), a system of units is used in which $c = 1$. Figure (**c**) shows the variation of the body's total energy with its momentum. Figure (**d**) also shows the relationships $E = mc^2 = m_0c^2 + K$, where K is the body's kinetic energy

energy of a particle with a charge equal to that of the proton, $|e|$, on moving between two points among which there exists a potential difference of one volt. It is $1\text{ eV} = |e|\Delta V = (1.602 \times 10^{-19}\text{ C}) \times (1\text{ V}) = 1.602 \times 10^{-19}\text{ J}$. Multiples of this unit which are used are the keV (10^3 eV), the MeV (10^6 eV), the GeV (10^9 eV) or even the TeV (10^{12} eV).

Given that the magnitude m_0c^2 has the dimensions of energy and, for relativistic particles, is measured in MeV or GeV, it is customary to use MeV/c^2 and GeV/c^2 as units of mass. So, if we find, for example, the numerical result $mc^2 = \text{X MeV}$, we express the mass as $m = \text{X MeV}/c^2$.

Similarly, for momentum, if we evaluate the quantity pc, which has the dimensions of energy, and we find the numerical result $pc = \text{X MeV}$, we express the momentum as $p = \text{X MeV}/c$.

Inversely, a body with mass $m = \text{X MeV}/c^2$ has $mc^2 = \text{X MeV}$. A body with momentum $p = \text{X MeV}/c$ has $pc = \text{X MeV}$.

The advantage in the use of these units is that we have simple numbers for the energy, the mass and the momentum of relativistic particles and the numerical values of these quantities for a certain particle are of the same order of magnitude. For example, for a very energetic particle, which moves with a speed very close to c, if its energy is $mc^2 = \text{X MeV}$, its momentum will be $p = m\upsilon \approx mc = \text{X MeV/c}$. The particle's mass expressed in MeV/c^2 has the same numerical value as its energy expressed in MeV. For a high-energy particle, its momentum, expressed in MeV/c, will almost have the same numerical value as well. A proton with a total energy of 10 GeV, has a relativistic mass of 10 GeV/c^2, a Lorentz factor of about 10, which corresponds to a reduced speed of $\beta = 0.995$, and a momentum p for which it is $pc = m\upsilon c = m\beta c^2 = 0.995mc^2 = 9.95\text{ GeV}$. It is, therefore, $p = 9.95\text{ GeV/c} \approx 10\text{ GeV/c}$.

In the Theory of Relativity, especially the General, a system of units is frequently used which takes the speed of light in vacuum as equal to unity, $c = 1$. This makes the symbol c 'disappear' from all equations, leading (according to some) to a simplification. This system will not be used in this book.

6.4 Classical Approximations

The relativistic equations must reduce to the classical ones for $c \to \infty$ or, equivalently, for $\beta \to 0$ and $\gamma \to 1$. The first approximations to the relativistic cases are found for $\gamma = \frac{1}{\sqrt{1-\upsilon^2/c^2}} \approx 1 + \frac{1}{2}\frac{\upsilon^2}{c^2}$. In this way, the kinetic energy is approximately equal to

$$K = \frac{m_0 c^2}{\sqrt{1-\upsilon^2/c^2}} - m_0 c^2 = m_0 c^2(\gamma - 1) \approx \frac{1}{2}\frac{\upsilon^2}{c^2} m_0 c^2 = \frac{1}{2} m_0 \upsilon^2, \tag{6.33}$$

the kinetic energy according to classical Mechanics. The total energy is, for a free body,

$$E = mc^2 = \gamma m_0 c^2 \approx \left(1 + \frac{1}{2}\frac{\upsilon^2}{c^2}\right) m_0 c^2 = m_0 c^2 + \frac{1}{2} m_0 \upsilon^2, \tag{6.34}$$

i.e. approximately equal to the rest energy of the body plus its classical kinetic energy.

The momentum of a body is [Eq. (6.11)]

$$\mathbf{p} = m\boldsymbol{\upsilon} = \frac{m_0 \boldsymbol{\upsilon}}{\sqrt{1-\upsilon^2/c^2}} \approx m_0 \boldsymbol{\upsilon}\left(1 + \frac{1}{2}\frac{\upsilon^2}{c^2}\right) \quad \text{or} \quad \mathbf{p} = m_0 \boldsymbol{\upsilon}, \tag{6.35}$$

to the approximation involving only powers lower than the second in the ratio υ/c.

It must be stressed that it is wrong to use these relations as exact in relativistic Mechanics!

Example 6.1 Speeds of Electrons and Protons From Modern Accelerators
The energy to which electrons are accelerated by modern synchrotrons has exceeded the value of 25 GeV, while protons are accelerated to 7 TeV at CERN, with the accelerator LHC. What are the speeds of the particles at these energies?

The rest energies of the electron and the proton are, respectively, $m_{e0}c^2 = 0.511$ MeV and $m_{p0}c^2 = 938$ MeV. At energies as high as the ones given, the Lorentz factor is so large that the kinetic energy may be taken as equal to the total. It is, therefore, $K = m_0 c^2(\gamma - 1) \approx \gamma m_0 c^2$ and thus $\gamma \approx K/m_0 c^2$, from which we have $\beta = \sqrt{1 - 1/\gamma^2} \approx 1 - 1/2\gamma^2 \approx 1 - (m_0 c^2)^2/2K^2$.

For an electron with $K_e = 25$ GeV $= 2.5 \times 10^{10}$ eV, it is

$$\gamma_e \approx K_e/m_{e0}c^2 = 2.5 \times 10^{10}/0.511 \times 10^6 = 49\,000.$$

We see that the approximation we made is justified. The reduced speed of the electron is

$$\beta_e \approx 1 - 1/2\gamma_e^2 = 1 - 2.1 \times 10^{-10} = 0.999\,999\,999\,8.$$

For a proton with $K_p = 7$ TeV $= 7 \times 10^{12}$ eV, it is

$$\gamma_p \approx K_p/m_{p0}c^2 = 7 \times 10^{12}/0.938 \times 10^9 = 7560.$$

We see that the approximation we made is, here, also justified. The reduced speed of the proton is

$$\beta_p \approx 1 - 1/2\gamma_p^2 = 1 - 8.7 \times 10^{-9} = 0.999\,999\,991\,3.$$

Example 6.2 Kinetic Energy in One-Dimensional Motion
Show that the kinetic energy of a body moving on a straight line is given by $K = mc^2 - m_0c^2$.

It is

$$K = \int_{\upsilon=0}^{\upsilon} \mathbf{F} \cdot d\mathbf{s} = \int_{\upsilon=0}^{\upsilon} F dx = \int_{\upsilon=0}^{\upsilon} \frac{d}{dt}(m\upsilon)dx = \int_{\upsilon=0}^{\upsilon} d(m\upsilon)\frac{dx}{dt} =$$
$$= \int_{\upsilon=0}^{\upsilon} (md\upsilon + \upsilon dm)\upsilon = \int_{\upsilon=0}^{\upsilon} (m\upsilon d\upsilon + \upsilon^2 dm)$$

From $m = \dfrac{m_0}{\sqrt{1-\upsilon^2/c^2}}$, it follows that $m^2c^2 - m^2\upsilon^2 = m_0c^2$.

Taking differentials, we find $2mc^2dm - m^2 2\upsilon d\upsilon - \upsilon^2 2mdm = 0$ or $m\upsilon d\upsilon + \upsilon^2 dm = c^2 dm$. Substituting in K,

$$K \int_{\upsilon=0}^{\upsilon} (m\upsilon d\upsilon + \upsilon^2 dm) = \int_{m=m_0}^{m} c^2 dm = c^2(m - m_0).$$

Problems
6.1 What is the speed of an electron whose kinetic energy is 2 MeV? What is the ratio of its relativistic mass to its rest mass? The rest mass of the electron is $m_0 = 0.511$ MeV/c^2. Ans.: $\upsilon = 0.98c$, $m/m_0 = 5$
6.2 The extremely rare event of the indirect observation, in cosmic radiation, of a particle with an energy of the order of 10^{20} eV (16 J!) occurred a few years ago [see J. Linsey, *Phys. Rev. Lett.* **10**, 146 (1963)]. Assuming the particle was a proton,

which has a rest energy of 1 GeV, approximately, evaluate its speed relative to the Earth. How much time would this proton need, in its own frame of reference, in order to cross our galaxy, whose diameter is 10^5 light years? What is the diameter of the Galaxy as seen by the proton? Ans.: $\upsilon = 0.999\,999\,999\,999\,999\,999\,999\,95\,c = c - 1.5 \times 10^{-14}$ m/s, 32 s, 10^{10} m approx.

6.3 The kinetic energy and the momentum of a particle were measured and found equal to 250 MeV and 368 MeV/c, respectively. Find the rest mass of the particle in MeV/c^2. Ans.: 270 MeV/c^2

6.4 At what value of the speed of a particle is its kinetic energy equal to (a) its rest energy and (b) 10 times its rest energy? Ans.: (a) $0.866c$, (b) $0.996c$

6.5 An electron with an energy of 100 MeV moves along a tube which is 5 m long. What is the length of the tube in the frame of reference of the electron? The rest energy of the electron is $E_0 = m_0c^2 = 0.511$ MeV, where m_0 is its rest mass. Ans.: 26 mm

6.6 A beam of identical particles with the same speed is produced by an accelerator. The particles of the beam travel, inside a tube, its full length of $l = 2400$ m in a time $\Delta t = 10\ \mu$s, as measured by an observer in the laboratory frame of reference.

(a) Find β and γ for the particles and the duration $\Delta t'$ of the trip as measured in their own frame of reference. Ans.: $\beta = 4/5$, $\gamma = 5/3$, $\Delta t' = 6\ \mu$s
(b) If the particles are unstable, with a mean lifetime of $\tau = 10^{-6}$ s, what proportion of the particles is statistically expected to reach the end of the tube? (Use: $e^3 \approx 20$). Ans.: 1/400
(c) If the rest energy of each particle is $m_0c^2 = 3$ GeV, find their kinetic energy. Ans.: $K = 2$ GeV
(d) Determine the quantity pc for a particle of the beam, where p is its momentum, and express p in units of GeV/c. Ans.: $p = 4$ GeV/c

6.7 Two particles, A and B, each having a rest mass of $m_0 = 1$ GeV/c^2, move, in the frame of reference of an accelerator, on the x-axis and in opposite directions, approaching each other. In this frame, particle A moves with a velocity of $\upsilon_{Ax} = -0.6\,c$ and particle B with a velocity of $\upsilon_{Bx} = 0.6\,c$. In the frame of reference of particle A,

(a) what is the speed of particle B? Ans.: $\upsilon'_{Bx} = 0.88c$
(b) what is the energy (in GeV) and the momentum (in GeV/c) of particle B? Ans.: $E'_B = 2.11$ GeV, $p'_{Bx} = 1.86$ GeV/c

6.8 A beam of π^+ particles, each with an energy of 1 GeV, has a total flow rate of 10^6 particles/s at the start of a trip that has a length of 10 m in the laboratory frame of reference. What is the flow rate of particles at the end of the trip? π^+ has a rest mass of $m_\pi = 140$ MeV/c^2 and a mean lifetime $\tau_\pi = 2.56 \times 10^{-8}$ s. Ans.: $0.83 \times 10^6\ \pi^+$/s

6.9 Show that, for a body of rest mass m_0, which moves with a speed υ and has momentum p and kinetic energy K, it is $\dfrac{p\upsilon}{K} = 1 + \dfrac{1}{1 + K/m_0c^2}$.

6.5 Particles with Zero Rest Mass

There exist in nature particles with zero rest mass. The better known of these is the photon, the carrier of the electromagnetic field. The graviton, the carrier of the gravitational field also has zero rest mass. The neutrino possibly has zero rest mass, although at present we can only set an upper limit to it, which is 0.3 eV/c^2. Strictly speaking, that is the situation with the other particles as well. The upper limit for the rest mass of the photon is known today, from measurements of Luo [1] and his co-workers, to be 1.2×10^{-54} kg or 7×10^{-21} eV/c^2.

If we put $m_0 = 0$ in the relativistic energy relations, we have for the energy, from Eq. (6.30)

$$E = pc. \tag{6.36}$$

Inversely, the momentum of a particle with zero rest mass which has an energy E is

$$p = E/c. \tag{6.37}$$

Since in general it is $p = \frac{E}{c^2}\upsilon$ [Eq. (6.32)], substituting in Eq. (6.37), we find that $\upsilon = c$. The same result is derived by letting $m_0 \to 0$ in the equation $E = \frac{m_0\, c^2}{\sqrt{1-\upsilon^2/c^2}}$, in which case it follows that $\upsilon \to c$. This important result tells us that *all particles having zero rest mass move with the maximum possible speed, which is the speed of light in vacuum, c.*

Also valid for the photon are the relations between its energy, its frequency f and its wavelength λ,

$$E = hf = \frac{hc}{\lambda}, \tag{6.38}$$

where h is Planck's constant $\left(h = 6.626 \times 10^{-34}\ \text{J} \cdot \text{s} = 4.136 \times 10^{-15} \text{eV} \cdot \text{s}\right)$. It follows that the momentum of a photon is also given by

$$p = \frac{E}{c} = \frac{hf}{c} = \frac{h}{\lambda}. \tag{6.39}$$

We have to be careful in the use of this expression, since, given that the energy of the photon is always positive, Eq. (6.39) only gives the magnitude of the momentum and not its algebraic value, which may, in some cases, be negative.

Problems

6.10 What is the mass m_ϕ which corresponds to the energy E_ϕ of a photon of wavelength 500 nm? Ans.: $m_\phi = 4.42 \times 10^{-36}$ kg

6.11 The wavelength of the photons from a laser is 633 nm. What is the momentum of one such photon? Ans.: $p = 1.96$ eV/c

6.6 The Conservation of Momentum and of Energy

Momentum was defined in such a manner so that, in an interaction of particles, both the mass and the momentum are conserved. Given that the total energy of a free body, not acted on by external forces, is equal to $E = mc^2$, the fact that in an interaction of n particles the conservation of relativistic mass applies,

$$\sum_{i=1}^{n} m_i = \text{const.} \tag{6.40}$$

means that the energy is also conserved:

$$\sum_{i=1}^{n} E_i = \text{const.} \tag{6.41}$$

This law is used, together with the law of conservation of momentum

$$\sum_{i=1}^{n} \mathbf{p}_i = \text{const.} \tag{6.42}$$

in the solution of problems in which particles interact in any way.

For n particles, the components of the total momentum and the energy are defined as

$$P_{\text{tot},x} \equiv \sum_{i=1}^{n} p_{ix} \quad P_{\text{tot},y} \equiv \sum_{i=1}^{n} p_{iy} \quad P_{\text{tot},z} \equiv \sum_{i=1}^{n} p_{iz} \quad E_{\text{tot}} \equiv \sum_{i=1}^{n} E_i \tag{6.43}$$

for frame S, with similar expressions for frame S′.

If, in frame S, during the interaction of n particles, the momentum and the energy are conserved, then, for the total values of the components of momentum and of energy, the following relations are true:

$$\left(P_{\text{tot},x}\right)_{\text{initially}} = \left(P_{\text{tot},x}\right)_{\text{finally}} \quad \left(P_{\text{tot},y}\right)_{\text{initially}} = \left(P_{\text{tot},y}\right)_{\text{finally}} \quad \left(P_{\text{tot},z}\right)_{\text{initially}} = \left(P_{\text{tot},z}\right)_{\text{finally}}$$
$$\left(E_{\text{tot}}\right)_{\text{initially}} = \left(E_{\text{tot}}\right)_{\text{finally}} \tag{6.44}$$

The conservation of these quantities is valid in general, in every physical process. The subject will be further discussed in the next section.

Example 6.3 The Photon Rocket

The use of photons in the propulsion of rockets has been proposed. While for chemical fuels there is a limit in the speed of ejection of mass of the order of 10 km/s, for photons this speed is 30 000 times greater. For this reason, the emission of photons by the rocket, with their high speed, was considered that it would give the rocket a greater speed per unit mass ejected. Find the speed attained by such a rocket, as a function of the fraction κ of its mass that has been ejected as photons.

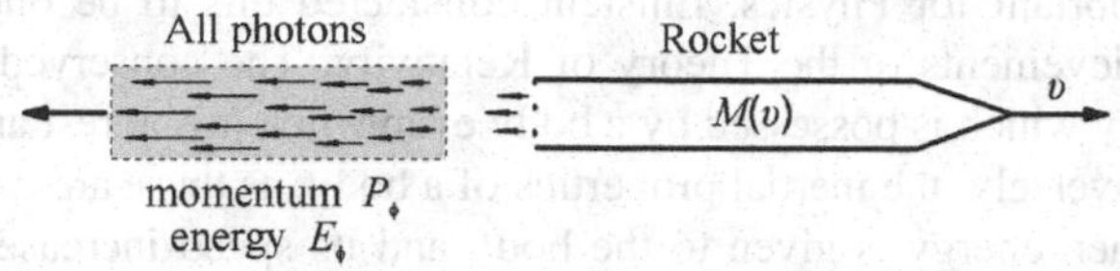

Let the initial rest mass of the rocket be M_0 and, at some time, a fraction κ of this has been emitted as photons. All the photons are considered to be emitted in the same direction. The emitted photons carry, in total, an energy E_ϕ and momentum $P_\phi = E_\phi/c$, as measured by an observer on the Earth. If at that moment the rocket is moving with a speed equal to υ and has a relativistic mass $M(\upsilon) = (1-\kappa)M_0\gamma$, where $\gamma = 1\big/\sqrt{1-\upsilon^2/c^2}$, the laws of conservation give:

Conservation of energy: $E_{\text{total}} = M_0c^2 = M(\upsilon)c^2 + E_\phi = (1-\kappa)M_0c^2\gamma + E_\phi$
Conservation of momentum: $P_{\text{total}} = 0 = M(\upsilon)\upsilon - E_\phi/c = (1-\kappa)M_0\gamma\upsilon - E_\phi/c$
Eliminating E_ϕ between these two relations, it follows that

$M_0c^2 = (1-\kappa)M_0c^2\gamma(1+\beta)$ and $(1-\kappa)\gamma(1+\beta) = 1$, $\quad (1-\kappa)\sqrt{\dfrac{1+\beta}{1-\beta}} = 1$,

$\dfrac{1-\beta}{1+\beta} = (1-\kappa)^2$, from which we find $\beta = \dfrac{1-(1-\kappa)^2}{1+(1-\kappa)^2}$ and

$\gamma = \dfrac{1}{2}\left(1-\kappa+\dfrac{1}{1-\kappa}\right).$

Some numerical values are:

β	0.5	0.9	0.95	0.99	0.995	0.999
γ	1.16	2.94	3.20	7.09	10	22.4
κ	0.42	0.77	0.84	0.93	0.95	0.98

We see that, in order to achieve a speed of $0.995c$, which corresponds to $\gamma = 10$, a proportion of 95 % of the rocket's total mass must be emitted as photons. The useful load is, therefore, only 5 % of the spaceship. Of course, if we wish to decelerate the spaceship in order to stop it, then accelerate it towards its starting point and finally decelerate it to a halt at its starting point, the useful load would be only a fraction equal to $(1-\kappa)^4$, which has a value smaller than 10^{-5} for maximum γ equal to 10. This makes the photon rockets of doubtful usefulness, without the situation being very much better for chemical fuels of course. The problem is examined in greater detail in Sect. 7.6.1.

6.7 The Equivalence of Mass and Energy

The relation $E = mc^2$ describes an equivalence of mass and energy which is extremely important for Physics. Einstein considered this to be one of the most important achievements of the Theory of Relativity. The conserved magnitude E includes energy which is possessed by a body even when it is at rest and is due to its rest mass. Conversely, the inertial properties of a body, as these are expressed by its mass, vary when energy is given to the body and its speed increases. An equivalence between mass and energy seems to exist, which is verified experimentally in many ways. This equivalence makes possible the creation or annihilation of particles in a reaction, with the assumption of the conservation of mass-energy, m or E, as well as of other physical quantities (momentum, angular momentum, charge, hypercharge, lepton numbers, baryon number, taste, color, charm and others, in addition to some that are still unknown!).

The relation $E = mc^2$ has a twofold meaning: it states that mass may be changed into energy, but also that to every form of energy E there corresponds a mass equal to $m = E/c^2$. Although the mass appearing in the equation $E = mc^2$ is the inertial mass, in the General Theory of Relativity the inertial mass is equal to the gravitational, and, hence, the relation also attributes a gravitational mass to every form of energy. We must, however, be careful. If the motion of the gravitational mass of a photon in a gravitational field is examined, a deflection of light will be predicted, as demonstrated experimentally. The angle of deflection will, however, be wrong by a factor of 2, because the calculation does not take into account the geometrical distortion or curving of space by the mass to which the gravitational field is due.

In general, a change of Δm in the mass, corresponds to a change ΔE in the energy, and vice versa, where the two quantities are related by

$$\Delta E = \Delta m\, c^2. \qquad (6.45)$$

The experimental evidence for the validity of this relation is indisputable.

It is worth making a comment, here, regarding this equation. The opinion is sometimes heard that Einstein discovered the equation $E = mc^2$, which showed that great amounts of energy may be released when mass is transformed into energy, something that led to the construction of the atomic bomb, initially, and the nuclear bomb subsequently. This point of view is wrong, both as regards History as well as Physics. It must be understood that this equation applies in all physical and chemical changes, not only in processes in nuclear Physics. It is valid when a spring is compressed or stretched (Problem 6.12) and in every chemical reaction (Example 6.4). It applies when we light a match, a fire and in all the chemical reactions happening in our bodies. Man has used fire for many millennia before the discovery of the equation. He did not need to know the origin of the energy he used for war purposes when he used incendiary bombs. The same was true in the construction of the atom bomb. From the moment the phenomenon of nuclear fission was discovered in the laboratory, between 1934 and 1939, and the enormous amount of energy released was measured, the construction of the atomic bomb was a matter of time. The only difference with the past was that, now, the amount of energy released is so great that the change in the nuclear masses is measurable. Neither Einstein nor the equation $E = mc^2$ should be incriminated. The equation simply gives us the ability to interpret the origin of the released energy, both in nuclear reactions and in many others.

6.7.1 The Validity of the Conservation of Momentum and Energy During the Transmutation of Nuclei and the Annihilation and Creation of Particles

The equation of equivalence of mass and energy is valid in all changes a system suffers and not only to 'classical' processes in which the reacting bodies retain their form unchanged. In the processes of nuclear Physics and the Physics of Elementary Particles, the nuclei may decay or suffer fission or fusion and the elementary particles may be created or annihilated. In these processes, the reacting bodies are not necessarily the same before and after the reaction. It has been established experimentally that, in all these changes, momentum and energy, as defined by the Special Theory of Relativity, are conserved.

It was found that relativistic definitions of mass and momentum make it possible to conserve both momentum and energy in all inertial frames of reference in the case of a decaying particle. This is shown in Fig. 6.6a and may be symbolized by $(1) \to (2) + (3)$.

If we reverse time, we have the creation of one body from the collision of two particles, as seen in Fig. 6.6b, and is symbolized by $(1) + (2) \to (3)$. The conservation of momentum and energy in case (a) makes it possible to conserve the two quantities in case (b) too. The laws of conservation also hold in more complex cases, such as the one shown in Fig. 6.6c where two bodies react to create two other bodies but with different rest masses. We may assume that, intermediately, a single body is formed (circle A), which then decays into two bodies. Thus, we have a process (b) followed by a process (a), again safeguarding the possibility of the conservation of momentum and energy at all stages. Symbolically, we may write the equation of the reaction as $(1) + (2) \to (A) \to (3) + (4)$. By the same method, more complex processes may be analyzed, such as the interaction of two bodies for the creation of three, as shown in Fig. 6.6d, with the equation of the reaction being $(1) + (2) \to (A) \to (3) + (B) \to (3) + (4) + (5)$. In this way it is possible to

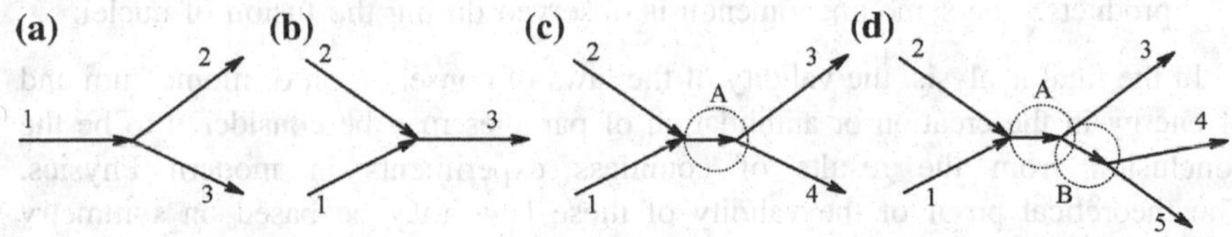

Fig. 6.6 Various collisions of point masses in order to form a single body or other bodies with different rest masses

conserve relativistic momentum and energy during the interaction of particles for the creation of other, possibly different, particles. It must be stressed that in these cases, the rest masses and the velocities of the particles created do not have uniquely determined values, as may be seen, for example, in the elastic collision of two identical particles which is examined in Sect. 6.11. Other laws determine which particles may be created and limit further the range of values their energies and momenta may have. There exist many such examples which have been studied experimentally, three of which will be discussed below.

The equivalence of mass and energy and the validity of the laws of conservation of momentum and energy are demonstrated in a dramatic way in the annihilation of matter-antimatter such as an electron-positron pair, in the creation of particles from a photon as in the phenomenon of pair production, as well as in the fission or fusion of atomic nuclei:

(a) *Annihilation of an electron-positron pair*. An electron and its antiparticle, the positron, annihilate completely producing two photons,

$$e^- + e^+ \to 2\gamma.$$

Since for 'thermal' particles the momentum of the pair of particles is initially almost zero, the two photons have the same energy (511 keV) and move in opposite directions, in order to conserve momentum. The energy available, which is entirely due to the rest masses of the two particles, is $2m_e c^2 = 1.022$ MeV, which is shared equally among the two photons.

(b) *Production of an electron-positron pair*. A photon may produce an electron-positron pair, provided it has an energy higher than $2m_e c^2 = 1.022$ MeV. For both momentum and energy to be conserved, the process can only take place in the vicinity of another body, such as the nucleus of an atom. The *threshold* of $2m_e c^2 = 1.022$ MeV in energy exists because this is the energy equivalent of the two electron masses. If the photon possesses just this amount of energy in a certain frame of reference, then the two particles will have zero kinetic energies, in this frame of reference, after their creation. Photons of higher energy also impart kinetic energy to the products of pair production.

(c) *Fission and fusion of atomic nuclei*. Another example of the conversion of mass to energy is observed during the phenomenon of nuclear fission, which will be examined in the next chapter. During fission, a small proportion of the rest mass of the nucleus is transformed into kinetic energy of the fission products. The same phenomenon is observed during the fusion of nuclei.

In the final analysis, the validity of the laws of conservation of momentum and of energy in the creation or annihilation of particles may be considered to be the conclusion from the results of countless experiments in modern Physics. The theoretical proof of the validity of these laws may be based on symmetry arguments [2]. The conservation of energy is derived from the invariance of the laws of Physics during a displacement in time and the law of the conservation of (vectorial) momentum follows from the invariance of the laws of Physics on the transportation along the three dimensions of space. If this invariance holds then the laws of conservation follow. Modern Physics in general and Nuclear Physics and the Physics of Elementary Particles in particular, are based to a large extent on the validity of these laws. If nothing else, therefore, we have to admit that the validity of the laws of the conservation of momentum and of energy in all physical processes are based on indisputable experimental evidence. In the examples and the problems that follow, the usefulness of these laws in the study of physical processes and phenomena will be demonstrated.

Example 6.4 Conversion of Mass into Energy in a Chemical Reaction
During the combination of 1 kg of hydrogen with 8 kg of oxygen for the production of 9 kg of water, an energy of approximately 10^8 J is released. Is it possible to detect the conversion of mass into energy in this case, using a balance which is capable in measuring a proportional change in mass equal to 1 part in 10^7?

The mass equivalent of the energy released is $\Delta m = \Delta E/c^2 = 10^8 \big/ (3 \times 10^8)^2 \approx 10^{-9}$ kg. This is approximately 1 part in 10^{10} of the mass involved in the chemical reaction and cannot be detected with the balance we have or with any other balance available today.

Example 6.5 The De-excitation of a Nucleus
A stationary excited nucleus returns to its ground state by liberating an energy ΔE and emitting a photon. The rest mass of the de-excited nucleus is M. What is the energy E_γ of the photon? What fraction of the excitation energy is given to the nucleus as kinetic energy?

The rest energy M^*c^2 of the excited nucleus is equal to the rest energy Mc^2 of the de-excited nucleus plus the excitation energy ΔE.
The laws of conservation give:

Conservation of energy $M^*c^2 = Mc^2 + \Delta E = E + E_\gamma$, where $E = Mc^2\gamma$.

Conservation of momentum $\dfrac{E_\gamma}{c} = Mc\gamma\beta$.

Therefore, $Mc^2 + \Delta E = Mc^2\gamma(1+\beta) = Mc^2\sqrt{\dfrac{1+\beta}{1-\beta}}$,

from which we get $\sqrt{\dfrac{1+\beta}{1-\beta}} = 1 + \dfrac{\Delta E}{Mc^2}$.

This relation may be written in terms of the ratio $\alpha \equiv \dfrac{\Delta E}{Mc^2}$ as $\dfrac{1+\beta}{1-\beta} = (1+\alpha)^2$.

Solving, $\beta = \dfrac{(1+\alpha)^2 - 1}{(1+\alpha)^2 + 1}$, $\quad \gamma = \dfrac{1 + (1+\alpha)^2}{2(1+\alpha)}$.

It follows that $E_\gamma = Mc^2\beta\gamma = Mc^2\left[\dfrac{(1+\alpha)^2-1}{(1+\alpha)^2+1}\right]\left[\dfrac{1+(1+\alpha)^2}{2(1+\alpha)}\right]$

or $E_\gamma = \dfrac{Mc^2}{2}\left(\dfrac{(1+\alpha)^2-1}{1+\alpha}\right) = \dfrac{Mc^2}{2}\left(\dfrac{\alpha^2+2\alpha}{1+\alpha}\right)$

and $E_\gamma = \dfrac{\Delta E}{2}\dfrac{\dfrac{\Delta E}{Mc^2}+2}{\dfrac{\Delta E}{Mc^2}+1}$, $\quad E_\gamma = \Delta E\left(\dfrac{Mc^2+\frac{1}{2}\Delta E}{Mc^2+\Delta E}\right)$.

The energy given to the de-excited nucleus as kinetic energy is

$$K = \Delta E - E_\gamma = \Delta E\left(1-\frac{Mc^2+\frac{1}{2}\Delta E}{Mc^2+\Delta E}\right) = \frac{\Delta E}{2}\left(\frac{\Delta E}{Mc^2+\Delta E}\right).$$

The fraction of the excitation energy which is given to the de-excited nucleus as kinetic energy is

$$\frac{K}{\Delta E} = \frac{1}{2}\frac{\Delta E}{Mc^2+\Delta E}.$$

If it is $\Delta E \ll Mc^2$, then the fraction is approximately equal to $\dfrac{K}{\Delta E} \approx \dfrac{\Delta E}{2Mc^2}$.

Example 6.6 The Creation of a Neutral Pion

What is the energy of the π^0 produced during the capture of an antiproton, which is moving with a low speed, by a stationary deuterium nucleus ($\overline{\mathrm{p}}+\mathrm{d} \to \mathrm{n}+\pi^0$)? The rest energies of the particles $\overline{\mathrm{p}}$, d, n and π^0 are, respectively, $E_{\mathrm{p}0} = 938.2$ MeV, $E_{\mathrm{d}0} = 1875.5$ MeV, $E_{\mathrm{n}0} = 939.5$ MeV and $E_{\pi 0} = 135.0$ MeV.

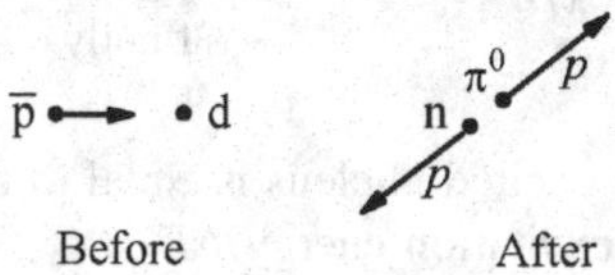

Since the initial momentum of the system is negligible, the momenta of n and π^0 must be equal and opposite. Let the momenta of the two particles have magnitude equal to p. The initial kinetic energy is also negligible. The conservation of energy gives

$$E_{\mathrm{p}0} + E_{\mathrm{d}0} = E_{\mathrm{n}} + E_\pi \tag{1}$$

where E_{n} and E_π are the energies of n and π^0, respectively, while, from the relation $E^2 = (m_0c^2)^2 + (pc)^2$ we have, for the particles n and π^0

$$(pc)^2 = E_n^2 - E_{n0}^2 = E_\pi^2 - E_{\pi 0}^2. \tag{2}$$

Equation (1) gives $\quad E_n^2 = (E_{p0}+E_{d0}-E_\pi)^2$

and Eq. (2) $\quad E_n^2 = E_\pi^2 + E_{n0}^2 - E_{\pi 0}^2$.

Equating, $$(E_{0\mathrm{p}} + E_{0\mathrm{d}} - E_\pi)^2 = E_\pi^2 + E_{0\mathrm{n}}^2 - E_{0\pi}^2$$

or $$(E_{\mathrm{p}0} + E_{\mathrm{d}0})^2 - 2E_\pi(E_{\mathrm{p}0} + E_{\mathrm{d}0}) + \cancel{E_\pi^2} = \cancel{E_\pi^2} + E_{\mathrm{n}0}^2 - E_{\pi 0}^2$$

and, finally, $$E_\pi = \frac{(E_{\mathrm{p}0} + E_{\mathrm{d}0})^2 + E_{\pi 0}^2 - E_{\mathrm{n}0}^2}{2(E_{\mathrm{p}0} + E_{\mathrm{d}0})}.$$

Substituting,

$$E_\pi = \frac{(938.2 + 1875.5)^2 + 135.0^2 - 939.5^2}{2(938.2 + 1875.5)} = 1253\ \mathrm{MeV} = 1.25\ \mathrm{GeV}.$$

Example 6.7 The Rest Mass of a Spherical Distribution of Charge due to Its Electrostatic Energy

Verify Eddington's statement that, if a quantity of electrons of mass 1 g could be compressed uniformly into a sphere of a radius of 10 cm, the rest mass of this sphere, which would correspond to its electrostatic energy, would be of the order of magnitude of 10 million tons.

We evaluate the electric charge of the spherical distribution. If m_e is the mass of the electron and e its charge, a mass of electrons equal to m will have a charge equal to $Q = em/m_\mathrm{e}$. From theory we know that the electrostatic energy of a charge Q distributed uniformly inside a sphere of radius R is $E_0 = \dfrac{3Q^2}{20\pi\varepsilon_0 R}$.

Therefore, $E_0 = \dfrac{3}{20\pi\varepsilon_0} \dfrac{m^2 e^2}{m_\mathrm{e}^2 R}$.

This energy corresponds to a rest mass $M_0 = \dfrac{E_0}{c^2} = \dfrac{3}{20\pi\varepsilon_0 c^2} \dfrac{m^2 e^2}{m_\mathrm{e}^2 R}$.

Substituting $\varepsilon_0 = 8.85 \times 10^{-12}$ F/m, $m_\mathrm{e} = 9.11 \times 10^{-31}$ kg, $e = -1.602 \times 10^{-19}$ C and $m = 0.001$ kg, $R = 0.1$ m, we find $M_0 = 1.86 \times 10^{10}$ kg or 20 million tons, approximately.

Problems

6.12 A spring has a constant equal to $k = 2 \times 10^4$ N/m. What is the increase in the mass of the spring when it is compressed by $x = 5$ cm from its natural length?
Ans.: $\Delta m = 2.8 \times 10^{-16}$ kg

6.13 A particle with rest mass m, is stationary in the laboratory. The particle splits into two others: one with rest mass m_1 which moves with a speed $V_1 = (3/5)c$, and another with rest mass m_2 which moves with a speed $V_2 = (4/5)c$.
(a) Find m_1 and m_2 as fractions of m. Ans.: $m_1 = (16/35)m$, $m_2 = (9/35)m$
(b) What are the kinetic energies, K_1 and K_2, of the two particles, in terms of m? Ans.: $K_1 = (4/35)mc^2$, $K_2 = (6/35)mc^2$

6.14 A particle with rest mass m moves in the laboratory with speed $V = (3/5)c$. The particle disintegrates into two others: one with rest mass m_1 which remains stationary and another with rest mass m_2 which moves with a speed of $\upsilon = (4/5)c$. Find the masses m_1 and m_2 in terms of m. Ans.: $m_1 = (5/16)m$, $m_2 = (9/16)m$

6.15 A particle of rest mass $m_1 = 1\ \text{GeV}/c^2$ moves with a speed of $\upsilon_1 = (4/5)c$ and collides with another particle with rest mass $m_2 = 10\ \text{GeV}/c^2$ which is stationary. After the collision, the two particles form a single body with rest mass M. Find:

(a) The total energy of the system, in GeV. Ans.: 11.7 GeV
(b) The total momentum of the system, in GeV/c. Ans.: 1.33 GeV/c
(c) The mass M, in GeV/c^2. Ans.: 11.6 GeV/c^2

6.16 Show that, if a photon could disintegrate into two photons, both the photons produced would have to be moving in the same direction as the original photon.

6.17 A neutral pion disintegrates into two photons, which move on the same straight line. The energy of the one photon is twice that of the other. Show that the speed of the pion was $c/3$.

6.18 A π particle, which is at rest in the laboratory frame of reference, disintegrates into a muon and a neutrino: $\pi \to \mu + \nu$. Show that the energy of μ is $E_\mu = \dfrac{c^2}{2m_\pi}\left(m_\pi^2 + m_\mu^2 - m_\upsilon^2\right)$, where m_π, m_μ and m_ν are the rest masses of the three particles, respectively. What is the energy of ν? Ans.: $E_\nu = \dfrac{c^2}{2m_\pi}\left(m_\nu^2 + m_\pi^2 - m_\mu^2\right)$

6.19 A stationary particle of rest mass M, disintegrates into a new particle of rest mass m and a photon. Find the energies of the new particle and the photon.

Ans.: $E = \dfrac{M^2 + m^2}{2M}c^2$, $E_\gamma = \dfrac{M^2 - m^2}{2M}c^2$

6.20 A particle decays, producing a π^+ and a π^-. Both pions have momenta equal to 530 MeV/c and move in directions which are normal to each other. Find the rest mass of the original particle. The rest mass of $\pi^\pm$ is $m_\pi = 140\ \text{MeV/c}^2$.

Ans.: 799 MeV/c^2

6.21 A neutral kaon decays into two pions: $\text{K}^0 \to \pi^+ + \pi^-$. If the negative pion produced is at rest, what is the energy of the positive pion? What was the energy of the kaon? Given are the rest masses of K^0, $m_\text{K} = 498\ \text{MeV}/c^2$ and the $\pi^\pm$, $m_\pi = 140\ \text{MeV}/c^2$. Ans.: $E_\pi = 753$ MeV, $E_\text{K} = 886$ MeV

6.22 A photon which has an energy equal to E, collides with an electron which is moving in the opposite direction to that of the photon. After the collision, the photon still has an energy E and reverses direction of motion. Show that, for this to happen, the electron must initially have a momentum of magnitude E/c. Also show that the final speed of the electron is $\upsilon = c \Big/ \sqrt{1 + (m_0c^2/E)^2}$, where m_0c^2 is its rest energy.

6.23 A particle which has rest mass m_1 and speed υ_1, collides with a particle at rest, which has a rest mass equal to m_2. The two particles are united into one body which has a rest mass M and moves with speed υ. Show that $\upsilon = (m_1\gamma_1\upsilon_1)/(m_1\gamma_1 + m_2)$ and $M^2 = m_1^2 + m_2^2 + 2\gamma_1 m_1 m_2$, where $\gamma_1 = 1\Big/\sqrt{1 - \upsilon_1^2/c^2}$.

6.24 Two identical particles with rest mass m_0 each move towards each other, in the laboratory frame of reference, with the same speed βc. Find the energy of the one particle in the frame of reference of the other. Ans.: $E_{\mathrm{BA}} = E_{\mathrm{AB}} = m_0 c^2 \dfrac{1+\beta^2}{1-\beta^2}$

6.25 A particle with rest mass M, moves with velocity $\boldsymbol{\upsilon}$. The particle disintegrates into two others, 1 and 2, with rest masses m_1 and m_2 respectively. If particle 2 moves in a direction perpendicular to $\boldsymbol{\upsilon}$, find the angle θ that the direction of motion of particle 1 makes with $\boldsymbol{\upsilon}$. Ans.: $\tan\theta = \dfrac{1}{2\beta\gamma^2}\sqrt{\left[1-\left(\dfrac{m_1}{M}\right)^2+\left(\dfrac{m_2}{M}\right)^2\right]^2-\left(2\gamma\dfrac{m_2}{M}\right)^2}$

6.26 A Σ particle disintegrates, in motion, into three charged pions, A, B and C. The rest mass of each pion is 140 MeV/c^2. Their kinetic energies are, respectively, $K_{\mathrm{A}} = 190$ MeV, $K_{\mathrm{B}} = 321$ MeV and $K_{\mathrm{C}} = 58$ MeV. The velocities of the pions form with the x-axis angles equal to $\theta_{\mathrm{A}} = 22.4^\circ$, $\theta_{\mathrm{B}} = 0^\circ$ and $\theta_{\mathrm{C}} = -12.25^\circ$, respectively. Evaluate the rest mass of the original particle and the direction of its motion. Ans.: $M_\Sigma = 495\ \mathrm{MeV}/c^2$, $\theta = 5.6^\circ$

6.27 A particle with rest mass M, disintegrates, in motion, into two other particles, 1 and 2. Particle 1 has rest mass m_1, momentum p_1 and energy E_1, while particle 2 has rest mass m_2, momentum p_2 and energy E_2. The angle formed by their directions of motion is θ. Show that $E_1E_2 - p_1p_2c^2\cos\theta = \frac{1}{2}(M^2 - m_1^2 - m_2^2)c^4$, a quantity that is an invariant.

6.28 A neutral pion with energy E disintegrates into two photons, $\pi^0 \to 2\gamma$, with energies E_1 and E_2. Express the angle between the directions of motion of the two photons, θ, in terms of E, the rest mass m_π of the pion and the variable $\varepsilon = (E/2) - E_1 = E_2 - (E/2) = (E_2 - E_1)/2$. What is the minimum value of this angle for $E = 10$ GeV? It is given that $m_\pi c^2 = 0.140$ GeV. Ans.: $\sin\dfrac{\theta}{2} = \dfrac{m_\pi c^2}{\sqrt{E^2 - 4\varepsilon^2}}$, $\theta_{\min} = 1.6^\circ$

6.29 A Λ hyperon disintegrates in motion into a proton and a pion ($\Lambda \to \mathrm{p} + \pi$). These have, in the laboratory frame of reference, momenta p_{p} and p_π, and energies E_{p} and E_π, respectively. The angle between p_{p} and p_π is equal to θ. Show that the energy released in this disintegration is $Q = \sqrt{m_{\mathrm{p}}^2c^4 + m_\pi^2c^4 + 2E_{\mathrm{p}}E_\pi - 2p_{\mathrm{p}}p_\pi c^2\cos\theta} - (m_{\mathrm{p}} + m_\pi)c^2$, where m_{p} and m_π are the rest masses of the proton and the pion, respectively.

6.30 In the disintegration $\mathrm{K}^0 \to \pi^+ + \pi^-$, both the momenta of the particles produced both have magnitude equal to $p_\pi = 360\ \mathrm{MeV}/c$ and form an angle of 70° between them. What is the rest mass m_{K} of K^0 in MeV/c^2? For the pions, it is $m_\pi c^2 = 140$ MeV. Ans.: $m_{\mathrm{K}} = 499\ \mathrm{MeV}/c^2$

6.31 A K^0 particle, having a rest energy $m_{\mathrm{K}}c^2 = 498$ MeV, disintegrates into two mesons, π^+ and π^-, which have rest masses equal to m_π. In the frame of reference of K^0 both the mesons move with speed $0.83\,c$.

(a) Find the ratio m_π/m_K of the rest masses and the rest mass m_π of the π particles. Ans.: $m_\pi/m_K = 0.28$, $m_\pi = 139\ \mathrm{MeV}/c^2$

(b) Let the K^0 particle move with a speed of $0.83\,c$ in the laboratory frame of reference and the two mesons move on the initial direction of motion of the K^0 particle. Find the kinetic energies of the mesons in the laboratory frame of reference. Ans.: 0 and 616 MeV

6.32 *Symmetric Elastic Collision.* A particle with rest mass m and kinetic energy K collides elastically with a particle at rest that has the same rest mass. After the collision, the two particles move in directions which form equal and opposite angles, $\pm\theta$, with the direction of motion of the initially moving particle. Find angle θ in terms of m and K. *Note:* In the Special Theory of Relativity, the term *elastic collision* implies that the reacting particles remain the same, with the same rest masses, respectively, before and after the collision. Ans.: $\tan\theta = \pm 1\Big/\sqrt{1+K/2mc^2}$

6.8 The Transformation of Momentum and Energy

The momentum and the energy of a particle in a frame of reference S is $\mathbf{p} = m_0\gamma_\mathrm{P}\boldsymbol{\upsilon}$ and $E = m_0\gamma_\mathrm{P}c^2$, where $\gamma_\mathrm{P} = \frac{1}{\sqrt{1-\upsilon^2/c^2}}$ is the Lorentz factor for the particle in the frame of reference in which it is being observed. Therefore,

$$p_x = m_0\gamma_\mathrm{P}\frac{dx}{dt}, \qquad p_y = m_0\gamma_\mathrm{P}\frac{dy}{dt}, \qquad p_z = m_0\gamma_\mathrm{P}\frac{dz}{dt}, \qquad \frac{E}{c^2} = m_0\gamma_\mathrm{P}. \tag{6.46}$$

If time dt elapses in the laboratory, the corresponding time interval in the frame of reference of the particle (the particle's rest time or proper time), is $d\tau = dt/\gamma_\mathrm{P}$. So, equivalently, Eq. (6.46) may be written as

$$p_x = m_0\frac{dx}{d\tau}, \qquad p_y = m_0\frac{dy}{d\tau}, \qquad p_z = m_0\frac{dz}{d\tau}, \qquad \frac{E}{c^2} = m_0\frac{dt}{d\tau}. \tag{6.47}$$

In frame S′,

$$p'_x = m_0\frac{dx'}{d\tau}, \qquad p'_y = m_0\frac{dy'}{d\tau}, \qquad p'_z = m_0\frac{dz'}{d\tau}, \qquad \frac{E'}{c^2} = m_0\frac{dt'}{d\tau}. \tag{6.48}$$

Given that the magnitude $d\tau$ remains invariant under a transformation from one inertial frame of reference to another and the same is true for m_0, these relations show that the magnitudes p_x, p_y, p_z and E/c^2 transform like the magnitudes x, y, z and t, respectively.

Alternatively, we may use the differential form of the Lorentz transformation for x, y, z and t from frame S to frame S′

$$dx' = \gamma(dx - \beta c dt), \qquad dy' = dy, \qquad dz' = dz, \qquad dt' = \gamma(dt - (\beta/c)dx) \tag{6.49}$$

where $\gamma = \dfrac{1}{\sqrt{1 - V^2/c^2}}$ is the Lorentz factor for the transformation from one frame to the other (with the relative speed V of the two frames). Multiplying all the terms by $m_0 \dfrac{1}{d\tau}$, we derive the relations

$$m_0 \frac{dx'}{d\tau} = \gamma\left(m_0 \frac{dx}{d\tau} - \beta c m_0 \frac{dt}{d\tau}\right), \qquad m_0 \frac{dy'}{d\tau} = m_0 \frac{dy}{d\tau}, \qquad m_0 \frac{dz'}{d\tau} = m_0 \frac{dz}{d\tau},$$
$$m_0 \frac{dt'}{d\tau} = \gamma\left(m_0 \frac{dt}{d\tau} - \frac{\beta}{c} m_0 \frac{dx}{d\tau}\right) \tag{6.50}$$

Using Eqs. (6.47), (6.48) and (6.50) gives:

$$p'_x = \gamma\left(p_x - \frac{\beta}{c}E\right), \qquad p'_y = p_y, \qquad p'_z = p_z, \qquad E' = \gamma(E - c\beta p_x) \tag{6.51}$$

We notice that the components of momentum p_x, p_y, p_z and E/c^2 do actually transform like the quantities x, y, z and t, respectively.

Vectorially, for two frames of reference S and S′ with corresponding axes parallel to each other and with S′ moving with a constant velocity **V** relative to S,

$$\mathbf{p}' = \mathbf{p} + (\gamma - 1)\frac{\mathbf{p} \cdot \mathbf{V}}{V^2}\mathbf{V} - \gamma \mathbf{V}\frac{E}{c^2}, \qquad E' = \gamma(E - \mathbf{p} \cdot \mathbf{V}) \tag{6.52}$$

and, inversely,

$$\mathbf{p} = \mathbf{p}' + (\gamma - 1)\frac{\mathbf{p}' \cdot \mathbf{V}}{V^2}\mathbf{V} + \gamma \mathbf{V}\frac{E'}{c^2}, \qquad E = \gamma(E' + \mathbf{p}' \cdot \mathbf{V}). \tag{6.53}$$

Example 6.8 The Doppler Effect

A light source moves at a constant speed V along the x-axis of the frame of reference S. The source emits in all directions photons, which, in the frame of reference S′ of the source, have energy E_0. At rest in frame S, and on the xy plane, is an observer A. At some moment, photons from the source reach the observer at an angle θ to the x-axis, having an energy E in the observer's frame of reference. Find the ratio of the frequencies, f/f_0, of the photons in the two frames of reference.

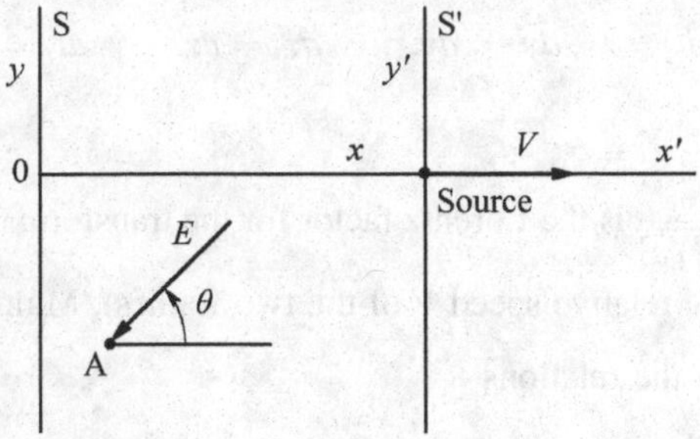

The magnitude of the momentum of a photon which has an energy E is $p = E/c$. Therefore, in frame S, the components of the momentum of the photon are:

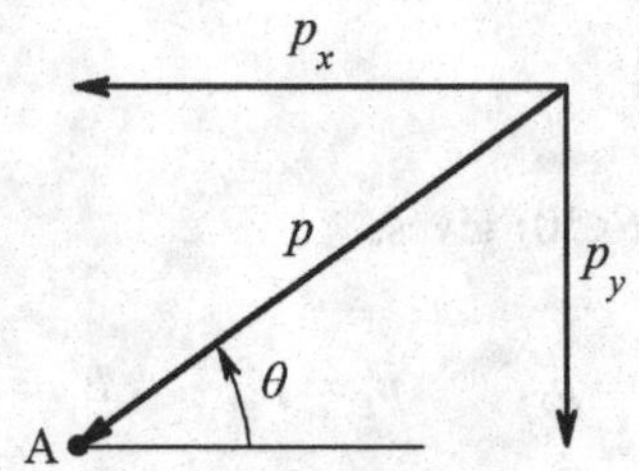

$$p_x = -\frac{E}{c}\cos\theta \qquad p_y = -\frac{E}{c}\sin\theta \qquad p_z = 0.$$

In frame S we have photons with an x-component of momentum equal to $p_x = -\frac{E}{c}\cos\theta$ and energy E. In the frame of reference of the source, S′, therefore, the energy of the photon is:

$$E_0 = E' = \gamma(E - p_x c\beta) = \gamma[E + (E/c)c\beta\cos\theta]$$

where $\beta = \frac{V}{c}$. Thus, $E_0 = E\gamma(1 + \beta\cos\theta)$ and, finally, since it is $E = hf$ and $E_0 = hf_0$ where h is Planck's constant, we have,

$$\frac{E}{E_0} = \frac{f}{f_0} = \frac{\sqrt{1-\beta^2}}{1+\beta\cos\theta}.$$

This is the general equation for the Doppler effect we found in Sect. 4.6.2.

Problems

6.33 Show that the quantity $E^2 - p_x^2c^2 - p_y^2c^2 - p_z^2c^2 = E^2 - p^2c^2 = m_0^2c^4$ is invariant under the special Lorentz transformation from one inertial frame of reference to another.

6.34 A particle S, which has rest mass M, is stationary in the laboratory frame of reference. The particle disintegrates into a particle with rest mass $M/2$ and a photon. Find:

(a) The speed V of the particle produced, in the laboratory frame of reference. Ans.: $V = (3/5)c$
(b) The energy of the photon, in terms of M,
   i. in the laboratory frame of reference, E_γ. Ans.: $E_\gamma = (3/8)Mc^2$
   ii. in the frame of reference of the particle produced, E'_γ. Ans.: $E'_\gamma = (3/16)Mc^2$

6.9 The Zero-Momentum Frame of Reference

The advantages of the *Center of Mass Frame of Reference* (CMFR) are well known from Classical Mechanics, in the cases where we examine the interactions of a system of many particles. One of the advantages of the CMFR is that in this system the total momentum of the system of particles is equal to zero. The conservation of momentum is ensured in a simplified manner in this frame. This and other advantages lead us to define a corresponding frame of reference in the Special Theory of Relativity. However, as the center of mass is not always easy to define uniquely in all frames of reference, we will use in our definition the property we mentioned above, namely the property of zero momentum:

> We define the *Zero-Momentum Frame of Reference* (ZMFR) of a system of particles as that inertial frame of reference in which the total momentum of the system is equal to zero.

Given the momenta of a system of N particles $(\mathbf{p}_1,\ \mathbf{p}_2, \ldots,\ \mathbf{p}_i, \ldots,\ \mathbf{p}_N)$ in a certain inertial frame of reference S, we assume the existence of another inertial frame of reference, the ZMFR S′, which moves with a velocity $\mathbf{V}$ relative to the first frame, such that the sum of the momenta $\left(\mathbf{p}'_1,\ \mathbf{p}'_2, \ldots,\ \mathbf{p}'_i, \ldots,\ \mathbf{p}'_N\right)$ of the particles in this frame is equal to zero: $\sum_N \mathbf{p}'_i = 0$. From this condition, the velocity $\mathbf{V}$ is found.

The Zero-Momentum Frame of Reference is also called *Center of Momentum Frame of Reference*. It should be understood, however, that this is not a geometrical point such as the center of mass in Classical Mechanics. The term describes the frame of reference in which the total momentum of a system vanishes.

Once we solve a given problem in the Zero-Momentum Frame of Reference, using the advantages that this frame of reference has to offer, we may then transform the quantities found to another frame of reference, such as, for example, the Laboratory Frame of Reference (LFR). Some examples and some problems will demonstrate the usefulness of the Zero-Momentum Frame of Reference in the solution of problems.

Example 6.9 Zero-Momentum Frame of Reference of a Photon and a Proton
Let a photon with energy E_γ move towards a proton which is stationary in the laboratory frame of reference.
(a) In the laboratory frame of reference, which is the speed V of the zero-momentum frame of reference of the photon-proton system?

(b) What are the energies of the photon and of the proton in the zero-momentum frame of reference?

(c) What are the momenta of the photon and of the proton in the zero-momentum frame of reference?

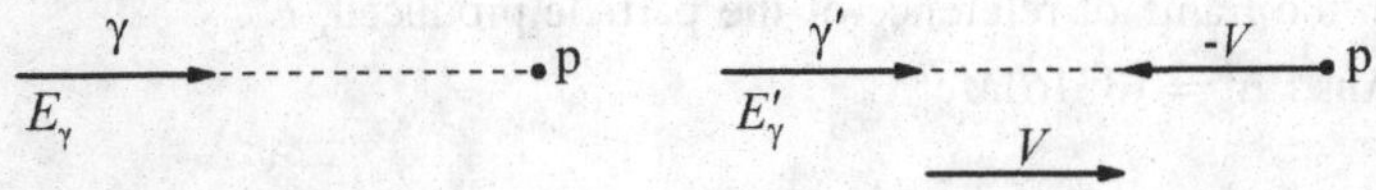

Laboratory frame of reference   Zero-momentum frame of reference

(a) We assume that the zero-momentum frame of reference has a speed equal to $V = \beta c$ relative to the laboratory frame of reference. Obviously, so that no transverse momentum should appear in ZMFR, the direction of V must be that of the original photon.

In ZMFR, the proton has a speed of $-V$, as can be found from the relation $\upsilon'_x = \dfrac{\upsilon_x - V}{1 - \upsilon_x V/c^2}$ for $\upsilon_x = 0$. Its momentum is $p'_{\text{p}} = \dfrac{-m_{\text{p}}\beta c}{\sqrt{1-\beta^2}}$, where m_{p} is its rest mass and $\beta = V/c$.

The photon has energy $E'_\gamma = \gamma(E_\gamma - Vp_\gamma) = \gamma\left(E_\gamma - V\dfrac{E_\gamma}{c}\right) = E_\gamma\gamma(1-\beta) = E_\gamma\sqrt{\dfrac{1-\beta}{1+\beta}}$ and momentum $p'_\gamma = \dfrac{E'_\gamma}{c}$.

Since in the ZMFR the total momentum is zero, it follows that $\left|p'_\gamma\right| = \left|p'_{\text{p}}\right|$. Therefore, $\dfrac{E'_\gamma}{c} = -p'_{\text{p}}$, which gives $\dfrac{E_\gamma}{c}\sqrt{\dfrac{1-\beta}{1+\beta}} = \dfrac{m_{\text{p}}\beta c}{\sqrt{1-\beta^2}}$ or $\dfrac{E_\gamma}{c}(1-\beta) = m_{\text{p}}\beta c$. Thus, $\dfrac{E_\gamma}{c} = \beta\left(\dfrac{E_\gamma}{c} + m_{\text{p}}c\right)$ and, finally, $\beta = \dfrac{E_\gamma}{E_\gamma + m_{\text{p}}c^2}$, which is the reduced speed of the ZMFR relative to the laboratory frame of reference.

(b) Using the value of β found, the energy of the photon in the ZMFR turns out to be

$$E'_\gamma = E_\gamma\sqrt{\frac{1-\beta}{1+\beta}} = E_\gamma\sqrt{\frac{m_{\text{p}}c^2}{2E_\gamma + m_{\text{p}}c^2}}.$$

As expected, it is $E'_\gamma < E_\gamma$. For the energy of the proton in the ZMFR, we find

$$E'_{\text{p}} = \frac{m_{\text{p}}c^2}{\sqrt{1-\beta^2}} = \frac{m_{\text{p}}c^2}{\sqrt{1 - \dfrac{E_\gamma^2}{\left(E_\gamma + m_{\text{p}}c^2\right)^2}}} = \frac{m_{\text{p}}c^2\left(E_\gamma + m_{\text{p}}c^2\right)}{\sqrt{m_{\text{p}}^2c^4 + 2E_\gamma m_{\text{p}}c^2}} = \frac{E_\gamma + m_{\text{p}}c^2}{\sqrt{1 + 2\dfrac{E_\gamma}{m_{\text{p}}c^2}}}.$$

(c) The momentum of a photon of energy E_γ is $p_\gamma = E_\gamma/c$. The momentum of the photon in the FMFR is, therefore,

$$p'_\gamma = \frac{E_\gamma}{c}\sqrt{\frac{m_p c^2}{2E_\gamma + m_p c^2}}.$$

The proton will have an equal and opposite momentum,

$$p'_p = -\frac{E_\gamma}{c}\sqrt{\frac{m_p c^2}{2E_\gamma + m_p c^2}}.$$

Example 6.10 Zero-Momentum Frame of Reference of Two Masses
Two bodies with rest masses m_1 and m_2 move, in the laboratory frame of reference S, on the same line, with speeds υ_1 and υ_2, respectively. Find, relative to the frame S, the speed V of the zero-momentum frame of reference S′ of the two particles. Find also the speeds of the two masses, υ'_1 and υ'_2, in frame S′.

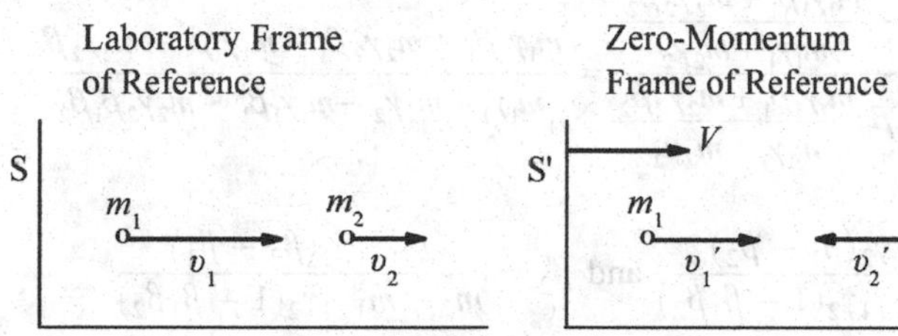

Since the two masses move on the same line, if the momentum in frame S′ is to be zero, V must be parallel to this line, as shown in the figure.

The speeds of the two masses in frame S′ are: $\upsilon'_1 = \dfrac{\upsilon_1 - V}{1 - \upsilon_1 V/c^2}$ and $\upsilon'_2 = \dfrac{\upsilon_2 - V}{1 - \upsilon_2 V/c^2}$.

Their reduced speeds are given by the relations: $\beta'_1 = \dfrac{\beta_1 - \beta}{1 - \beta\beta_1}$ and $\beta'_2 = \dfrac{\beta_2 - \beta}{1 - \beta\beta_2}$.

The Lorentz factor corresponding to β'_1 is

$$\gamma'_1 = \frac{1}{\sqrt{1 - \left(\dfrac{\beta_1 - \beta}{1 - \beta\beta_1}\right)^2}} = \frac{1 - \beta\beta_1}{\sqrt{1 - \cancel{2\beta\beta_1} + \beta^2\beta_1^2 - \beta_1^2 + \cancel{2\beta\beta_1} - \beta^2}} = \frac{1 - \beta\beta_1}{\sqrt{(1-\beta^2)(1-\beta_1^2)}}$$

and, by analogy, $\gamma'_2 = \dfrac{1 - \beta\beta_2}{\sqrt{(1-\beta^2)(1-\beta_2^2)}}$.

The total momentum in frame S′ is zero and so $m_1\gamma'_1\beta'_1 + m_2\gamma'_2\beta'_2 = 0$. Substituting, we find the relation

$$m_1 \frac{1-\beta\beta_1}{\sqrt{(1-\beta^2)(1-\beta_1^2)}}\left(\frac{\beta_1-\beta}{1-\beta\beta_1}\right) + m_2 \frac{1-\beta\beta_2}{\sqrt{(1-\beta^2)(1-\beta_2^2)}}\left(\frac{\beta_2-\beta}{1-\beta\beta_2}\right) = 0,$$

which simplifies to $m_1 \dfrac{\beta_1-\beta}{\sqrt{(1-\beta_1^2)}} + m_2 \dfrac{\beta_2-\beta}{\sqrt{(1-\beta_2^2)}} = 0$

or

$$m_1\gamma_1(\beta_1-\beta) + m_2\gamma_2(\beta_2-\beta) = 0.$$

This gives for the reduced speed of the zero-momentum frame of reference the value of

$$\beta = \frac{m_1\gamma_1\beta_1 + m_2\gamma_2\beta_2}{m_1\gamma_1 + m_2\gamma_2}.$$

The reduced speeds of the two masses in frame S′ may now be found:

$$\beta'_1 = \frac{\beta_1 - \dfrac{m_1\gamma_1\beta_1 + m_2\gamma_2\beta_2}{m_1\gamma_1 + m_2\gamma_2}}{1-\beta_1\dfrac{m_1\gamma_1\beta_1 + m_2\gamma_2\beta_2}{m_1\gamma_1 + m_2\gamma_2}} = \frac{\cancel{m_1\gamma_1\beta_1} + m_2\gamma_2\beta_1 - \cancel{m_1\gamma_1\beta_1} - m_2\gamma_2\beta_2}{m_1\gamma_1 + m_2\gamma_2 - m_1\gamma_1\beta_1^2 - m_2\gamma_2\beta_1\beta_2}$$

i.e. $\beta'_1 = \dfrac{m_2\gamma_1\gamma_2(\beta_1-\beta_2)}{m_1 + m_2\gamma_1\gamma_2(1-\beta_1\beta_2)}$ and $\beta'_2 = \dfrac{m_1\gamma_1\gamma_2(\beta_2-\beta_1)}{m_2 + m_1\gamma_1\gamma_2(1-\beta_1\beta_2)}$.

In the special case in which particle 2 is initially at rest ($\upsilon_2 = 0,\ \beta_2 = 0,\ \gamma_2 = 1$), we have

$$\beta = \frac{\beta_1}{1+m_2/(m_1\gamma_1)},\quad \beta'_1 = \frac{\beta_1}{1+m_1/(m_2\gamma_1)} \quad \text{and} \quad \beta'_2 = \frac{-\beta_1}{1+m_2/(m_1\gamma_1)}.$$

Problems

6.35 Explain why the following cannot happen:

(a) A photon collides with an electron at rest and imparts all its energy to it.

(b) An isolated photon is transformed into an electron-positron pair. (The positron is the antiparticle of the electron.)

(c) A moving positron collides with an electron at rest and both annihilate into a single photon.

6.36 A positron, e^+, with rest mass m and speed υ in the laboratory frame of reference, collides with a stationary electron, e^- (which has the same rest mass m). The two particles annihilate, producing two photons, γ_1 and γ_2, which, in the zero-momentum frame of reference, move in directions perpendicular to the straight line on which the positron and the electron moved (figure a). Let **P**, $\mathbf{P}_1$ and $\mathbf{P}_2$ be the momentum vectors of the positron and the two photons, respectively, in the laboratory frame of reference, and θ_1 and θ_2 be the angles formed by vectors $\mathbf{P}_1$ and $\mathbf{P}_2$ with the vector **P** (figure (b)).

(a) Show that $|\mathbf{P}_1| = |\mathbf{P}_2|$ and $\theta_1 = \theta_2$.
(b) Find the angle θ between vectors $\mathbf{P}_1$ and $\mathbf{P}_2$ in terms of υ.

Ans.: $\tan\theta = \dfrac{\beta\sqrt{1-\beta^2}}{\beta - 1/2}$

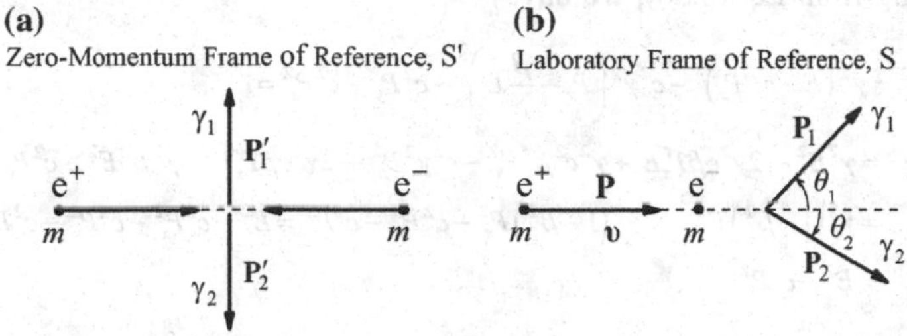

6.10 The Transformation of the Total Momentum and the Total Energy of a System of Particles

We will now examine a system consisting of n particles. The momentum-energy transformation holds for each of the particles separately. For the i-th particle, we have the relations:

$$p'_{ix} = \gamma\left(p_{ix} - \frac{\beta}{c}E_i\right), \quad p'_{iy} = p_{iy}, \quad p'_{iz} = p_{iz}, \quad E'_i = \gamma(E_i - c\beta p_{ix}). \qquad (6.54)$$

The components of the total momentum and the total energy of the system of particles in the two frames of reference are:

$$P_x = \sum_{i=1}^{n} p_{ix} \quad P_y = \sum_{i=1}^{n} p_{iy} \quad P_z = \sum_{i=1}^{n} p_{iz} \quad E = \sum_{i=1}^{n} E_i \qquad (6.55)$$

with corresponding expressions for the primed magnitudes.

The transformations (6.54) are linear and so, summing the corresponding equations over the n particles and using Eq. (6.55) and those corresponding to frame S′, we have for the components of the total momentum and for the total energy of the system

$$P'_x = \gamma\left(P_x - \frac{\beta}{c}E\right), \qquad P'_y = P_y, \qquad P'_z = P_z, \qquad E' = \gamma(E - c\beta P_x). \quad (6.56)$$

These quantities are conserved, being the sums of conserved quantities.

6.10.1 The Invariance of the Quantity $E^2 - c^2P^2$ for a System of Particles

We will find the effect of the Lorentz transformation on the quantity $E'^2 - c^2P'^2$. Substituting from Eq. (6.56), we have

$$\begin{aligned} E'^2 - c^2P'^2 &= \gamma^2(E - c\beta P_x)^2 - c^2\gamma^2\left(P_x - \frac{\beta}{c}E\right)^2 - c^2P_y^2 - c^2P_z^2 = \\ &= \gamma^2E^2 - \cancel{2\gamma^2 c\beta P_x E} + \gamma^2c^2\beta^2P_x^2 - c^2\gamma^2P_x^2 + \cancel{2\gamma^2 c\beta P_x E} - \gamma^2\beta^2E^2 - c^2P_y^2 - c^2P_z^2 = \\ &= \gamma^2(1-\beta^2)E^2 - c^2\gamma^2(1-\beta^2)P_x^2 - c^2P_y^2 - c^2P_z^2 = E^2 - c^2P_x^2 - c^2P_y^2 - c^2P_z^2 = \\ &= E^2 - c^2P^2 \end{aligned} \tag{6.57}$$

The result shows that the quantity $E^2 - c^2P^2$ is invariant under the Lorentz transformation. We conclude that the quantity $E^2 - c^2P^2$ is not only conserved during the changes that happen to a system of n particles, but it also has the same value in all inertial frames of reference. By analogy to the relation which holds for one particle, $E^2 - c^2p^2 = (m_0c^2)^2$, where m_0 is the rest mass of the particle, we define the quantity M_0 for the system of n particles, in which case it is

$$E^2 = (M_0c^2)^2 + c^2P^2. \tag{6.58}$$

We will consider the quantity M_0 to be the rest mass of the system, on the understanding that M_0 is not the sum of the rest masses of the n particles, but it includes the internal energy of the system due to the interaction of the particles with each other and the kinetic energies of the particles with respect to their common center of mass. The term 'rest mass' does not, therefore, mean that the particles are at rest in the system. It is called rest mass in the sense that the rest energy of the system is given, for $P = 0$, as $E_0 = M_0c^2$. In the zero-momentum frame of reference of the system of the particles, in which by definition it is $P = 0$, the magnitude $E_0 = M_0c^2$ is the total energy of the system. The examples that follow will show the use of these ideas in the solution of problems.

Example 6.11 The Velocity of the Zero-Momentum Frame of Reference of Two Particles, One of Which Is at Rest

A particle A, with rest mass m_A, is at rest in the inertial frame of reference S. A second particle, B, with rest mass m_B, moves with velocity $\upsilon_B\hat{\mathbf{x}}$ in frame S. Find the speed V of the zero-momentum frame of reference of the particles A and B in frame S.

Let the zero-momentum frame of reference S′ of the two particles, move with velocity $V\hat{\mathbf{x}}$ relative to the frame S, to which there corresponds a Lorentz factor $\gamma = 1\big/\sqrt{1 - V^2/c^2}$. In frame S we have,

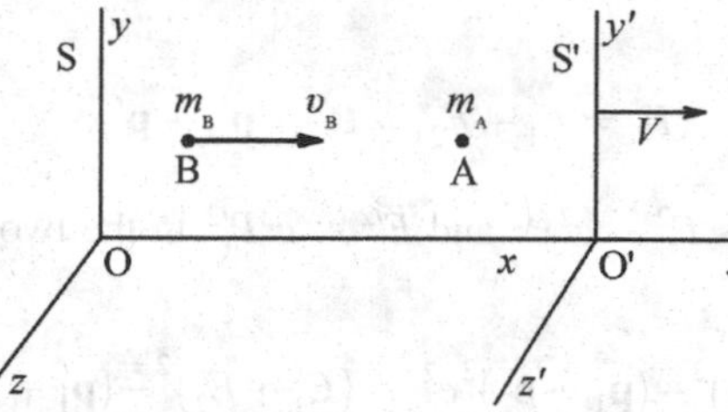

particle A: momentum $p_{Ax} = 0$,
energy $E_A = m_A c^2$,
particle B: momentum $p_{Bx} = m_B \gamma_B \upsilon_B$,
energy $E_B = m_B c^2 \gamma_B$,

where $\gamma_B = 1\Big/\sqrt{1 - \upsilon_B^2/c^2}$. The total momentum and the total energy of the two particles are, respectively, $P_x = p_{Ax} + p_{Bx} = m_B \gamma_B \upsilon_B$ and $E = E_A + E_B = m_A c^2 + m_B c^2 \gamma_B$.
Transforming in order to find the total momentum in frame S′, we have

$$P'_x = \gamma\left(P_x - VE/c^2\right) = \gamma(m_B \gamma_B \upsilon_B - m_A V + m_B \gamma_B V).$$

For the total momentum in frame S′ to be zero, it must be $P'_x = 0$ or

$$-m_A \gamma V + \gamma\left(m_B \gamma_B \upsilon_B - m_B \gamma_B V\right) = 0.$$

Therefore, $V = \dfrac{m_B \gamma_B \upsilon_B}{m_A + m_B \gamma_B}$ and, finally, $V = \dfrac{\upsilon_B}{1 + m_A/m_B \gamma_B}$.

This is the same result as that obtained in Example 6.10 for $\upsilon_A = 0$.

Example 6.12 Energy Available for the Production of Particles During the Collision of a Proton with Another, Stationary Proton
A moving proton collides with another, stationary proton. What is the total energy which is available for the production of particles?

In the laboratory frame of reference S, let the momentum and the total energy of the impinging proton be $\mathbf{p}_1$ and E_1, respectively. In another frame, S′, these quantities have the corresponding values of $\mathbf{p}'_1$ and E'_1. For the other proton, which is initially at rest in the laboratory frame of reference S, the corresponding quantities are $\mathbf{p}_2 = 0$ and $E_2 = m_p c^2$, and $\mathbf{p}'_2$ and E'_2, where m_p is the rest mass of the proton.

For the system of the two particles, it is,
in frame S:

$$E = E_1 + E_2, \quad \mathbf{P} = \mathbf{p}_1 + \mathbf{p}_2 \tag{1}$$

and in frame S′:

$$E' = E'_1 + E'_2, \quad \mathbf{P}' = \mathbf{p}'_1 + \mathbf{p}'_2. \tag{2}$$

Equating the quantities $E^2 - c^2P^2$ and $E'^2 - c^2P'^2$ in the two frames of reference, we have

$$(E_1 + E_2)^2 - (\mathbf{p}_1 + \mathbf{p}_2)^2c^2 = (E'_1 + E'_2)^2 - (\mathbf{p}'_1 + \mathbf{p}'_2)^2c^2. \tag{3}$$

If frame S′ is the zero-momentum frame of reference of the two particles, we have $\mathbf{p}'_1 + \mathbf{p}'_2 = 0$. Putting also $\mathbf{p}_2 = 0$ and $E_2 = m_\mathrm{p}c^2$ in Eq. (3), it follows that

$$E_1^2 + 2m_\mathrm{p}c^2E_1 + m_\mathrm{p}^2c^4 - c^2p_1^2 = E'^2, \tag{4}$$

where $E' = E'_1 + E'_2$ is the total energy of the two protons in the zero-momentum frame of reference after the collision. For the impinging proton, it is also true that

$$E_1^2 = m_\mathrm{p}^2c^4 + c^2p_1^2, \tag{5}$$

in which case Eq. (4) gives

$$2m_\mathrm{p}c^2E_1 + 2m_\mathrm{p}^2c^4 = E'^2. \tag{6}$$

Since the total energy in the laboratory frame of reference S is $E = E_1 + m_\mathrm{p}c^2$, Eq. (6) gives

$$2Em_\mathrm{p}c^2 = E'^2. \tag{7}$$

The available energy for the production of particles in the zero-momentum frame of reference is the whole energy $E' = E'_1 + E'_2$, since, at the threshold energy, i.e. at the energy with which the production of the specific particles is just possible, the particles produced must be at rest in the zero-momentum frame of reference. We notice that the fraction of the initial energy, in the laboratory frame of reference, which is available for the production of particles, is

$$\kappa \equiv \frac{E'}{E} = \frac{2m_\mathrm{p}c^2}{E'}. \tag{8}$$

If we want, for example, to have an available energy of $E' = 10$ GeV for the production of particles during the collision of the two protons, since it is $m_\mathrm{p}c^2 \approx 1$ GeV, this fraction is $\kappa = 0.2$. Thus, the energy required in the laboratory frame of reference will be $E = E'/\kappa = 50$ GeV. In general, it is

$$E = \frac{E'^2}{2m_\mathrm{p}c^2}. \tag{9}$$

The low proportion of the energy used led to the construction of accelerators in which the collisions happen not between moving particles and particles at rest, but between particles having equal and opposite momenta. This is the case, for example, in the LHC at CERN, in which, when in full operation, two beams, each with protons of 7 TeV energy, will collide, making available the entire energy of 14 TeV for the production of particles.

6.11 The Collision of Two Identical Particles

We will examine the collision of a moving particle with another, identical one, in which the two particles remain unchanged. The particles have rest mass m_0. In the laboratory frame of reference (LFR), the one particle, particle 1, moves with a speed υ_0 towards the other, particle 2, which is at rest. In this frame of reference, particle 1 has, before the collision, momentum p_0, energy E_0 and Lorentz factor $\gamma_0 = 1\Big/\sqrt{1-\upsilon_0^2/c^2}$. After the collision, the particles 1 and 2 have, in frame S, respectively, momenta $\bar{p}_1$ and $\bar{p}_2$, and move in directions which form angles θ_1 and θ_2 with the initial direction of motion of particle 1, as shown in Fig. 6.7a. To avoid problems with the signs, we will consider θ_1 and θ_2 positive as drawn in the figure. For momentum to be conserved, the vectors of p_0, $\bar{p}_1$ and $\bar{p}_2$ lie on a plane, which we will take to be plane xy with the x-axis in the direction of p_0.

We will initially solve the problem in the zero-momentum frame of reference (ZMFR) S′, which we will assume to be moving with velocity $\mathbf{V} = V\hat{\mathbf{x}}$ with respect to frame S. The direction of motion of frame S′ relative to frame S is, obviously, the original direction of motion of particle 1. Particle 2, which in the LFR is initially at rest, in the ZMFR must move with speed V, in a direction which is opposite of that

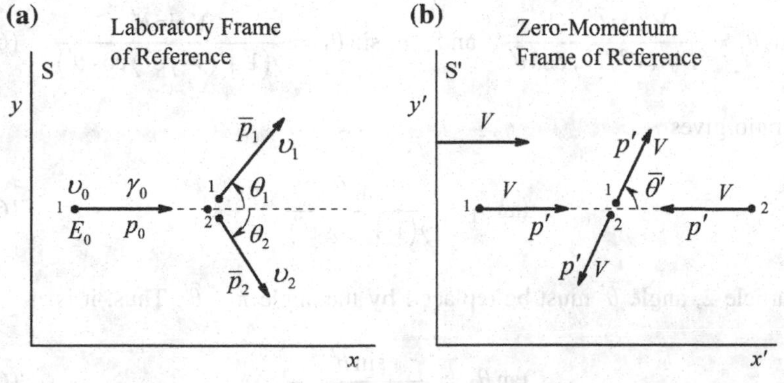

Fig. 6.7 The collision of two identical particles, as seen **a** from the Laboratory Frame of Reference, in which one of the particles is initially at rest and **b** from the Zero-Momentum Frame of Reference of the particles

of the ZMFR's motion (Fig. 6.7b). Since the two particles are identical, in order for the initial momentum to be zero, particle 1 must have an equal and opposite velocity. After the collision, in the ZMFR, the particles must have equal and opposite momenta and, therefore, equal and opposite velocities. For energy to be conserved, the magnitude of these velocities must be equal to V. We will assume that after the collision, in the ZMFR, particle 1 moves in a direction that forms an angle $\bar{\theta}'$ with the initial line of motion of the particles. The direction of motion of particle 2 after the collision forms an angle $\pi - \bar{\theta}'$ with the same line.

To find the value of V, to which there corresponds a Lorentz factor $\gamma = 1\big/\sqrt{1 - V^2/c^2}$, we use the transformation of the total momentum $\mathbf{P}$ and of the total energy E of the two particles from frame S to frame S′. For the x-component of momentum we have

$$P'_x = \gamma\left(P_x - \frac{V}{c^2}E\right). \tag{6.59}$$

Because it must be $\mathbf{P}' = 0$ and $P'_x = 0$, we find

$$V = \frac{c^2 P_x}{E}. \tag{6.60}$$

Using the relations $P_x = p_0 = m_0\gamma_0\upsilon_0$ and $E = (\gamma_0 + 1)m_0c^2$, we find

$$V = \frac{c^2 m_0\gamma_0\upsilon_0}{(\gamma_0 + 1)m_0c^2} \quad \text{or} \quad V = \frac{\gamma_0}{\gamma_0 + 1}\upsilon_0. \tag{6.61}$$

We may now transform and find the angles in the LFR. The x and y components of the velocity of particle 1 in the ZMFR are $V\cos\bar{\theta}'$ and $V\sin\bar{\theta}'$, respectively. Transforming into frame S, we find

$$\upsilon_1\cos\theta_1 = \frac{V\cos\bar{\theta}' + V}{1 + (V^2/c^2)\cos\bar{\theta}'} \quad \text{and} \quad \upsilon_1\sin\theta_1 = \frac{V\sin\bar{\theta}'}{\gamma(1 + (V^2/c^2)\cos\bar{\theta}')}. \tag{6.62}$$

Their ratio gives

$$\tan\theta_1 = \frac{\sin\bar{\theta}'}{\gamma(1 + \cos\bar{\theta}')}. \tag{6.63}$$

For particle 2, angle $\bar{\theta}'$ must be replaced by the angle $\pi - \bar{\theta}'$. Thus, it is

$$\tan\theta_2 = \frac{\sin\bar{\theta}'}{\gamma(1 - \cos\bar{\theta}')}. \tag{6.64}$$

Multiplying the last two equations with each other, it follows that it is

$$\tan\theta_1 \tan\theta_2 = \frac{1}{\gamma^2}. \tag{6.65}$$

Now, $1/\gamma^2$ may be expressed in terms of γ_0, making use of Eq. (6.61). Thus, it is

$$\frac{1}{\gamma^2} = 1 - \frac{V^2}{c^2} = 1 - \left(\frac{\gamma_0}{\gamma_0+1}\right)^2 \frac{v_0^2}{c^2} = 1 - \left(\frac{\gamma_0}{\gamma_0+1}\right)^2 \left(1 - \frac{1}{\gamma_0^2}\right) = 1 - \frac{\gamma_0^2 - 1}{(\gamma_0+1)^2}, \tag{6.66}$$

in which case it is

$$\frac{1}{\gamma^2} = \frac{2}{\gamma_0 + 1}. \tag{6.67}$$

and, finally,

$$\tan\theta_1 \tan\theta_2 = \frac{2}{\gamma_0 + 1}. \tag{6.68}$$

We note that this relation differs from the classical one, $\tan\theta_1 \tan\theta_2 = 1$, which of course results on letting $\gamma_0 \to 1$ in Eq. (6.68). The classical relation means that $\theta_1 + \theta_2 = \pi/2$ and that the two particles move, after the collision, in directions which are normal to each other in the laboratory frame of reference, in which one of the particles was initially at rest (one may observe this by watching a game of billiards.) The relativistic analysis gives $\tan\theta_1 \tan\theta_2 < 1$, since it is $\gamma_0 > 1$, which in turn means that $\theta_1 + \theta_2 < \pi/2$. In the extreme relativistic case, for very large γ_0, both the particles move after the collision in directions which are very close to that of the originally moving particle. This difference was used, as we will see below, in order to test the predictions of the Special Theory of Relativity regarding the dependence of the inertial mass and the momentum of the body on its speed.

The angles θ_1 and θ_2 may be found in terms of $\overline{\theta}'$, by using Eq. (6.67) in Eqs. (6.63) and (6.64). We find

$$\tan\theta_1 = \sqrt{\frac{2}{1+\gamma_0}}\left(\frac{\sin\overline{\theta}'}{1+\cos\overline{\theta}'}\right) \quad \text{and} \quad \tan\theta_2 = \sqrt{\frac{2}{1+\gamma_0}}\left(\frac{\sin\overline{\theta}'}{1-\sin\overline{\theta}'}\right). \tag{6.69}$$

The data of the problem do not permit the determination of the angle $\overline{\theta}'$. In order to find $\overline{\theta}'$ more information must be given, such as the nature of the force of interaction between the two particles and the collision parameter. With the information we are given, $\overline{\theta}'$ is just a parameter which can take any value between 0 and 2π.

6.11.1 *Experimental Test of the Special Theory of Relativity with Colliding Electrons*

In 1932, Champion [3] put under experimental test the relativistic theory of the collision of two identical particles, by observing in a cloud chamber (Wilson chamber) the orbits of the two electrons after the collision of moving electrons (β particles) with the almost stationary electrons of the gas in the chamber. The angles θ_1 and θ_2 were measured in photographs of the orbits of the electrons and the validity of Eq. (6.68) was verified. The experiment was an experimental verification of the relativistic prediction regarding the variation of the inertial mass and of the momentum of a body with its speed, as well as of the validity of the laws of conservation of momentum and of energy at relativistic energies and speeds. In addition, the theory proposed in 1903 by Abraham [4] for the interpretation of the results of Kaufmann's experiments regarding the variation of the inertial mass with speed (see Sect. 1.12) was proved wrong.

6.12 The Transformation of Force

We have already preserved Newton's second law of motion in the form

$$\mathbf{F} = \frac{d\mathbf{p}}{dt} \tag{6.70}$$

which it has in classical Mechanics, with the momentum still being defined as $\mathbf{p} = m\boldsymbol{\upsilon}$. However, since the relativistic mass m depends on speed, the force is no longer equal to mass times acceleration. In place of the momentum we now have the relativistic momentum, in which case it is

$$\mathbf{F} = \frac{d}{dt}\left(\frac{m_0\boldsymbol{\upsilon}}{\sqrt{1-\upsilon^2/c^2}}\right). \tag{6.71}$$

With this definition, the acceleration is, in general, no longer in the direction of the force. This definition has, however, certain advantages: it makes it possible to preserve the law of action-reaction and it simplifies the interpretation of electromagnetic phenomena.

We have also kept the definition of the work dW produced by the force $\mathbf{F}$ on moving its point of application by a displacement $d\mathbf{r}$ as

$$dW = \mathbf{F}\cdot d\mathbf{r} \tag{6.72}$$

in which case the rate of production of work by the force (power) is given by

$$\frac{dW}{dt} = \mathbf{F} \cdot \frac{d\mathbf{r}}{dt} = \mathbf{F} \cdot \boldsymbol{\upsilon}. \tag{6.73}$$

We consider a particle with rest mass m_0, velocity $\boldsymbol{\upsilon}$ and momentum $\mathbf{p}$ in the frame of reference S, and velocity $\boldsymbol{\upsilon}'$ and momentum $\mathbf{p}'$ in the frame of reference S′, which moves relative to S with a velocity $V\hat{\mathbf{x}}$. The x component of the force in frame S′ is given by the relation

$$F'_x = \frac{dp'_x}{dt'}. \tag{6.74}$$

From the transformation of momentum-energy, it is

$$p'_x = \gamma\left(p_x - \frac{\beta}{c}E\right) \tag{6.75}$$

where $\beta = \frac{V}{c}$ and $\gamma = 1\Big/\sqrt{1-\beta^2}$. Thus

$$F'_x = \gamma \frac{d}{dt'}\left(p_x - \frac{\beta}{c}E\right). \tag{6.76}$$

From the transformation $t' = \gamma\left(t - \frac{V}{c^2}x\right)$, we have

$$\frac{dt'}{dt} = \frac{d}{dt}\gamma\left(t - \frac{V}{c^2}x\right) = \gamma\left(1 - \frac{V\upsilon_x}{c^2}\right) \tag{6.77}$$

and, since it is $\frac{d}{dt'} = \frac{dt}{dt'}\frac{d}{dt}$, it follows that

$$F_x = \gamma \frac{d}{dt'}\left(p_x - \frac{\beta}{c}E\right) = \frac{\gamma}{\gamma\left(1 - \frac{V\upsilon_x}{c^2}\right)}\frac{d}{dt}\left(p_x - \frac{\beta}{c}E\right)\frac{1}{1 - \frac{V\upsilon_x}{c^2}}\left(F_x - \frac{V}{c^2}\frac{dE}{dt}\right). \tag{6.78}$$

However

$$\frac{dE}{dt} = \frac{dW}{dt} = \mathbf{F} \cdot \boldsymbol{\upsilon} = F_x\upsilon_x + F_y\upsilon_y + F_z\upsilon_z, \tag{6.79}$$

in which case, finally,

$$F'_x = F_x - \frac{V\upsilon_y}{c^2(1 - V\upsilon_x/c^2)}F_y - \frac{V\upsilon_z}{c^2(1 - V\upsilon_x/c^2)}F_z. \tag{6.80}$$

For the y component of the force, because it is

$$p'_y = p_y, \tag{6.81}$$

it follows that

$$F'_y = \frac{dp'_y}{dt'} = \frac{dp_y}{dt'} = \frac{dt}{dt'}\frac{dp_y}{dt} = \frac{dt}{dt'}F_y \tag{6.82}$$

and from Eq. (6.77) we have

$$F'_y = \frac{F_y}{\gamma(1 - V\upsilon_x/c^2)} = \frac{\sqrt{1 - V^2/c^2}}{1 - V\upsilon_x/c^2}F_y. \tag{6.83}$$

Similarly, from $p'_z = p_z$, it is found that

$$F'_z = \frac{F_z}{\gamma(1 - V\upsilon_x/c^2)} = \frac{\sqrt{1 - V^2/c^2}}{1 - V\upsilon_x/c^2}F_z. \tag{6.84}$$

Summarizing:

$$\begin{aligned} F'_x &= F_x - \frac{V\upsilon_y/c^2}{1 - V\upsilon_x/c^2}F_y - \frac{V\upsilon_z/c^2}{1 - V\upsilon_x/c^2}F_z \\ F'_y &= \frac{F_y}{\gamma(1 - V\upsilon_x/c^2)} = \frac{\sqrt{1 - V^2/c^2}}{1 - V\upsilon_x/c^2}F_y \quad F'_z = \frac{F_z}{\gamma(1 - V\upsilon_x/c^2)} = \frac{\sqrt{1 - V^2/c^2}}{1 - V\upsilon_x/c^2}F_z. \end{aligned} \tag{6.85}$$

The inverse transformation is found (by putting $V \to -V$, $\mathbf{\upsilon} \to \mathbf{\upsilon}'$ and $\mathbf{F} \to \mathbf{F}'$) to be:

$$\begin{aligned} F_x &= F'_x + \frac{V\upsilon'_y/c^2}{1 + V\upsilon'_x/c^2}F'_y + \frac{V\upsilon'_z/c^2}{1 + V\upsilon'_x/c^2}F'_z \\ F_y &= \frac{F'_y}{\gamma(1 + V\upsilon'_x/c^2)} = \frac{\sqrt{1 - V^2/c^2}}{1 + V\upsilon'_x/c^2}F'_y \quad F_z = \frac{F'_z}{\gamma(1 + V\upsilon'_x/c^2)} = \frac{\sqrt{1 - V^2/c^2}}{1 + V\upsilon'_x/c^2}F'_z. \end{aligned} \tag{6.86}$$

In vector form, for the general Lorentz transformation without rotation of the axes, we have

$$\mathbf{F}' = \frac{\mathbf{F}}{\gamma(1 - \mathbf{V}\cdot\mathbf{\upsilon}/c^2)} + \frac{\mathbf{V}}{1 - \mathbf{V}\cdot\mathbf{\upsilon}/c^2}\left[\frac{\mathbf{V}\cdot\mathbf{F}}{V^2}\left(1 - \frac{1}{\gamma}\right) - \frac{\mathbf{\upsilon}\cdot\mathbf{F}}{c^2}\right]. \tag{6.87}$$

and the inverse transformation

$$\mathbf{F} = \frac{\mathbf{F}'}{\gamma(1+\mathbf{V}\cdot\boldsymbol{\upsilon}'/c^2)} + \frac{\mathbf{V}}{1+\mathbf{V}\cdot\boldsymbol{\upsilon}'/c^2}\left[\frac{\mathbf{V}\cdot\mathbf{F}'}{V^2}\left(1-\frac{1}{\gamma}\right)+\frac{\boldsymbol{\upsilon}'\cdot\mathbf{F}'}{c^2}\right]. \tag{6.88}$$

These transformations were derived for a force which is exerted on a particle whose velocity it changes. For a steady point of application of the force ($\upsilon' = 0$), the relations (6.86) reduce to

$$F_x = F'_x \quad F_y = \frac{F'_y}{\gamma} \quad F_z = \frac{F'_z}{\gamma}. \tag{6.89}$$

There is a trap here we could fall into if we are not careful. Putting $\boldsymbol{\upsilon} = 0$ in (6.85) we find

$$F'_x = F_x \quad F'_y = \frac{F_y}{\gamma} \quad F'_z = \frac{F_z}{\gamma} \tag{6.90}$$

which seem to disagree with Eq. (6.89). We will see that there is no paradox and that Eqs. (6.89) and (6.90) are all correct, if we notice that it is not possible that we are referring to the same point in the two cases. A point cannot be at rest in two frames of reference which are in relative motion to each other. If a point has velocity $\upsilon' = 0$ in the frame of reference S′, it will have a speed equal to $\upsilon_x = V$ in frame S′, where $V\hat{\mathbf{x}}$ is the velocity of S′ relative to S. Putting $\upsilon_x = V,\ \upsilon_y = 0,\ \upsilon_z = 0$ in Eq. (6.86), we find

$$F'_x = F_x \quad F'_y = \gamma F_y \quad F'_z = \gamma F_z,$$

which are Eq. (6.89).

The components of the force have their maximum values in the frame of reference in which the point of application of the force is stationary.

6.13 Motion Under the Influence of a Constant Force. The Motion of a Charged Particle in a Constant Uniform Electric Field

Let a particle with rest mass m_0 and charge q enter with velocity $\boldsymbol{\upsilon}_0$ a space where there exists a constant homogeneous electric field $\boldsymbol{\mathcal{E}}$. We choose the x-axis to be in the direction of the field, in which case it is $\boldsymbol{\mathcal{E}} = \mathcal{E}\hat{\mathbf{x}}$. We also choose the y-axis in such a way that $\boldsymbol{\upsilon}_0$ lies in the zy plane, in which case we may write $\boldsymbol{\upsilon}_0 = \upsilon_{0x}\hat{\mathbf{x}} + \upsilon_{0y}\hat{\mathbf{y}}$. The motion that will follow is confined to the xy plane.

The equation of motion. If the instantaneous momentum of the particle is $\mathbf{p}$, the equation of motion of the charged particle is

$$\frac{d\mathbf{p}}{dt} = q\mathcal{E}\hat{\mathbf{x}}. \tag{6.91}$$

Momentum. If the momentum of the particle at $t = 0$ is $\mathbf{p}_0 = p_{0x}\hat{\mathbf{x}} + p_{0y}\hat{\mathbf{y}}$, the integration of Eq. (6.91) gives

$$\mathbf{p} - \mathbf{p}_0 = q\mathcal{E}t\hat{\mathbf{x}} \quad \text{or} \quad \mathbf{p} = q\mathcal{E}t\hat{\mathbf{x}} + p_{0x}\hat{\mathbf{x}} + p_{0y}\hat{\mathbf{y}} \tag{6.92}$$

and the momentum $\mathbf{p} = p_x\hat{\mathbf{x}} + p_y\hat{\mathbf{y}}$ has components

$$p_x = p_{0x} + q\mathcal{E}t, \qquad p_y = p_{0y}. \tag{6.93}$$

The magnitude of the momentum is

$$p = \frac{m_0 \upsilon}{\sqrt{1 - \upsilon^2/c^2}} = \sqrt{(q\mathcal{E}t + p_{0x})^2 + p_{0y}^2}. \tag{6.94}$$

We define the parameters

$$\tau \equiv \frac{p_{0x}}{q\mathcal{E}}, \quad \lambda \equiv \frac{p_{0y}}{q\mathcal{E}}, \quad \mu \equiv \frac{m_0 c}{q\mathcal{E}} \quad \text{and} \quad \chi^2 \equiv \frac{m_0^2 c^2 + p_{0y}^2}{q^2 \mathcal{E}^2} = \mu^2 + \lambda^2, \tag{6.95}$$

in which case it is

$$p = q\mathcal{E}\sqrt{\lambda^2 + (t + \tau)^2}. \tag{6.96}$$

Energy. The energy of the particle is

$$E = \sqrt{(m_0 c^2)^2 + p^2 c^2} = \sqrt{(m_0 c^2)^2 + p_{0y}^2 c^2 + (p_{0x} c + qc\mathcal{E}t)^2}, \tag{6.97}$$

which is also written as

$$E = \frac{m_0 c^2}{\mu}\sqrt{\mu^2 + \lambda^2 + (t + \tau)^2} = \frac{m_0 c^2}{\mu}\sqrt{\chi^2 + (t + \tau)^2}. \tag{6.98}$$

The Lorentz factor. The Lorentz factor corresponding to the speed of the particle is

$$\gamma = \frac{E}{m_0 c^2} = \frac{1}{\mu}\sqrt{\chi^2 + (t + \tau)^2}\,. \tag{6.99}$$

The components of velocity. From Eq. (6.92) we have $\mathbf{p} = (q\mathcal{E}t + p_{0x})\,\hat{\mathbf{x}} + p_{0y}\hat{\mathbf{y}}$. Equating to $\mathbf{p} = m_0\gamma\boldsymbol{\upsilon} = m_0\gamma\upsilon_x\hat{\mathbf{x}} + m_0\gamma\upsilon_y\hat{\mathbf{y}}$ and using Eq. (6.99), we have

$$(q\mathcal{E}t + p_{0x})\hat{\mathbf{x}} + p_{0y}\hat{\mathbf{y}} = \frac{m_0}{\mu}\sqrt{\chi^2 + (t+\tau)^2}\,(\upsilon_x\hat{\mathbf{x}} + \upsilon_y\hat{\mathbf{y}}), \tag{6.100}$$

from which we find

$$\upsilon_x = c\frac{t+\tau}{\sqrt{\chi^2 + (t+\tau)^2}} \quad \text{and} \quad \upsilon_y = c\frac{\lambda}{\sqrt{\chi^2 + (t+\tau)^2}}. \tag{6.101}$$

The speed. The speed of the particle is

$$\upsilon = \sqrt{\upsilon_x^2 + \upsilon_y^2} = c\sqrt{\frac{\lambda^2 + (t+\tau)^2}{\chi^2 + (t+\tau)^2}} = c\sqrt{1 - \frac{\mu^2}{\chi^2 + (t+\tau)^2}}. \tag{6.102}$$

An interesting conclusion that follows from this relation is that an infinite amount of time is needed for a body on which a constant force is exerted to reach a speed equal to c.

The displacement. The displacement may be found by the integration of υ_x and υ_y. Assuming that at $t = 0$ it is $x = 0$ and $y = 0$, we find

$$\begin{aligned} x &= c\int_0^t \frac{t+\tau}{\sqrt{\chi^2 + (t+\tau)^2}}\,dt = c\left[\sqrt{\chi^2 + (t+\tau)^2}\right]_0^t \\ &= c\left(\sqrt{\chi^2 + (t+\tau)^2} - \sqrt{\chi^2 + \tau^2}\right) \end{aligned} \tag{6.103}$$

and

$$y = c\int_0^t \frac{\lambda}{\sqrt{\chi^2 + (t+\tau)^2}}\,dt = \lambda c\left[\ln\left(t + \tau + \sqrt{\chi^2 + (t+\tau)^2}\right)\right]_0^t$$

or

$$y = \lambda c\left[\ln\left(t + \tau + \sqrt{\chi^2 + (t+\tau)^2}\right) - \ln\left(\tau + \sqrt{\chi^2 + \tau^2}\right)\right]. \tag{6.104}$$

6.14 The Motion of a Charged Particle in a Constant Homogeneous Magnetic Field

Let a particle with rest mass m_0 and charge q enter with velocity $\boldsymbol{\upsilon}_0$ a space where there is a constant homogeneous magnetic field $\mathbf{B}$. We choose the z-axis in the direction of the magnetic field, in which case it is $\mathbf{B} = B\hat{\mathbf{z}}$. We also choose the y-axis in such a way as to make the velocity of the particle, $\boldsymbol{\upsilon}$, have initial value $\boldsymbol{\upsilon}_0$

which lies in the yz plane, in which case we may write $\boldsymbol{\upsilon}_0 = \upsilon_{0y}\hat{\mathbf{y}} + \upsilon_{0z}\hat{\mathbf{z}}$. Let also the initial position of the charge be $\mathbf{r}_0 = x_0\hat{\mathbf{x}} + y_0\hat{\mathbf{y}} + z_0\hat{\mathbf{z}}$.

The equation of motion. The force exerted on the charge is $\mathbf{F} = q\boldsymbol{\upsilon} \times \mathbf{B}$. If the instantaneous relativistic momentum of the particle is $\mathbf{p} = \gamma m_0 \boldsymbol{\upsilon}$, the equation of motion of the charge is

$$\frac{d\mathbf{p}}{dt} = q\boldsymbol{\upsilon} \times \mathbf{B}. \tag{6.105}$$

For $\mathbf{B} = B\hat{\mathbf{z}}$, the components of this equation are

$$\frac{d(\gamma m_0 \upsilon_x)}{dt} = qB\upsilon_y, \qquad \frac{d(\gamma m_0 \upsilon_y)}{dt} = -qB\upsilon_x, \qquad \frac{d(\gamma m_0 \upsilon_z)}{dt} = 0. \tag{6.106}$$

These will be solved for the velocity and the position of the charge as functions of time.

Momentum and energy. Given that the force exerted on the charge is always normal to its velocity, the force does not produce work and the kinetic energy of the particle remains constant, and so do the magnitudes of its momentum and its velocity. This also follows from

$$\frac{d(p^2)}{dt} = 2p\frac{dp}{dt} = 2\mathbf{p} \cdot \frac{d\mathbf{p}}{dt} = 2q\mathbf{p} \cdot \boldsymbol{\upsilon} \times \mathbf{B} = 2qm(\boldsymbol{\upsilon} \cdot \boldsymbol{\upsilon} \times \mathbf{B}) = 0. \tag{6.107}$$

The speed of the particle and the Lorentz factor γ which corresponds to it are, therefore, constant.

The velocity. The components of the equation of motion may now be written as

$$\frac{d\upsilon_x}{dt} = \omega_c \upsilon_y, \qquad \frac{d\upsilon_y}{dt} = -\omega_c \upsilon_x, \qquad \frac{d\upsilon_z}{dt} = 0, \tag{6.108}$$

where $\omega_c = \dfrac{qB}{\gamma m_0}$ is a constant, known as the *cyclotron angular frequency*. It differs from the corresponding classical Larmor angular frequency $\omega_0 = \dfrac{qB}{m_0}$ only by the constant factor γ. It is

$$\omega_c = \frac{\omega_0}{\gamma}. \tag{6.109}$$

Because it is $(\omega_c/\omega_0)^2 = 1 - (\upsilon_0/c)^2$, the plot of the ratio ω_c/ω_0 as a function of the reduced speed of the charge, υ_0/c, is a circle of unit radius and center at the point (0, 0), as shown in Fig. 6.8.

The last of Eqs. (6.108) may be integrated to give $\upsilon_z = \upsilon_{0z}$ and $z = z_0 + \upsilon_{0z}t$. To solve the other two, we multiply the second with i and add to the first so as to get

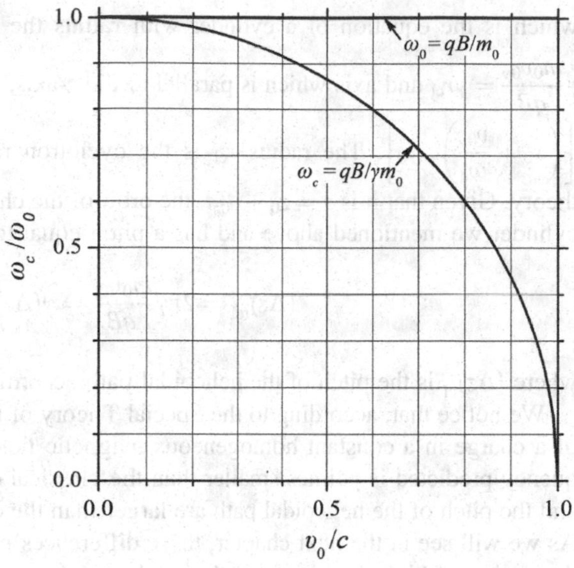

Fig. 6.8 The ratio of the relativistic cyclotron angular frequency of a charge q to its classical one, ω_c/ω_0, as a function of the charge's reduced speed, υ_0/c. In the classical case, the cyclotron angular frequency of the charge, ω_0, does not vary with its speed

$$\frac{d}{dt}(\upsilon_x + i\upsilon_y) = -i\omega_c(\upsilon_x + i\upsilon_y). \tag{6.110}$$

The integration of this equation with the given initial conditions gives

$$\upsilon_x + i\upsilon_y = i\upsilon_{0y}e^{-i\omega_c t}, \tag{6.111}$$

from which it finally follows that

$$\upsilon_x = \upsilon_{0y}\sin(\omega_c t), \qquad \upsilon_y = \upsilon_{0y}\cos(\omega_c t). \tag{6.112}$$

We see that the magnitude of the velocity remains constant and equal to

$$\upsilon = \sqrt{\upsilon_x^2 + \upsilon_y^2 + \upsilon_z^2} = \sqrt{\upsilon_{0y}^2 + \upsilon_{0z}^2} = \upsilon_0. \tag{6.113}$$

The position of the charge. The displacements may be found by integrating υ_x, υ_y and υ_z. We find them to be

$$x = x_0 + \frac{\upsilon_{0y}}{\omega_c}[1 - \cos(\omega_c t)], \quad y = y_0 + \frac{\upsilon_{0y}}{\omega_c}\sin(\omega_c t), \quad z = z_0 + \upsilon_{0z}t. \tag{6.114}$$

The first two give

$$\left[x - \left(x_0 + \frac{\upsilon_{0y}}{\omega_c}\right)\right]^2 + (y - y_0)^2 = \left(\frac{\upsilon_{0y}}{\omega_c}\right)^2, \tag{6.115}$$

which is the equation of a cylinder with radius the *cyclotron radius* $r_c = \frac{\upsilon_{0y}}{\omega_c} = \gamma \frac{m_0 \upsilon_{0y}}{qB} = \gamma r_{cl}$ and axis which is parallel to the z-axis, and passes through the point $\left[\left(x_0 + \frac{\upsilon_{0y}}{\omega_c}\right),\ y_0\right]$. The radius r_{cl} is the cyclotron radius predicted by classical theory. Given that it is $z = z_0 + \upsilon_{0z} t$, the orbit of the charge is helicoidal, lies on the cylinder we mentioned above and has a pitch equal to

$$(\Delta z)_{\text{rel}} = 2\pi\gamma \frac{m_0 \upsilon_{0z}}{qB} = \gamma (\Delta z)_{\text{cl}} \tag{6.116}$$

where $(\Delta z)_{\text{cl}}$ is the pitch of the helicoidal path according to non-relativistic theory.

We notice that, according to the Special Theory of Relativity, during the motion of a charge in a constant homogeneous magnetic field, the cyclotron angular frequency predicted is γ times smaller than the classical one and the cyclotron radius and the pitch of the helicoidal path are larger than the classical values by a factor γ. As we will see in the next chapter, these differences must be taken into account in the design of high-energy charged particle accelerators.

Example 6.13 Charged Particle in a Magnetic Field
Use Newton's second law of motion to determine the relationship between the radius of the circular orbit and the speed of a charged particle moving at right angles to a uniform magnetic field **B**.

Newton's second law gives for the equation of motion of the charge

$$\mathbf{F} = \frac{d}{dt}(m\boldsymbol{\upsilon}) = \frac{d}{dt}\left(\frac{m_0 \boldsymbol{\upsilon}}{\sqrt{1 - \upsilon^2/c^2}}\right).$$

According to Eq. (6.123), the speed υ of the charge is constant and so is the Lorentz factor corresponding to it. The velocity of the charge, $\boldsymbol{\upsilon}$, just changes direction. Therefore,

$$\mathbf{F} = \frac{m_0}{\sqrt{1 - \upsilon^2/c^2}} \frac{d\boldsymbol{\upsilon}}{dt}.$$

It is also $F_r = q\upsilon B$ and $\left|\frac{d\boldsymbol{\upsilon}}{dt}\right| = \frac{\upsilon^2}{R}$, where R is the radius of the circular orbit on which the charge moves. It follows that

$$q\upsilon B = \frac{m_0}{\sqrt{1 - \upsilon^2/c^2}} \frac{\upsilon^2}{R},$$

which may be solved to give $\upsilon = \dfrac{c}{\sqrt{1+(m_0 c/qBR)^2}}$.

Example 6.14 The Relativistic Oscillator [5]
A body with rest mass m_0 moves on the x-axis under the influence of a restoring force $-kx$ provided by a spring. The total energy of the body,

(rest energy + kinetic energy + potential energy in the spring),

remains constant and equal to W and the body executes oscillations with amplitude a. What is the period of the oscillations?

Let $x = 0$ be the equilibrium point of the oscillator. The potential energy stored in the spring when the body is at a distance x from its equilibrium position is $U = \frac{1}{2}kx^2$. The total energy of the body at this point is

$$W = E + U = m_0 c^2 \gamma + \frac{1}{2}kx^2 = \frac{m_0 c^2}{\sqrt{1-\upsilon^2/c^2}} + \frac{1}{2}kx^2. \quad (1)$$

At maximum displacement, $x = a$, the speed of the body is zero, in which case it follows that

$$W = m_0 c^2 + \frac{1}{2}ka^2. \quad (2)$$

Substituting in Eq. (1) and placing $\omega = \sqrt{k/m_0}$, the non-relativistic angular frequency, we have

$$c^2 + \frac{1}{2}\omega^2 a^2 = \frac{c^2}{\sqrt{1-\upsilon^2/c^2}} + \frac{1}{2}\omega^2 x^2. \quad (3)$$

Thus,

$$1 + \frac{\omega^2}{2c^2}(a^2 - x^2) = \frac{1}{\sqrt{1-\upsilon^2/c^2}}. \quad (4)$$

To find the relationship between x and t, we define the variable $x = a\cos\theta$, in which case Eq. (4) gives

$$\upsilon = \pm\, \omega a \,\sin\theta \,\frac{\sqrt{1+\dfrac{\omega^2 a^2}{4c^2}\sin^2\theta}}{1+\dfrac{\omega^2 a^2}{2c^2}\sin^2\theta}. \quad (5)$$

Because it is $\upsilon = -a\sin\theta \dfrac{d\theta}{dt}$, it follows that

$$\frac{d\theta}{dt} = \pm\,\omega\,\frac{\sqrt{1+\dfrac{\omega^2 a^2}{4c^2}\sin^2\theta}}{1+\dfrac{\omega^2 a^2}{2c^2}\sin^2\theta}, \tag{6}$$

from which we have

$$\omega t = \pm \int \frac{1+\dfrac{\omega^2 a^2}{2c^2}\sin^2\theta}{\sqrt{1+\dfrac{\omega^2 a^2}{4c^2}\sin^2\theta}}\,d\theta. \tag{7}$$

The evaluation of this integral requires the use of elliptic integrals but an approximate solution may be found by expanding the denominator in powers up to and including $\omega^2 a^2/c^2$

$$\begin{aligned} \omega t &\approx \int \left(1+\frac{\omega^2 a^2}{2c^2}\sin^2\theta\right)\left(1-\frac{\omega^2 a^2}{8c^2}\sin^2\theta\right) d\theta \\ &\approx \int \left(1+\frac{3\omega^2 a^2}{8c^2}\sin^2\theta\right) d\theta \approx \int \left(1+\frac{3\omega^2 a^2}{8c^2}-\frac{3\omega^2 a^2}{8c^2}\cos 2\theta\right) d\theta \end{aligned} \tag{8}$$

where we have taken the solution for positive values of t. Integrating with $\theta = 0$ and $x = a$ when it is $t = 0$,

$$t \approx \left(1+\frac{3\omega^2 a^2}{16c^2}\right)\frac{\theta}{\omega} - \frac{3\omega a^2}{16c^2}\sin\theta\cos\theta. \tag{9}$$

Together with

$$x = a\cos\theta, \tag{10}$$

the two equations give the relationship between t and x in parametric form.

Because it is $x = a$ and $t = 0$ for $\theta = 0$, Eq. (9) gives the period of the oscillation for $\theta = 2\pi$,

$$T \approx \left(1+\frac{3\omega^2 a^2}{16c^2}\right) T_0 \approx T_0 + \frac{3\pi^2}{4}\frac{(a/c)^2}{T_0}, \tag{11}$$

where $T_0 = 2\pi/\omega$ is the period in the non-relativistic solution.

If the quantity $\frac{3\omega^2 a^2}{16c^2}$ is small compared to unity, Eqs. (9) and (10) give, to a first approximation, $\theta = \omega t$. To the same approximation, we may substitute this in the product $\sin\theta\cos\theta$ of Eq. (9) and obtain

$$\left(1+\frac{3\omega^2a^2}{16c^2}\right)\theta = \omega t - \frac{3\omega^2a^2}{16c^2}\sin(\omega t)\cos(\omega t) = \omega t - \frac{1}{2}\frac{3\omega^2a^2}{16c^2}\sin(2\omega t). \quad (12)$$

Using $\left(1+\frac{3\omega^2a^2}{16c^2}\right) \approx \frac{T}{T_0}$, it follows that, to a first approximation, it is

$$\frac{T}{T_0}\theta = \frac{2\pi}{T_0}t - \frac{1}{2}\left(\frac{T}{T_0}-1\right)\sin\left(4\pi\frac{t}{T_0}\right), \quad \text{or} \quad \theta = \frac{2\pi}{T}t - \left(\frac{T-T_0}{2T}\right)\sin\left(4\pi\frac{t}{T_0}\right). \quad (13)$$

Finally,

$$x = a\cos\left[2\pi\frac{t}{T} - \left(\frac{T-T_0}{2T}\right)\sin\left(4\pi\frac{t}{T_0}\right)\right], \quad (14)$$

to a first approximation. If $\frac{T-T_0}{2T}$ is very small compared to unity, we may expand cos[] and have

$$x = a\cos\left(2\pi\frac{t}{T}\right) + a\frac{T-T_0}{2T}\sin\left(4\pi\frac{t}{T_0}\right)\sin\left(2\pi\frac{t}{T}\right). \quad (15)$$

References

1. J Luo, L-C Tu, Z-K Hu, E-J Luan, Phys. Rev. Lett. **90**, 081801–1 (2003)
2. L.D. Landau, E.M. Lifshitz, *Mechanics* (Pergamon Press, London, 1960), sec. 7
3. F.C. Champion, Proc. R. Soc. A **136**, 630 (1932)
4. M. Abraham, Annln. Phys. **10**, 105 (1903)
5. P.H. Penfield, H. Zatzkis, Jour. Franklin Inst. **262**, 121 (1956)

Chapter 7
Applications of Relativistic Dynamics

7.1 The Compton Effect

The corpuscular nature of light was established not only by the interpretation Einstein gave to the photoelectric effect, but also by the success of the application of Relativistic Mechanics in the explanation of the observations of A.H. Compton, in 1922, of the scattering of X rays by light elements [1]. Compton's main observation was that, apart from those X rays that had the same wavelength as the impinging rays, there was also a component in the spectrum of the scattered rays consisting of X rays of a slightly greater wavelength.

Compton's experimental arrangement is shown in Fig. 7.1. Monoenergetic electrons falling on a molybdenum target produced X rays, in a chamber shielded with lead. The X rays were scattered by a carbon target, which was kept small in order to avoid multiple scattering. The X rays that had been scattered at an angle θ, were selected by an arrangement of slits and fell on the monocrystal (*Crystal* in the figure) of a Bragg spectrometer. The spectrometer diffracted the X rays by angles α which depended on their wavelengths or energies. The intensity of the X rays diffracted in a certain direction was measured by an ionization chamber, in which the X rays produced ionization in a gas. By recording the ionization current as a function of the diffraction angle α, the energy spectrum of the X rays scattered by the carbon target by an angle θ was determined. The results of Compton's initial experiment are shown in Fig. 7.2 for scattering by carbon of the molybdenum K_α line X rays.

The results shown in Fig. 7.2 are for scattering angles $\theta = 45°$, $90°$ and $135°$. The peak due to the X rays with the same wavelength as that of the impinging rays is seen at a constant position on the left. The peak appearing in each case at a greater diffraction angle α (on the right) is due to X rays of a greater wavelength. The *change* $\Delta\lambda$ in the wavelength λ was found to be positive and to vary from zero for $\theta = 0°$ up to a maximum value for $\theta = 180°$. The change in wavelength for a given scattering angle was found experimentally to be independent of the

© Springer International Publishing Switzerland 2016
C. Christodoulides, *The Special Theory of Relativity*,
Undergraduate Lecture Notes in Physics, DOI 10.1007/978-3-319-25274-2_7

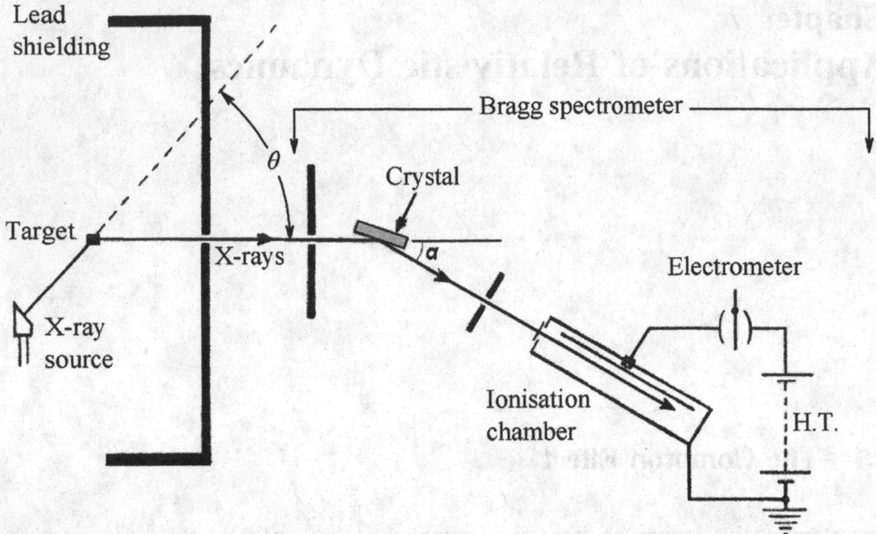

Fig. 7.1 The experimental arrangement used by Compton in the study of the scattering of X rays from light elements

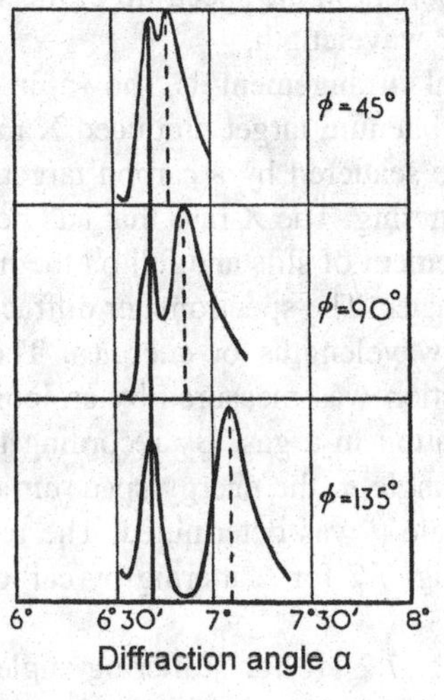

Fig. 7.2 Spectrum of the K_{α} X rays from molybdenum that have been scattered by a carbon target in Compton's experiments, for three scattering angles (angle ϕ in this figure is angle θ in Fig. 7.1 and in the equations). The X rays that are scattered elastically and, therefore, lose no energy, appear at a diffraction angle of $\alpha = 6^{\circ}40'$, approximately. Scattered X rays with a greater wavelength appear, in each case, as a peak at a higher diffraction angle

wavelength of the scattered X rays and to be the same for all materials. For the explanation of this change, the use of the Special Theory of Relativity was necessary.

Compton's success in the interpretation of the observations is due to the fact that he considered the X rays to consist of photons, with corpuscular properties, which collide with the free electrons according to relativistic dynamics. The scattering is represented symbolically in Fig. 7.3. A photon with wavelength λ, or frequency f, has an energy of $E = hf = hc/\lambda$, where h is Planck's constant. According to the theory of relativity, the photon, as a particle with zero rest mass, will have a momentum equal to $p = E/c = hf/c = h/\lambda$. Initially, the photon has an energy E and momentum $p = E/c$. After scattering by an angle θ, the photon has an energy E' and momentum $p' = E'/c$. The electron, initially at rest, has an energy of $E_e = m_0c^2$, where m_0 is its rest mass. After the scattering, the electron moves in a direction forming an angle ϕ with the initial direction of motion of the photon. It has speed υ, momentum $p'_e = \gamma m_0 \upsilon = m\upsilon$ and energy $E'_e = \gamma m_0 c^2 = mc^2$.

The laws of conservation give:

Conservation of energy: $$m_0c^2 + E = mc^2 + E' \tag{7.1}$$

Conservation of longitudinal momentum: $$\frac{E}{c} = \frac{E'}{c}\cos\theta + m\upsilon\cos\phi \tag{7.2}$$

Conservation of transverse momentum: $$0 = \frac{E'}{c}\sin\theta - m\upsilon\sin\phi \tag{7.3}$$

We will eliminate angle ϕ between Eqs. (7.2) and (7.3). These two equations give

$$m^2c^2\upsilon^2\cos^2\phi = (E - E'\cos\theta)^2 \quad \text{and} \quad m^2c^2\upsilon^2\sin^2\phi = E'^2\sin^2\theta. \tag{7.4}$$

Adding, $$m^2c^2\upsilon^2 = (E - E'\cos\theta)^2 + E'^2\sin^2\theta = E^2 + E'^2 - 2EE'\cos\theta. \tag{7.5}$$

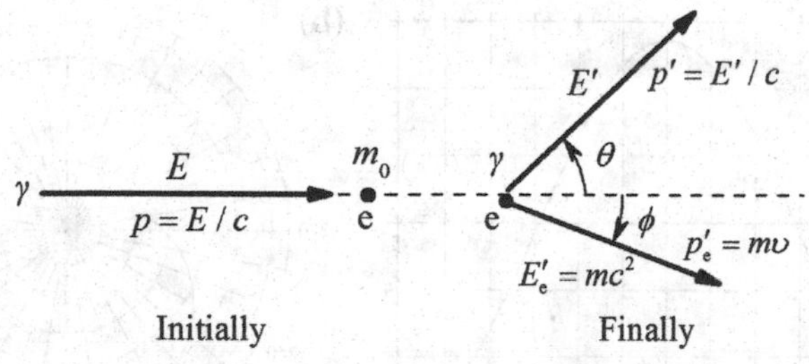

Fig. 7.3 The Compton effect. Scattering of a photon, γ, by a free electron, e

However, $m^2 = \dfrac{m_0^2}{1 - \upsilon^2/c^2}$ or $m^2\upsilon^2 = c^2(m^2 - m_0^2)$. (7.6)

This, together with Eq. (7.1), gives

$$\begin{aligned} m^2\upsilon^2c^2 &= (m_0c^2 + E - E')^2 - m_0^2c^4 \\ &= \cancel{m_0^2c^4} + E^2 + E'^2 + 2m_0c^2E - 2m_0c^2E' - 2EE' - \cancel{m_0^2c^4} \end{aligned} \tag{7.7}$$

which, with Eq. (7.5), results in

$$\cancel{E^2} + \cancel{E'^2} + 2m_0c^2E - 2m_0c^2E' - 2EE' = \cancel{E^2} + \cancel{E'^2} - 2EE'\cos\theta. \tag{7.8}$$

Finally, $$2m_0c^2(E - E') = 2EE'(1 - \cos\theta). \tag{7.9}$$

Dividing throughout by $2EE'm_0c^2$, we find that

$$\frac{1}{E'} - \frac{1}{E} = \frac{1}{m_0c^2}(1 - \cos\theta). \tag{7.10}$$

Because it is $E = hc/\lambda$ and $E' = hc/\lambda'$, we get the relation

$$\Delta\lambda = \lambda' - \lambda = \frac{h}{m_0c}(1 - \cos\theta) \quad \text{or} \quad \Delta\lambda = \lambda' - \lambda = \frac{2h}{m_0c}\sin^2\frac{\theta}{2}. \tag{7.11}$$

The variation of $\Delta\lambda$ with the angle of scattering θ is shown in Fig. 7.4.

This relation explains the change in the wavelength of the scattered X rays as a function of the scattering angle θ. It constitutes yet another experimental verification of the predictions of the Special Theory of Relativity and of the corpuscular nature of light. The relation also explains the presence in the spectrum of X rays

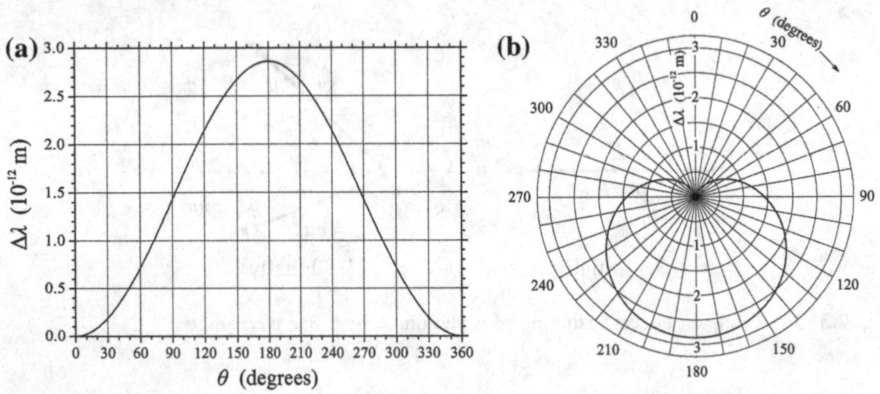

Fig. 7.4 The variation of $\Delta\lambda$ with the scattering angle θ, **a** in orthogonal coordinates and **b** in polar coordinates

that retain (almost) all their initial energy. If in their scattering the whole atom participates, and not just an electron, then in Eq. (7.11), instead of the mass of the electron, the mass appearing will be that of the whole atom, which is about four orders of magnitude larger and, as a consequence, the change in the wavelength of the X rays, $\Delta\lambda$, will be smaller by the same factor.

Substituting, we find that

$$\frac{h}{m_e c} = 2.43 \times 10^{-12}\ \mathrm{m},$$

where m_e is the rest mass of the electron. The quantity

$$\frac{\hbar}{m_e c} = 3.86 \times 10^{-13}\ \mathrm{m}, \tag{7.12}$$

where $\hbar \equiv h/2\pi$, is called *Compton wavelength of the electron.*

Problems

7.1 A photon of energy E_γ collides with an electron at rest, which has a rest mass of m_0. After the collision, the photon moves in a direction opposite to its initial direction of motion. Find, from first principles:

(a) The fraction of the energy E_γ that is given to the electron as kinetic energy. Ans.: $\dfrac{K}{E_\gamma} = \dfrac{2E_\gamma}{m_0 c^2 + 2E_\gamma}$

(b) The change in the wavelength λ of the photon. Ans.: $\lambda' - \lambda = \dfrac{2h}{m_0 c}$

7.2 Show that, during Compton scattering, the scattering angles of the photon and the electron satisfy the relation $\cot\phi = \left(1 + \dfrac{E}{m_0 c^2}\right) \tan\dfrac{\theta}{2}$.

7.2 The Inverse Compton Effect

With the term *inverse Compton effect* we refer to that process by which very high-energy photons are produced from photons of low energy, through their collision with fast electrons. The phenomenon is important in Astrophysics as well as in the production, in the laboratory, of photons with high energies through the collision of visible photons from a laser with fast electrons from an accelerator. The general process is shown in Fig. 7.5a. A fast-moving electron e, with momentum p_e and energy E_e, collides with a low-energy photon ϕ, with momentum p_ϕ and energy E_ϕ, which is moving at an angle α relative to the direction of motion of the electron. The photon is backscattered as a high-energy gamma ray, γ, at an angle θ, with momentum p_γ and energy E_γ. The electron moves at an angle ϕ, with momentum p'_e and energy E'_e.

The laws of conservation give:

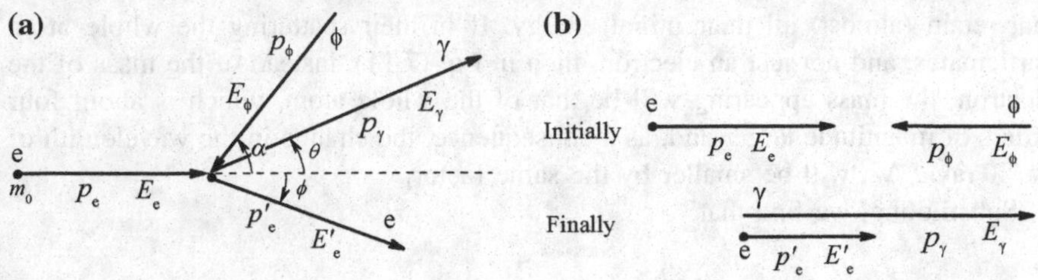

Fig. 7.5 The inverse Compton effect: **a** the general case and **b** the case of head-on collision

Longitudinal momentum: $$p_e - p_\phi \cos\alpha = p'_e \cos\phi + p_\gamma \cos\theta \tag{7.13}$$

Transverse momentum: $$-p_\phi \sin\alpha = -p'_e \sin\phi + p_\gamma \sin\theta \tag{7.14}$$

Energy: $$E_e + E_\phi = E'_e + E_\gamma \tag{7.15}$$

Solving Eqs. (7.13) and (7.14) for $p'_e \cos\phi$ and $p'_e \sin\phi$, squaring and adding, we get

$$p'^2_e = p_e^2 + p_\phi^2 + p_\gamma^2 - 2p_e p_\phi \cos\alpha - 2p_e p_\gamma \cos\theta + 2p_\phi p_\gamma \cos(\alpha - \theta) \tag{7.16}$$

Expressing the momenta in terms of the energies, we have

$$E'^2_e \cancel{-m_0^2c^2} = E_e^2 \cancel{-m_0^2c^2} + E_\phi^2 + E_\gamma^2 - 2E_e\beta E_\phi \cos\alpha - 2E_e\beta E_\gamma \cos\theta + 2E_\phi E_\gamma \cos(\alpha - \theta) \tag{7.17}$$

Substituting for E'^2_e from Eq. (7.15), results in

$$\cancel{E_e^2} + \cancel{E_\phi^2} + \cancel{E_\gamma^2} + 2E_e E_\phi - 2E_\phi E_\gamma - 2E_e E_\gamma =$$
$$= \cancel{E_e^2} + \cancel{E_\phi^2} + \cancel{E_\gamma^2} - 2E_e\beta E_\phi \cos\alpha - 2E_e\beta E_\gamma \cos\theta + 2E_\phi E_\gamma \cos(\alpha - \theta)$$

or

$$E_e E_\phi (1 + \beta\cos\alpha) = E_\gamma \left[E_e(1 - \beta\cos\theta) + E_\phi[1 + \cos(\alpha - \theta)]\right] \tag{7.18}$$

which can be solved for the ratio E_γ / E_e to give

$$\frac{E_\gamma}{E_e} = \frac{1 + \beta\cos\alpha}{1 + \cos(\alpha - \theta) + (1 - \beta\cos\theta)(E_e/E_\phi)}. \tag{7.19}$$

For the case of the head-on collision of Fig. 7.5b, we substitute $\alpha = 0$ and $\theta = 0$ in Eq. (7.19) to get:

$$\frac{E_\gamma}{E_e} = \frac{1+\beta}{2+(1-\beta)(E_e/E_\phi)} \tag{7.20}$$

For $\gamma \gg 1$, it is $1+\beta \approx 2$ and $1-\beta = \dfrac{1-\beta}{\gamma^2(1-\beta^2)} = \dfrac{1}{\gamma^2(1+\beta)} \approx \dfrac{1}{2\gamma^2}$ and Eq. (7.20) becomes

$$\frac{E_\gamma}{E_e} = \frac{1}{1+(E_e/E_\phi)/4\gamma^2} \quad \text{or} \quad \frac{E_\gamma}{E_e} = \frac{1}{1+m_0^2c^4/4E_eE_\phi}. \tag{7.21}$$

The behaviour of this function depends on the relative magnitudes of $m_0^2c^4$ and E_eE_ϕ, the latter being the product of a low value of energy, E_ϕ, and a high value, E_e. For the case of the head-on collision, the fraction E_γ/E_e of the initial electron energy, E_e, which is given to the backscattered photon, E_γ, is given in Fig. 7.6 (full lines), as a function of E_e, for some values of the initial photon energy E_ϕ between 0.1 and 100 eV.

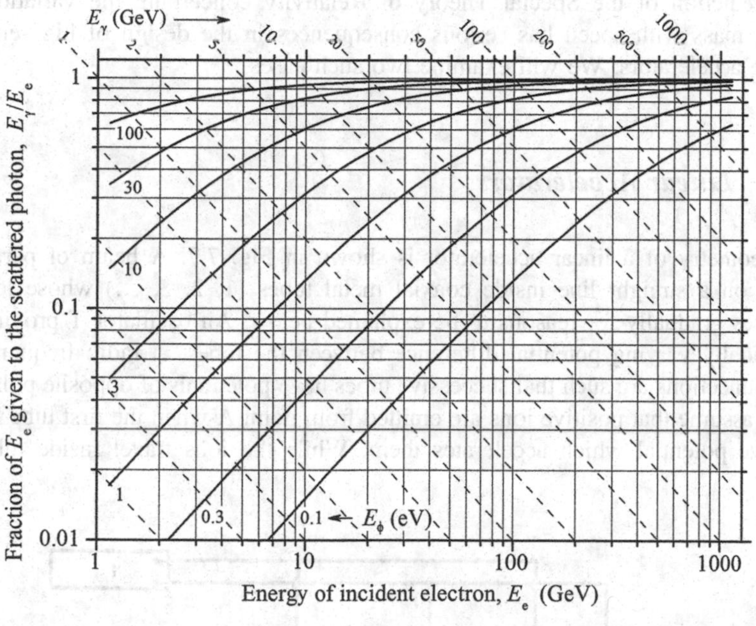

Fig. 7.6 The inverse Compton effect—case of the head-on collision. The fraction E_γ/E_e of the initial electron energy, E_e, which is given to the backscattered photon, E_γ, is given in the figure (*full lines*), as a function of E_e, for seven values of the initial photon energy E_ϕ between 0.1 and 100 eV. Also shown, in *dashed lines*, are the lines of constant photon energy E_γ, the values of which are given at the top of the figure

Example: An electron with energy $E_e = 100$ GeV colliding with a photon with $E_\phi = 1$ eV, results in a photon with $E_\gamma / E_e = 0.6$. This gives a γ ray with $E_\gamma = 60$ GeV.

We notice that the fraction of the electron energy which is transferred to the photon is higher for higher electron energy, E_e, and for higher energy of the impinging photon, E_ϕ. It is possible for almost all the electron energy to be transferred to the photon. A 3 eV photon of visible light colliding with a 30 GeV electron, will recoil as a gamma ray with energy 18 GeV.

The inverse Compton effect plays a very important role in Astrophysics, for the production of X and γ rays. In interstellar or intergalactic space, photons of visible light are very abundant. Electrons with high energies also form one of the components of cosmic radiation. The collision of the two leads, through the inverse Compton effect, to the production of very energetic photons, with energies which classify them as γ rays.

7.3 The Consequences of the Special Theory of Relativity on the Design of Particle Accelerators

The prediction of the Special Theory of Relativity concerning the variation of inertial mass with speed has serious consequences in the design of high-energy particle accelerators. We will examine two such cases.

7.3.1 *Linear Accelerators*

The geometry of a linear accelerator is shown in Fig. 7.7. A beam of particles moves on a straight line inside coaxial metal tubes (1, 2, 3, …) whose length increases gradually for reasons to be explained below. An oscillator T provides a sinusoidally varying potential difference between the tubes, at radio frequencies. The connections are such that successive tubes have potentials of opposite polarity. Let us assume that positive ions are emitted from point A when the first tube has a negative potential which accelerates them. While the ions travel inside tube 1,

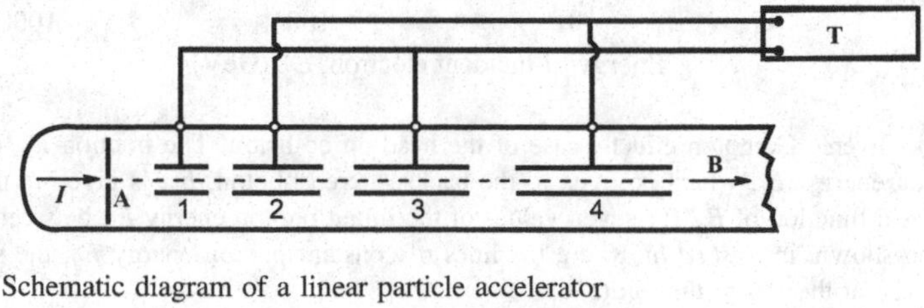

Fig. 7.7 Schematic diagram of a linear particle accelerator

where they are not accelerated, the potential difference between successive tubes change polarity, so that, when the ions come out of tube 1 the potential of tube 2 relative to tube 1 is negative and the ions are accelerated again. The same is repeated between tubes 2 and 3 etc.

If the charge of the ions is q and the potential difference between successive tubes is $\Delta\phi$, the kinetic energy of an ion, when it comes out of the n-th tube, will be, according to non-relativistic Mechanics, equal to $K_n = \frac{1}{2}mv_n^2 = nq\Delta\phi$, where m is the mass of the ion and v_n its speed at that moment. Now, the time needed for the ion to travel through the n-th tube must be equal to half the period of the oscillator, $T/2$. The length of the n-th tube must, therefore, be

$$l_n = \frac{1}{2}Tv_n = \frac{1}{f}\sqrt{\frac{nq\Delta\phi}{2m}} \tag{7.22}$$

where f is the frequency of the oscillator. Therefore,

$$l_n = \sqrt{n}\,l_1, \quad \text{where} \quad l_1 = \frac{1}{f}\sqrt{\frac{q\Delta\phi}{2m}} \tag{7.23}$$

is the length of the first tube. We see that the lengths of the tubes must gradually increase.

For a final kinetic energy of the ions equal to $K_N = \frac{1}{2}mv_N^2 = Nq\Delta\phi$, the number of tubes required is $N = K_N/q\Delta\phi$. The total length of the linear accelerator would then be

$$L(N) = \left(1 + \sqrt{2} + \sqrt{3} + \sqrt{4} + \cdots + \sqrt{N}\right)l_1 \approx \frac{2}{3}N^{3/2}l_1 \tag{7.24}$$

where the approximation is better the higher the number N is.

The Stanford linear accelerator (SLAC, Stanford Linear Accelerator Center), which first operated in 1961, accelerates electrons to energies of 25 GeV. It has a total length of 2 miles (3.2 km), and achieves an electron energy of 8 MeV per meter of accelerator length. For $N = 50\,000$, the energy given to the electrons per tube is estimated to be $q\Delta\phi = 25$ GeV/50 000 = 0.5 MeV. For a frequency of 5 GHz, it is $l_1 = 4.2$ cm. Since $\frac{2}{3}N^{3/2} = 7.45 \times 10^6$, the total length of the accelerator should have been, according to classical Mechanics, equal to $L(N) = \frac{2}{3}N^{3/2}l_1 = 7.45 \times 10^6 \times 0.042 = 3.13 \times 10^5$ m = 313 km. In fact, the accelerator is only 3.2 km long.

The explanation is simple. According to the Special Theory of Relativity, the speed of the particles does not increase indefinitely at the rate predicted by classical Mechanics, but tends, asymptotically, to the value of c. In fact, at the SLAC the electrons are first accelerated to energies equal to 30 MeV before they enter the linear accelerator. This means that their speed already corresponds to $\beta = 0.99986$. Therefore, the length of the tubes remains constant and equal to about

$l = c/f = 6$ cm. We see that, as the particles are given more energy, their speed remains almost constant, a fact that must be taken into account in the construction of a linear accelerator.

7.3.2 Cyclotron, Synchrocyclotron and Synchrotron

7.3.2.1 Cyclotron

In the *cyclotron*, the particles move in a magnetic field which is perpendicular to the plane of their orbits, describing circular orbits of ever increasing radius, as their energy increases. Most of the time the particles move inside two cavities in the shape of a D, as shown in Fig. 7.8 (D_1 and D_2). These two electrodes are connected to the poles of an oscillator T, which applies between them a potential difference at a high frequency. The particles enter the cyclotron at its center C and start moving in a circular orbit. If, when the particles are in the gap EE between the two D's, the potential difference $\Delta\phi$ between these is such as to accelerate the particles, the energy of the particles increases. The frequency of alternating the polarity between the two D's must be such that the particles, moving with the speed they have acquired, travel the semicircle and are found again in the gap between the electrodes half a period later, so that they can be accelerated again.

The condition for this to happen is found as follows: Let the magnetic field in which the particles move be **B**, normal to the section shown in the figure. If the

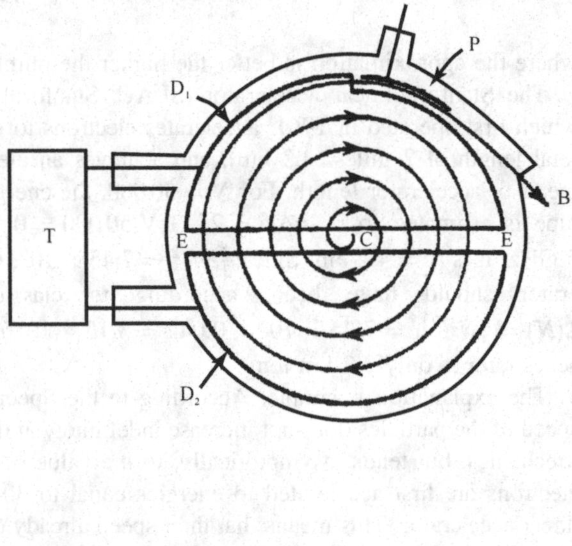

Fig. 7.8 Plan of a cyclotron

mass of the particle is m, its charge q and its speed υ, then the radius of the orbit it describes is equal to r, which satisfies the condition

$$\frac{m\upsilon^2}{r} = Bq\upsilon, \tag{7.25}$$

from which we find the speed of the particle in terms of the radius of its orbit as

$$\upsilon = \frac{qB}{m}r. \tag{7.26}$$

The frequency at which the particle revolves around the center C is

$$f = \frac{\upsilon}{2\pi r} = \frac{qB}{2\pi m}. \tag{7.27}$$

This is called *cyclotron frequency* and is independent of the speed υ or the radius r.

When the particles acquire enough energy so that they move in an orbit of radius R equal to that of the cyclotron, a deflection electrode P pushes them so that they exit electrode D_1 and move on a straight line as a beam, B. Theoretically, the maximum kinetic energy the particles may gain is

$$K_{\max} = \frac{1}{2}m\upsilon_{\max}^2 = \frac{q^2B^2R^2}{2m}. \tag{7.28}$$

In practice, however, with conventional cyclotrons we achieve maximum energies only of the order of 5–25 MeV, lower than the theoretical limit. This happens because the condition of Eq. (7.27) presupposes that the mass m of each particle remains constant during the acceleration process. At relativistic speeds this is not true and m increases with the speed or the energy of the particles. The limit to the maximum energy that can be achieved with a cyclotron is due to this effect.

There are two alternative solutions to the problem. These are implemented in the *synchrocyclotron* and the *synchrotron*, as we will explain below.

7.3.2.2 Synchrocyclotron

In the synchrocyclotron, the increase of the mass m is offset by a suitable change in the frequency. In the relativistic case, Eq. (7.27) is written as

$$f = \frac{qB}{2\pi m} = \frac{qB}{2\pi(m_0 + K/c^2)} \tag{7.29}$$

where m_0 is the rest mass of the particles and K is their kinetic energy. As the energy of the particles increases, the frequency f must be suitably decreased.

The 184 inch synchrocyclotron at Berkeley, California, which was constructed in 1942, was the first to accelerate by this method protons to the energy of 340 MeV, which was considered as very high at that time. In order to achieve this, a change of the frequency by 35 % was necessary. For higher energies with heavy ions or for the acceleration of electrons, synchrotrons are generally preferred.

7.3.2.3 Synchrotron

In the case of the synchrotron, the relativistic effects are faced by keeping the frequency constant but increasing the magnetic field B as the energy of the particles increases. Equation (7.29) is also valid in this case. The particles are introduced into the synchrotron after they have been accelerated to relativistic energies by other means. In this way they have an almost constant speed very near the value of c and, for constant frequency f, they move, on the basis of Eq. (7.27), on an almost constant radius, equal to $R = c/2\pi f$. This means that a magnetic field is needed only in a small region around this radius R, with an ensuing decrease in energy consumption. Electrons are accelerated to energies of the order of 500 MeV or higher with synchrotrons.

7.4 Mass Defect and Binding Energy of the Atomic Nucleus

An isotope, symbolized by ${}^{A}_{Z}\mathrm{X}$, where X is the chemical symbol of the atom, has atomic number Z and mass number A. The nucleus of the isotope consists of Z protons, of A nucleons in total and, consequently, of $(A-Z)$ neutrons. If we add the rest masses of Z protons and $(A-Z)$ neutrons, we will find a total mass which is greater than the rest mass of the nucleus of the isotope ${}^{A}_{Z}\mathrm{X}$. If m_{p} is the rest mass of the proton, m_{n} is the rest mass of the neutron and m_{X} is the rest mass of the nucleus of the isotope ${}^{A}_{Z}\mathrm{X}$, the quantity

$$\Delta m = Zm_{\mathrm{p}} + (A-Z)m_{\mathrm{n}} - m_{\mathrm{X}} \tag{7.30}$$

is defined as the *true mass defect* of the isotope's nucleus. Due to the fact that tables usually give the masses of the atoms of the isotopes and not of their nuclei, taking into account the masses of the electrons around the nucleus, this quantity may also be defined as

$$\Delta M = ZM_{\mathrm{H}} + (A-Z)m_{\mathrm{n}} - M_{\mathrm{X}} \tag{7.31}$$

where M_{H} is the mass of the neutral *atom* of hydrogen and M_{X} the mass of the neutral *atom* of the isotope ${}^{A}_{Z}\mathrm{X}$. This mass may be transformed into energy by multiplying with c^2. In this case we refer to the *binding energy* (*B.E.*) of the nucleus of the isotope ${}^A_Z$X:

$$B.E. = \Delta M\, c^2 = [ZM_{\rm H} + (A - Z)m_{\rm n} - M_{\rm X}]c^2. \tag{7.32}$$

The binding energy is a quantity that has been measured experimentally many times, during the fission of nuclei and during nuclear reactions. According to the interpretation given by the Special Theory of Relativity, it is the energy which is released during the 'formation' of a nucleus by protons and neutrons. It is mainly due to the attractive strong nuclear forces between the nucleons forming the particular nucleus. This is the energy that must be given to the nucleus in order to move its nucleons at infinite distances from each other. It is equivalent to the difference in the masses and is measured experimentally. Obviously, in the terms $ZM_{\rm H} + (A - Z)m_{\rm n}$ of Eq. (7.32) it was not taken into account that the binding energies of the Z electrons to the nucleus of the isotope ${}^A_Z$X are greater than that of the electron of the hydrogen atom. These energies are, however, small compared to the nuclear binding energies and may be ignored. Besides, as we will see below, we will usually evaluate *differences* between binding energies of nuclei, a fact that makes the remaining error even smaller.

The magnitude usually evaluated is the average binding enegy per nucleon of the nucleus examined,

$$\frac{B.E.}{A} = \frac{\Delta M\, c^2}{A}. \tag{7.33}$$

Figure 7.9 gives the binding energies per nucleon as a function of the mass number A. Obviously, to a given A there correspond many different isotopes. We notice that the B.E. per nucleon starts at low values for low values of A, it reaches a maximum for the isotope ${}^{56}_{26}$Fe and then it becomes smaller for heavier isotopes. Greater binding energy per nucleon means greater stability of the nucleus, since a greater amount of energy would be required in order to dissolve the nucleus in its constituents. We notice two important characteristics of the figure:

(a) The nucleus of the atom of helium-4, ${}^4_2$He, i.e. an α particle, has a greater stability than those of the neighboring isotopes. The same is true for the nuclei of isotopes than may be thought of as being formed by 2, 3, 4 etc. α particles, such as the nuclei of ${}^8_4$Be, ${}^{12}_6$C and ${}^{16}_8$O.

(b) Because the nuclei in the region of $A = 56$ are more stable, energy is released whenever a heavy nucleus is split into two nuclei of approximately equal mass numbers (nuclear fission) or whenever two lighter nuclei combine to form a heavier one (nuclear fusion).

The unit of measurement of the masses of atoms is the *atomic mass unit*, u. It is defined so that the mass of a neutral atom of the isotope ${}^{12}_6$C is exactly equal to 12 u.

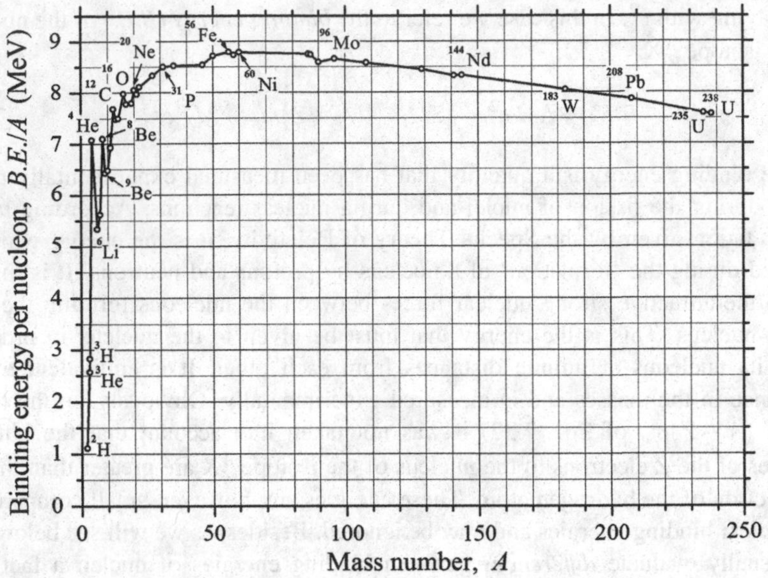

Fig. 7.9 The binding energy per nucleon of isotopes as a function of their mass number A

Knowing that one mol of ${}^{12}_{6}\mathrm{C}$ has a mass of 12 g and consists of a number of atoms equal to Avogadro's constant $N_{\mathrm{A}} = 6.022\ 142 \times 10^{23}\ \mathrm{mol}^{-1}$, we find that

$$1\ \mathrm{u} = 1.660\ 539 \times 10^{-27}\ \mathrm{kg}. \tag{7.34}$$

From the relation $E = mc^2$ we find the energy equivalent of 1 u to be

$$1\ \mathrm{u} \equiv 1.492\ 418 \times 10^{-10}\ \mathrm{J} = 931.494\ \mathrm{MeV}. \tag{7.35}$$

The masses of some particles and some neutral atoms are given in the table below, together with their energy equivalents. The binding energies per nucleon are also given for the nuclei (Table 7.1).

7.4.1 Nuclear Reactions and Binding Energy

A(a, b)B is another way to symbolize the nuclear reaction $\mathrm{A} + \mathrm{a} \to \mathrm{B} + \mathrm{b} + Q$, where Q is the energy released (+) or absorbed (–) in the reaction. Nucleus A is bombarded with a particle a, producing a new nucleus B and a particle b. Obviously, it is $Q = \Delta M c^2$. In Table 7.2, the values of Q are given for some nuclear reactions.

Table 7.1 Mass, energy equivalent and binding energy per nucleon for some particles and atoms

Particle or atom	Mass (u)	(kg)	Energy equivalent (MeV)	Binding energy per nucleon (MeV/nucleon)
1 u	1	$1.660\ 539 \times 10^{-27}$	931.494	–
e	0.000 548 6	$9.109\ 382 \times 10^{-31}$	0.511	–
p	1.007 276	$1.672\ 622 \times 10^{-27}$	938.272	–
n	1.008 665	$1.674\ 927 \times 10^{-27}$	939.565	–
α	4.001 506	$6.644\ 656 \times 10^{-27}$	3727.379	7.074
$^{1}_{1}H$	1.007 825	$1.673\ 533 \times 10^{-27}$	938.783	–
$^{2}_{1}H$ (D)	2.014 102	$3.344\ 495 \times 10^{-27}$	1877.124	1.112
$^{3}_{1}H$ (T)	3.016 049	$5.008\ 267 \times 10^{-27}$	2809.432	2.827
$^{4}_{2}He$	4.002 603	$6.646\ 478 \times 10^{-27}$	3728.401	7.074
$^{12}_{6}C$	12	$19.92\ 6468 \times 10^{-27}$	11 177.93	7.680
$^{16}_{8}O$	15.994 915	$26.560\ 180 \times 10^{-27}$	14 899.17	7.976
$^{56}_{26}Fe$	55.934 939	$92.882\ 148 \times 10^{-27}$	52 103.06	8.790
$^{58}_{28}Ni$	57.935 347	$96.203\ 903 \times 10^{-27}$	53 967.43	8.732
$^{60}_{28}Ni$	59.930 786	$99.517\ 378 \times 10^{-27}$	55 825.17	8.781
$^{90}_{36}Kr$	89.919 517	$149.314\ 865 \times 10^{-27}$	83 759.49	8.591
$^{141}_{56}Ba$	140.914 411	$233.993\ 875 \times 10^{-27}$	131 260.9	8.326
$^{208}_{82}Pb$	207.976 64	$345.353\ 322 \times 10^{-27}$	193 729.0	7.867
$^{235}_{92}U$	235.043 930	$390.299\ 612 \times 10^{-27}$	218 942.0	7.591
$^{238}_{92}U$	238.050 786	$395.292\ 61 \times 10^{-27}$	221 742.9	7.570

Table 7.2 Energy released in nuclear reactions

Exoergic Reaction	Q (MeV)	Endoergic Reaction	Q (MeV)
$^{2}H(n,\gamma)^{3}H$	+6.26	$^{7}Li(p,n)^{7}Be$	−1.65
$^{6}Li(d,\alpha)^{4}He$	+22.17	$^{9}Be(\gamma,n)^{8}Be$	−1.67
$^{9}Be(p,\alpha)^{6}Li$	+2.25	$^{14}N(\alpha,p)^{17}O$	−1.15
$^{10}B(d,n)^{11}C$	+6.38	$^{18}O(p,n)^{18}F$	−2.45
$^{14}N(n,\gamma)^{15}N$	+10.83	$^{28}Si(\alpha,p)^{31}P$	−2.25

Example 7.1 The Binding Energy of the Deuteron
The deuteron (d) is the nucleus of deuterium (D), an isotope of hydrogen (H) and consists of a proton and a neutron. Find its binding energy.

The reaction of 'formation' of the deuteron is $p + n \rightarrow d + \Delta E$, where $\Delta E = Q$ is the binding energy of the deuteron and is released on its formation.

The mass defect of the deuteron is $\Delta m = m_p + m_n - m_d = 0.002\ 388$ u (equal to 4.4 rest masses of the electron).

The conservation of mass-energy gives: $m_{\rm p}c^2 + m_{\rm n}c^2 = m_{\rm d}c^2 + \Delta E$
The masses of the particles are:

$$m_{\rm p} = 1.007\,276\ {\rm u} = 938.272\ {\rm MeV}/c^2$$

$$m_{\rm n} = 1.008\,665\ {\rm u} = 939.565\ {\rm MeV}/c^2$$

Summing, $m_{\rm p} + m_{\rm n} = 1877.837\ {\rm MeV}/c^2.$

Also, $m_{\rm d} = 2.013\,553\ {\rm u} = 1875.613\ {\rm MeV}/c^2.$

The difference gives $\Delta E = \Delta m\, c^2 = 2.224\ {\rm MeV}$

which corresponds to

$$1.112\ {\rm MeV/nucleon}.$$

[We find the same value from
$\Delta E = \Delta mc^2 = (0.002\,388\ {\rm u}) \times (931.5\ {\rm MeV/u}) = 2.224\ {\rm MeV}.$]

The binding energy of the deuteron, depending only on the strong nuclear forces, gives us an estimate of the strength of this force between a proton and a neutron. It might be worth making two observations here:

If the strong nuclear force was 5 % weaker, the deuteron would not have been stable. Without deuterons, the synthesis of heavier nuclei in the interior of the stars would have been impossible and so would the appearence of life [2].

Had the nuclear force been 2 % stronger, the formation of the *diproton* (pp) inside the stars would have been possible, at a huge rate. The decay of excited diprotons into deuterons would convert ^{1}H into ^{2}H at too fast a rate. The production of energy in stars would not occur at a slow enough rate for life to have time to evolve!

It appears that, at least in the universe we live in, the intensity of the strong nuclear force has been fine-tuned so that the existence of life would not be impossible!

Example 7.2 Nuclear Fission
A nucleus of $^{235}_{92}$U, bombarded by a thermal neutron, suffers fission into $^{141}_{56}$Ba, $^{92}_{36}$Kr and 3 neutrons. Find the energy released.

The nuclear reaction is

$$^{235}_{92}{\rm U} + {}^1_0{\rm n} \rightarrow {}^{141}_{56}{\rm Ba} + {}^{92}_{36}{\rm Kr} + 3\,{}^1_0{\rm n} + Q.$$

From Fig. 7.9 we see that the binding energy per nucleon for $^{235}_{92}$U is about 7.6 MeV/nucleon, while for $^{141}_{56}$Ba and $^{92}_{36}$Kr it is about 8.5 MeV/nucleon. The total energy released during the fission is

Q = (Number of nucleons, A)×(Difference in the binding energies per nucleon)
Q = 235 ×(8.5–7.6) = 212 MeV per fission, approximately.

Example 7.3 Nuclear fusion. Energy production in the stars
The reactions of nuclear fusion taking place in the interior of a star are the following:

(1) ${}^1_1\mathrm{p} + {}^1_1\mathrm{p} \rightarrow {}^2_1\mathrm{d} + \mathrm{e}^+ + \nu_\mathrm{e} + 0.420\ \mathrm{MeV}$
(2) ${}^2_1\mathrm{d} + {}^1_1\mathrm{p} \rightarrow {}^3_2\mathrm{He}^{2+} + \gamma + 5.493\ \mathrm{MeV}$
(3) ${}^3_2\mathrm{He}^{2+} + {}^3_2\mathrm{He}^{2+} \rightarrow {}^4_2\mathrm{He}^{2+} + 2{}^1_1\mathrm{p} + 12.860\ \mathrm{MeV}$

In a complete proton–proton cycle, consisting of 2 reactions (1), 2 reactions (2) and one reaction (3) how much energy is released?

A complete proton–proton cycle consists of the following reactions:

$$
\begin{array}{ll}
2\times(1) & 2{}^1_1\mathrm{p} + 2{}^1_1\mathrm{p} \rightarrow 2{}^2_1\mathrm{d} + 2\mathrm{e}^+ + 2\nu_\mathrm{e} + 0.840\ \mathrm{MeV} \\
2\times(2) & 2{}^2_1\mathrm{d} + 2{}^1_1\mathrm{p} \rightarrow 2{}^3_2\mathrm{He}^{2+} + 2\gamma + 10.986\ \mathrm{MeV} \\
1\times(3) & 2{}^3_2\mathrm{He}^{2+} \rightarrow {}^4_2\mathrm{He}^{2+} + 2{}^1_1\mathrm{p} + 12.860\ \mathrm{MeV}
\end{array}
$$

Adding,

$$4{}^1_1\mathrm{p} \rightarrow {}^4_2\mathrm{He}^{2+} + 2\mathrm{e}^+ + 2\nu_\mathrm{e} + 2\gamma + 24.686\ \mathrm{MeV}$$

i.e. the net result is the fusion of four protons for the formation of a helium nucleus, 2 positrons, 2 neutrinos and 2 γ rays, with a total release of energy equal to 24.7 MeV. In this way, the hydrogen of a star is converted into helium, producing energy.

The change in the mass is

$$
\begin{aligned}
\Delta m &= 4m({}^1_1\mathrm{p}) - m({}^4_2\mathrm{He}^{2+}) - 2m_\mathrm{e} = 4m({}^1_1\mathrm{p}) - m({}^4_2\mathrm{He} - 2\mathrm{e}) - 2m_\mathrm{e} \\
&= 4m({}^1_1\mathrm{p}) - M({}^4_2\mathrm{He}) = 4 \times 1.007\,276 - 4.002\,603 = 0.0265\ \mathrm{u}
\end{aligned}
$$

where it was taken into account that the mass of the neutrino is negligible and that the nucleus of helium, ${}^4_2\mathrm{He}^{2+}$, has a mass which is equal to that of the neutral atom minus the mass of 2 electrons.

This change in mass is equivalent to a released energy of $\Delta m\, c^2 = 24.7$ MeV.

Problems

7.3 *The binding energy of an* α *particle*. The α particle is a helium nucleus, consisting of two protons and two neutrons. Find the binding energy of the α particle and its binding energy per nucleon. Ans.: 28.3 MeV, 7.07 MeV/nucleon

7.4 *Neutron decay.* Free neutrons are unstable and decay with a mean lifetime of $\tau = 898\ \mathrm{s} \approx 15$ minutes, according to $\mathrm{n} \rightarrow \mathrm{p} + \mathrm{e}^- + \bar{\nu}_\mathrm{e} + \Delta E$. Find the energy released during the decay. Ans.: $\Delta E = 0.79$ MeV

7.5 *Photofission of uranium-235.* A photon with energy $E_\phi = 6$ MeV collides with a ${}^{235}_{92}\mathrm{U}$ nucleus at rest and causes the fission ${}^{235}_{92}\mathrm{U} + \gamma \rightarrow {}^{90}_{36}\mathrm{Kr} + {}^{142}_{56}\mathrm{Ba} + 3{}^1_0\mathrm{n}$. What is the total kinetic energy K of the products of this fission? The isotopic masses are given as $M({}^{235}_{92}\mathrm{U}) = 235.043\,930$ u, $M({}^{90}_{36}\mathrm{Kr}) = 89.919\,72$ u, $M({}^{142}_{56}\mathrm{Ba}) = 141.916\,35$ u and $m_\mathrm{n} = 1.008\,665$ u. Ans.: $K = 175$ MeV

7.6 (a) How much energy is released during the fusion of two deuterium nuclei to form a helium nucleus? The nuclear rest mass of deuterium is $m_D = 2.0136$ u and of helium $m_{He} = 4.0015$ u. (b) How much energy is released in the fusion of a mass of deuterium equal to 1 kg for the formation of helium? Express your answer in J and in kWh. Ans.: (a) $\Delta E = 23.9$ MeV, (b) $W = 5.7 \times 10^{14}$ J $= 1.6 \times 10^8$ kWh

7.7 The radium isotope ${}^{226}_{88}$Ra decays into radon ${}^{222}_{86}$Rn emitting an α particle, ${}^{226}_{88}\mathrm{Ra} \to {}^{222}_{86}\mathrm{Rn} + {}^{4}_{2}\mathrm{He}$. How much energy is released in this decay? The atomic masses $M({}^{226}_{88}\mathrm{Ra}) = 226.02541$ u, $M({}^{222}_{86}\mathrm{Rn}) = 222.01758$ u and $M({}^{4}_{2}\mathrm{He}) =$ 4.00260 u are given. Ans.: $\Delta E = 4.87$ MeV

7.8 The β decay of ${}^{55}_{24}$Cr takes place according to ${}^{55}_{24}\mathrm{Cr} \to {}^{55}_{25}\mathrm{Mn}^{+} + \mathrm{e}^{-} + \bar{\nu}_{\mathrm{e}}$. What kinetic energy is given to the electron and the neutrino if the two nuclei are considered stationary? The atomic masses are given as $M_{\text{Cr-55}} = 54.940\,840$ u and $M_{\text{Mn-55}} = 54.938\,045$ u. By comparison, the mass of the neutrino is negligible, if not zero. Ans.: $\Delta E = 2.6$ MeV

7.9 (a) What is the binding energy per nucleon of the isotope ${}^{12}_{6}$C? (b) A ${}^{12}_{6}$C nucleus is to be split into three ${}^{4}_{2}$He nuclei. How much energy is needed for this? It is given that the mass of the ${}^{12}_{6}$C atom is $M_C = 12$ u exactly, of the atom of ${}^{4}_{2}$He $M_{He} = 4.002603$ u, of the atom of hydrogen $M_H = 1.007825$ u and of the neutron $M_n = 1.008665$ u. Ans.: (a) 7.68 MeV/nucleon, (b) Energy 7.27 MeV is absorbed

7.10 What mass of ${}^{235}_{92}$U must undergo fission for the production of thermal energy of 1 GW for a whole day? During the fission of a ${}^{235}_{92}$U nucleus, an energy of about 220 MeV is released. Ans.: $M = 0.976$ kg

7.11 In the *carbon cycle* or the *Bethe cycle*, energy is produced in the stars by the following reactions:

$$
\begin{aligned}
&{}^{12}_{6}\mathrm{C} + \mathrm{p} \to {}^{13}_{7}\mathrm{N} \to {}^{13}_{6}\mathrm{C} + \mathrm{e}^{+} + \nu_{\mathrm{e}} \\
&{}^{13}_{6}\mathrm{C} + \mathrm{p} \to {}^{14}_{7}\mathrm{N} \\
&{}^{14}_{7}\mathrm{N} + \mathrm{p} \to {}^{15}_{8}\mathrm{O} \to {}^{15}_{7}\mathrm{N} + \mathrm{e}^{+} + \nu_{\mathrm{e}} \\
&{}^{15}_{7}\mathrm{N} + \mathrm{p} \to {}^{12}_{6}\mathrm{C} + {}^{4}_{2}\mathrm{He}
\end{aligned}
$$

Find the total energy released in the cycle. Assume that no orbital electrons are present. Ans.: 24.7 MeV

7.4.2 *The Experimental Test of the Equivalence of Mass and Energy in Nuclear Reactions*

The development of mass spectrographs, which led to the accurate determination of atomic masses, made it possible to test experimentally the equivalence of mass and energy and their conservation during nuclear reactions. One of the first examples was the experiment of Cockcroft and Walton of 1932 [3]. In it, a lithium target was bombarded with fast protons, resulting in the formation of an excited nucleus, which immediately splitted in two α particles, according to

$$^{7}_{3}\mathrm{Li} + {}^{1}_{1}\mathrm{H} \rightarrow \left({}^{8}_{4}\mathrm{Be}^{*}\right) \rightarrow 2{}^{4}_{2}\mathrm{He}$$

We will consider the nuclei of the atoms in the reaction as having all their orbital electrons, so that we can use the masses of neutral atoms which are given in the tables. As many electrons actually missing from the left-hand-side of the equation (i.e. one), will also be missing from the products of the reaction. The masses of the atoms involved in this reaction are, to 4 decimal digits available at the time:

$$M\left({}^{7}_{3}\mathrm{Li}\right) = 7.0160\ \mathrm{u},\quad M\left({}^{1}_{1}\mathrm{H}\right) = 1.0078\ \mathrm{u},\quad M\left({}^{4}_{2}\mathrm{Be}\right) = 8.0053\ \mathrm{u},\quad M\left({}^{4}_{2}\mathrm{He}\right) = 4.0026\ \mathrm{u}$$

The original mass is equal to 8.0238 u, while that of the de-excited Be atom produced is smaller by 0.0185 u. The excitation energy of this nucleus is, therefore, equal to $\Delta E = (0.0185\ \mathrm{u}) \times (931.494\ \mathrm{MeV/u}) = 17.23\ \mathrm{MeV}$. The two helium atoms have a total mass of 8.0052 u, which is smaller than the original mass by 0.0186 u. This is equivalent to an energy of $Q = (0.0186\ \mathrm{u}) \times\ (931.494\ \mathrm{MeV/u}) = 17.33$ MeV, which is given as kinetic energy to the α particles produced.

The kinetic energies of the α particles could be determined from their known curves of energy-range in air. Accurate measurements performed later (1939) by Smith [4], gave the difference between the kinetic energies of the α particles and the impinging protons as $\Delta K = 17.28 \pm 0.03$ MeV. The agreement of experiment with theory is very good and may be considered as verifying the equivalence of mass and energy to an accuracy of better than 1 %. More recent measurements have given even better agreement.

In 1944, Dushman [5] compared the theoretical prediction for the liberated energy to the measured one, for some then well known nuclear reactions. The results are presented in Table 7.3. Again, the agreement between experiment and theory was very good.

Table 7.3 Dashman's comparison of the theoretical to the measured energies liberated during nuclear reactions

Nuclear reaction	Mass defect ΔM (u)	Liberated energy	
		Theoretical ΔMc^2 (MeV)	Experimental Q (MeV)
${}^{9}_{4}\mathrm{Be} + {}^{1}_{1}\mathrm{H} \rightarrow {}^{6}_{3}\mathrm{Li} + {}^{4}_{2}\mathrm{He}$	0.00242	2.25	2.28
${}^{6}_{3}\mathrm{Li} + {}^{2}_{1}\mathrm{H} \rightarrow {}^{4}_{2}\mathrm{He} + {}^{4}_{2}\mathrm{He}$	0.02381	22.17	22.20
${}^{10}_{5}\mathrm{B} + {}^{2}_{1}\mathrm{H} \rightarrow {}^{11}_{6}\mathrm{C} + {}^{1}_{0}\mathrm{n}$	0.00685	7.38	7.08
${}^{14}_{7}\mathrm{N} + {}^{2}_{1}\mathrm{H} \rightarrow {}^{12}_{6}\mathrm{C} + {}^{4}_{2}\mathrm{He}$	0.01436	13.37	13.40
${}^{14}_{7}\mathrm{N} + {}^{4}_{2}\mathrm{He} \rightarrow {}^{17}_{8}\mathrm{O} + {}^{1}_{1}\mathrm{H}$	–0.00124	–1.15	–1.16
${}^{28}_{14}\mathrm{Si} + {}^{4}_{2}\mathrm{He} \rightarrow {}^{31}_{15}\mathrm{P} + {}^{1}_{1}\mathrm{H}$	–0.00242	–2.25	–2.23

7.5 Threshold Energy

The threshold energy of a nuclear reaction is the lowest possible energy that must be available for the reaction to take place. Naturally, a reaction may be allowed by the conservation of mass-energy, but be forbidden to happen for other reasons. For example, the capture of a proton by a nucleus for the production of another nucleus may be an exoergic reaction and thus allowed purely on the grounds of the examination of the initial masses and that of the final nucleus. The reaction may be, however, impossible to happen if the proton does not have at least certain energy. The reason is the existence of a *Coulomb barrier* which the proton must overcome if it is to enter the nucleus. Being positive, it is repelled by the also positive nucleus until it gets close enough to it for the strong attractive nuclear forces to come into action. The nucleus produced is in an excited state and is de-excited in some manner. For example, in the reaction

$$^{12}\text{C} + {^1\text{p}} \to (^{13}\text{N}^*) \to {^{13}\text{N}} + \gamma$$

the ^{12}C atom together with the proton have the rest mass necessary to create the ^{13}N plus 1.43 MeV. This does not mean that the reaction can take place with zero proton energy. The proton has to overcome a Coulomb barrier of a height equal to 2.70 MeV in order to enter the ^{12}C nucleus and this is the minimum kinetic energy the proton must have for the reaction to take place. With the proton incorporated, the ^{12}C nucleus is transmuted to a ^{13}N nucleus but in an excited state (symbolized by ^{13}N*). Part of the excess energy is given to the nucleus as recoil energy and the rest is expelled via the emission of a γ ray photon. The nucleus finally settles in its ground energy state. The total energy available after the reaction is at least equal to 2.70 + 1.43 = 4.13 MeV. On the other hand, for the reaction

$$^{16}_{8}\text{O} + {^1_1\text{p}} \to {^{15}_{8}\text{O}} + {^2_1\text{H}^+}$$

the threshold energy is 12.9 MeV and the Coulomb barrier is 1.38 MeV. If the threshold energy is available, the proton can easily overcome the Coulomb barrier. Naturally, a Coulomb barrier only exists, in nuclei, for positively charged incident particles.

The threshold energy resulting solely from the requirements of the conservation laws is evaluated taking into account the conservation of energy and momentum in a reaction. For example, in the reaction

$$\gamma + \text{p} \to \pi^0 + \text{p}$$

by which a photon collides with an initially stationary proton for the production of a π^0 particle and a proton, the photon must not only have enough energy equivalent to the mass of the π^0 particle, i.e. 135 MeV. In order for momentum to be conserved, the products of the reaction must be moving and, therefore, they must have some kinetic energy. It turns out (Example 7.5) that the photon must have an energy of at least 150 MeV if it is to create a π^0 and give enough energy to the p and π^0 for momentum to be conserved. This energy is called *threshold energy for the production of a π^0 particle during the collision of a photon with a proton at rest.*

The arguments are significantly simplified if the whole analysis is performed in the zero-momentum frame of reference of the reacting particles. In this frame, the total momentum must be zero. This condition is satisfied also in the case of the products of a reaction being at rest. As a consequence, the energy available is maximum in this case. In the zero-momentum frame of reference, if the products of a reaction are stationary, we can be sure that the available energy just managed to create them. This will, therefore, be the threshold energy for the particular reaction in this frame of reference. The conclusion reached is that, in the laboratory frame of reference, if the available energy is just equal to the threshold energy, the products of the reaction must all move with the same velocity. Summarizing:

1. *In the zero-momentum frame of reference,* when the available energy is just equal to the threshold energy, the products of the reaction will be stationary. Consequently,
2. *In the laboratory frame of reference*, when the available energy is just equal to the threshold energy, the products of the reaction will all be moving with the same velocity.

These ideas are clarified further in the following examples and problems.

Example 7.4 Threshold Energy for the Production of an Antiproton
During the fall of a proton on another proton, at rest, show that the threshold energy (minimum kinetic energy) for the reaction of the production of an additional proton-antiproton pair, $\mathrm{p}+\mathrm{p} \to \mathrm{p}+\mathrm{p}+\mathrm{p}+\bar{\mathrm{p}}$, is equal to $6m_\mathrm{p}c^2$, where m_p is the rest mass of the proton and of the antiproton.

We will solve the problem in two ways. First in the zero-momentum frame of reference (ZMFR) and then in the laboratory frame of reference (LFR).

(a) In the zero-momentum frame of reference

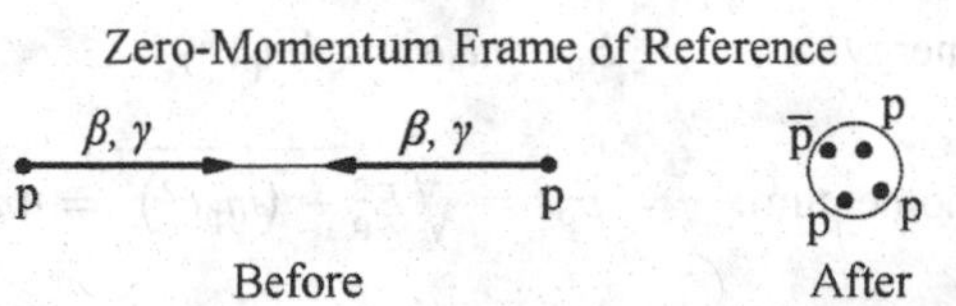

In the zero-momentum frame of reference the two initial protons have equal and opposite momenta and, therefore, equal and opposite velocities, to which there corresponds the same Lorentz factor, γ. If the available energy is just the threshold energy, the particles produced after the collision will all be at rest in the ZMFR.

The conservation of energy gives:

$$2m_\mathrm{p}c^2\gamma = 4m_\mathrm{p}c^2.$$

Therefore, $\gamma = 2$ and $|\beta| = \sqrt{1-1/\gamma^2} = \sqrt{1-1/4} = \sqrt{3}/2$.

Since the proton which was initially at rest in the LFR has in the ZMFR a speed equal to $-(\sqrt{3}/2)c$, then the speed of the LFR relative to the ZMFR is $V = -(\sqrt{3}/2)c$. Transforming the speed of the proton which moves in the LFR and has in the ZMFR a speed equal to $\upsilon = (\sqrt{3}/2)c$, from the ZMFR to the LFR, we get

$$\upsilon' = \frac{\upsilon - V}{1-(\upsilon V)/c^2} = c\frac{\sqrt{3}/2 - (-\sqrt{3}/2)}{1+(\sqrt{3}/2)^2} = c\frac{4\sqrt{3}}{7} \quad \Rightarrow \quad \beta' = \frac{4\sqrt{3}}{7}.$$

In the LFR, the Lorentz factor of the moving proton is, therefore,

$$\gamma' = \frac{1}{\sqrt{1-(4\sqrt{3}/7)^2}} = \frac{7}{\sqrt{49-48}} = 7.$$

The energy of the impinging proton is $E_\mathrm{p} = m_\mathrm{p}c^2\gamma' = 7m_\mathrm{p}c^2$ and its kinetic energy, or the threshold energy, is $K_\mathrm{p} = 6m_\mathrm{p}c^2$.

(b) In the laboratory frame of reference

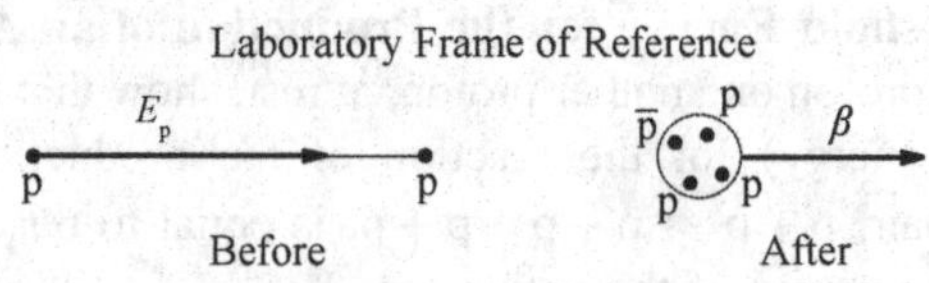

For the available energy being equal to the threshold energy, the particles produced must be at rest in the ZMFR. In the LFR, therefore, all the particles produced move with the same velocity, say βc, as shown in the figure.

Thus, we have,

conservation of energy: $$E_\mathrm{p} + m_\mathrm{p}c^2 = 4m_\mathrm{p}c^2\gamma \qquad (1)$$

conservation of momentum: $$p_\mathrm{p}c = \sqrt{E_\mathrm{p}^2 - (m_\mathrm{p}c^2)^2} = 4m_\mathrm{p}c^2\beta\gamma \qquad (2)$$

where E_p and p_p are the energy and the momentum, respectively, of the impinging proton.

Equation (1) gives

$$E_p^2 = (m_p c^2)^2 (4\gamma - 1)^2$$

and Eq. (2)

$$E_p^2 = 16(m_p c^2)^2 \beta^2 \gamma^2 + (m_p c^2)^2.$$

Equating, $$(m_p c^2)^2 (4\gamma - 1)^2 = 16(m_p c^2)^2 \beta^2 \gamma^2 + (m_p c^2)^2$$

and

$$(4\gamma - 1)^2 = 16\beta^2\gamma^2 + 1, \quad 16\gamma^2 - 8\gamma + \not{1} = 16\beta^2\gamma^2 + \not{1}, \quad \gamma = 2\gamma^2(1 - \beta^2), \quad \gamma = 2.$$

From Eq. (1) it follows that

$$E_p = 7m_p c^2$$

and the threshold (kinetic) energy is $K_p = 6m_p c^2$.

Example 7.5 Threshold Energy for the Production of a π^0 by the Reaction $\gamma + p \rightarrow p + \pi^0$

What is the threshold energy for the production of a pion during the collision of a photon with a proton at rest, $\gamma + p \rightarrow p + \pi^0$? The rest masses of the particles p and π^0 are 938 and 140 MeV/c^2, respectively.

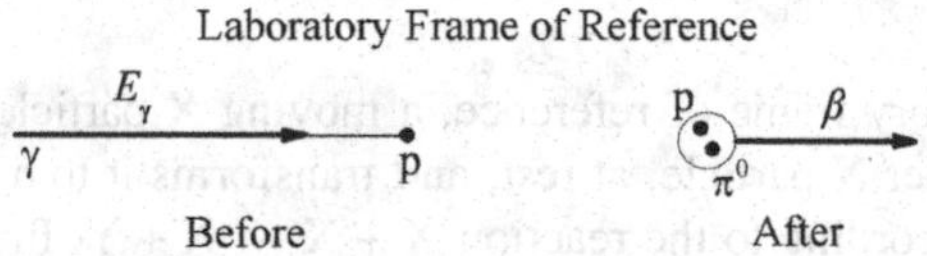

If the energy available is equal to the threshold energy, in the LFR all the particles produced will move with the same velocity, say βc, as shown in the figure.

We therefore have,

conservation of energy: $$E_\gamma + m_p c^2 = (m_p c^2 + m_\pi c^2)\gamma \quad (1)$$

conservation of momentum: $$\frac{E_\gamma}{c} = \frac{m_p c^2 + m_\pi c^2}{c}\beta\gamma, \quad (2)$$

where E_γ is the energy of the impinging photon.

From these two equations, it follows that

$$E_\gamma = (m_p c^2 + m_\pi c^2)\gamma - m_p c^2 = (m_p c^2 + m_\pi c^2)\beta\gamma$$

or

$$(m_{\mathrm{p}}c^2 + m_{\pi}c^2)\gamma(1-\beta) = m_{\mathrm{p}}c^2 \quad \Rightarrow \quad \left(1+\frac{m_{\pi}}{m_{\mathrm{p}}}\right)\sqrt{\frac{1-\beta}{1+\beta}} = 1.$$

Putting $\lambda \equiv 1+\dfrac{m_{\pi}}{m_{\mathrm{p}}}$ we find $\sqrt{\dfrac{1+\beta}{1-\beta}} = \lambda \quad \Rightarrow \quad \dfrac{1+\beta}{1-\beta} = \lambda^2 \quad \Rightarrow \quad \beta = \dfrac{\lambda^2-1}{\lambda^2+1}.$

Dividing Eq. (2) by Eq. (1), we get $\beta = \dfrac{E_{\gamma}}{E_{\gamma}+m_{\mathrm{p}}c^2}$.

Therefore, $E_{\gamma} = \dfrac{\beta m_{\mathrm{p}}c^2}{1-\beta}$ and $E_{\gamma} = \dfrac{m_{\mathrm{p}}c^2}{1/\beta - 1} = \dfrac{m_{\mathrm{p}}c^2}{\frac{\lambda^2+1}{\lambda^2-1}-1} = \dfrac{m_{\mathrm{p}}c^2}{2}(\lambda^2-1).$

So, $$E_{\gamma} = \frac{c^2}{2}m_{\mathrm{p}}\left[2\frac{m_{\pi}}{m_{\mathrm{p}}} + \left(\frac{m_{\pi}}{m_{\mathrm{p}}}\right)^2\right] \quad \Rightarrow \quad E_{\gamma} = m_{\pi}c^2\left(1+\frac{m_{\pi}}{2m_{\mathrm{p}}}\right).$$

Substituting,

$$E_{\gamma} = 140 \times \left(1+\frac{140}{2\times 938}\right) = 150\ \text{MeV}.$$

This is the the threshold energy required. Since the rest energy of the π^0 particle is 140 MeV, the remaining energy of 10 MeV is given as kinetic energy to the products of the reaction.

Problems

7.12 In the laboratory frame of reference, a moving X particle (with rest mass m), collides with another X particle, at rest, and transforms it to a Y particle (with rest mass $M = 3m$), according to the reaction $\mathrm{X}+\mathrm{X} \to \mathrm{X}+\mathrm{Y}$. In the laboratory frame of reference, how much is the threshold energy (kinetic energy) of the moving X particle for this to happen? Ans.: $K = 6mc^2$

7.13 What is the threshold energy for the production of an electron-positron pair (e^-, e^+) in the collision of a photon, γ, with a stationary electron, according to the reaction $\gamma + \mathrm{e}^- \to \mathrm{e}^- + \mathrm{e}^- + \mathrm{e}^+$? For the e^- and e^+, it is given that $m_{\mathrm{e}}c^2 = 0.511$ MeV. Ans.: 2.044 MeV

7.14 Which is the threshold energy for the production of a proton-antiproton pair (p, $\overline{\mathrm{p}}$) during the collision of an electron with a proton at rest, according to the reaction $\mathrm{e}+\mathrm{p} \to \mathrm{e}+\mathrm{p}+\mathrm{p}+\overline{\mathrm{p}}$? The rest energies $m_{\mathrm{e}}c^2 = 0.511$ MeV and $m_{\mathrm{p}}c^2 =$ 938 MeV are given. Ans.: 3.75 GeV

7.15 Show that, in the laboratory frame of reference, the threshold (kinetic) energy for the production of n pions in the collision of protons with a hydrogen target,

$\mathrm{p+p \to p+p}+n\pi$, is $K = 2nm_{\pi}c^2(1+nm_{\pi}/4m_{\mathrm{p}})$, where m_{π} and m_{p} are, respectively, the rest masses of the pion and the proton. Which is the threshold energy for the reactions $\mathrm{p+p \to p+p+\pi}$ and $\mathrm{p+p \to p+p+2\pi}$, if $m_{\mathrm{p}}c^2 =$ 938 MeV and $m_{\pi}c^2 = 140$ MeV? Ans.: $K_{\mathrm{p1}} = 290$ MeV, $K_{\mathrm{p2}} = 602$ MeV

7.16 Evaluate the threshold energy for the reaction $\pi^- + \mathrm{p} \to \Xi^- + \mathrm{K}^+ + \mathrm{K}^0$, during the incidence of a pion on a proton at rest. The rest energies of the particles involved, π^-, p, Ξ^-, K^+ and K^0, are 140, 938, 1321, 494 and 498 MeV, respectively. Ans.: $K_{\pi} = 2.23$ GeV

7.17 Find the threshold energy for the general case $\mathrm{A+B \to X_1+X_2+\ldots+X}_i+\ldots+\mathrm{X}_N$, in which a particle A falls on a particle B at rest, producing a number N of different particles X_i. The rest masses of the particles are $M_{\mathrm{A}}, M_{\mathrm{B}}, M_1, \ldots, M_i, \ldots, M_N$, respectively. Ans.: $E_{\mathrm{T}} = \dfrac{M^2-(M_{\mathrm{A}}+M_{\mathrm{B}})^2}{2M_{\mathrm{B}}}c^2$, where $M = \sum_i M_i$

7.6 The General Equations for the Motion of a Relativistic Rocket

We will examine, on the basis of the Special Theory of Relativity, the general problem of a rocket which is propelled by ejecting mass. We will assume that the initial rest mass of the rocket is M_0 and that the gases it ejects backwards have a speed of $-V_0$ relative to the rocket itself. We will consider that at any moment in time the rocket is at rest in an inertial frame of reference, which we will call the frame of reference of the rocket at that particular moment. At some moment, the rocket has rest mass equal to M. In an infinitesimal time interval, the rest mass of the rocket changes from M to $M+dM$. Naturally, the quantity dM is negative here. The mass $-dM$, which is ejected, moves with a speed $-V_0$ in the frame of reference of the rocket, as shown in Fig. 7.10.

At that moment, the rocket moves with a speed V relative to the frame of reference of the Earth, in which the rocket was initially at rest. The mass of the rocket will, therefore, be equal to

$$M' = \frac{M}{\sqrt{1-\dfrac{V^2}{c^2}}}. \tag{7.36}$$

In the same frame of reference, the speed of the gases being ejected is

$$V_0' = \frac{-V_0+V}{1-\dfrac{VV_0}{c^2}}. \tag{7.37}$$

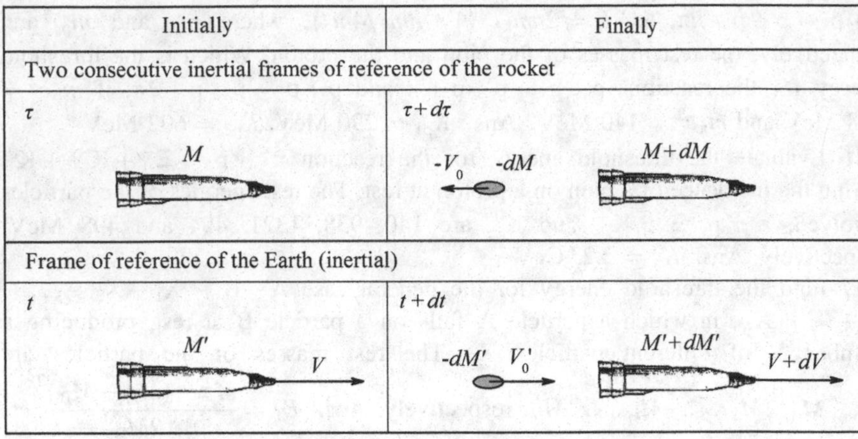

Fig. 7.10 The acceleration of a rocket by ejection of mass. The process is examined in two consecutive inertial frames of reference of the rocket, with an infinitesimal time difference between them, and in the frame of reference of the Earth. The figures show the masses and the velocities in these two frames of reference at two successive moments of time, τ and $\tau + d\tau$, and t and $t + dt$, respectively

The whole system is isolated and, therefore, its momentum remains constant during the ejection of mass $-dM$. The conservation of momentum in the frame of reference of the Earth gives

$$M'V = (V + dV)(M' + dM') + V_0'(-dM') \tag{7.38}$$

or

$$\cancel{M'V} = \cancel{M'V} + M'dV + VdM' + dVdM' - V_0'dM'. \tag{7.39}$$

Since the term $dVdM'$ may be ignored by comparison to the other differentials, it follows that

$$M'dV = -(V - V_0')dM' \tag{7.40}$$

and, finally,
$$\frac{dM'}{M'} = -\frac{dV}{V - V_0'}. \tag{7.41}$$

Substituting in Eq. (7.41) the quantities $V - V_0' = V_0 \dfrac{1 - V^2/c^2}{1 - VV_0/c^2}$ [from Eq. (7.37)] and $\dfrac{dM'}{M'} = \dfrac{dM}{M} + \dfrac{V/c^2}{1 - V^2/c^2} dV$ [from Eq. (7.36)], it follows that

$$\frac{dM}{M} + \left(\frac{V/c^2}{1 - V^2/c^2}\right) dV = -\left(\frac{1 - VV_0/c^2}{1 - V^2/c^2}\right) \frac{dV}{V_0}. \tag{7.42}$$

By rearranging terms, we have

$$\frac{dM}{M} = -\frac{dV}{V_0(1 - V^2/c^2)}, \tag{7.43}$$

which is integrated, with the condition that $M = M_0$ for $V = 0$, to give

$$\ln\left(\frac{M}{M_0}\right) = \frac{c}{2V_0}\ln\left(\frac{1 - V/c}{1 + V/c}\right) \quad \text{or} \quad \left(\frac{M}{M_0}\right)^{2V_0/c} = \frac{1 - V/c}{1 + V/c}. \tag{7.44}$$

Finally,

$$\frac{V}{c} = \frac{1 - (M/M_0)^{2V_0/c}}{1 + (M/M_0)^{2V_0/c}}. \tag{7.45}$$

We notice that, in contrast to the non-relativistic case, when all the mass of the rocket has been exhausted, the speed of the rocket (or, better, of its infinitesimal remnants!) is not infinite, but it is equal to c.

The Lorentz factor corresponding to the speed found is

$$\gamma = \frac{1}{2}\left[\left(\frac{M}{M_0}\right)^{-V_0/c} + \left(\frac{M}{M_0}\right)^{V_0/c}\right]. \tag{7.46}$$

If we define the parameters $\alpha = V_0/c$ and $\mu = M/M_0$, the two last equations are also written as

$$\beta = \frac{1 - \mu^{2\alpha}}{1 + \mu^{2\alpha}} \quad \text{and} \quad \gamma = \frac{1}{2}(\mu^{-\alpha} + \mu^{\alpha}). \tag{7.47}$$

The variation of the reduced speed of the rocket, $\beta = V/c$, as a function of M/M_0, is shown in Fig. 7.11, for various values of the parameter $\alpha = V_0/c$.

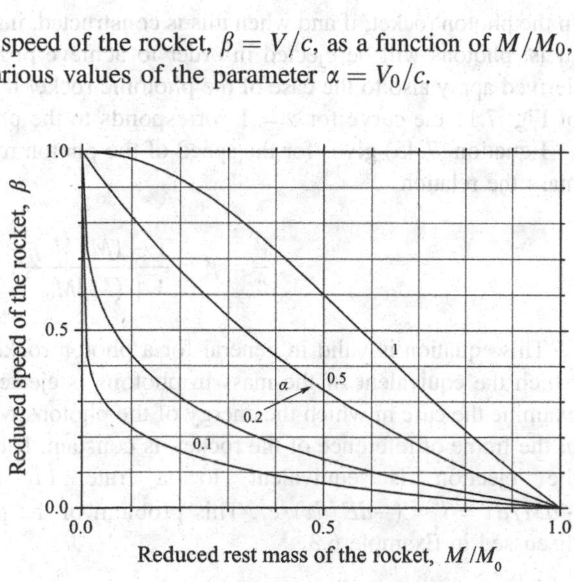

Fig. 7.11 The variation of the reduced speed of the rocket, β, for various values of the parameter $\alpha = V_0/c$, as a function of the ratio M/M_0 of the rest mass of the rocket to its initial rest mass

Equation (7.45) gives the speed attained by the rocket in the frame of reference of the Earth, as a function of its (rest) mass in its own momentary inertial frame of reference, independently of the rate of ejection of mass, which could possibly be varying with time. The relativistic mass of the rocket in the frame of the Earth, M', is given by the relation

$$\frac{M'}{M_0} = \gamma\frac{M}{M_0} = \frac{1}{2}\left(\mu^{1-\alpha} + \mu^{1+\alpha}\right). \tag{7.48}$$

We notice that as $\mu \to 0$ the mass M' of the rocket, as measured by an observer on the Earth, tends to zero in all cases except in the case when it is $\alpha = 1$ (photon rocket), when it tends to the value of $M' = M_0/2$!

Problems

7.18 For the case of the rocket which expels mass in its own frame of reference at a constant rate equal to $dM/d\tau = -k$, find the rocket's acceleration in the Earth's frame of reference. Express the result in terms of the variable $\mu = M/M_0$, where M and M_0 are the instantaneous and the initial rest mass of the rocket, respectively.

Ans.: $\dfrac{d(V/c)}{d(t/\tau_0)} = \dfrac{8\alpha}{\mu(\mu^{\alpha} + \mu^{-\alpha})^3}$, where $\tau_0 = \dfrac{M_0}{k}$

7.19 Examine the motion of a rocket which moves with an exponential decrease of mass in its own frame of reference.

7.6.1 The Photon Rocket

In the photon rocket, if and when this is constructed, instead of gases from chemical fuels, photons will be ejected in order to achieve propulsion. The results already derived apply also to the case of the photonic rocket if we place $V_0 = c$. In the plot of Fig. 7.11 the curve for $\alpha = 1$ corresponds to the photon rocket.

Equation (7.45) gives for the speed of the photon rocket as a function of its rest mass the relation

$$\frac{V}{c} = \beta = \frac{1 - (M/M_0)^2}{1 + (M/M_0)^2}. \tag{7.49}$$

This equation is valid in general for a photon rocket, irrespective of the rate at which the equivalent of the mass in photons is ejected. In what follows, we will examine the case in which the energy of the photons which is ejected per unit time, in the frame of reference of the rocket, is constant. If this rate is equal to $-dE/d\tau$, the ejection is equivalent to a rate of mass ejection equal to $-dM/d\tau = k = (-dE/d\tau)/c^2$. This problem of the photon rocket has also been discussed in Example 6.3.

The relationship between time in the frame of reference of the Earth and the rest mass of the rocket. The relationship between the time as this is measured in the frame of reference of the Earth, t, and the proper time of the rocket τ, is $dt = \gamma d\tau$. Substituting for γ from Eq. (7.46), with $V_0 = c$, we have

$$dt = -\frac{1}{2k}\left(\frac{M}{M_0} + \frac{M_0}{M}\right) dM. \tag{7.50}$$

This relation may be integrated, with the condition $M = M_0$ for $t = 0$, and with $\tau_0 = M_0/k$, to give

$$\frac{t}{\tau_0} = \frac{1}{4}\left(1 - \frac{M^2}{M_0^2}\right) - \frac{1}{2}\ln\left(\frac{M}{M_0}\right). \tag{7.51}$$

We notice that for $M \to 0$ we have $t \to \infty$. The finite time τ_0 which is required in the frame of the rocket for all its mass to be ejected, is infinite in the frame of the Earth.

From Eq. (7.49) it follows that it is $\dfrac{M}{M_0} = \sqrt{\dfrac{1-\beta}{1+\beta}}$. This combines with Eq. (7.51) to give

$$\frac{t}{\tau_0} = \frac{\beta}{2(1+\beta)} - \frac{1}{4}\ln\left(\frac{1-\beta}{1+\beta}\right). \tag{7.52}$$

We notice that $\beta \to 1$ for $t \to \infty$ or that an infinite time is needed for the rocket to reach a speed equal to c.

The relativistic mass of the rocket in the frame of reference of the Earth. By analogy with the reduced mass of the rocket in its own frame of reference, $\mu = M/M_0$, we define the reduced relativistic mass in the frame of reference of the Earth as $\mu' = M'/M_0$. Because it is $M' = \gamma M$ and from Eq. (7.47) with $\alpha = 1$ it follows that $\gamma = \dfrac{1}{2}\left(\dfrac{1}{\mu} + \mu\right)$, we will have

$$\mu' = \frac{M'}{M_0} = \gamma \frac{M}{M_0} = \frac{1}{2}(1 + \mu^2). \tag{7.53}$$

We notice the interesting result that for $\mu \to 0$, it is $\mu' \to \frac{1}{2}$ or $M' = M_0/2$. In other words, as the speed of the rocket tends to c, the rest mass of the rocket tends to zero but its relativistic mass in the frame of the Earth tends to the value $M' = M_0/2$. Essentially, we end up with a particle with zero rest mass and energy $M_0c^2/2$. The only disadvantage in using this mechanism to create very energetic zero rest mass particles is that, in the frame of reference of the Earth, the process needs an infinite time to be completed.

The displacement of the rocket as a function of its speed. If we define the *Doppler factor* as $D = \sqrt{\frac{1-\beta}{1+\beta}}$, the relations $\beta = \frac{1-D^2}{1+D^2}$ and $M = M_0 D$ hold. Therefore,

$$dx = Vdt = c\beta dt = c\beta\gamma d\tau = -\frac{cM_0}{2k}\left(\frac{1-D^2}{1+D^2}\right)\left(\frac{1}{D}+D\right)dD = \frac{cM_0}{2k}\left(D-\frac{1}{D}\right)dD, \tag{7.54}$$

which may be integrated with the condition that it is $x = 0$ for $t = 0$ (when it is also $\beta = 0$, $D = 1$) to give

$$\frac{x}{x_0} = -\frac{\beta}{2(1+\beta)} - \frac{1}{4}\ln\left(\frac{1-\beta}{1+\beta}\right), \tag{7.55}$$

where $x_0 = cM_0/k$.

The Doppler shift of the photons of the rocket, for an observer on the Earth. If the photons emitted by the rocket have a wavelength of λ_0 in its own frame of reference, as it moves away from the Earth with a speed of $V = c\beta$, the photons will be seen by an observer on the Earth to have a wavelength equal to

$$\lambda = \lambda_0\sqrt{\frac{1+\beta}{1-\beta}} = \lambda_0\frac{M_0}{M} = \frac{\lambda_0\tau_0}{\tau_0-\tau}. \tag{7.56}$$

We notice that the wavelength of the photons increases with time and as the rocket is accelerated. Thus, due to the dilation of time, the observer on the Earth sees the number of the photons emitted per unit time to grow smaller and smaller, while, due to the Doppler effect, the photons have momentum which is gradually reduced. This, together with the fact that the relativistic mass of the rocket tends to the constant value of $M' = M_0/2$, make the acceleration of the rocket in the frame of reference of the Earth to tend to zero as the rocket's speed tends to c.

References

1. A.H. Compton, Bulletin Nat. Res. Council, No.20, 16 (1922); Phys. Rev. **21**, 715 and **22**, 409 (1922)
2. J.D. Barrow, F.J. Tipler, *The Anthropic Cosmological Principle* (Clarendon Press, Oxford, 1986) or C.P.W. Davies, *The Accidental Universe* (C.U.P., 1982)
3. J.D. Cockcroft, G.T.S. Walton, Proc. R. Soc. A **137**, 229 (1932)
4. N.M. Smith Jr, Phys. Rev. **56**, 548 (1939)
5. S. Dushman, General Electric Review **47**, 6–13 (1944)

Chapter 8
Minkowski's Spacetime and Four-Vectors

8.1 The 'World' of Minkowski

In three-dimensional Euclidean space, the square of the distance between the two points (x_1, y_1, z_1) and (x_2, y_2, z_2) is given by the Pythagorean theorem as

$$s^2 = (x_2 - x_1)^2 + (y_2 - y_1)^2 + (z_2 - z_1)^2. \tag{8.1}$$

For infinitesimal variations in the coordinates of the two points, this relation is written as

$$ds^2 = dx^2 + dy^2 + dz^2, \tag{8.2}$$

where ds^2 stands for $(ds)^2$ and not $d(s^2)$, dx^2 for $(dx)^2$ etc. The transformation $(x, y, z) \Rightarrow (x', y', z')$ from the coordinate system (x, y, z) to any other coordinate system (x', y', z'), in which relative displacements and rotations of the axes are involved, leave the form of ds^2 invariant. By analogy, in a four-dimensional space with mutually perpendicular axes (x, y, z, w) the corresponding relation for ds^2 would be:

$$ds^2 = dx^2 + dy^2 + dz^2 + dw^2. \tag{8.3}$$

In the Special Theory of Relativity, the place of points in space are taken by points in space *and* time. Instead of geometrical points we have events, which are determined by three spatial coordinates and one temporal, (x, y, z, t). We are tempted, after we change time to length by multiplying t by c, to define $w = ct$ as the fourth dimension, in which case we would expect the infinitesimal distance between two geometric points

© Springer International Publishing Switzerland 2016

C. Christodoulides, *The Special Theory of Relativity*,
Undergraduate Lecture Notes in Physics, DOI 10.1007/978-3-319-25274-2_8

$$ds = \sqrt{dx^2 + dy^2 + dz^2}, \tag{8.4}$$

to be replaced by the 'distance' between two events,

$$ds = \sqrt{dx^2 + dy^2 + dz^2 + c^2dt^2}. \tag{8.5}$$

This quantity, however, does not remain invariant under the Lorentz transformation, i.e. in going from one inertial frame of reference to another. On the other hand, it is easily seen that the quantity which does remain invariant under the Lorentz transformation is

$$ds = \sqrt{dx^2 + dy^2 + dz^2 - c^2dt^2}. \tag{8.6}$$

In this case, in Eq. (8.4), it is $w \equiv ict$, an imaginary variable, and the square of the magnitude of the infinitesimal distance between two events, ds^2, may be positive, zero or even negative. This fact shows us that time is of a different nature than the three spatial coordinates. This notation is adopted in a significant proportion of the literature on the Special Theory of Relativity and is particularly convenient in the analysis of electromagnetism. An alternative solution, in order to avoid the use of an imaginary coordinate, is to take $w \equiv ct$ and *define* the square of the magnitude of the infinitesimal distance between two events as

$$ds^2 = c^2dt^2 - dx^2 - dy^2 - dz^2. \tag{8.7}$$

This notation (namely the choice of the signs $+, -, -, -$) is useful in the General Theory of Relativity. For uniformity, we define the coordinates x^μ $(\mu = 0, 1, 2, 3)$ as

$$x^0 \equiv ct, \quad x^1 \equiv x, \quad x^2 \equiv y, \quad x^3 \equiv z, \tag{8.8}$$

where the numbers 0, 1, 2 and 3 are upper indices and not exponents. Equation (8.7) is then written as

$$ds^2 = (dx^0)^2 - (dx^1)^2 - (dx^2)^2 - (dx^3)^2. \tag{8.9}$$

In the General Theory of Relativity, the expression for ds^2 is more complex, and is written as

$$ds^2 = \sum_{\mu=0}^{3}\sum_{\nu=0}^{3} g_{\mu\nu}dx^\mu dx^\nu \quad \text{or} \quad ds^2 = g_{\mu\nu}dx^\mu dx^\nu, \tag{8.10}$$

where we adopted Einstein's suggestion for the notation, i.e. that an index appearing up and down in one side of an equation implies summation with respect to this index for all its possible values. The coefficients $g_{\mu\nu}$ $(\mu, \nu = 0, 1, 2, 3)$ are called *metric coefficients* or *components* and together they form the *metric tensor* $g_{\mu\nu}$. In the General Theory of Relativity, these coefficients in general depend on the curvature of space, which is due to gravitational fields. In the Special Theory of Relativity, it is

$$g_{00} = 1, \quad g_{11} = g_{22} = g_{33} = -1, \tag{8.11}$$

with all the other coefficients equal to zero. The space or, better, spacetime defined in this manner is called *pseudo-Euclidean*: 'Euclidean' because the space is 'flat' and 'pseudo-' because the distance between two points in this space is given by the modified form of the Pythagorean theorem given by Eq. (8.9).

Minkowski (Hermann Minkowski, 1864-1909) proposed in 1908, a very illustrative way of representing this space, which he called *World*. An event is represented by a point in the four-dimensional spacetime, (ct, x, y, z), which is called *world point*. The evolution of this event, in space and in time, is generally represented by a curve, which is called *world line*. A world line represents the *history* of a phenomenon and may be given in parametric form, with a parameter θ, say, as $[ct(\theta),\ x(\theta),\ y(\theta),\ z(\theta)]$ or in our familiar form $[x(t),\ y(t),\ z(t)]$ in a *Minkowski diagram*. Of course, the representation of a four-dimensional space on a plane figure is impossible. The qualitative description of a phenomenon may be achieved by drawing two or, at most, three coordinates in the plane figure. In Fig. 8.1, for example, a world point P and its history is represented in two or three dimensions of *Minkowski's world*.

Figure 8.2a shows the history of a point that is at rest at the position $x = x_P$, in the inertial frame of reference S. The other two coordinates of the point are not drawn. Although the point is stationary, its history develops in time, a fact that is described by a straight line parallel to the ct-axis. Figure 8.2b shows the history of a point moving with a constant speed $dx/dt = V$ and passing through point $(x = x_P;\ t = 0)$. Its history is represented by a straight line with a slope of $d(ct)/dx = c/V = 1/\beta$. In Fig. 8.2c are shown the world lines of a point moving with a speed V and passing through the point $(x = 0,\ t = 0)$ and of a photon that

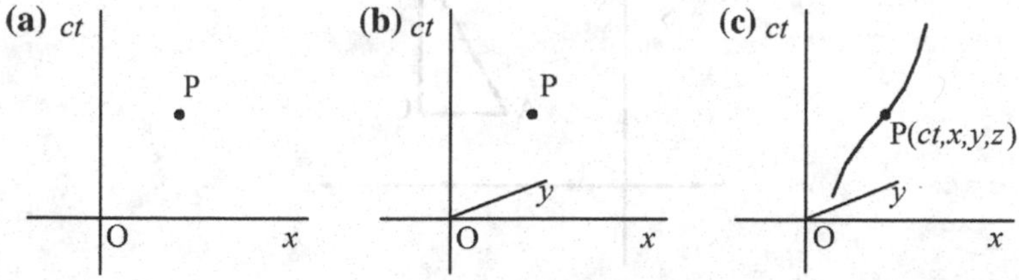

Fig. 8.1 Minkowski's spacetime. A world point P is represented in a Minkowski diagram in (**a**) two dimensions (*x*, *ct*) and (**b**) three dimensions (*x*, *y*, *ct*). Figure (**c**) shows the history of an event in three dimensions

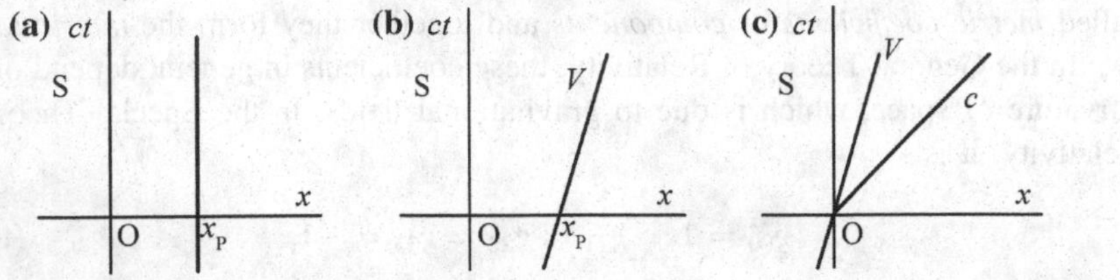

Fig. 8.2 **a** The world line, in only two dimensions, of a point at rest at the position $x = x_P$. **b** The (straight) world line, in two dimensions, of a point that moves with a constant speed $dx/dt = V$ and passes through the point ($x = x_P$, $t = 0$). **c** The world lines of a point that moves with a constant speed V and passes through the point ($x = 0$, $t = 0$) and of a photon which moves with speed c and is emitted at ($x = 0$, $t = 0$)

starts from the point ($x = 0,\ t = 0$) and moves with speed equal to c. In the case of a photon, the slope of the straight line is equal to unity

In all cases, we must not forget that, in Minkowski's world, the Pythagorean theorem, used in the evaluation of distances between world points, takes the pseudo-Euclidean form of Eq. (8.7) (Fig. 8.3).

Let us now assume that a spherical wave front starts propagating from the point ($x = 0$, $y = 0$, $z = 0$) at the moment $t = 0$ in all directions with speed c. In a diagram in xyz space, we could draw successive views that show the positions of the wave front at different moments in time (Fig. 8.4a). Alternatively, we could ignore one of the spatial dimensions (say z) and draw the diagram with axes those of x, y and ct (Fig. 8.4b). In this second figure, the history of a luminous wave front is represented by a conical surface, $x^2 + y^2 = c^2t^2$. Simplifying the diagram even further, we could draw the figure with only two axes, x and ct (Fig. 8.4c). In the last figure, the history of the light wave front is represented by two straight lines, $x = \pm ct$. Identical figures can be drawn for the other two spatial dimensions, y and z.

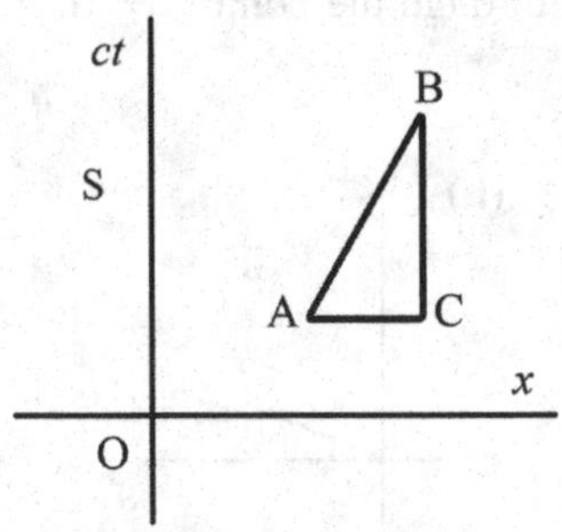

Fig. 8.3 Pythagoras' theorem in Minkowski's spacetime. In two dimensions, the lengths of the three straight world lines of the figure, which form a right angled triangle ABC, satisfy the relation: $(AB)^2 = (BC)^2 - (AC)^2$

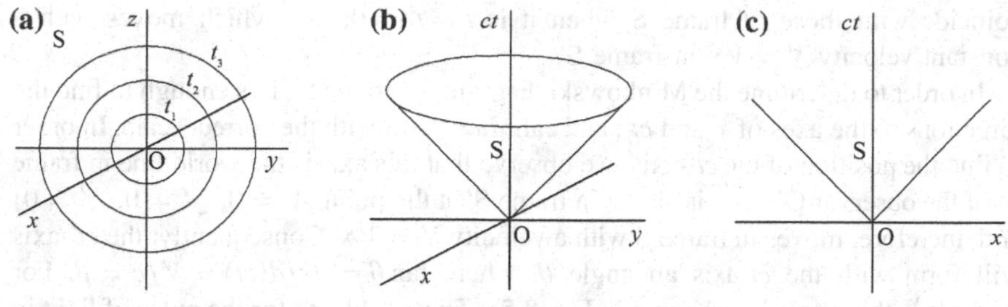

Fig. 8.4 (a) The spherical wave fronts, at different values of time, $0 < t_1 < t_2 < t_3$, for a pulse of light that was emitted from the point ($x = 0$, $y = 0$, $z = 0$) at time $t = 0$. Minkowski diagrams for the history of the pulse are shown in figure (**b**) in only three dimensions (x, y, ct) and in figure (**c**) in only two dimensions (x, ct)

8.1.1 The Minkowski Diagram of the Lorentz Transformation

Let us assume that we have a Minkowski diagram of the events in an inertial frame of reference S, as shown in Fig. 8.5a. Due to the limitations of the figure, we only plotted axes x and ct. Point O shows the position of an observer at the origin of the spatial axes ($x = 0$, $y = 0$, $z = 0$) at time $t = 0$. In this figure, the world line of a pulse of light emitted from the origin of the spatial axes ($x = 0$, $y = 0$, $z = 0$) at the moment $t = 0$, is given by the two straight lines $x = \pm ct$, which have slopes equal to ± 1 respectively. Which is the corresponding Minkowski diagram for an observer O′, who has as his frame of reference frame S′, whose spatial axes

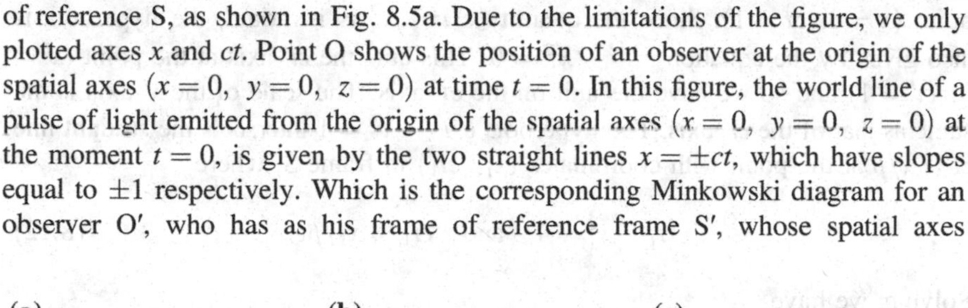

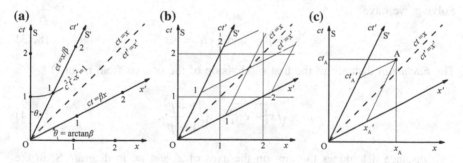

Fig. 8.5 **a** The ct-x Minkowski diagram for an observer O in the inertial frame of reference S and the ct'-x' axes of the Minkowski diagram for observer O′ in the inertial frame of reference S′ whose spatial axes coincide with those of S at the moment $t = t' = 0$, and which moves with a constant velocity $\mathbf{V} = V\hat{\mathbf{x}}$ in frame S. The *dashed line* is the world line of a pulse of light which is emitted from the origin ($x = 0$, $y = 0$, $z = 0$) at the moment $t = 0$ (light cone). This world line is common to both diagrams. The calibration of the axes of diagram S′ is performed with the aid of the hyperbolas $c^2t^2 - x^2 = \lambda$, where λ can take various values. **b** The calibration of the axes of diagram S′. **c** The Lorentz transformation between frames S and S′, for an event A

coincide with those of frame S when it is $t = t' = 0$, and which moves with a constant velocity $\mathbf{V} = V\hat{\mathbf{x}}$ in frame S?

In order to determine the Minkowski diagram for frame S′, it is enough to find the directions of the axes of x' and ct', and calibrate them with the correct scale. In order to find the position of the ct'-axis, we observe that this axis is the world line in frame S, of the observer O′, who is at rest in frame S′ at the point $(x' = 0,\ y' = 0,\ z' = 0)$ and, therefore, moves in frame S with a velocity $\mathbf{V} = V\hat{\mathbf{x}}$. Consequently, the ct'-axis will form with the ct-axis an angle θ, where $\tan\theta = dx/d(ct) = V/c = \beta$. For positive β, this angle is as shown in Fig. 8.5a. The world line for the pulse of light in frame S is given by the equation $c^2t^2 - x^2 = 0$. This is an invariant quantity and, therefore, it is also $c^2t'^2 - x'^2 = 0$. The dashed straight line in Fig. 8.5a, which bisects the angle between the x and ct axes, also bisects the angle between the x' and ct' axes, since it is also given by the equation $c^2t'^2 - x'^2 = 0$. The x'-axis forms, therefore, an angle θ with the x-axis. With the value of θ given, the equations of the axes of x' and of ct' in the S diagram are $ct = \beta x$ and $ct = x/\beta$, respectively, and have been marked on Fig. 8.5a.

In order to calibrate the scale of the axis of ct', we may use (for example) the unit hyperbola $c^2t^2 - x^2 = 1$, which has been drawn in Fig. 8.5a. This passes through point $(x = 0,\ ct = 1)$. Since the quantity $c^2t^2 - x^2$ is invariant, the hyperbola is also given by the equation $c^2t'^2 - x'^2 = 0$. This cuts the ct'-axis at the point $(x' = 0,\ ct' = 1)$ and so we have the unit on the ct'-axis. The scale of the x'-axis is the same as that of the ct'-axis. The hyperbola $c^2t^2 - x^2 = 1$ intersects the straight line $ct = x/\beta$ at the point with coordinates $(x_1,\ ct_1)$ in frame S, where

$$c^2t_1^2 - x_1^2 = 1 \quad \text{and} \quad ct_1 = x_1/\beta. \tag{8.12}$$

Solving, we have

$$x_1 = \beta\gamma \quad \text{and} \quad ct_1 = \gamma. \tag{8.13}$$

The *Euclidean* distance of the first subdivision of the ct'-axis from point O is

$$\mu = \sqrt{x_1^2 + c^2t_1^2} = \sqrt{\frac{1+\beta^2}{1-\beta^2}}. \tag{8.14}$$

This distance determines the unit on the axes of x' and ct' in diagram S, in the following sense: If a length a corresponds in the figure to the unit of measurement of the axes of x and ct, then the unit for the axes of x' and ct' will have a length of μa.

We note that, for positive values of the velocity V, the greater this velocity is the more acute is the angle between axes of x' and ct' in diagram S. For negative velocities V, angles θ are in the opposite direction of those of Fig. 8.5a. In all cases, the straight lines with slopes ± 1, being the world lines of the cone of light, bisect the angles between the axes x' and ct' as well as the right angle between the axes of x and ct.

The divisions of the axes of the two diagrams have been drawn in Fig. 8.5b. In Fig. 8.5c the coordinates of event A are shown in the Minkowski diagram for the two frames of reference.

The transformation of the coordinates of an event P from the inertial frame of reference S to frame S′ is seen in Fig. 8.6. The angle θ formed by the axes, was found to be given by the relation $\tan\theta = V/c = \beta$. From the figure, we find the following relations

$$\begin{aligned}(\mathrm{OB}) &= (\mathrm{OA}) + (\mathrm{AB}) = (\mathrm{OE})\cos\theta + (\mathrm{EP})\sin\theta = (\mathrm{OE})\cos\theta + (\mathrm{OF})\sin\theta \\ (\mathrm{OD}) &= (\mathrm{OC}) + (\mathrm{CD}) = (\mathrm{OF})\cos\theta + (\mathrm{FP})\sin\theta = (\mathrm{OE})\sin\theta + (\mathrm{OF})\cos\theta\end{aligned} \tag{8.15}$$

which are solved for

$$(\mathrm{OE}) = \frac{(\mathrm{OB})\cos\theta - (\mathrm{OD})\sin\theta}{\cos^2\theta - \sin^2\theta} \quad \text{and} \quad (\mathrm{OF}) = \frac{(\mathrm{OD})\cos\theta - (\mathrm{OB})\sin\theta}{\cos^2\theta - \sin^2\theta}. \tag{8.16}$$

Taking into account the scales of the frame of reference S' in relation to those of frame S [Eq. (8.14)], we have

$$x' = \frac{(\mathrm{OE})}{\mu} = (\mathrm{OE})\sqrt{\frac{1-\beta^2}{1+\beta^2}}, \qquad ct' = \frac{(\mathrm{OF})}{\mu} = (\mathrm{OF})\sqrt{\frac{1-\beta^2}{1+\beta^2}}. \tag{8.17}$$

Because it is $\tan\theta = \beta$, it also is $\sin\theta = \beta/\sqrt{1+\beta^2}$ and $\cos\theta = 1/\sqrt{1+\beta^2}$. Also, $(\mathrm{OB}) = x$ and $(\mathrm{OD}) = ct$, in which case from Eq. (8.16) the following relations follow

Fig. 8.6 The Minkowski diagram for the Lorentz transformation from frame S to frame S′ for an event P

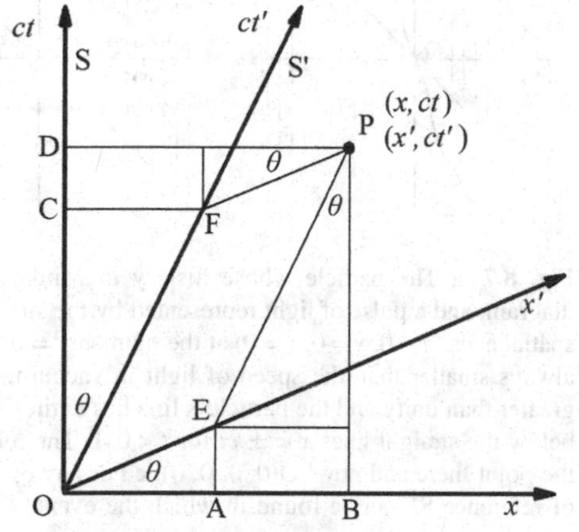

$$x' = \frac{x - \beta ct}{\sqrt{1-\beta^2}}, \quad ct' = \frac{ct - \beta x}{\sqrt{1-\beta^2}}, \tag{8.18}$$

which express the Lorentz transformation.

8.1.2 *Causality. Past, Present and Future*

The history of a particle in Minkowski's spacetime is given by a curve, which must have certain properties. One such curve was drawn in Fig. 8.7a. The particle is at the origin of the spatial axes ($x = 0,\ y = 0,\ z = 0$) at the time $t = 0$. A pulse of light emitted from this point at $t = 0$, moves with speed c in all directions. In the ct-x diagram of the figure, the history of the pulse is described by the straight lines $x = \pm ct$. The particle cannot move with a speed greater than that of light in vacuum, c, a fact which means that the slope of the curve giving the history of the particle must be everywhere greater than unity $[d(ct)/dx = c/V = 1/\beta > 1]$. Given that both the particle and the pulse of light pass through the point O, this means that the particle's line must lie entirely between the straight lines $x = \pm ct$ of Fig. 8.7a. More generally, the particle will lie, at all times, within the corresponding spherical wave front of the pulse and inside the cone of light in Fig. 8.4a. All these follow from the condition $c^2t^2 - x^2 - y^2 - z^2 > 0$.

Relative to a certain event, ($x = 0$, $y = 0$, $z = 0$, $t = 0$) say, all other events are classified according to the square of their distance from it. This quantity is defined as:

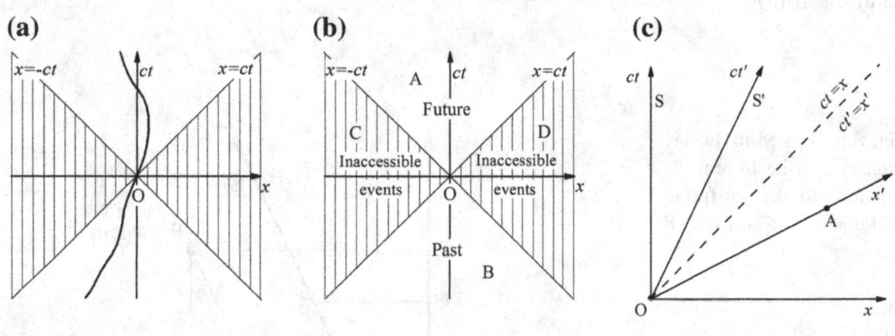

Fig. 8.7 **a** The particle whose history in Minkowski's spacetime is given by the line in the diagram, and a pulse of light represented by the straight lines $x = \pm\, ct$, are both at the origin of the spatial axes ($x = 0$, $y = 0$, $z = 0$) at the moment $t = 0$. Due to the fact that the speed of the particle is always smaller than the speed of light in vacuum, the slope of the particle's line is everywhere greater than unity and the particle's line lies entirely above the straight lines $x = \pm\, ct$ for $t > 0$ and below the straight lines $x = \pm\, ct$ for $t < 0$. **b** The relationships of various events in spacetime and the point 'here and now' O(0, 0, 0, 0). **c** For any event A in the regions C and D, an inertial frame of reference S′ can be found in which the events O and A are simultaneous

$$s^2 = c^2t^2 - x^2 - y^2 - z^2. \tag{8.19}$$

If the distance of the world point or event (x, y, z, t) from $(0, 0, 0, 0)$ is such that it is $s^2 > 0$, the interval between the two points is called *time-like*. For $s^2 < 0$, the interval between the two points is called *space-like*. Finally, for $s^2 = 0$ the interval between the two points is called *light-like* or *null interval*. The corresponding vectors in Minkowski space are called time-like, space-like and light-like or null. In a time-like vector, the term corresponding to the time component (c^2t^2) predominates compared to those corresponding to the space components ($x^2 + y^2 + z^2$). In a space-like vector the opposite is true. For a null or light-like vector, it is $c^2t^2 = x^2 + y^2 + z^2$, an equation which is satisfied on the wave front of a pulse of light which is emitted from the point ($x = 0,\ y = 0,\ z = 0$) at the moment $t = 0$. Time-like and light-like vectors are collectively called *causal* vectors, for reasons that will become apparent below.

Two world points whose coordinates differ by Δt, Δx, Δy and Δz, are at a distance between them equal to $\Delta s = \sqrt{(c\Delta t)^2 - (\Delta x)^2 - (\Delta y)^2 - (\Delta z)^2}$. If it is $(\Delta s)^2 > 0$, the two points are separated in a time-like manner, if it is $(\Delta s)^2 < 0$, the two points are separated in a space-like manner and if it is $(\Delta s)^2 = 0$, the two points are separated in a light-like manner.

In the *ct-x* diagram of Fig. 8.7b, Minkowski's space is separated in different regions on the basis of the criteria mentioned above. Consequently, spaces A and B consist of points which are separated by a time-like interval from point $(0, 0, 0, 0)$. Space A, for which it is $t > 0$, consists of events which are in the *future* of event $(0, 0, 0, 0)$. The transfer from point $(0, 0, 0, 0)$ to a point (x, y, z, t) of space A is possible with speeds smaller than c. Space B, for which it is $t < 0$, consists of events which constitute the *past* of event $(0, 0, 0, 0)$. Again, the transfer from a point (x, y, z, t) of space B to the point $(0, 0, 0, 0)$ is possible with speeds smaller than c. Points situated on the four-dimensional light-like cone $s^2 = c^2t^2 - x^2 - y^2 - z^2 = 0$, which in Fig. 8.7b is represented by the straight lines $x = \pm ct$, are accessible from point $(0, 0, 0, 0)$ only with motion at a speed equal to c. Points in regions C and D are *inaccessible* from point $(0, 0, 0, 0)$, unless we accept the existence of superluminal speeds ($> c$). These points of space are said to be *elsewhere*, i.e. neither in the past or future of point $(0, 0, 0, 0)$. At least with the laws of Physics as we know them today, it is impossible for a relationship of cause-effect to exist between the event $(0, 0, 0, 0)$ and any other event represented by a point in regions C and D, which actually form a unique region in spacetime. As seen in Fig. 8.7c, for any event A in the regions C and D, an inertial frame of reference S′ may be found, in which events O and A are simultaneous. The x'-axis of this frame is the straight line OA, for which it is $ct' = 0$. The ct'-axis is symmetrical to the x'-axis relative to the straight line $ct' = x'$ (or $ct = x$), which represent the light cone. In this sense, the events of regions C and D may be thought of as constituting the *present* of event $(0, 0, 0, 0)$.

8.1.3 The Minkowski Diagram for the Effect of Length Contraction

Figure 8.8a shows a Minkowski diagram with the aid of which the relativistic effect of length contraction may be described. A rod AB is at rest along the x-axis of the inertial frame of reference S. In this frame, the rest length of the rod is L_0. The world lines of the two ends of the rod (A and B) have been drawn in the figure. Their intersection points with a straight line of a given value of ct determines the length L_0 of the rod in frame S as measured at that particular moment in time. In order to find the length of the rod in frame S′, which is moving with a velocity $\mathbf{V} = V\hat{\mathbf{x}}$ relative to frame S, we must find the positions of the two ends of the rod simultaneously in this frame. The intersections of the world lines of the two ends of the rod with a straight line of given value of ct' gives us the positions of the two ends of the rod, A and B′, at the particular value of t'. The phenomenon of length contraction is demonstrated by the fact that the length of the rod in frame S′, $L' = \text{AB}'$, is smaller than the rest length, if we take into account the difference in the scales of the axes of the two diagrams (if a unit in the scale of the x's has length α in the figure, a unit in the scale of the x''s has a length $\alpha\sqrt{(1+\beta^2)/(1-\beta^2)}$, according to what has been said in Section 8.1.1).

Given that the angle between the axes of x and x' has a tangent equal to $\tan\theta = \beta$, it follows that

$$\frac{(\text{AB}')}{(\text{AB})} = \frac{1}{\cos\theta} = \sqrt{1+\tan^2\theta} = \sqrt{1+\beta^2}. \tag{8.20}$$

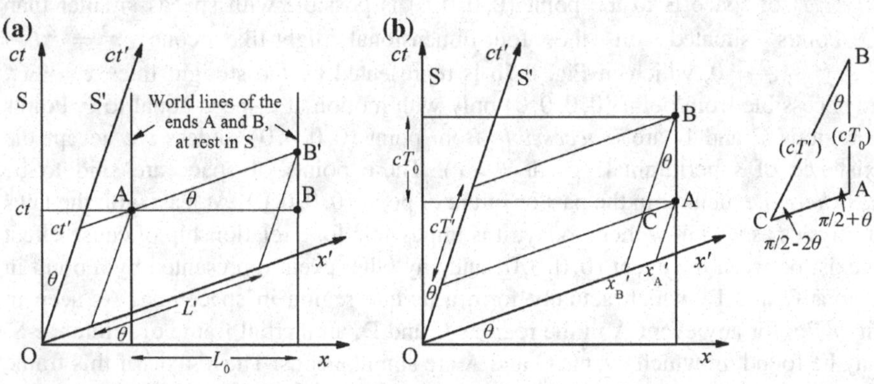

Fig. 8.8 Minkowski diagrams for the effects of (**a**) contraction of length and (**b**) dilation of time

Taking into account also the difference in the scales of the two axes, it is

$$\frac{L'}{L_0} = \frac{(AB')}{(AB)}\sqrt{\frac{1-\beta^2}{1+\beta^2}} = \sqrt{1+\beta^2}\sqrt{\frac{1-\beta^2}{1+\beta^2}}, \tag{8.21}$$

from which it follows that

$$L' = \sqrt{1-\beta^2}\, L_0. \tag{8.22}$$

This expression describes the contraction of length. It should be noted that, although in the figure the line representing L' has a greater length than the line representing L_0, when the difference in the scales is taken into account, it is $L' < L_0$.

8.1.4 The Minkowski Diagram for the Effect of Time Dilation

Figure 8.8b shows a Minkowski diagram with the use of which the relativistic effect of the dilation of time is described. The world line of a clock which is at rest in a frame of reference S is given by the straight line AB, which is normal to the *x*-axis. The time difference between events A and B is T_0 in frame S and T' in frame S′. To find the relationship between T' and T_0, we examine the geometry of triangle ABC, which is redrawn on the right of Fig. 8.8b. The quantities cT_0 and cT' are placed on the sides AB and BC of the triangle in parentheses, as they are represented by these sides but their ratio is not equal to the ratio of the lengths of the two lines; the difference in scales must also be taken into account. The angles of the triangle ABC are as shown in the figure on the right. Using the law of sines, we have:

$$\frac{(\mathrm{AB})}{\sin(\pi/2-2\theta)} = \frac{(\mathrm{BC})}{\sin(\pi/2+\theta)} \quad \Rightarrow \quad \frac{(\mathrm{AB})}{\cos 2\theta} = \frac{(\mathrm{BC})}{\cos\theta}. \tag{8.23}$$

$$\frac{(\mathrm{AB})}{(\mathrm{BC})} = \frac{\cos 2\theta}{\cos\theta} = \frac{2\cos^2\theta - 1}{\cos\theta} = \left(\frac{2}{1+\beta^2} - 1\right)\sqrt{1+\beta^2} = \frac{1-\beta^2}{\sqrt{1+\beta^2}}. \tag{8.24}$$

Taking into account the difference in scales, we have

$$\frac{cT_0}{cT'} = \frac{(\mathrm{AB})}{(\mathrm{BC})}\sqrt{\frac{1+\beta^2}{1-\beta^2}} = \frac{1-\beta^2}{\sqrt{1+\beta^2}}\sqrt{\frac{1+\beta^2}{1-\beta^2}} = \sqrt{1-\beta^2} \tag{8.25}$$

$$\text{or} \quad T' = \frac{T_0}{\sqrt{1-\beta^2}}. \tag{8.26}$$

This expression describes the dilation of time.

Example 8.1 The Minkowski diagram for the twin paradox
An observer, P, remains at rest at point $(0, 0, 0)$ of the inertial frame of reference S. A second observer, Q, moves away from P, moving along the x-axis with a constant velocity $\mathbf{V} = V\hat{\mathbf{x}}$ in frame S. Q, after moving in this way for time $t_P/2$, as measured by P, reverses the direction of its motion and returns to P after a total time t_P, as measured by P. For how long did the journey last, as measured by Q? We will evaluate this time, τ_Q, with the aid of a Minkowski diagram describing the history of the two observers.

Observer P is at rest in frame S, at the point $(0, 0, 0)$. The world line he will describe will, therefore, be the axis of ct. Let it be that in time t_P he moves along the straight line OA [see figure (a)]. Observer Q moves with velocity $\mathbf{V} = V\hat{\mathbf{x}}$ for a time $t_P/2$, as measured by P, and for the rest of the time, $t_P/2$, he moves with velocity $\mathbf{V} = -V\hat{\mathbf{x}}$, so that he meets P again at point A. OBA will, therefore, be the world line of Q. The straight line OB forms with the ct-axis an angle $\theta = \arctan(V/c)$, in the direction of positive x's and line BA forms an angle $-\theta$.

The square of the distance OB is found from the right-angled triangle which has OB as its hypotenuse and the other two sides with lengths equal to $ct_P/2$ and $Vt_P/2$, respectively. According to the 'Pythagorean theorem' of the Minkowski space, it is

$$(\mathrm{OB})^2 = \left(\frac{ct_P}{2}\right)^2 - \left(\frac{Vt_P}{2}\right)^2$$

$$\text{and} \quad (\mathrm{OB}) = \frac{ct_P}{2}\sqrt{1 - \frac{V^2}{c^2}}.$$

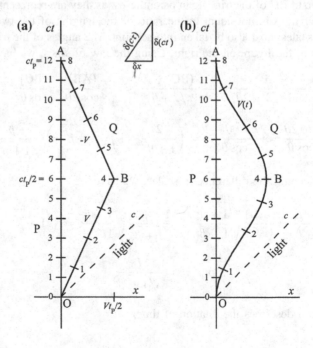

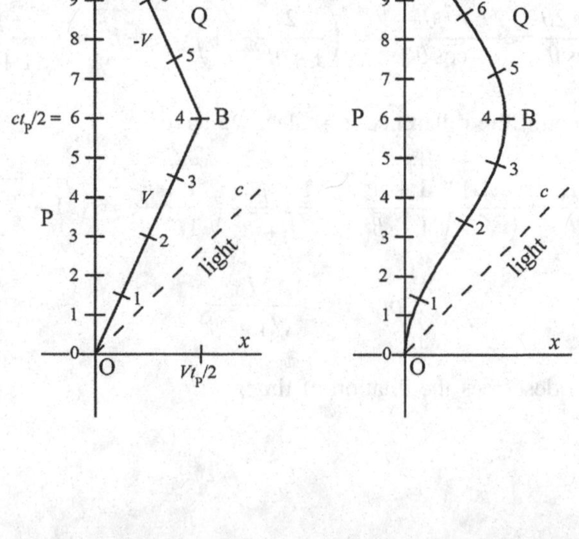

The total length of OBA is (OBA) $= s_Q = ct_P\sqrt{1-\frac{V^2}{c^2}}$.

The proper time that elapsed for observer Q is $\tau_Q = \frac{s_Q}{c} = t_P\sqrt{1-\frac{V^2}{c^2}}$.
The proper time elapsed for P during the path OA is t_P. We notice that it is

$$\frac{\tau_Q}{t_P} = \sqrt{1-\frac{V^2}{c^2}} \quad \text{or} \quad \tau_Q = \frac{t_P}{\gamma}.$$

Obviously the paths of the two observers are not symmetrical. Q jumps from one inertial frame to another at the point B. The total proper times drawn in figure (a) are 12 units for P and 8 units for Q (corresponding to $\gamma = 3/2$, $\beta = \sqrt{5}/3$). The time interval along the path of Q is evaluated with the aid of the relation $[\delta(c\tau)]^2 = [\delta(ct)]^2 - [\delta x]^2$ of the right-angled triangle in the figure.

Figure (b) shows another path for observer Q, with speed which varies continuously with time, $V(t)$, but with the same total time τ_Q. In this case, the line lengths corresponding to equal intervals of proper time on the curve are different. Time passes at a slower rate for Q when the absolute value of the slope of its world line is smaller. Observer P finds that the clock which is at rest relative to him registers a largest interval of time.

Example 8.2 Wheeler's single electron universe
In his acceptance lecture for the Nobel prize for Physics in 1965, Richard Feynman mentioned how, when he was a postgraduate student of John Archibald Wheeler, at Princeton, Wheeler telephoned him on a day in the spring of 1940 and announced: 'I know why all electrons in the universe have the same mass, charge and all their other properties the same'. What did the great teacher have in mind?

Wheeler's argument was based on the idea that the antiparticle of the electron, the positron, is simply an electron moving backwards in time. In the Minkowski diagram shown in the figure, a positron would be the particle moving from A to B. At point B it is created, for the positive direction of time, an electron-positron pair from a photon (dotted line) which has the energy required for this purpose. The *Electron* then moves at times towards positive values of time until it meets with (itself as) a positron at C and the 'two' annihilate into two photons. This process is repeated again and again at various points in spacetime. In this way, the one and only electron in the universe reacts with itself forward and backward in time. This electron also represents all the positrons in the universe.

If we look at the picture at a particular moment of time (horizontal line in the figure), we will see a universe full with many electrons, positrons and photons moving in various directions.

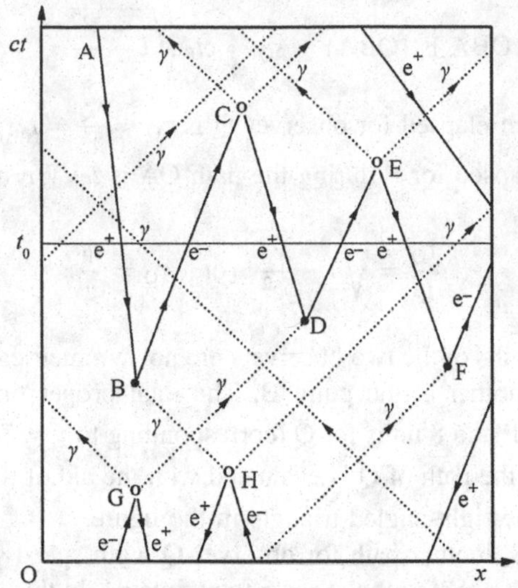

Naturally, there are numerous objections to Wheeler's idea. For example, the question arises of why the number of positrons in the universe is not equal to that of the electrons. Wheeler would answer that 'they may be hiding inside protons, or somewhere.' Wheeler enjoyed very much to raise unusual questions such as this one, apparently for the purpose of amusement. However, the godfather of 'black hole' was in the habit of making very important statements through such paradoxical comments [1].

8.1.5 A Minkowski Diagram for the Doppler Effect

Figure 8.9 shows the Minkowski diagram of an observer and a source of light moving relative to him with a velocity equal to V along the x-axis. The world line of the source in the frame of reference of the observer is a straight line with a slope greater than that of the world line of a light wave front. At regular intervals, the source emits light pulses at the points marked in the figure by the dots.

In order to find the moments at which these pulses reach the observer, we draw straight lines with a slope of $+1$ from the emission points at negative x's and -1 from points at positive values of x. These straight lines define the world lines of the wave front of each pulse, from the point of its emission.

From the geometry of the figure, the time between the arrivals of successive pulses to the observer, T_-, is smaller when the source moves at negative values of x, approaching the observer, as compared to the time between the arrival, at the observer, of successive pulses, T_+, when the source is moving at positive values of x, receding from the observer.

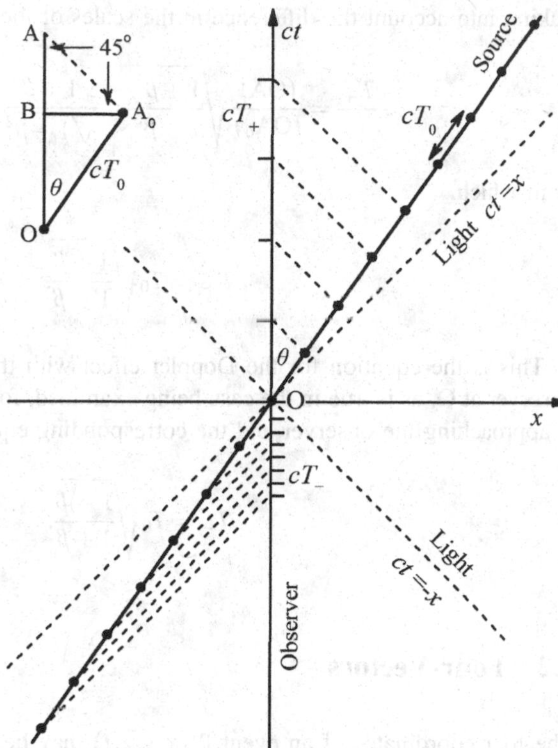

Fig. 8.9 A Minkowski diagram for the Doppler effect

In order to find the quantitative relation between the two periods, the difference in the scales of the two frames of reference in the Minkowski diagram must be taken into account. If the speed of the source is $V = \beta c$, then the angle formed by the world line with the ct-axis is θ, where it is $\tan\theta = \beta$. The relation between the periods of the signal in the two frames of reference is described by the triangle OAA_0 in Fig. 8.9. In this triangle, straight line OA_0 is a section of the world line of the source between the emissions of two successive pulses (the two dots), while OA is the corresponding section for the observer between the arrivals of the two pulses. The straight line A_0A is the world line of the light signal from the point A_0 towards the observer. The angles of the triangle are, therefore, as shown in the figure. Thus,

$$(\mathrm{OA}) = (\mathrm{OB}) + (\mathrm{BA}) = (\mathrm{OB}) + (\mathrm{BA_0}) = (\mathrm{OA_0})\cos\theta + (\mathrm{OA_0})\sin\theta. \quad (8.27)$$

Because it is $\sin\theta = \beta\big/\sqrt{1+\beta^2}$ and $\cos\theta = 1/\sqrt{1+\beta^2}$, this relation gives

$$(\mathrm{OA}) = (\mathrm{OA_0})\frac{1+\beta}{\sqrt{1+\beta^2}}. \quad (8.28)$$

Taking into account the difference in the scales of the two time axes, it is

$$\frac{cT_+}{cT_0} = \frac{(\mathrm{OA})}{(\mathrm{OA_0})}\sqrt{\frac{1+\beta^2}{1-\beta^2}} = \frac{1+\beta}{\sqrt{1+\beta^2}}\sqrt{\frac{1+\beta^2}{1-\beta^2}}, \tag{8.29}$$

from which,

$$T_+ = T_0\sqrt{\frac{1+\beta}{1-\beta}}. \tag{8.30}$$

This is the equation for the Doppler effect with the source receding from the observer at O, as is true in the case being examined, for $t > 0$. For $t < 0$, the source is approaching the observer and the corresponding equation is

$$T_- = T_0\sqrt{\frac{1-\beta}{1+\beta}}. \tag{8.31}$$

8.2 Four-Vectors

The four coordinates of an event P (x, y, z, t) may be considered to constitute the position vector of the event in the four-dimensional spacetime. The origin of the vector is the origin of the axes $(0, 0, 0, 0)$ and the point (x, y, z, t) is the end of the vector. We will refer to this vector as a *four-vector* and we will denote it with a bold underlined letter, such as $\underline{\mathbf{R}}$. We may also give the components of the four-vector: $\underline{\mathbf{R}} = (ct, x, y, z)$ or $\underline{\mathbf{R}} = (ct, \mathbf{r})$. Coordinate t is called the *temporal* component and coordinates (x, y, z) or the vector $\mathbf{r}$ are called the *spatial* components of the position four-vector. As we have already mentioned [Eq. (8.8)], the components of the position four-vector are also symbolized by x^μ $(\mu = 0, 1, 2, 3)$, where

$$x^0 \equiv ct, \quad x^1 \equiv x, \quad x^2 \equiv y, \quad x^3 \equiv z. \tag{8.32}$$

The transformation from one inertial frame of reference S to another, S′, which is moving with a velocity $\mathbf{V} = V\hat{\mathbf{x}}$ relative to frame S, is performed using the Lorentz transformation

$$x'^0 = \gamma\left(x^0 - \beta x^1\right), \quad x'^1 = \gamma\left(x^1 - \beta x^0\right), \quad x'^2 = x^2, \quad x'^3 = x^3, \tag{8.33}$$

$$\text{where} \quad \beta \equiv V/c \quad \text{and} \quad \gamma \equiv 1/\sqrt{1-\beta^2}.$$

The square of the magnitude of the position four-vector, its *norm*, is defined as

$$\underline{\mathbf{R}}^2 = s^2 = (x^0)^2 - (x^1)^2 - (x^2)^2 - (x^3)^2 = c^2t^2 - x^2 - y^2 - z^2 \tag{8.34}$$

and is invariant.

In a similar manner, we define, in the inertial frame of reference S, any four-vector $\underline{\mathbf{A}} = (A^0, A^1, A^2, A^3)$, the norm of which is defined as

$$\underline{\mathbf{A}}^2 \equiv (A^0)^2 - (A^1)^2 - (A^2)^2 - (A^3)^2. \tag{8.35}$$

The spatial part of the four vector may be written as a vector in three dimensions, $\mathbf{a}$, and the four-vector also be denoted by $\underline{\mathbf{A}} = (A^0, \mathbf{a})$. The norm of the four-vector $\underline{\mathbf{A}}$ is defined as

$$\underline{\mathbf{A}}^2 = (A^0)^2 - \mathbf{a}^2. \tag{8.36}$$

Depending on whether the norm of a four-vector is positive, negative or zero, it is called *time-like*, *space-like* or *null*, respectively.

The inner product of two four-vectors, $\underline{\mathbf{A}} = (A^0, \mathbf{a})$ and $\underline{\mathbf{B}} = (B^0, \mathbf{b})$, is defined as

$$\underline{\mathbf{A}} \cdot \underline{\mathbf{B}} = A^0 B^0 - \mathbf{a} \cdot \mathbf{b}. \tag{8.37}$$

The transformation of a four-vector from one inertial frame of reference S to another inertial frame of reference S′, which is moving with velocity $\mathbf{V} = V\hat{\mathbf{x}}$ relative to frame S, is performed according to the Lorentz transformation

$$A'^0 = \gamma(A^0 - \beta A^1), \quad A'^1 = \gamma(A^1 - \beta A^0), \quad A'^2 = A^2, \quad A'^3 = A^3, \tag{8.38}$$

which may also be written, with the use of matrices (see Appendix 6), as:

$$\begin{pmatrix} A'^0 \\ A'^1 \\ A'^2 \\ A'^3 \end{pmatrix} = \begin{pmatrix} \gamma & -\beta\gamma & 0 & 0 \\ -\beta\gamma & \gamma & 0 & 0 \\ 0 & 0 & 1 & 0 \\ 0 & 0 & 0 & 1 \end{pmatrix} \begin{pmatrix} A^0 \\ A^1 \\ A^2 \\ A^3 \end{pmatrix}. \tag{8.39}$$

In tensor form, the transformation is written as

$$A'^{\mu} = a'_{\mu\nu} A^{\nu}, \tag{8.40}$$

where $(\mu, \nu = 0, 1, 2, 3)$ and

$$d'_{\mu\nu} = \begin{pmatrix} \gamma & -\beta\gamma & 0 & 0 \\ -\beta\gamma & \gamma & 0 & 0 \\ 0 & 0 & 1 & 0 \\ 0 & 0 & 0 & 1 \end{pmatrix}. \tag{8.41}$$

Equations (3.24) express the general Lorentz transformation without rotation of the axes for the (position) four-vector $(t, x, y, z) = (t, \mathbf{r})$. By analogy, we have the general Lorentz transformation without rotation of the axes for any four-vector $\underline{\mathbf{A}} = (A^0, A^1, A^2, A^3) = (A^0, \mathbf{a})$,

$$A'^0 = \gamma\left(A^0 - \frac{\mathbf{V}\cdot\mathbf{a}}{c^2}\right), \qquad \mathbf{a}' = \mathbf{a} + (\gamma - 1)\mathbf{V}\frac{\mathbf{V}\cdot\mathbf{a}}{V^2} - \gamma\mathbf{V}A^0 \tag{8.42}$$

where $\gamma = 1/\sqrt{1 - V^2/c^2}$. The inverse transformation is

$$A^0 = \gamma\left(A'^0 + \frac{\mathbf{V}.\mathbf{a}'}{c^2}\right), \qquad \mathbf{a} = \mathbf{a}' + (\gamma - 1)\mathbf{V}\frac{\mathbf{V}.\mathbf{a}'}{V^2} + \gamma\mathbf{V}A'^0. \tag{8.43}$$

Problem

8.1 If $\underline{\mathbf{A}}$ and $\underline{\mathbf{B}}$ are two four-vectors, show that: (a) $\underline{\mathbf{A}}\cdot\underline{\mathbf{B}} = \underline{\mathbf{B}}\cdot\underline{\mathbf{A}}$, (b) $\underline{\mathbf{A}}\cdot(\underline{\mathbf{B}}+\underline{\mathbf{C}}) = \underline{\mathbf{A}}\cdot\underline{\mathbf{B}}+\underline{\mathbf{A}}\cdot\underline{\mathbf{C}}$, (c) $\delta(\underline{\mathbf{A}}\cdot\underline{\mathbf{B}}) = \delta\underline{\mathbf{A}}\cdot\underline{\mathbf{B}}+\underline{\mathbf{A}}\cdot\delta\underline{\mathbf{B}}$ and, in the limit, $d(\underline{\mathbf{A}}\cdot\underline{\mathbf{B}}) = d\underline{\mathbf{A}}\cdot\underline{\mathbf{B}}+\underline{\mathbf{A}}\cdot d\underline{\mathbf{B}}$.

8.2.1 The Position Four-Vector

As we have already mentioned, the four-vector of the position of an event is defined as $\underline{\mathbf{R}} = (ct, \mathbf{r}) = (ct, x, y, z)$ and has the (invariant) square of its magnitude or norm

$$\underline{\mathbf{R}}^2 = s^2 = c^2t^2 - x^2 - y^2 - z^2. \tag{8.44}$$

In differential form, for infinitesimal changes in the coordinates, it is

$$(ds)^2 = c^2(dt)^2 - (dx)^2 - (dy)^2 - (dz)^2. \tag{8.45}$$

Let the spatial coordinates refer to the position of a particle in three-dimensional space as a function of time. The last quantity is invariant, and, therefore, in two inertial frames of reference, frame S and another inertial frame of reference S′ which is moving with velocity $\mathbf{V} = V\hat{\mathbf{x}}$ relative to S, it is true that

$$(ds)^2 = c^2(dt)^2 - (dx)^2 - (dy)^2 - (dz)^2 = c^2(dt')^2 - (dx')^2 - (dy')^2 - (dz')^2. \tag{8.46}$$

If now we suppose that the particle is at rest in frame S′, it will be $dx' = dy' = dz' = 0$ and

$$(ds)^2 = c^2(dt)^2 - (dx)^2 - (dy)^2 - (dz)^2 = c^2(dt')^2. \tag{8.47}$$

Time dt' is the time as measured by the particle, the *proper time* or *rest time* of the particle, which we denote by τ. It is, therefore,

$$d\tau = dt' = \frac{ds}{c} = \sqrt{1 - \frac{(dx)^2 + (dy)^2 + (dz)^2}{c^2(dt)^2}}\, dt. \tag{8.48}$$

However, $\sqrt{\left[(dx)^2 + (dy)^2 + (dz)^2\right] / (dt)^2} = \upsilon$ is the magnitude of the three-vector of the velocity of the particle in frame S. Therefore,

$$d\tau = \frac{ds}{c} = \sqrt{1 - \frac{\upsilon^2}{c^2}}\, dt = \frac{dt}{\gamma}, \tag{8.49}$$

where $\beta \equiv \upsilon/c$ and $\gamma \equiv 1 / \sqrt{1 - \beta^2}$ refer to the motion of the particle in frame S. This quantity is invariant. If we have two frames of reference, frame S and another, S′, which moves with a velocity $\mathbf{V} = V\hat{\mathbf{x}}$ relative to the frame S, we will denote by $\beta \equiv V/c$ and $\gamma \equiv 1 / \sqrt{1 - \beta^2}$ the quantities corresponding to the relative velocity of the two frames of reference. The particle will have a speed υ in frame S and υ' in frame S′. To these speeds will correspond the quantities $\beta_P \equiv \upsilon/c$, $\gamma_P \equiv 1 / \sqrt{1 - \beta_P^2}$ and $\beta'_P \equiv \upsilon'/c$, $\gamma'_P \equiv 1 / \sqrt{1 - \beta'^2_P}$.

The speed of the particle in frame S may be a function of time t, $\beta(t) = \upsilon(t)/c$, and Eq. (8.49) then gives

$$\tau = \int \sqrt{1 - \frac{\upsilon^2}{c^2}}\, dt \tag{8.50}$$

with integration between the appropriate limits.

8.2.2 The Four-Vector of Velocity

For the definition of the four-vector of velocity, we cannot divide the variation of the position four-vector, $d\underline{\mathbf{R}}$, by dt, because dt is not an invariant quantity under the Lorentz transformation. The proper time of the particle, τ, is, however, such an invariant quantity. We thus define the four-vector of velocity as

$$\underline{\mathbf{U}} \equiv \frac{d\underline{\mathbf{R}}}{d\tau}. \tag{8.51}$$

$\underline{\mathbf{U}}$ is a four-vector because $d\underline{\mathbf{R}}$ is a four-vector and $d\tau$ is invariant. From the components of $\underline{\mathbf{R}} = (t, x, y, z)$, we find the components of $\underline{\mathbf{U}}$:

$$\begin{aligned} U^0 &= \frac{dx^0}{d\tau} = \gamma_P \frac{d(ct)}{dt} = \gamma_P c, \qquad & U^1 &= \frac{dx^1}{d\tau} = \gamma_P \frac{dx}{dt} = \gamma_P \upsilon_x \\ U^2 &= \frac{dx^2}{d\tau} = \gamma_P \frac{dy}{dt} = \gamma_P \upsilon_y, \qquad & U^3 &= \frac{dx^3}{d\tau} = \gamma_P \frac{dz}{dt} = \gamma_P \upsilon_z \end{aligned} \tag{8.52}$$

where the Lorentz factor γ_P corresponds to the speed of the particle in frame S, $\gamma_P \equiv 1/\sqrt{1-\upsilon^2/c^2}$. The four-vector of velocity is, therefore,

$$\underline{\mathbf{U}} = (\gamma_P c,\ \gamma_P \upsilon_x,\ \gamma_P \upsilon_y,\ \gamma_P \upsilon_z), \quad \underline{\mathbf{U}} = (\gamma_P c,\ \gamma_P \boldsymbol{\upsilon}) \quad \text{or} \quad \underline{\mathbf{U}} = \gamma_P (c,\ \boldsymbol{\upsilon}). \tag{8.53}$$

The norm of the velocity four-vector is

$$\underline{\mathbf{U}}^2 = \gamma_P^2 c^2 - \gamma_P^2 \boldsymbol{\upsilon}^2 = c^2, \tag{8.54}$$

which is invariant, as expected for a four-vector. This means that the four-velocity can never be zero, since a particle can be at rest in space, $\boldsymbol{\upsilon} = 0$, but the temporal component of the four-velocity, $\gamma_P c$, can never be zero, as expected since nothing can stand still in time!

The Lorentz transformation for the components of the velocity four-vector gives, according to Eq. (8.38),

$$U'^0 = \gamma(U^0 - \beta U^1), \quad U'^1 = \gamma(U^1 - \beta U^0), \quad U'^2 = U^2, \quad U'^3 = U^3, \tag{8.55}$$

where, here, β and γ correspond to the relative speed V of the two frames of reference. Substituting from Eq. (8.52), we have

$$\gamma'_P c = \gamma(\gamma_P c - \beta\gamma_P \upsilon_x), \quad \gamma'_P \upsilon'_x = \gamma(\gamma_P \upsilon_x - \beta\gamma_P c), \quad \gamma'_P \upsilon'_y = \gamma_P \upsilon_y, \quad \gamma'_P \upsilon'_z = \gamma_P \upsilon_z. \tag{8.56}$$

The first equation gives

$$\frac{\gamma_P}{\gamma'_P} = \frac{1}{\gamma\left(1 - \dfrac{V\upsilon_x}{c^2}\right)}, \tag{8.57}$$

which is the transformation of the Lorentz factor of the particle, γ_P, from the one frame of reference to the other. By substitution of this relation in the other three, we get

$$v'_x = \frac{v_x - V}{1 - Vv_x/c^2}, \quad v'_y = \frac{v_y}{\gamma(1 - Vv_x/c^2)}, \quad v'_z = \frac{v_z}{\gamma(1 - Vv_x/c^2)}, \tag{8.58}$$

i.e. the transformation of the three-vector of velocity.

8.2.3 The Four-Vector of Acceleration

The four-vector of the acceleration of a particle is defined as the derivative of the four-vector of its velocity, $\underline{\mathbf{U}}$, with respect to the proper time τ of the particle or as the second derivative of the position four-vector of the particle, $\underline{\mathbf{R}}$, with respect to its proper time:

$$\underline{\mathbf{A}} \equiv \frac{d\underline{\mathbf{U}}}{d\tau} \equiv \frac{d^2\underline{\mathbf{R}}}{d\tau^2}. \tag{8.59}$$

From the components of $\mathbf{v}_P$, given by Eq. (8.52), it follows that

$$A^0 = \frac{dU^0}{d\tau} = \frac{d(\gamma_P c)}{dt}\frac{dt}{d\tau}, \qquad A^1 = \frac{dU^1}{d\tau} = \frac{d(\gamma_P v_x)}{dt}\frac{dt}{d\tau},$$
$$A^2 = \frac{dU^2}{d\tau} = \frac{d(\gamma_P v_y)}{dt}\frac{dt}{d\tau}, \qquad A^3 = \frac{dU^3}{d\tau} = \frac{d(\gamma_P v_z)}{dt}\frac{dt}{d\tau}. \tag{8.60}$$

However,

$$\frac{dt}{d\tau} = \gamma_P \quad \text{and} \quad \frac{d\gamma_P}{dt} = \dot{\gamma}_P = \gamma_P^3 \beta\dot{\beta} = \frac{\gamma_P^3 v\dot{v}}{c^2}. \tag{8.61}$$

Equations (8.60) give, therefore,

$$A^0 = \frac{\gamma_P^3 v\dot{v}}{c}\gamma_P = \frac{\gamma_P^4 v\dot{v}}{c} = \frac{v\dot{v}}{c(1 - v^2/c^2)^2},$$
$$A^1 = \frac{d(\gamma_P v_x)}{dt}\frac{dt}{d\tau} = \gamma_P \frac{dv_x}{dt}\gamma_P + v_x \frac{d\gamma_P}{dt}\gamma_P = \frac{\dot{v}_x}{1 - v^2/c^2} + \frac{v_x v\dot{v}}{c^2(1 - v^2/c^2)^2},$$
$$A^2 = \frac{d(\gamma_P v_y)}{dt}\frac{dt}{d\tau} = \gamma_P \frac{dv_y}{dt}\gamma_P + v_y \frac{d\gamma_P}{dt}\gamma_P = \frac{\dot{v}_y}{1 - v^2/c^2} + \frac{v_y v\dot{v}}{c^2(1 - v^2/c^2)^2},$$
$$A^3 = \frac{d(\gamma_P v_z)}{dt}\frac{dt}{d\tau} = \gamma_P \frac{dv_z}{dt}\gamma_P + v_z \frac{d\gamma_P}{dt}\gamma_P = \frac{\dot{v}_z}{1 - v^2/c^2} + \frac{v_z v\dot{v}}{c^2(1 - v^2/c^2)^2}. \tag{8.62}$$

The four-vector of the acceleration may also be written as

$$\underline{\mathbf{A}} = \left(\gamma_{\mathrm{P}} \frac{d(\gamma_{\mathrm{P}} c)}{dt},\ \gamma_{\mathrm{P}} \frac{d(\gamma_{\mathrm{P}} \boldsymbol{\upsilon})}{dt} \right) = \left(\frac{\gamma_{\mathrm{P}}}{m_0 c} \frac{dE}{dt},\ \gamma_{\mathrm{P}} \frac{d(\gamma_{\mathrm{P}} \boldsymbol{\upsilon})}{dt} \right), \tag{8.63}$$

where $E = m_0 c^2 \gamma_{\mathrm{P}}$ is the energy of the particle in frame S.

Problems

8.2 If $\underline{\mathbf{U}}$ is the four-vector of the velocity of a particle and $\underline{\mathbf{A}}$ the four-vector of its acceleration, show that the two four-vectors are orthogonal, i.e. that it is $\underline{\mathbf{U}} \cdot \underline{\mathbf{A}} = 0$.
8.3 Show that the four-vector of acceleration is $\underline{\mathbf{A}} = \gamma_{\mathrm{P}}(\dot{\gamma}_{\mathrm{P}} c,\ \dot{\gamma}_{\mathrm{P}} \boldsymbol{\upsilon} + \gamma_{\mathrm{P}} \boldsymbol{a})$, where $\dot{\gamma}_{\mathrm{P}} = d\gamma_{\mathrm{P}}/dt$ and $\boldsymbol{a} = d\boldsymbol{\upsilon}/dt$ is the three-vector of the acceleration.
8.4 Use the result of Problem 8.3 in order to find the four-acceleration of a particle in its own frame of reference, i.e. in the inertial frame of reference in which the particle in momentarily at rest, and show that this is equal to zero only if the particle's proper acceleration, $\boldsymbol{\alpha}$, is zero. Ans.: $\underline{\mathbf{A}} = (0,\ \boldsymbol{\alpha})$
8.5 Let $\underline{\mathbf{U}}$ and $\underline{\mathbf{V}}$ be the velocity four-vectors of two particles. Evaluating their inner product in the frame of reference of the particle with four-velocity $\underline{\mathbf{U}}$, show that it is $\underline{\mathbf{U}} \cdot \underline{\mathbf{V}} = c^2 \gamma(\upsilon)$, where $\gamma(\upsilon)$ is the Lorentz factor corresponding to the relative speed υ of the particles.

8.2.3.1 The Proper Acceleration of a Particle Moving with a General Velocity

As a four-vector, $\underline{\mathbf{A}}$ has an invariant norm $\underline{\mathbf{A}}^2$. Evaluating this in the instantaneously co-moving inertial reference frame of the particle, and using the result of Problem 8.4, we have $\underline{\mathbf{A}}^2 = -\boldsymbol{\alpha}^2 = -\alpha^2$. This result is valid in every inertial frame of reference.

The general result of Problem 8.3 is $\underline{\mathbf{A}} = \gamma_{\mathrm{P}}(\dot{\gamma}_{\mathrm{P}} c, \dot{\gamma}_{\mathrm{P}} \boldsymbol{\upsilon} + \gamma_{\mathrm{P}} \boldsymbol{a})$.
We will use result (iii) of Example 3.3, for the particle, i.e. $d\gamma_{\mathrm{P}}/dt = (\gamma_{\mathrm{P}}^3/c^2)\boldsymbol{\upsilon} \cdot d\boldsymbol{\upsilon}/dt$, or $\dot{\gamma}_{\mathrm{P}} = (\gamma_{\mathrm{P}}^3/c^2)\boldsymbol{\upsilon} \cdot \boldsymbol{a}$. We also have the general relations $\boldsymbol{\upsilon}^2 = \upsilon^2$ and $(\boldsymbol{\upsilon} \cdot \boldsymbol{a})^2 = \upsilon^2 a^2 - (\boldsymbol{\upsilon} \times \boldsymbol{a})^2$. Therefore, it is

$$\begin{aligned} \underline{\mathbf{A}}^2 &= \gamma_{\mathrm{P}}^2(\dot{\gamma}_{\mathrm{P}}^2 c^2 - \dot{\gamma}_{\mathrm{P}}^2 \upsilon^2 - 2\,\gamma_{\mathrm{P}} \dot{\gamma}_{\mathrm{P}} \boldsymbol{\upsilon} \cdot \boldsymbol{a} - \gamma_{\mathrm{P}}^2 a^2) \\ &= \gamma_{\mathrm{P}}^8 (\boldsymbol{\upsilon} \cdot \boldsymbol{a})^2 (c^2 - \upsilon^2)/c^4 - 2\gamma_{\mathrm{P}}^6 (\boldsymbol{\upsilon} \cdot \boldsymbol{a})^2/c^2 - \gamma_{\mathrm{P}}^4 a^2 \\ &= -\gamma_{\mathrm{P}}^4 a^2 + \gamma_{\mathrm{P}}^6 (\boldsymbol{\upsilon} \cdot \boldsymbol{a})^2/c^2 - 2\gamma_{\mathrm{P}}^6 (\boldsymbol{\upsilon} \cdot \boldsymbol{a})^2/c^2 \end{aligned} \tag{8.64}$$

$$\begin{aligned} \underline{\mathbf{A}}^2 = -\gamma_{\mathrm{P}}^4 a^2 - \gamma_{\mathrm{P}}^6 (\boldsymbol{\upsilon} \cdot \boldsymbol{a})^2/c^2 &= -\gamma_{\mathrm{P}}^4 a^2 - \frac{\gamma_{\mathrm{P}}^6}{c^2}\left[\upsilon^2 a^2 - (\boldsymbol{\upsilon} \times \boldsymbol{a})^2\right] \\ &= -\frac{\gamma_{\mathrm{P}}^6}{c^2} a^2 \left(\frac{c^2}{\gamma_{\mathrm{P}}^2} + \upsilon^2 \right) + \frac{\gamma_{\mathrm{P}}^6}{c^2} (\boldsymbol{\upsilon} \times \boldsymbol{a})^2 \end{aligned} \tag{8.65}$$

and, finally,

$$\underline{\mathbf{A}}^2 = -\gamma_{\mathrm{P}}^6 a^2 + \frac{\gamma_{\mathrm{P}}^6}{c^2}(\boldsymbol{\upsilon} \times \boldsymbol{a})^2. \tag{8.66}$$

Equating with $\underline{\mathbf{A}}^2 = -\alpha^2$, we find

$$\alpha^2 = \gamma_{\mathrm{P}}^6 \left[a^2 - (\boldsymbol{\upsilon} \times \boldsymbol{a})^2/c^2\right]. \tag{8.67}$$

The magnitude of the proper acceleration of the particle is

$$\alpha = \gamma_{\mathrm{P}}^3 \sqrt{a^2 - (\boldsymbol{\upsilon} \times \boldsymbol{a})^2/c^2}. \tag{8.68}$$

If θ is the angle between the vectors of the velocity $\boldsymbol{\upsilon}$ and the acceleration $\boldsymbol{a}$ of the particle, then it is $(\boldsymbol{\upsilon} \times \boldsymbol{a})^2 = \upsilon^2 a^2 \sin^2\theta$ and

$$\alpha = \gamma_{\mathrm{P}}^3 a \sqrt{1 - \beta^2 \sin^2\theta}. \tag{8.69}$$

For rectilinear motion, $\boldsymbol{\upsilon}$ and $\boldsymbol{\alpha}$ are parallel to each other, and $\alpha = \gamma_{\mathrm{P}}^3 a$, (8.70)

as we have found in Sect. 3.3.1, Eq. (3.63).

For mutually perpendicular $\boldsymbol{\upsilon}$ and $\boldsymbol{\alpha}$, it is $\alpha = \gamma_{\mathrm{P}}^2 a$, (8.71)

as we have shown in Example 3.11.

8.2.4 The Energy-Momentum Four-Vector

We define the four-vector of momentum of a particle as the product of its rest mass with its velocity four-vector:

$$\underline{\mathbf{P}} \equiv m_0 \underline{\mathbf{U}} = (\gamma_{\mathrm{P}} m_0 c,\ \gamma_{\mathrm{P}} m_0 \boldsymbol{\upsilon}). \tag{8.72}$$

According to Eq. (8.36), the square of the magnitude or the norm of $\underline{\mathbf{P}}$ is

$$\underline{\mathbf{P}}^2 = m_0^2 c^2. \tag{8.73}$$

Obviously, for a photon or any other particle with zero rest mass, it is $\underline{\mathbf{P}}^2 = 0$.

The four-vector $\underline{\mathbf{P}}$ may also be written as

$$\underline{\mathbf{P}} = \left(\frac{E}{c},\ \mathbf{p}\right), \tag{8.74}$$

where $E = \gamma_P m_0 c^2$ is the energy and $\mathbf{p} = \gamma_P m_0 \boldsymbol{\upsilon}$ is the three-momentum of the particle. For this reason, $\underline{\mathbf{P}}$ is called the energy-momentum four-vector. Evaluating the norm of $\underline{\mathbf{P}}$ from the two forms of Eqs. (8.72) and (8.74), and equating, we get

$$\underline{\mathbf{P}}^2 = m_0^2 c^2 = E^2/c^2 - p^2, \tag{8.75}$$

which is the well known relation between the quantities E, p and m_0.

The momentum-energy transformation is found by applying the Lorentz transformation to the four-vector $\underline{\mathbf{P}} = (P^0, P^1, P^2, P^3)$,

$$P'^0 = \gamma(P^0 - \beta P^1), \quad P'^1 = \gamma(P^1 - \beta P^0), \quad P'^2 = P^2, \quad P'^3 = P^3, \tag{8.76}$$

with

$$P^0 = E/c, \quad P^1 = p_x, \quad P^2 = p_y, \quad P^3 = p_z. \tag{8.77}$$

There follow the known relations

$$E'/c = \gamma(E/c - \beta p_x), \quad p'_x = \gamma(p_x - \beta E/c), \quad p'_y = p_y, \quad p'_z = p_z. \tag{8.78}$$

8.2.4.1 Conservation of the Energy-Momentum Four-Vector

The energy-momentum four-vector has four components, each of which is conserved separately. We may, therefore, state that the energy-momentum four-vector or the four-momentum is conserved:

$$\underline{\mathbf{P}}_{\text{final}} = \underline{\mathbf{P}}_{\text{initial}} \quad \text{or} \quad \left(\frac{E_{\text{final}}}{c},\ \mathbf{p}_{\text{final}}\right) = \left(\frac{E_{\text{initial}}}{c},\ \mathbf{p}_{\text{initial}}\right). \tag{8.79}$$

In Sect. 6.10.1 we proved [Eq. (6.58)] that the total energy E and the total momentum P of a system satisfy the relation $E^2 - c^2P^2 = (M_0c^2)^2$. By Eqs. (6.55), it is $E = \sum_{i=1}^{n} E_i$ and $\mathbf{P} = \sum_{i=1}^{n} \mathbf{p}_i$. These imply that, for a system of particles, the energy-momentum four-vector satisfies the equation

$$\underline{\mathbf{P}}^2 = \underline{\mathbf{P}} \cdot \underline{\mathbf{P}} = M_0^2 c^2. \tag{8.80}$$

The quantity M_0 is an invariant of the system of particles.

It is convenient to use the law of conservation of the energy-momentum four-vector and the invariant property of $\underline{\mathbf{P}}^2$ in the solution of problems.

Example 8.3 The inverse Compton effect
An ultrarelativistic electron collides with a low energy photon. Find the final energy of the photon in the case of the head-on collision.

Let the electron initially move with velocity $\boldsymbol{\upsilon}$ and have momentum p_{e} and energy E_{e} as shown in Fig. (a). Its four-momentum is, therefore, $\underline{\mathbf{P}}_{\mathrm{e}} = \gamma m_0(c,\ \boldsymbol{\upsilon})$, where $\gamma = 1\big/\sqrt{1-\upsilon^2/c^2}$. After the collision with the photon, its four-momentum becomes $\underline{\mathbf{P}}'_{\mathrm{e}}$.

The photon initially has an energy E_ϕ and direction of propagation given by the unit vector $\hat{\mathbf{k}}$. Its four-momentum is $\underline{\mathbf{P}}_\phi = \dfrac{E_\phi}{c}(1,\ \hat{\mathbf{k}})$. After the collision, these properties of the photon have the values of E_γ, $\hat{\mathbf{k}}'$ and $\underline{\mathbf{P}}_\gamma = \dfrac{E_\gamma}{c}(1,\ \hat{\mathbf{k}}')$, respectively.

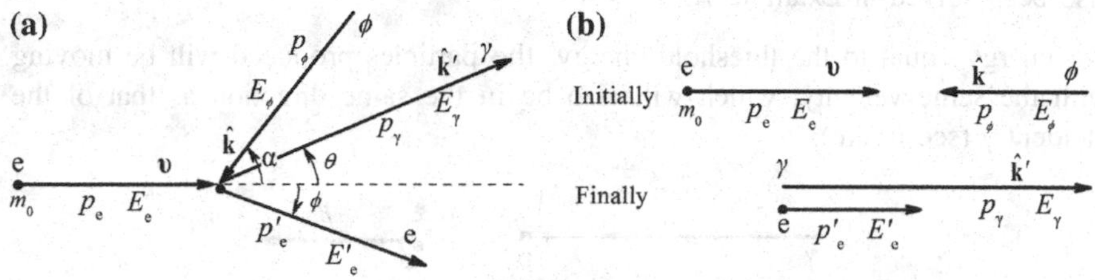

Conservation of four-momentum gives: $\underline{\mathbf{P}}_\phi + \underline{\mathbf{P}}_{\mathrm{e}} = \underline{\mathbf{P}}_\gamma + \underline{\mathbf{P}}'_{\mathrm{e}}$,
from which we get $\underline{\mathbf{P}}'^2_{\mathrm{e}} = \left(\underline{\mathbf{P}}_\phi + \underline{\mathbf{P}}_{\mathrm{e}} - \underline{\mathbf{P}}_\gamma\right)^2$.

Expanding, $\underline{\mathbf{P}}'^2_{\mathrm{e}} = \underline{\mathbf{P}}^2_\phi + \underline{\mathbf{P}}^2_{\mathrm{e}} + \underline{\mathbf{P}}^2_\gamma + 2\underline{\mathbf{P}}_\phi \cdot \underline{\mathbf{P}}_{\mathrm{e}} - 2\underline{\mathbf{P}}_\phi \cdot \underline{\mathbf{P}}_\gamma - 2\underline{\mathbf{P}}_\phi \cdot \underline{\mathbf{P}}_\gamma$ and using $\underline{\mathbf{P}}^2_{\mathrm{e}} = \underline{\mathbf{P}}'^2_{\mathrm{e}} = m_0^2c^2$ for the electron and $\underline{\mathbf{P}}^2_\phi = \underline{\mathbf{P}}^2_\gamma = 0$ for the photons, we have

$$\underline{\mathbf{P}}_\phi \cdot \underline{\mathbf{P}}_{\mathrm{e}} - \underline{\mathbf{P}}_\phi \cdot \underline{\mathbf{P}}_\gamma - \underline{\mathbf{P}}_{\mathrm{e}} \cdot \underline{\mathbf{P}}_\gamma = 0.$$

For a head-on collision [Fig. (b)], it is $\boldsymbol{\upsilon} \cdot \hat{\mathbf{k}} = -\upsilon$, $\boldsymbol{\upsilon} \cdot \hat{\mathbf{k}}' = \upsilon$ and $\hat{\mathbf{k}} \cdot \hat{\mathbf{k}}' = -1$. Substituting in the last equation,

$$\gamma m_0 \frac{E_{\mathrm{e}}}{c}(c+\upsilon) - 2\frac{E_\phi E_\gamma}{c^2} - \gamma m_0 \frac{E_\gamma}{c}(c-\upsilon)$$

$$\text{or} \quad E_\gamma = \frac{\gamma m_0 E_\phi c(c+\upsilon)}{2E_\phi + \gamma m_0 c(c-\upsilon)}.$$

Now, for $\gamma \gg 1$, it is $\upsilon \approx c$, $c+\upsilon \approx 2c$ and $\gamma(c-\upsilon) = \dfrac{c}{\gamma}\dfrac{1-\upsilon/c}{1-\upsilon^2/c^2} = \dfrac{c}{\gamma}\dfrac{1}{1+\upsilon/c} \approx \dfrac{c}{2\gamma}$.

Using these results, we find $E_\gamma \approx \dfrac{2\gamma m_0 E_\phi c^2}{2E_\phi + m_0 c^2/2\gamma}$

and, finally, $E_\gamma \approx \dfrac{\gamma m_0 c^2}{1 + m_0 c^2/4\gamma E_\phi}$, which, with $E_e = \gamma m_0 c^2$, is the result of Eq. (7.21),

$$\frac{E_\gamma}{E_e} = \frac{1}{1 + (E_e/E_\phi)/4\gamma^2} \quad \text{or} \quad \frac{E_\gamma}{E_e} = \frac{1}{1 + m_0^2 c^4/4E_e E_\phi}.$$

Example 8.4 Threshold energy for production of a π^0 by a photon and a proton

What is the threshold energy of the photon for the production of a π^0 during the scattering of a photon on a stationary proton, $\gamma + \mathrm{p} \to \mathrm{p} + \pi^0$? (This problem has also been solved in Example 7.5.)

For energy equal to the threshold energy, the particles produced will be moving with the same velocity, which will also be in the same direction as that of the incident γ (see figure).

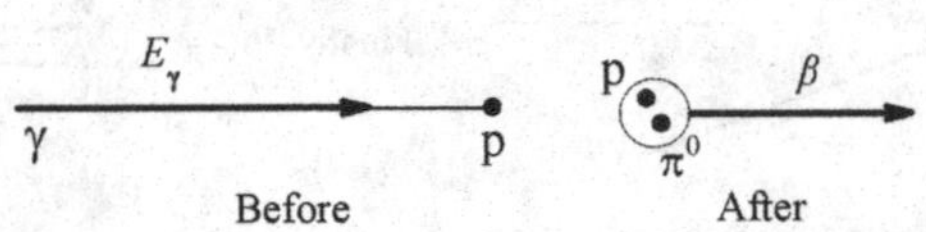

Let $\underline{\mathbf{P}}_\gamma$ and $\underline{\mathbf{P}}_\mathrm{p}$ be the initial energy-momentum four-vectors of the photon and the proton, respectively. The energy-momentum four-vectors of the pion and the proton created are $\underline{\mathbf{P}}'_\mathrm{p}$ and $\underline{\mathbf{P}}_\pi$. The law of conservation of the *total* energy-momentum four-vector gives

$$\left(\underline{\mathbf{P}}_\gamma + \underline{\mathbf{P}}_\mathrm{p}\right)^2 = \left(\underline{\mathbf{P}}_\pi + \underline{\mathbf{P}}'_\mathrm{p}\right)^2 = M_0^2 c^2. \tag{1}$$

M_0 is the rest mass of the system, as explained in Sect. 6.10.1. We are not interested in evaluating M_0 and it is used in order to stress the fact that the quantities in the parentheses are invariant.

We evaluate the energy-momentum four-vectors involved in the reaction:

$$\underline{\mathbf{P}}_\gamma = \frac{1}{c}(E_\gamma, E_\gamma, 0, 0), \quad \underline{\mathbf{P}}_\mathrm{p} = (m_\mathrm{p} c, 0, 0, 0). \tag{2}$$

Since $\left(\underline{\mathbf{P}}_\pi + \underline{\mathbf{P}}'_\mathrm{p}\right)^2$ is invariant, we may evaluate it in any frame of reference we wish. We choose to evaluate $\underline{\mathbf{P}}'_\mathrm{p}$ and $\underline{\mathbf{P}}_\pi$ in the zero-momentum frame of reference, since we know that, for threshold energy, the two particles are at rest in this frame. Thus, denoting these quantities in the zero-momentum frame of reference by $\underline{\mathbf{P}}'_{\mathrm{p},0}$ and $\underline{\mathbf{P}}_{\pi,0}$, we may write

$$\underline{\mathbf{P}}'_{\mathrm{p},0} = (m_{\mathrm{p}}c, 0, 0, 0), \qquad \underline{\mathbf{P}}_{\pi,0} = (m_\pi c, 0, 0, 0) \tag{3}$$

and

$$\left(\underline{\mathbf{P}}'_{\mathrm{p}} + \underline{\mathbf{P}}_\pi\right)^2 = \left(\underline{\mathbf{P}}'_{\mathrm{p},0} + \underline{\mathbf{P}}_{\pi,0}\right)^2. \tag{4}$$

Equation (1) now gives

$$\underline{\mathbf{P}}_\gamma^2 + \underline{\mathbf{P}}_{\mathrm{p}}^2 + 2\underline{\mathbf{P}}_\gamma \cdot \underline{\mathbf{P}}_{\mathrm{p}} = \underline{\mathbf{P}}'^2_{\mathrm{p},0} + \underline{\mathbf{P}}^2_{\pi,0} + 2\underline{\mathbf{P}}'_{\mathrm{p},0} \cdot \underline{\mathbf{P}}_{\pi,0} \tag{5}$$

or

$$0 + m_{\mathrm{p}}^2c^2 + 2E_\gamma m_{\mathrm{p}} = m_{\mathrm{p}}^2c^2 + m_\pi^2c^2 + 2m_{\mathrm{p}}m_\pi c^2 \tag{6}$$

which is solved to give $E_\gamma = \dfrac{m_\pi^2c^2 + 2m_{\mathrm{p}}m_\pi c^2}{2m_{\mathrm{p}}}$ or $E_\gamma = m_\pi c^2\left(1 + \dfrac{m_\pi}{2m_{\mathrm{p}}}\right)$, (7)

in agreement with the result of Example 7.5.

Problems

8.6 A photon and a particle have four-momenta $\underline{\mathbf{P}}$ and $\underline{\mathbf{Q}}$, respectively. Show that, in the frame of reference of the particle, it is $\underline{\mathbf{P}} \cdot \underline{\mathbf{Q}} = hfm_0$, where f is the frequency of the photon and m_0 the rest mass of the particle.

8.7 Two particles have four-momenta $\underline{\mathbf{P}}_1$ and $\underline{\mathbf{P}}_2$, respectively, and relative speed between them υ. If m_{10} and m_{20} are the rest masses of the particles, m_1 is the mass of the first particle in the frame of reference of the second and m_2 is the mass of the second particle in the frame of reference of the first, show that $\underline{\mathbf{P}}_1 \cdot \underline{\mathbf{P}}_2 = m_{10}m_{20}c^2\gamma(\upsilon) = m_{10}m_2c^2 = m_{20}m_1c^2$.

8.8 Consider an elastic collision between two particles. If the four-momenta of the two particles are, respectively, $\underline{\mathbf{P}}_1$ and $\underline{\mathbf{P}}_2$ before, and $\underline{\mathbf{P}}'_1$ and $\underline{\mathbf{P}}'_2$ after the collision, show that $\underline{\mathbf{P}}_1 \cdot \underline{\mathbf{P}}_2 = \underline{\mathbf{P}}'_1 \cdot \underline{\mathbf{P}}'_2$.

8.9 Show that for two photons with energies E_1 and E_2 and four-momenta $\underline{\mathbf{P}}_1$ and $\underline{\mathbf{P}}_2$, it is $\underline{\mathbf{P}}_1 \cdot \underline{\mathbf{P}}_2 = (E_1E_2/c^2)(1 - \cos\theta)$, where θ is the angle between the directions of motion of the two photons.

8.10 Use four-vectors to prove Eq. (7.10) for the Compton effect.

8.11 A pion decays into a muon and a neutrino: $\pi^- \to \mu^- + \bar{\nu}_\mu$. Taking into account the fact that the neutrino has a negligible rest mass, show that $\underline{\mathbf{P}}_\pi \cdot \underline{\mathbf{P}}_\mu = \frac{1}{2}\left(m_\pi^2 + m_\mu^2\right)c^2$, where $\underline{\mathbf{P}}_\pi$ and $\underline{\mathbf{P}}_\mu$ are the energy-momentum four-vectors of the pion and the muon, respectively.

8.12 A rocket is propelled by emitting photons backwards. The rocket is initially at rest and has a rest mass equal to M_{0i}. At time t, the rocket has a speed of υ and its rest mass has been reduced to M_0. Find the ratio M_{0i}/M_0 as a function of υ.

Ans.: $\dfrac{M_{0i}}{M_0} = \sqrt{\dfrac{1+\beta}{1-\beta}}$

8.13 A particle in a magnetic field moves on a helical orbit given in parametric form by

$$x = R\sin\omega t, \quad y = R\cos\omega t, \quad z = ut$$

where t is the time. Find the proper acceleration of the particle.

$$\text{Ans.: } \alpha = \frac{c^2\omega^2 R}{c^2 - \omega^2 R^2 - u^2}$$

8.14 A particle P has four-momentum $\underline{\mathbf{P}}$ and velocity $\boldsymbol{\upsilon}_{\mathrm{P}}$ as observed in the frame of reference S. In the same frame of reference, a second observer, S′, has velocity $\boldsymbol{\upsilon}_{\mathrm{S}'}$ and four-velocity $\underline{\mathbf{U}}_{\mathrm{S}'}$. Show that $\underline{\mathbf{P}} \cdot \underline{\mathbf{U}}_{\mathrm{S}'}$ is the energy of the particle in the frame of reference S′.

8.2.5 The Four-Vector of Force

We define the four-vector of force as the rate of change of the energy-momentum four-vector $\underline{\mathbf{P}} = m_0 \underline{\mathbf{U}}$ with respect to the particle's proper time τ,

$$\underline{\mathbf{F}} = \frac{d\underline{\mathbf{P}}}{d\tau} = m_0 \frac{d\underline{\mathbf{U}}}{d\tau}. \tag{8.81}$$

From Eq. (8.63),

$$\frac{d\underline{\mathbf{U}}}{d\tau} = \underline{\mathbf{A}} = \left(\gamma_{\mathrm{P}} \frac{d(\gamma_{\mathrm{P}} c)}{dt},\ \gamma_{\mathrm{P}} \frac{d(\gamma_{\mathrm{P}} \boldsymbol{\upsilon})}{dt} \right) = \left(\frac{\gamma_{\mathrm{P}}}{m_0 c} \frac{dE}{dt},\ \gamma_{\mathrm{P}} \frac{d(\gamma_{\mathrm{P}} \boldsymbol{\upsilon})}{dt} \right), \tag{8.82}$$

it follows that it is

$$\underline{\mathbf{F}} = \left(\gamma_{\mathrm{P}} \frac{d(\gamma_{\mathrm{P}} m_0 c)}{dt},\ \gamma_{\mathrm{P}} \frac{d(\gamma_{\mathrm{P}} m_0 \boldsymbol{\upsilon})}{dt} \right) = \left(\frac{\gamma_{\mathrm{P}}}{c} \frac{dE}{dt},\ \gamma_{\mathrm{P}} \frac{d(\gamma_{\mathrm{P}} m_0 \boldsymbol{\upsilon})}{dt} \right). \tag{8.83}$$

The three-vector of the force is given by the equation $\mathbf{F} = \dfrac{d(\gamma_{\mathrm{P}} m_0 \boldsymbol{\upsilon})}{dt}$, while $\dfrac{dE}{dt} = \mathbf{F} \cdot \boldsymbol{\upsilon}$ is the rate of production of work by the force $\mathbf{F}$ on the particle, which is moving with velocity $\boldsymbol{\upsilon}$ (power).

Therefore,

$$\underline{\mathbf{F}} = \left(\frac{\gamma_{\mathrm{P}}}{c} \mathbf{F} \cdot \boldsymbol{\upsilon},\ \gamma_{\mathrm{P}} \mathbf{F} \right). \tag{8.84}$$

The transformation of force is found by applying the Lorentz transformation to the four-vector $\underline{\mathbf{F}} = (F^0, F^1, F^2, F^3)$,

$$F'^0 = \gamma(F^0 - \beta F^1), \quad F'^1 = \gamma(F^1 - \beta F^0), \quad F'^2 = F^2, \quad F'^3 = F^3. \tag{8.85}$$

Let us first examine the simple case in which the point of application of the force **F** is stationary in the frame of reference S, ($\mathbf{v} = 0$ and $\gamma_P = 1$). The four-vector of the force is, according to Eq. (8.84), $\underline{\mathbf{F}} = (0,\ \mathbf{F})$

and

$$F^0 = 0, \quad F^1 = F_x, \quad F^2 = F_y, \quad F^3 = F_z. \tag{8.86}$$

From Eqs. (8.84) and (8.85) there follow the relations

$$\frac{\gamma'_P}{c}\mathbf{F}' \cdot \mathbf{v}' = -\gamma\beta F_x, \qquad \gamma'_P F'_x = \gamma F_x, \qquad \gamma'_P F'_y = F_y, \qquad \gamma'_P F'_z = F_z. \tag{8.87}$$

However, the point of application of the force is moving with the frame of reference S with speed $-V$ relative to the frame S′, and $\gamma'_P = \gamma$. The transformation of the force is, therefore,

$$\mathbf{F}' \cdot \mathbf{v}' = -VF_x, \qquad F'_x = F_x, \qquad F'_y = F_y/\gamma, \qquad F'_z = F_z/\gamma. \tag{8.88}$$

We will now assume that the force is exerted on a particle which is moving with instantaneous velocity $\mathbf{v}$ in the frame S and that frame S′ moves with velocity $\mathbf{V} = V\hat{\mathbf{x}}$ relative to frame S. The four-vector of the force is given by Eq. (8.84) and

$$F^0 = \frac{\gamma_P}{c}\mathbf{F} \cdot \mathbf{v}, \qquad F^1 = \gamma_P F_x, \qquad F^2 = \gamma_P F_y, \qquad F^3 = \gamma_P F_z. \tag{8.89}$$

Equations (8.85) now give

$$\begin{aligned} &\frac{\gamma'_P}{c}\mathbf{F}' \cdot \mathbf{v}' = \gamma\left(\frac{\gamma_P}{c}\mathbf{F} \cdot \mathbf{v} - \gamma_P \beta F_x\right), \\ &\gamma'_P F'_x = \gamma\left(\gamma_P F_x - \beta\frac{\gamma_P}{c}\mathbf{F} \cdot \mathbf{v}\right), \qquad \gamma'_P F'_y = \gamma_P F_y, \qquad \gamma'_P F'_z = \gamma_P F_z. \end{aligned} \tag{8.90}$$

These simplify to

$$\begin{aligned} &\mathbf{F}' \cdot \mathbf{v}' = \frac{\gamma_P}{\gamma'_P}\gamma(\mathbf{F} \cdot \mathbf{v} - VF_x), \\ &F'_x = \frac{\gamma_P}{\gamma'_P}\gamma\left(F_x - \frac{V}{c^2}\mathbf{F} \cdot \mathbf{v}\right), \quad F'_y = \frac{\gamma_P}{\gamma'_P}F_y, \quad F'_z = \frac{\gamma_P}{\gamma'_P}F_z. \end{aligned} \tag{8.91}$$

Taking into account Eq. (8.57), $\dfrac{\gamma_P}{\gamma'_P}\gamma = \left(1 - \dfrac{V v_x}{c^2}\right)^{-1}$, we find

$$\mathbf{F}' \cdot \mathbf{\upsilon}' = \frac{\mathbf{F} \cdot \mathbf{\upsilon} - VF_x}{1 - \dfrac{V\upsilon_x}{c^2}},$$

$$F'_x = \frac{F_x - \dfrac{V}{c^2}\mathbf{F} \cdot \mathbf{\upsilon}}{1 - \dfrac{V\upsilon_x}{c^2}}, \quad F'_y = \frac{F_y\sqrt{1 - V^2/c^2}}{1 - \dfrac{V\upsilon_x}{c^2}}, \quad F'_z = \frac{F_z\sqrt{1 - V^2/c^2}}{1 - \dfrac{V\upsilon_x}{c^2}} \tag{8.92}$$

for the transformation of the force when the point of its application is moving with an instantaneous velocity $\mathbf{\upsilon}$ in the frame of reference S. It should be noted that $\mathbf{F} \cdot \mathbf{\upsilon}$ is the rate of production of work by the force (power) in the frame of reference S.

8.2.6 The Four-Vectorial Equation of Motion

The definition of the four-vector of force [Eq. (8.81)], constitutes the four-vectorial equation of motion:

$$\frac{d\underline{\mathbf{P}}}{d\tau} = \underline{\mathbf{F}}. \tag{8.93}$$

Substituting, this gives

$$\frac{d\underline{\mathbf{P}}}{d\tau} = m_0 \frac{d\underline{\mathbf{U}}}{d\tau} = m_0 \left(\gamma_{\mathrm{P}} \frac{d(\gamma_{\mathrm{P}} c)}{dt}, \ \gamma_{\mathrm{P}} \frac{d(\gamma_{\mathrm{P}} \mathbf{\upsilon})}{dt} \right) = \underline{\mathbf{F}} = \left(\frac{\gamma_{\mathrm{P}}}{c} \mathbf{F} \cdot \mathbf{\upsilon}, \ \gamma_{\mathrm{P}} \mathbf{F} \right) \tag{8.94}$$

and, therefore,

$$\frac{d(\gamma_{\mathrm{P}} m_0 c^2)}{dt} = \mathbf{F} \cdot \mathbf{\upsilon}, \qquad \frac{d(\gamma_{\mathrm{P}} m_0 \mathbf{\upsilon})}{dt} = \mathbf{F}, \tag{8.95}$$

where $\mathbf{F}$ is the three-vector of force. The first equation gives the rate of change of the particle's energy as equal to the rate of production of work by the force. The second equation is Newton's second law, $d\mathbf{p}/dt = d(m\mathbf{\upsilon})/dt = \mathbf{F}$, with the mass being the relativistic mass $m = \gamma_{\mathrm{P}} m_0$.

Reference

1. J.R. Klauder (ed.), *Magic without Magic: John Archibald Wheeler* (W.H. Freeman, New York, 1973)

Chapter 9
Electromagnetism

9.1 Introduction

The Special Theory of Relativity was originally formulated in order to solve problems which appeared in electromagnetic theory. The electromagnetic waves and light are the most relativistic entities in nature, so it was only natural that the first disagreements with classical Physics should involve phenomena related to electromagnetic waves and light. These disagreements appeared after the strict mathematical formulation of the laws of Electromagnetism by Maxwell, in 1862. The mathematical forms of the laws of electromagnetism are given by Maxwell's equations, which are the following:

$$\nabla \cdot \mathbf{E} = \frac{\rho}{\varepsilon_0}, \qquad \nabla \cdot \mathbf{B} = 0, \qquad \nabla \times \mathbf{E} = -\frac{\partial \mathbf{B}}{\partial t}, \qquad \nabla \times \mathbf{B} = \varepsilon_0 \mu_0 \frac{\partial \mathbf{E}}{\partial t} + \mu_0 \mathbf{J}. \tag{9.1}$$

In vacuum, without free charges, $\rho = 0$, or currents, $\mathbf{J} = 0$, the equations simplify to:

$$\nabla \cdot \mathbf{E} = 0, \qquad \nabla \cdot \mathbf{B} = 0, \qquad \nabla \times \mathbf{E} = -\frac{\partial \mathbf{B}}{\partial t}, \qquad \nabla \times \mathbf{B} = \varepsilon_0 \mu_0 \frac{\partial \mathbf{E}}{\partial t}, \tag{9.2}$$

which may be combined to give one equation for the electric field $\mathbf{E}$

$$\frac{\partial^2 \mathbf{E}}{\partial x^2} + \frac{\partial^2 \mathbf{E}}{\partial y^2} + \frac{\partial^2 \mathbf{E}}{\partial z^2} = \frac{1}{c^2}\frac{\partial^2 \mathbf{E}}{\partial t^2} \qquad \text{or} \qquad \nabla^2 \mathbf{E} = \frac{1}{c^2}\frac{\partial^2 \mathbf{E}}{\partial t^2}, \tag{9.3}$$

© Springer International Publishing Switzerland 2016

C. Christodoulides, *The Special Theory of Relativity*,
Undergraduate Lecture Notes in Physics, DOI 10.1007/978-3-319-25274-2_9

and one equation for the magnetic field **B**

$$\frac{\partial^2 \mathbf{B}}{\partial x^2} + \frac{\partial^2 \mathbf{B}}{\partial y^2} + \frac{\partial^2 \mathbf{B}}{\partial z^2} = \frac{1}{c^2}\frac{\partial^2 \mathbf{B}}{\partial t^2} \qquad \text{or} \qquad \nabla^2 \mathbf{B} = \frac{1}{c^2}\frac{\partial^2 \mathbf{B}}{\partial t^2}. \tag{9.4}$$

These equations represent an electromagnetic field which moves with a speed of

$$c = \frac{1}{\sqrt{\varepsilon_0 \mu_0}} = 299\,792\,458 \text{ m/s}. \tag{9.5}$$

Maxwell's equations and the wave equation describe the behavior of the fields of **E** and **B**, as this was observed in the laboratory. Unless there is a reason for us to assume that the Earth is a privileged frame of reference, these equations must have the same form for all inertial observers. In addition, the speed of the electromagnetic waves, as it is obtained from the equations, is not referred to any privileged observer, but it must be considered to be the same for all observers. The propagation of electromagnetic waves required, according to pre-twentieth-century Physics, the existence of a medium, the aether, in which the waves propagate. In this case, however, the speed of the waves would be different for an observer moving relative to the aether, according to the Galilean transformation:

$$x' = x - Vt, \qquad y' = y, \qquad z' = z, \qquad t' = t. \tag{9.6}$$

The fact that the transformation of Galileo does not leave Maxwell's equations invariant has already been mentioned. The wave equation for the electric field vector, for example, is given by the Galilean transformation the form

$$\nabla'^2 \mathbf{E} = \frac{1}{c^2}\frac{\partial^2 \mathbf{E}}{\partial t'^2} + \frac{1}{c^2}\left(V^2 \frac{\partial^2 \mathbf{E}}{\partial x'^2} - 2V \frac{\partial^2 \mathbf{E}}{\partial x' \partial t'} \right). \tag{9.7}$$

On the other hand, experiments show that the speed of light in vacuum is independent of the motion of either the source or the observer. The Michelson-Morley experiment of 1887, showed that the motion of the Earth through the aether could not be observed. A possible explanation for this could be that the aether was dragged by the Earth as the latter moves through the aether. This, however, is precluded by the observation of the aberration of light. The final conclusion that had to be reached was that the aether does not exist.

Lorentz established that transformation from the inertial frame of reference S:(x, y, z, t) to the inertial frame S$'$: (x', y', z', t') which moves with a velocity $\mathbf{V} = V\hat{\mathbf{x}}$ relative to S, which must replace that of Galileo and which leaves

Maxwell's equations and the wave equation invariant. The transformation is the one known today by his name:

$$x' = \gamma(x - Vt) \quad y' = y \quad z' = z \quad t' = \gamma\left(t - \frac{V}{c^2}x\right). \tag{9.8}$$

In Appendix 4, it is shown that this transformation leaves Maxwell's equations invariant, provided that the fields transform in a particular way. The transformation, therefore, also leaves the wave equation invariant.

In this chapter we will follow a procedure which is different than that by which, historically, the Special Theory of Relativity was developed. Specifically, we will take for granted the experimental fact that, by contrast to mass, the electric charge is invariant with speed. We will accept that the Lorentz force $\mathbf{F} = q\mathbf{E} + q\mathbf{v} \times \mathbf{B}$ has the same form in all inertial frames of reference and we will make use of the transformation of force, as this has already been formulated. In this way, we will find the transformations of the fields **E** and **B** which leave Maxwell's equations invariant.

9.2 The Invariance of Electric Charge

By contrast to mass, the electric charge remains invariant during its motion. This is an experimental fact. It is known that the electrons of an atom move at speeds which are very different to those of the protons of the atom's nucleus. For example, the speed of the electron in the ground state of the hydrogen atom is about $c/137$, while the nucleus, i.e. a proton, is virtually stationary. Had there been a variation of charge with speed, the hydrogen atom would not have been neutral when it had its electron in orbit about the nucleus. The neutrality of the atoms and of the molecules proves the absence of a change of the electric charge with speed.

A simple experiment, consisting of observing that a solid remains neutral when heated, in which case its electrons move with speeds of the order of $0.001c$, places an upper limit to the variation of the electronic charge which is lower than $10^{-10}|e|$ in absolute value, where e is the electronic charge.

With experiments performed in 1960, King [1] showed, with very great accuracy, that the molecules of hydrogen and the atoms of helium are neutral. King used a vessel, electrically isolated from its surroundings, which contained hydrogen gas (Fig. 9.1). From a small hole the gas could be allowed to escape, in such a way that prevented the escape of ionized molecules. The vessel contained a total of 17 grams of gas, or about 5×10^{24} molecules. Each molecule has two protons and two electrons. If the molecules had a certain charge, it would have been possible to detect the gradual charging of the initially neutral vessel, above a certain limit, as the charged molecules escaped. For example, if the proton charge differs by 1 part in 10^{20} compared to that of the electron, the escape of the whole quantity of the gas

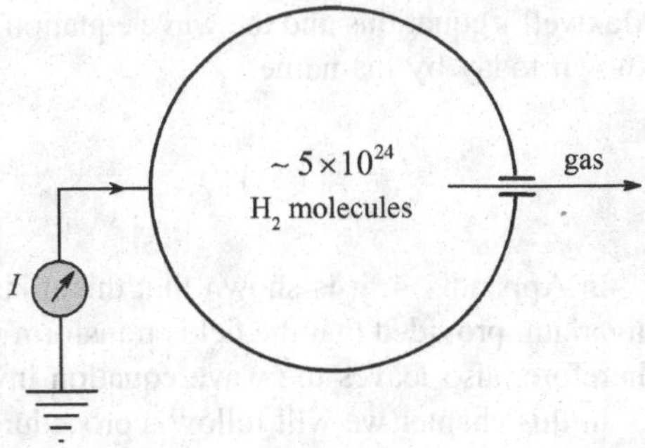

Fig. 9.1 The arrangement used by J.G. King in order to test the invariance of the electric charge

from the vessel would lead to it acquiring a charge equal, in absolute value, to $Q = 2 \times (5 \times 10^{24}) \times (10^{-20}|e|) = 10^5|e|$, where $|e|$ is the absolute value of the charge of the electron. This was, approximately, the limit of sensitivity of the experiment and, at this level, no charging of the vessel was observed. Consequently, the experiment showed that any charge present in the 'neutral' molecule of H_2 is smaller than 10^{-19} $|e|$ in absolute value. The same experiment was repeated with He gas, with the same results. The electrons of the atoms of hydrogen move with speeds of the order of 0.007 c, while in the He atoms the electrons have twice that speed. At speeds up to 0.01 c–0.02 c, no variation of the electric charge was observed by more than 1 part in 10^{19}.

Apart from the electrons, the protons and the neutrons inside the nuclei have large enough energies, corresponding in some cases to speeds of the order of 0.2 c–0.3 c. King's experiment with He essentially tests the invariance of the proton charge, to the accuracy mentioned, for speeds up to a significant fraction of the speed of light in vacuum. In subsequent experiments, King [2] improved the sensitivity of the experiment and measurements with H_2, He and SF_6 gave, as upper limits to the absolute values of possible charges per molecule of these gases the values of $(1.8 \pm 5.4) \times 10^{-21}$ $|e|$, $(-0.7 \pm 4.7) \times 10^{-21}$ $|e|$ and $(0 \pm 4.3) \times 10^{-21}$ $|e|$, respectively.

Hughes et al. [3] tried to detect the electrostatic deflection of the 'neutral' atoms in a beam of cesium atoms. The same experiment was performed with a beam of potassium atoms. No charge was detected which would mean a difference greater than 3.5×10^{-19} $|e|$ in the absolute values of the charges of the electron and of the proton. In potassium atoms, which has an atomic number $Z = 55$, the electrons of the K shell have speeds of the order of 0.4 c, in which case the neutrality of these atoms constitutes significant evidence for the invariance of the electric charge.

Very important additional evidence for the invariance of the electric charge is provided by the magnetic forces exerted on moving charges. The force exerted on an electric charge q which moves with velocity $\mathbf{\upsilon}$ in a space where there exist an electric field **E** and a magnetic field **B**, known as the Lorentz force, is given by the relation $\mathbf{F} = q\mathbf{E} + q\mathbf{\upsilon} \times \mathbf{B}$. This relation actually also constitutes the definition of the magnitude **B**. The equation of motion of a particle with rest mass m_0 and charge

q in a space where the fields are $\mathbf{E}$ and $\mathbf{B}$ will be, according to the Special Theory of Relativity,

$$\frac{d}{dt}\left(\frac{m_0\mathbf{v}}{\sqrt{1-v^2/c^2}}\right)=q\mathbf{E}+q\mathbf{v}\times\mathbf{B}. \tag{9.9}$$

This equation has been verified by experiments such as those of Kaufmann and Bucherer, which were described in Chap. 1. Since then, it has been tested countless times in applications as, for example, the construction and operation of high-energy accelerators, such as synchrotrons. In these, the variation with speed of the inertial mass is taken into account (see Sect. 7.3). Magnetic fields are used for the deflection of the particles being accelerated and Eq. (9.9) is used with constant electric charge q. Naturally, the emission of electromagnetic radiation by accelerated charges moving at high speeds must be accounted for. The energies to which electrons are accelerated by today's synchrotrons exceed 25 GeV, while protons will eventually be accelerated up to 7 TeV at CERN, with the LHC accelerator. It should be noted that an electron with kinetic energy of 25 GeV moves with a speed equal to 0.999 999 999 8 c, and a 7 TeV proton moves with a speed of 0.999 999 991 3 c (see Example 6.1). Even at such speeds of the moving charge, Eq. (9.9) is adequate, assuming q remains constant, with the correction we mentioned for the emission of electromagnetic radiation.

In what will follow, the invariance of the electric charge will be taken as an experimental fact. Alternatively, if Maxwell's equations are considered as being in agreement with the Special Theory of Relativity, the invariance of the electric charge follows as a consequence. This is proved in Section A4.8 of Appendix 4, as a consequence of the transformation of the charge density ρ and of the current density $\mathbf{J}$ which leave invariant the equation of continuity of charge, and which, in its turn, follows from the conservation of electric charge.

Problem

9.1 A capacitor with parallel plane plates is at rest, in vacuum, in a frame of reference S.The capacitor's plates are parallel to the xz plane and are rectangular, with sides a and b which are parallel to the x and z axes, respectively. The distance between the plates is d. The two plates have charges $+Q$ and $-Q$, respectively. The potential difference across the plates is ΔU and the electric field in the region between them has magnitude E.

Given the invariance of the electric charge, find the capacitance of the capacitor, C', the strength of the electric field in the space between its plates, E', the potential difference across the plates of the capacitor, $\Delta U'$, the total electrostatic energy stored in the capacitor, W', and the density of electrostatic energy per unit volume between the plates of the capacitor, w', as these are measured by an observer in another frame of reference, S$'$, which is moving relative to frame S with a speed V in the direction of positive values of x. Express the results in terms of V and the corresponding values in frame S. Ans.: $C'=C/\gamma$, $E'=\gamma E$, $\Delta U'=\gamma\,\Delta U$, $W'=\gamma W$, $w'=\gamma^2 w$.

9.3 The Transformations of the Electric Field and the Magnetic Field

We assume that the electric field and the magnetic field in the inertial frame of reference S : (x, y, z, t) are **E** and **B**, respectively, and in the inertial frame of reference S′ : (x', y', z', t'), which moves with a velocity $\mathbf{V} = V\hat{\mathbf{x}}$ relative to the frame S, they are $\mathbf{E}'$ and $\mathbf{B}'$. A particle with charge q moves in frame S with a velocity equal to $\boldsymbol{\upsilon}$ and with a velocity $\boldsymbol{\upsilon}'$ in frame S′. The Lorentz force exerted on it in the two frames of reference are, respectively,

$$\mathbf{F} = q\mathbf{E} + q\boldsymbol{\upsilon} \times \mathbf{B} \tag{9.10}$$

and

$$\mathbf{F}' = q\mathbf{E}' + q\boldsymbol{\upsilon}' \times \mathbf{B}'. \tag{9.11}$$

The y and y' components of the forces **F** and **F′** are found to be

$$F_y = q\left(E_y + \upsilon_z B_x - \upsilon_x B_z\right) \quad \text{and} \quad F'_y = q\left(E'_y + \upsilon'_z B'_x - \upsilon'_x B'_z\right). \tag{9.12}$$

The y component of the force transforms according to the relation [Eq. (6.83)]

$$F'_y = \frac{F_y}{\gamma(1 - V\upsilon_x/c^2)}, \tag{9.13}$$

where $\gamma = 1/\sqrt{1 - V^2/c^2}$. This equation may be rewritten as

$$F_y = \gamma(1 - V\upsilon_x/c^2)F'_y. \tag{9.14}$$

Substituting from Eqs. (9.12), it follows that

$$q\left(E_y + \upsilon_z B_x - \upsilon_x B_z\right) = \gamma(1 - V\upsilon_x/c^2)q\left(E'_y + \upsilon'_z B'_x - \upsilon'_x B'_z\right). \tag{9.15}$$

From the transformation of velocity, it is

$$\upsilon'_x = \frac{\upsilon_x - V}{1 - V\upsilon_x/c^2}, \quad \upsilon'_z = \frac{\upsilon_z}{\gamma(1 - V\upsilon_x/c^2)}. \tag{9.16}$$

Substituting in Eq. (9.15), we find

$$E_y + \upsilon_z B_x - \upsilon_x B_z = \gamma(1 - V\upsilon_x/c^2)E'_y + \upsilon_z B'_x - \gamma(\upsilon_x - V)B'_z \tag{9.17}$$

or

$$\left[E_y - \gamma\left(E'_y + VB'_z\right)\right] - \left[B_z - \gamma\left(B'_z + VE'_y/c^2\right)\right]\upsilon_x + \left[B_x - B'_x\right]\upsilon_z = 0. \tag{9.18}$$

Equation (9.18) must hold for every value of the velocity $\mathbf{\upsilon}$. Equating the coefficients of the components of $\mathbf{\upsilon}$ to zero, we find

$$E_y = \gamma\left(E'_y + VB'_z\right), \qquad B_x = B'_x, \qquad B_z = \gamma\left(B'_z + VE'_y/c^2\right). \tag{9.19}$$

Similarly, starting from the equation

$$F_z = F'_z\gamma(1 - V\upsilon_x/c^2), \tag{9.20}$$

we end up with the relations

$$E_z = \gamma\left(E'_z - VB'_y\right), \qquad B_x = B'_x, \qquad B_y = \gamma\left(B'_y - VE'_z/c^2\right). \tag{9.21}$$

Finally, from the equation

$$F_x = F'_x + \frac{V\upsilon'_y/c^2}{1 + V\upsilon'_x/c^2}F'_y + \frac{V\upsilon'_z/c^2}{1 + V\upsilon'_x/c^2}F'_z, \tag{9.22}$$

and substituting (only) for F_x and F'_x from Eqs. (9.10) to (9.11), we find

$$q(E_x + \upsilon_y B_z - \upsilon_z B_y) = q\left(E'_x + \upsilon'_y B'_z - \upsilon'_z B'_y\right)_x + \frac{V\upsilon'_y/c^2}{1 + V\upsilon'_x/c^2}F'_y + \frac{V\upsilon'_z/c^2}{1 + V\upsilon'_x/c^2}F'_z. \tag{9.23}$$

Because it is

$$\upsilon'_y = \frac{\upsilon_y}{\gamma(1 - V\upsilon_x/c^2)} \quad \text{and} \quad \upsilon'_z = \frac{\upsilon_z}{\gamma(1 - V\upsilon_x/c^2)}, \tag{9.24}$$

the only terms in Eq. (9.23) which are independent of the components of the velocity $\mathbf{\upsilon}$ are qE_x and qE'_x. Equating these two terms, we find that it is

$$E_x = E'_x. \tag{9.25}$$

Summarizing, the results found are

$$\begin{aligned} E'_x &= E_x & E_x &= E'_x \\ E'_y &= \gamma(E_y - VB_z) & E_y &= \gamma\left(E'_y + VB'_z\right) \\ E'_z &= \gamma(E_z + VB_y) & E_z &= \gamma\left(E'_z - VB'_y\right) \end{aligned} \tag{9.26}$$

$$\begin{aligned} B'_x &= B_x & B_x &= B'_x \\ B'_y &= \gamma(B_y + VE_z/c^2) & B_y &= \gamma\left(B'_y - VE'_z/c^2\right) \\ B'_z &= \gamma(B_z - VE_y/c^2) & B_z &= \gamma\left(B'_z + VE'_y/c^2\right) \end{aligned} \tag{9.27}$$

where the inverse transformations have also been given.

The transformations correlate the components of the electric field **E** and the magnetic field **B** at the point (x, y, z, t) of the inertial frame of reference S, with those of $\mathbf{E}'$ and $\mathbf{B}'$ at the corresponding point (x', y', z', t') of the inertial frame of reference S′, which moves with velocity $\mathbf{V} = V\hat{\mathbf{x}}$ relative to frame S. Denoting by the indices $\|$ the components of the fields which are parallel to the velocity **V** and by the indices $\perp$ the components of the fields which are normal to the velocity **V**, the equations can also be stated in vector form as

$$\mathbf{E}'_{\|} = \mathbf{E}_{\|} \qquad \mathbf{E}'_{\perp} = \gamma(\mathbf{E} + \mathbf{V} \times \mathbf{B})_{\perp} \tag{9.28}$$

$$\mathbf{B}'_{\|} = \mathbf{B}_{\|} \qquad \mathbf{B}'_{\perp} = \gamma\left(\mathbf{B} - \mathbf{V} \times \mathbf{E}/c^2\right)_{\perp} \tag{9.29}$$

and the inverse transformation:

$$\mathbf{E}_{\|} = \mathbf{E}'_{\|} \qquad \mathbf{E}_{\perp} = \gamma(\mathbf{E}' - \mathbf{V} \times \mathbf{B}')_{\perp} \tag{9.30}$$

$$\mathbf{B}_{\|} = \mathbf{B}'_{\|} \qquad \mathbf{B}_{\perp} = \gamma\left(\mathbf{B}' + \mathbf{V} \times \mathbf{E}'/c^2\right)_{\perp}. \tag{9.31}$$

The general transformation without rotation of the coordinate axes is, in vector form,

$$\mathbf{E}' = \gamma\mathbf{E} - (\gamma - 1)\frac{\mathbf{V}}{V^2}(\mathbf{V} \cdot \mathbf{E}) + \gamma(\mathbf{V} \times \mathbf{B}) \tag{9.32}$$

$$\mathbf{B}' = \gamma\mathbf{B} - (\gamma - 1)\frac{\mathbf{V}}{V^2}(\mathbf{V} \cdot \mathbf{B}) - \frac{\gamma}{c^2}(\mathbf{V} \times \mathbf{E}). \tag{9.33}$$

with the inverse transformation being:

$$\mathbf{E} = \gamma\mathbf{E}' - (\gamma - 1)\frac{\mathbf{V}}{V^2}(\mathbf{V}\cdot\mathbf{E}') - \gamma(\mathbf{V}\times\mathbf{B}') \tag{9.34}$$

$$\mathbf{B} = \gamma\mathbf{B}' - (\gamma - 1)\frac{\mathbf{V}}{V^2}(\mathbf{V}\cdot\mathbf{B}') + \frac{\gamma}{c^2}(\mathbf{V}\times\mathbf{E}'), \tag{9.35}$$

In the transformations of the electric and the magnetic field the interdependence of one field with the other is obvious, on the transition from one inertial frame of reference to another. This explains why the Special Theory of Relativity is considered to have unified these two fields into one, the electromagnetic. This was, historically, the second unification of forces, after the unification, by Newton, of the gravitational force, as observed on the Earth, with the force that keeps the celestial bodies in their orbits.

Problems

9.2 Using the transformations of the fields, find the electric field and the magnetic field in the region between the plates of the capacitor of Problem 9.1, as measured by an observer in the frame of reference S′. Express the results in terms of the components of the fields in frame S and the relative speed of the two frames of reference.
Ans.: $E'_x = 0$, $E'_y = \gamma E$, $E'_z = 0$, $B'_x = 0$, $B'_y = 0$, $B'_z = -\gamma VE/c^2$

9.3 A capacitor with parallel plane plates is at rest in the frame of reference S. The capacitor's plates are rectangular with sides a and b, have a distance d between them, and are oriented so that they are parallel to the yz plane. The charges on the plates are $\pm Q$, which create an electric field $\mathbf{E} = E\hat{\mathbf{x}}$ in the space between the plates. The magnetic field in this region is zero. The potential difference across the plates is ΔU. Another frame of reference, S′, moves with a velocity $V\hat{\mathbf{x}}$ relative to S. The axes of the two frames are parallel and coincided at $t = t' = 0$.

(a) Without using the transformations of the electromagnetic field, find, in terms of C, $\mathbf{E}$, ΔU and the speed V, the capacitance of the capacitor, C', the strength of the electric field in the space between the plates of the capacitor, $\mathbf{E}'$, and the potential difference across the plates of the capacitor, $\Delta U'$, as measured by an observer in the frame of reference S′. Ans.: $C' = \gamma C$, $\mathbf{E}' = \mathbf{E}$, $\Delta U' = \Delta U/\gamma$

(b) Use now the transformations of the fields, in order to find the electric field again and the magnetic field in the space between the plates of the capacitor, in the frame of reference S′. Ans.: $\mathbf{E}' = E\hat{\mathbf{x}}$, $\mathbf{B} = 0$

9.4 Show that the magnitudes $\mathbf{E}\cdot\mathbf{B}$ and $E^2 - c^2B^2$ remain invariant under the Lorentz transformations from one inertial system of reference S to another, S′, which is moving with a velocity $V\hat{\mathbf{x}}$ relative to the frame S.

9.5 A rectilinear charge distribution of linear charge density λ', extending to infinity on both sides, lies on the x'-axis of an inertial frame of reference S′. The strength of the electric field it produces is given by the well known equations

$$E'_x = 0, \qquad E'_y = \frac{\lambda' y'}{2\pi\varepsilon_0 (y'^2 + z'^2)}, \qquad E'_z = \frac{\lambda' z'}{2\pi\varepsilon_0 (y'^2 + z'^2)}.$$

In another frame of reference, S, in which frame S′ moves with a velocity $\mathbf{V} = V\hat{\mathbf{x}}$ (V positive), the charge distribution appears as a current I in the direction of positive x's. What are the magnetic and the electric fields in frame S? Compare the magnitudes of the electric and the magnetic forces exerted on a charge q which moves with velocity $\mathbf{V}$ in frame S and is at point (x, y, z).

Ans.: $\mathbf{E} = \dfrac{\lambda}{2\pi\varepsilon_0} \dfrac{y\hat{\mathbf{y}} + z\hat{\mathbf{z}}}{y^2 + z^2}$, $\mathbf{B} = \dfrac{\mu_0}{2\pi} I \dfrac{z\hat{\mathbf{y}} - y\hat{\mathbf{z}}}{y^2 + z^2}$, $\dfrac{\text{Electric Force}}{\text{Magnetic Force}} = -\dfrac{c^2}{V^2}$

9.4 Fields of a Moving Electric Charge

We will apply the Lorentz transformation in order to evaluate the fields produced by an electric charge which moves with a constant velocity. In the inertial frame of reference S, the charge Q is moving with a constant velocity $\mathbf{V} = V\hat{\mathbf{x}}$ on the x-axis and is situated at the point $x = 0,\ \ y = 0,\ \ z = 0$ at the moment $t = 0$. We assume that the charge is stationary at the origin of the axes of an inertial frame of reference S′ (Fig. 9.2). In S′, the charge produces the fields

$$\mathbf{E}' = \frac{Q\,\mathbf{r}'}{4\pi\varepsilon_0 r'^3} \quad \text{and} \quad \mathbf{B}' = 0. \tag{9.36}$$

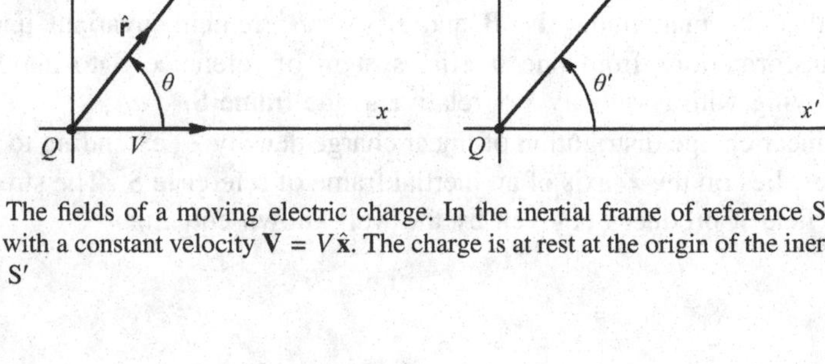

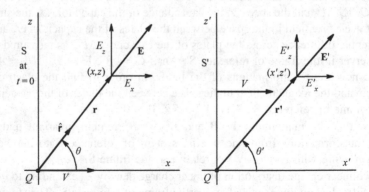

Fig. 9.2 The fields of a moving electric charge. In the inertial frame of reference S, the charge Q moves with a constant velocity $\mathbf{V} = V\hat{\mathbf{x}}$. The charge is at rest at the origin of the inertial frame of reference S′

The fields have axial symmetry about the x'-axis. We evaluate the fields on the $x'z'$ plane, on which the components of the electric field are

$$E'_x = \frac{1}{4\pi\varepsilon_0}\frac{Q}{r'^2}\cos\theta' = \frac{Q}{4\pi\varepsilon_0}\frac{x'}{(x'^2+z'^2)^{3/2}}, \tag{9.37}$$

$$E'_z = \frac{1}{4\pi\varepsilon_0}\frac{Q}{r'^2}\sin\theta' = \frac{Q}{4\pi\varepsilon_0}\frac{z'}{(x'^2+z'^2)^{3/2}}. \tag{9.38}$$

The transformation of coordinates of the charge gives

$$x' = \gamma(x - Vt), \qquad y' = y, \qquad z' = z, \qquad t' = \gamma(t - Vx/c^2), \tag{9.39}$$

where t is the time that has passed since the origins of the two frames coincided. The transformation of the components of the electric field, Eqs. (9.26), give, for $\mathbf{B}' = 0$, that $E_x = E'_x$ and $E_z = \gamma\left(E'_z - VB'_y\right)$. Therefore, at $t = 0$, when the charge Q is at the origin of the axes of frame S, it is

$$E_x = E'_x = \frac{Q}{4\pi\varepsilon_0}\frac{x'}{(x'^2+z'^2)^{3/2}} = \frac{\gamma Q x}{4\pi\varepsilon_0(\gamma^2x^2+z^2)^{3/2}} \tag{9.40}$$

and

$$E_z = \gamma E'_z = \frac{\gamma Q}{4\pi\varepsilon_0}\frac{z'}{(x'^2+z'^2)^{3/2}} = \frac{\gamma Q z}{4\pi\varepsilon_0(\gamma^2x^2+z^2)^{3/2}}. \tag{9.41}$$

Since it is $E_x/E_z = x/z$, the field $\mathbf{E}$ is in the direction of $\mathbf{r}$.

The magnitude of the vector $\mathbf{E}$ is found from the relation

$$E^2 = E'^2_x + E'^2_y = \frac{\gamma^2Q^2(x^2+z^2)}{(4\pi\varepsilon_0)^2(\gamma^2x^2+z^2)^3} = \frac{Q^2(x^2+z^2)}{(4\pi\varepsilon_0)^2\gamma^4(x^2+z^2-\beta^2z^2)^3} \tag{9.42}$$

$$E^2 = \frac{Q^2(1-\beta^2)^2}{(4\pi\varepsilon_0)^2(x^2+z^2)^2\left(1-\frac{\beta^2z^2}{x^2+z^2}\right)^3}, \quad \text{as} \quad E = \frac{Q(1-\beta^2)}{4\pi\varepsilon_0(x^2+z^2)\left(1-\frac{\beta^2z^2}{x^2+z^2}\right)^{3/2}}. \tag{9.43}$$

In terms of the distance r of the point (x, z) from the charge, which is momentarily situated at the origin of the axes in S, and the angle θ that the position vector $\mathbf{r}$ of the point forms with the velocity $\mathbf{V}$, we finally have

$$E = \frac{Q}{4\pi\varepsilon_0 r^2}\frac{1-\beta^2}{(1-\beta^2\sin^2\theta)^{3/2}}. \tag{9.44}$$

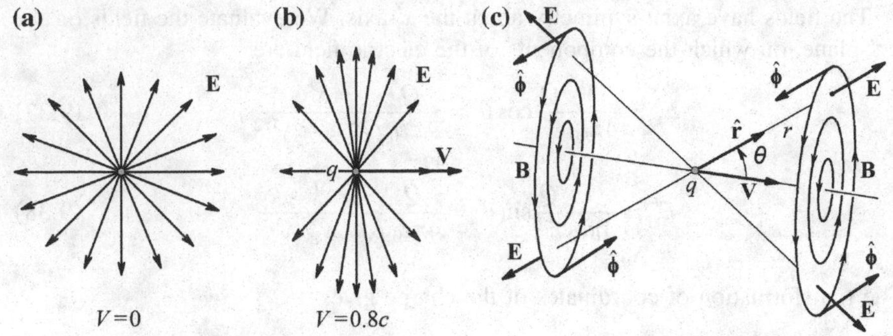

Fig. 9.3 The fields of a moving positive charge q. **a** The charge is stationary and the field it produces is spherically symmetric. **b** At speeds approaching the speed of light in vacuum ($V = 0.8c$ here), the electric field is radial but is more intense in directions normal to the velocity vector and weaker in directions forming small angles with the straight line on which the charge moves. **c** The lines of the magnetic field of the moving charge are circles with their centers on the straight line on which the charge moves

Vectorially, the electric field may be written as

$$\mathbf{E} = \frac{Q\hat{\mathbf{r}}}{4\pi\varepsilon_0 r^2}\frac{1-\beta^2}{\left(1-\beta^2\sin^2\theta\right)^{3/2}} \tag{9.45}$$

where $\hat{\mathbf{r}}$ is the unit vector from the charge towards the point at which the field is evaluated. The field is radial but not spherically symmetric. For large speeds of the charge, the field is intensely concentrated near the plane which passes through the charge and is normal to its velocity (Fig. 9.3b).

In order to find the magnetic field, we notice that, for $\mathbf{B}' = 0$, Eq. (9.34) becomes

$$\mathbf{E} = \gamma\mathbf{E}' - (\gamma - 1)\frac{\mathbf{V}}{V^2}(\mathbf{V}\cdot\mathbf{E}') \tag{9.46}$$

and Eq. (9.35),

$$\mathbf{B} = \frac{\gamma}{c^2}(\mathbf{V}\times\mathbf{E}'). \tag{9.47}$$

Therefore,

$$\mathbf{V}\times\mathbf{E} = \gamma\mathbf{V}\times\mathbf{E}' \tag{9.48}$$

and, finally,

$$\mathbf{B} = \frac{1}{c^2}\mathbf{V}\times\mathbf{E}. \tag{9.49}$$

From the last equation and using the relation $c^2 = 1/\varepsilon_0\mu_0$, we find that it is:

$$\mathbf{B} = \frac{\mu_0 Q}{4\pi r^2} \frac{1-\beta^2}{\left(1-\beta^2 \sin^2\theta\right)^{3/2}} \mathbf{V} \times \hat{\mathbf{r}}. \tag{9.50}$$

The lines of the magnetic field are circles which have their centers on the straight line on which the charge moves (Fig. 9.3c). Since it is $\mathbf{V}\times\hat{\mathbf{r}} = V\hat{\mathbf{V}}\times\hat{\mathbf{r}} = V\sin\theta\,\widehat{\boldsymbol{\phi}} = c\beta\sin\theta\,\widehat{\boldsymbol{\phi}}$, where $\widehat{\boldsymbol{\phi}}$ is the unit vector in the direction of increasing azimuth angle, this last equation may also be written as

$$\mathbf{B} = \frac{\mu_0 c Q}{4\pi r^2} \frac{\beta\left(1-\beta^2\right)\sin\theta}{\left(1-\beta^2 \sin^2\theta\right)^{3/2}} \widehat{\boldsymbol{\phi}}. \tag{9.51}$$

9.4.1 Ionization Caused by a Relativistic Charged Particle

When a charged particle moves in a material, it causes ionization of its atoms due to the forces it exerts on their electrons. The rate of energy loss per unit of length of the path, due to ionization, by a charge moving at relativistic speeds, is given by the relation

$$\frac{dE}{dx} = \frac{z^2 e^4 \rho N Z}{4\pi\varepsilon_0^2 m_0 c^2}\left[\frac{1}{\beta^2}\ln\left(\frac{2m_0c^2}{I}\right) - \frac{1}{\beta^2}\ln\left(\frac{1}{\beta^2}-1\right) - 1\right], \tag{9.52}$$

where m_0 is the rest mass of the particle, ze its electric charge and βc its speed. The material's density is ρ, N is the density of atoms of the material per unit volume, Z is their atomic number and $I = kZ$, with $k \approx 11.5$ keV, is a mean ionization potential of the particular atoms. The function $-dE/dx$ appears to depend on two constants which are determined by the nature of the material and the particle and is simply a function of the speed of the particle, βc. The quantity $-dE/\rho dx$, in units of $\text{MeV}/(\text{g/cm}^2)$, is plotted in Fig. 9.4 for a particle of unit charge (e.g. e, p) moving in lead, as a function of the momentum p of the particle in units of $m_0 c$, i.e. as a function of the variable $p/m_0 c = \beta\gamma$. In fact, the magnitude $-dE/\rho dx$ is plotted as a function of the variable $\beta\gamma$ of the particle in order to demonstrate better its behavior at values of β near unity. The corresponding values of β are given in the upper scale of the figure. Also given is the scale of kinetic energy for electrons, K_e, corresponding to the scales of β and $\beta\gamma$.

The *classical region* of low particle speeds may be defined, approximately, by the condition $p/m_0 c < 1$. In the classical region, $-dE/dx$ is inversely proportional to the square of the speed or to $1/\beta^2$. Ionization decreases with speed, because the higher the speed of the particle, the smaller the time it spends in the vicinity of an

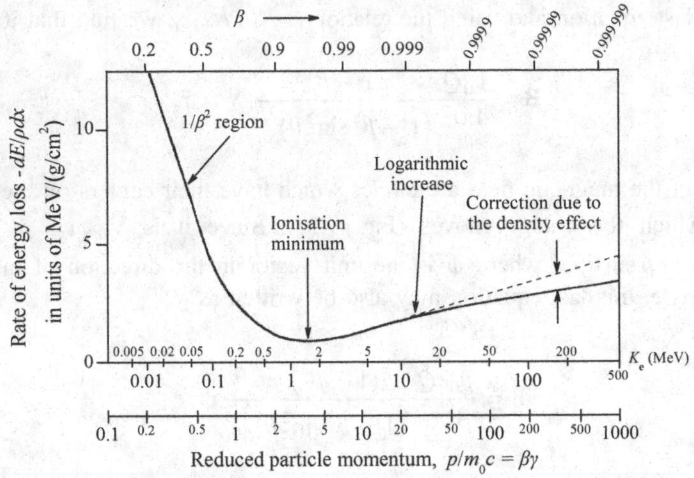

Fig. 9.4 The variation of the rate of energy loss by a charged particle per unit of path length, due to ionization, as a function of its momentum, for a particle with unit charge in lead

atom and, therefore, the smaller the impulse it gives to the atom's electrons in order to ionize them. As the reduced speed approaches unity, the value of this term is stabilized to the value corresponding to $\beta \approx 0.95$. We see that $-dE/dx$ stops decreasing with increasing speed at a value of $p/m_0c = \beta\gamma$ in the region of 3, it passes through a minimum and then starts increasing with speed.

The *relativistic region* may be thought of as extending to values of $p/m_0c = \beta\gamma$ greater than 4, where the so-called *logarithmic increase* is observed. The increase of $-dE/dx$ with speed is due to the concentration of the lines of the electric field of the moving charge in directions which are perpendicular to its velocity, as shown in Fig. 9.3b. The greater intensity of the field causes more intense ionization and at greater distances from the particle's path. This is a purely relativistic effect and is expressed by the second logarithmic term in Eq. (9.52).

The increase does not continue at the same rate for very high speeds of the particle. As the ionization extends to atoms at greater distances from the particle's path, the disturbance in the charge distribution of the atoms causes the appearance of a kind of shielding of the atoms at greater distances. This is known as the *density effect*.

The important phenomenon appearing here is the relativistic increase of the electric field in directions which are perpendicular to the particle's velocity at the point of the particle and the consequent increase of the ionization the charged particle causes. These observations provide evidence supporting the relativistic predictions concerning the electric field of a moving electric charge.

Example 9.1 The force between two identical charges moving together with the same velocity

In an inertial frame of reference S, two identical charges Q are at a distance r between them on a straight line parallel to the y-axis. The two charges move with the same velocity $\mathbf{V} = V\hat{\mathbf{x}}$. Use the results and the methods of the present chapter to find the force exerted by one charge on the other by three different methods.

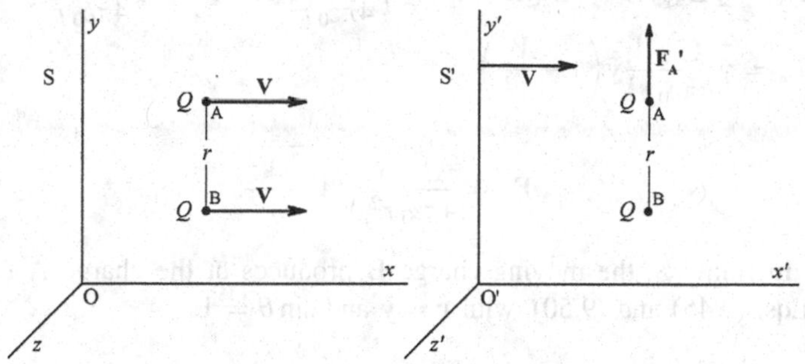

Method 1 Let the frame of reference S′ move with a velocity $\mathbf{V} = V\hat{\mathbf{x}}$ relative to the frame S. In this frame, the two charges are at rest and the force exerted by charge B on charge A is just the Coulomb force, equal to

$$\mathbf{F}'_{\mathrm{A}} = \frac{Q^2}{4\pi\varepsilon_0}\frac{\hat{\mathbf{y}}'}{r^2},$$

where $\hat{\mathbf{y}}'$ is the unit vector in the direction of positive y''s. Transforming the force to frame S, we have $\mathbf{F}_{\mathrm{A}} = F_{\mathrm{A}x}\hat{\mathbf{x}} + F_{\mathrm{A}y}\hat{\mathbf{y}} + F_{\mathrm{A}z}\hat{\mathbf{z}} = F'_{\mathrm{A}x}\hat{\mathbf{x}} + (F'_{\mathrm{A}y}/\gamma)\hat{\mathbf{y}} + (F'_{\mathrm{A}z}/\gamma)\hat{\mathbf{z}} = 0\hat{\mathbf{x}} + \frac{Q^2}{4\pi\varepsilon_0}\frac{\hat{\mathbf{y}}}{\gamma r^2} + 0\hat{\mathbf{z}}$ or $\mathbf{F}_{\mathrm{A}} = \frac{Q^2}{4\pi\varepsilon_0}\frac{\hat{\mathbf{y}}}{r^2}\sqrt{1-\frac{V^2}{c^2}}$.

Method 2 The electric and the magnetic fields produced by charge B at the charge A in frame S′ are $\mathbf{E}'_{\mathrm{A}} = \frac{Q}{4\pi\varepsilon_0}\frac{\hat{\mathbf{y}}'}{r^2}$ and $\mathbf{B}'_{\mathrm{A}} = 0$.

The transformation equations for the electromagnetic field are:

Equations (9.26) $\quad E_x = E'_x \quad E_y = \gamma\left(E'_y + VB'_z\right) \quad E_z = \gamma\left(E'_z - VB'_y\right),$

Equations (9.27) $\quad B_x = B'_x \quad B_y = \gamma\left(B'_y - VE'_z/c^2\right) \quad B_z = \gamma\left(B'_z + VE'_y/c^2\right).$

The fields in frame S at the position of charge A are:

$$E_{\mathrm{A}x} = 0 \qquad E_{\mathrm{A}y} = \gamma\left(\frac{Q}{4\pi\varepsilon_0}\frac{1}{r^2} + 0\right) = \gamma\frac{Q}{4\pi\varepsilon_0}\frac{1}{r^2} \qquad E_{\mathrm{A}z} = \gamma(0-0) = 0,$$

$$B_{Ax} = 0 \qquad B_{Ay} = \gamma(0 - 0) = 0 \qquad B_{Az} = \gamma\left(0 + \frac{V}{c^2}\frac{Q}{4\pi\varepsilon_0}\frac{1}{r^2}\right) = \gamma\frac{V}{c^2}\frac{Q}{4\pi\varepsilon_0}\frac{1}{r^2}.$$

The total force on charge A is the Lorentz force $\mathbf{F}_A = Q\mathbf{E}_A + Q\mathbf{V}\times\mathbf{B}_A$. This is

$$\mathbf{F}_A = QE_{Ay}\hat{\mathbf{y}} + QVB_{Az}\hat{\mathbf{x}}\times\hat{\mathbf{z}} = Q\gamma\frac{Q}{4\pi\varepsilon_0}\frac{1}{r^2}\hat{\mathbf{y}} - QV\gamma\frac{V}{c^2}\frac{Q}{4\pi\varepsilon_0}\frac{1}{r^2}\hat{\mathbf{y}}$$
$$= \gamma\frac{Q^2}{4\pi\varepsilon_0}\frac{1}{r^2}\hat{\mathbf{y}}\left(1 - \frac{V^2}{c^2}\right)$$

or
$$\mathbf{F}_A = \frac{Q^2}{4\pi\varepsilon_0}\frac{\hat{\mathbf{y}}}{r^2}\sqrt{1 - \frac{V^2}{c^2}}.$$

Method 3 In frame S, the moving charge B produces at the charge A the fields given by Eqs. (9.45) and (9.50), with $\hat{\mathbf{r}} = \hat{\mathbf{y}}$ and $\sin\theta = 1$:

$$\mathbf{E}_A = \frac{Q\hat{\mathbf{y}}}{4\pi\varepsilon_0 r^2}\frac{1}{\sqrt{1 - V^2/c^2}} \qquad \mathbf{B}_A = \frac{\mu_0 Q}{4\pi r^2}\frac{V}{\sqrt{1 - V^2/c^2}}\hat{\mathbf{z}} = \frac{Q}{4\pi\varepsilon_0 r^2}\frac{V/c^2}{\sqrt{1 - V^2/c^2}}\hat{\mathbf{z}},$$

where we used the fact that $c^2 = 1/\varepsilon_0\mu_0$. These are the same fields as those found in the solution by the second method and will result in a force on charge A equal to $\mathbf{F}_A = \frac{Q^2}{4\pi\varepsilon_0}\frac{\hat{\mathbf{y}}}{r^2}\sqrt{1 - \frac{V^2}{c^2}}$ again.

Example 9.2 The Poynting vector of a charge in uniform motion
Evaluate the Poynting vector for a charge moving with a constant velocity. Show that the total radiated power is equal to zero.

The fields produced by a charge Q moving with a constant velocity $\mathbf{V}$ were found to be

$$\mathbf{E} = \frac{Q\hat{\mathbf{r}}}{4\pi\varepsilon_0 r^2}\frac{1 - \beta^2}{(1 - \beta^2\sin^2\theta)^{3/2}} \quad \text{and} \quad \mathbf{B} = \frac{1}{c^2}\mathbf{V}\times\mathbf{E}.$$

The Poynting vector is: $\mathbf{S} = \frac{1}{\mu_0}\mathbf{E}\times\mathbf{B} = \frac{1}{\mu_0 c^2}\mathbf{E}\times(\mathbf{V}\times\mathbf{E}) = \varepsilon_0[\mathbf{V}(\mathbf{E}\cdot\mathbf{E}) - \mathbf{E}(\mathbf{E}\cdot\mathbf{V})]$.

We consider a sphere with its center at Q and with radius r. The electromagnetic energy crossing in unit time the element of surface $d\mathbf{a} = \hat{\mathbf{r}}\,da$ of the sphere is equal to $dP = \mathbf{S}\cdot d\mathbf{a} = \mathbf{S}\cdot\hat{\mathbf{r}}\,da$. However, vectors $\mathbf{E}$ and $\hat{\mathbf{r}}$ are parallel. We therefore may write: $\hat{\mathbf{r}} = \hat{\mathbf{E}} = \mathbf{E}/E$. Thus, the rate of flow of energy through any element of surface of the sphere is

$$dP = \mathbf{S} \cdot \hat{\mathbf{r}}\, da = (\varepsilon_0/E)[\mathbf{V}(\mathbf{E} \cdot \mathbf{E}) - \mathbf{E}(\mathbf{E} \cdot \mathbf{V})] \cdot \mathbf{E} da$$
$$= (\varepsilon_0/E)[(\mathbf{V} \cdot \mathbf{E})(\mathbf{E} \cdot \mathbf{E}) - (\mathbf{E} \cdot \mathbf{E})(\mathbf{E} \cdot \mathbf{V})]\, da = 0$$

i.e. equal to zero. The total flux of energy through the entire area of the sphere enclosing the charge will also be equal to zero.

9.5 The Derivation of the Differential Form of the Biot-Savart Law from Coulomb's Law

Equation (9.50) resulted from the application of the Lorentz transformations for position and for force, and from the law of Coulomb. According to this equation, a charge dQ, moving with velocity $\mathbf{V}$, produces at a point P which has position vector $\mathbf{r}$ relative to the charge (Fig. 9.5), a magnetic field

$$d\mathbf{B} = \frac{\mu_0 dQ}{4\pi r^2} \frac{1-\beta^2}{\left(1-\beta^2 \sin^2\theta\right)^{3/2}} \mathbf{V} \times \hat{\mathbf{r}}. \tag{9.53}$$

If we assume that this charge is distributed on an element of length $d\mathbf{s}$ oriented in the direction of $\mathbf{V}$, it will be equivalent to a current I which flows through the length element, where $I d\mathbf{s} = dQ\mathbf{V}$. Equation (9.53) becomes

$$d\mathbf{B} = \frac{\mu_0 I}{4\pi r^2} \frac{1-\beta^2}{\left(1-\beta^2 \sin^2\theta\right)^{3/2}} d\mathbf{s} \times \hat{\mathbf{r}}. \tag{9.54}$$

This is the exact formulation of the differential form of the law of Biot-Savart.

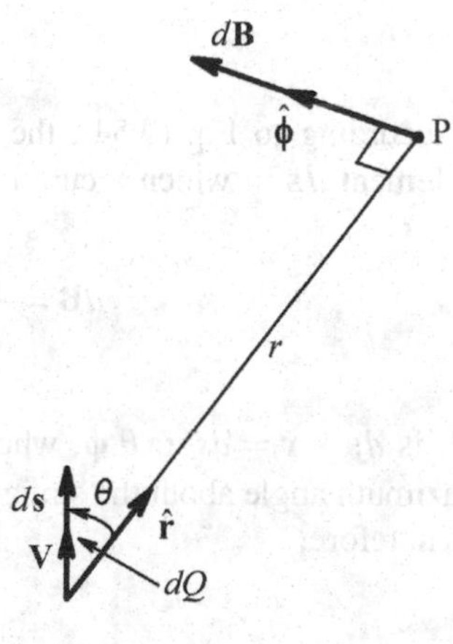

Fig. 9.5 The magnetic field of a charge dQ which moves with a constant velocity $\mathbf{V}$

The electrons in conductors have random velocities which at common temperatures are of the order of $c/250$, but the current is due to a slow drift of the electrons with a speed of the order of 10^{-4} m/s in a direction opposite to that of the current. Since these two speeds are much smaller than c, it is $\beta \ll 1$. Equation (9.54) may, therefore, be written as

$$d\mathbf{B} = \frac{\mu_0 I}{4\pi r^2} d\mathbf{s} \times \hat{\mathbf{r}}. \tag{9.55}$$

This is the classical formulation of the differential form of the law of Biot-Savart, which appears to be an approximation valid for low drift velocities of the charges to which the current is due.

Example 9.3 The magnetic field of a constant current flowing in an infinite straight conductor

Find the magnetic field produced by a constant current I flowing in a straight conductor extending to infinity in both directions.

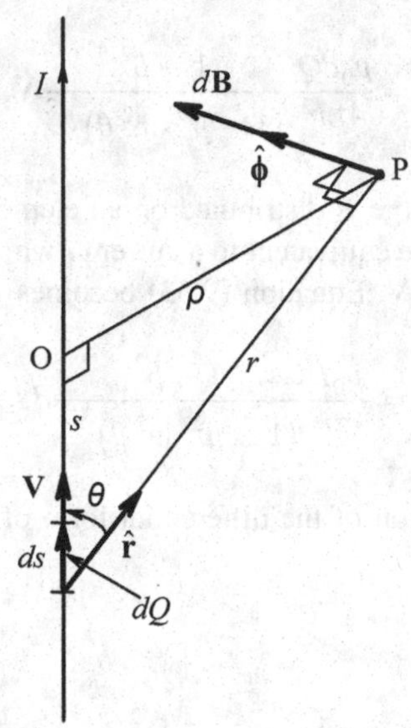

According to Eq. (9.54), the magnetic field at the point P, caused by the length element $d\mathbf{s}$ in which a current I is flowing, is

$$d\mathbf{B} = \frac{\mu_0 I}{4\pi r^2} \frac{1-\beta^2}{\left(1-\beta^2 \sin^2\theta\right)^{3/2}} d\mathbf{s} \times \hat{\mathbf{r}}.$$

It is $d\mathbf{s} \times \hat{\mathbf{r}} = ds \sin\theta\, \widehat{\boldsymbol{\phi}}$, where $\widehat{\boldsymbol{\phi}}$ is the unit vector in the direction of increasing azimuth angle about the straight conductor. Also, it is $r^2 = s^2 + \rho^2$ and $\sin\theta = \rho/r$. Therefore,

$$d\mathbf{B} = \frac{\mu_0 I(1-\beta^2)}{4\pi}\,\widehat{\boldsymbol{\phi}}\,\frac{(\rho/r)ds}{r^2(1-\beta^2\rho^2/r^2)^{3/2}}$$

or

$$d\mathbf{B} = \frac{\mu_0 I}{4\pi}\rho(1-\beta^2)\widehat{\boldsymbol{\phi}}\frac{ds}{\left[s^2+(1-\beta^2)\rho^2\right]^{3/2}}.$$

We will take β to be the same for all the charges to which the current is due. To find the total magnetic field at point P, we integrate from $s=-\infty$ to $s=+\infty$:

$$\mathbf{B} = \frac{\mu_0 I}{4\pi}\rho(1-\beta^2)\,\widehat{\boldsymbol{\phi}}\int_{-\infty}^{+\infty}\frac{ds}{[s^2+(1-\beta^2)\rho^2]^{\frac{3}{2}}}$$

$$= \frac{\mu_0 I}{4\pi}\widehat{\boldsymbol{\phi}}\frac{\rho(1-\beta^2)}{\rho^2(1-\beta^2)}\left[\frac{s}{\sqrt{s^2+(1-\beta^2)\rho^2}}\right]_{-\infty}^{+\infty} = \frac{\mu_0 I}{4\pi\rho}\widehat{\boldsymbol{\phi}}\left[\frac{s/|s|}{\sqrt{1+(1-\beta^2)\rho^2/s^2}}\right]_{-\infty}^{+\infty}$$

For $s=\pm\infty$ the numerator tends to the values ± 1, respectively, while the denominator tends to unity. So,

$$\mathbf{B} = \frac{\mu_0 I}{2\pi\rho}\widehat{\boldsymbol{\phi}},$$

which is, according the law of Biot-Savart, the magnetic field at a distance ρ from a straight conductor, through which flows a constant current I.

In practice, all the charges do not have the same β. We can, however, divide the charges dQ in groups, $\delta(dQ)$, to which currents δI are due, and which have values of β in narrow limits, between β and $\beta+\delta\beta$. The last equation gives the field produced by each of the groups. Summing, we have the magnetic field due to all the charges and this is proportional to the total current I. The next step would be to take into account the random motion of the charges in all directions [4].

Problem

9.6 A constant current I flows in a thin conductor in the shape of a circle of radius ρ. Assume that all charges to which the current is due move with the same speed βc. Using the relativistic version of the differential form of the Biot-Savart law, Eq. (9.54), find the magnetic field at the center of the circle.

Ans.: $\mathbf{B} = \dfrac{\mu_0 I}{2\rho}\dfrac{\widehat{\boldsymbol{\phi}}}{\sqrt{1-\beta^2}}$

9.6 The Force Exerted on a Moving Charge by an Electric Current

We will evaluate the force exerted by an electric current on a moving charge, with the aid of the Special Theory of Relativity. We will assume that, in the Laboratory Frame of Reference S (LFR), there exists a stationary straight neutral conductor. The positive ions of the conductor are at rest and have a (positive) linear charge density $+\lambda_0$. The electrons of the conductor have a (negative) linear charge density $-\lambda_0$ and all move with the same speed υ_0 in the positive x direction. A charge q is at a distance r from the conductor and moves in the positive x direction with a speed υ in the LFR (Fig. 9.6a). In the LFR, the conductor, being neutral, does not create an electric field.

In the Frame of Reference of Charge q (FRCq) the charge is stationary, the positive ions move in the negative x direction with speed $-\upsilon$ and the electrons move now with a speed υ_0' in the direction of the x-axis (Fig. 9.6b). We define the quantities:

$$\beta \equiv \frac{\upsilon}{c}, \quad \beta_0 \equiv \frac{\upsilon_0}{c}, \quad \beta_0' \equiv \frac{\upsilon_0'}{c},$$
$$\gamma \equiv 1/\sqrt{1-\beta^2}, \quad \gamma_0 \equiv 1/\sqrt{1-\beta_0^2}, \quad \gamma_0' \equiv 1/\sqrt{1-\beta_0'^2}. \tag{9.56}$$

From the relativistic law for the addition of velocities, we find that the reduced speed of the electrons in the FRCq is

$$\beta_0' = \frac{\beta_0 - \beta}{1 - \beta\beta_0}. \tag{9.57}$$

From the transformation of the Lorentz factor we have

$$\gamma_0' = 1/\sqrt{1-\beta_0'^2} = \gamma\gamma_0(1-\beta\beta_0). \tag{9.58}$$

Since the positive ions now move in the FRCq, the distances between them contract by a factor $1/\gamma$ and so the linear charge density of the positive charge is greater than its value in the LFR and equal to $\gamma\lambda_0$. The electrons move with a speed equal to υ_0 in the LFR. The linear charge density of the electrons in the LFR is $-\lambda_0$. In their own frame of reference, without length contraction, their linear charge density is $-\lambda_0/\gamma_0$. Next, in the FRCq, in which the electrons have a speed υ_0', the distances between them will be smaller by a factor of $1/\gamma_0'$ compared to those in their own frame of reference, and the linear density of the negative charge is $-\lambda_0(\gamma_0'/\gamma_0)$. The total linear density of the charge is the algebraic sum of the two densities and equal to

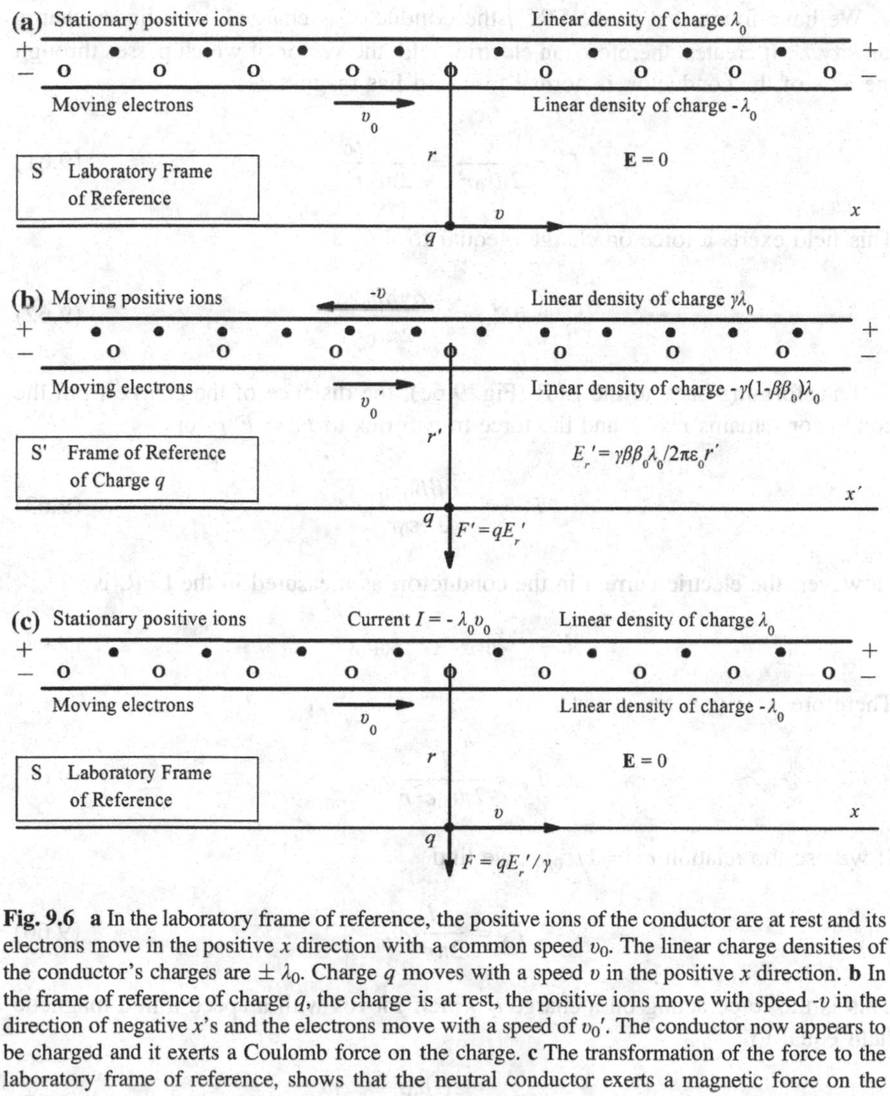

Fig. 9.6 **a** In the laboratory frame of reference, the positive ions of the conductor are at rest and its electrons move in the positive x direction with a common speed v_0. The linear charge densities of the conductor's charges are $\pm\,\lambda_0$. Charge q moves with a speed v in the positive x direction. **b** In the frame of reference of charge q, the charge is at rest, the positive ions move with speed $-v$ in the direction of negative x's and the electrons move with a speed of v_0'. The conductor now appears to be charged and it exerts a Coulomb force on the charge. **c** The transformation of the force to the laboratory frame of reference, shows that the neutral conductor exerts a magnetic force on the moving charge q, due to the current flowing through it

$$\lambda' = \gamma\lambda_0 - \lambda_0\frac{\gamma_0'}{\gamma_0}. \tag{9.59}$$

Substituting for γ_0' from Eq. (9.58), we find that it is

$$\lambda' = \gamma\lambda_0 - \lambda_0\frac{1}{\gamma_0}\gamma\gamma_0(1-\beta\beta_0) \quad \text{or} \quad \lambda' = \gamma\beta\beta_0\lambda_0. \tag{9.60}$$

We have found that in the FRCq the conductor is charged, with linear charge density λ'. It creates, therefore, an electric field, the vector of which passes through the axis of the conductor, is normal to it and has magnitude

$$E'_r = \frac{\lambda'}{2\pi\varepsilon_0 r'} = \frac{\gamma\beta\beta_0\lambda_0}{2\pi\varepsilon_0 r'}. \tag{9.61}$$

This field exerts a force on charge q equal to

$$F'_y = qE'_y = -\frac{q\gamma\beta\beta_0\lambda_0}{2\pi\varepsilon_0 r'}. \tag{9.62}$$

Transforming back to the LFR (Fig. 9.6c), the distance of the charge from the conductor remains $r = r'$ and the force transforms to $F_y = F'_y/\gamma$ or

$$F_y = -\frac{q\beta\beta_0\lambda_0}{2\pi\varepsilon_0 r}. \tag{9.63}$$

However, the electric current in the conductor, as measured in the LFR, is

$$I = -\lambda_0 \upsilon_0 = -\lambda_0\beta_0 c. \tag{9.64}$$

Therefore,

$$F_y = \frac{I}{2\pi\varepsilon_0 c^2 r} q\upsilon. \tag{9.65}$$

If we use the relation $c^2 = 1/\varepsilon_0\mu_0$, we find

$$F_y = \frac{\mu_0 I}{2\pi r} q\upsilon. \tag{9.66}$$

This is the force acting on a charge q which moves with a speed υ in a magnetic field equal to

$$B_r(r) = \frac{\mu_0 I}{2\pi r}, \tag{9.67}$$

and which is normal to the plane defined by the conductor and the charge.

We have seen that an electrically neutral conductor, in which an electric current flows, appears charged in the frame of reference of a moving charge and exerts a force on it. Transformed to the laboratory frame of reference, this force appears as a magnetic force.

Problem

9.7 A positive charge Q is at rest in the inertial frame of reference S (figure (a)). In the same frame there is a rectilinear conductor, extending to infinity in both directions, which contains stationary positive ions, all with charge $+e$, with a linear charge density λ_+, and electrons, with charge $-e$ and linear charge density λ_-. All the electrons move with a common speed υ_- in frame S. In the same frame, the distances between the positive ions are all equal to d_+ and those between the electrons are all equal to d_-. It is $d_- > d_+$. The distance of the charge Q from the conductor is r.

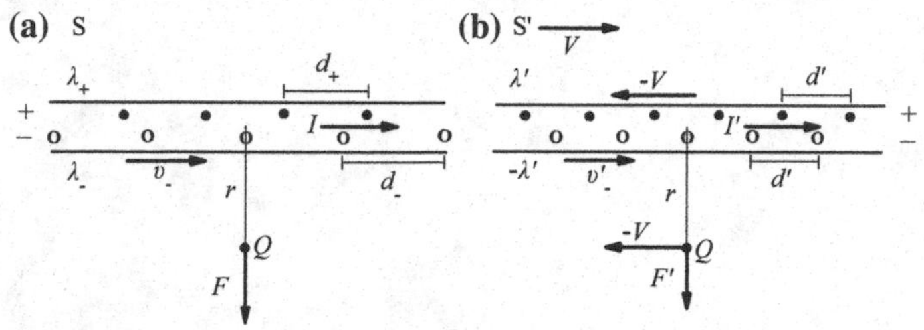

(a) Find the electrostatic force F exerted on charge Q in the frame of reference S.
(b) Find the speed V of a frame of reference S′ at which the distances between the positive charges and the negative charges are all equal to d' and, therefore, both the linear charge densities are equal to λ' in absolute values (figure (b)).
(c) With the value of V found in (b), find the force exerted on charge Q in frame S′. Since the conductor is neutral in frame S′, this force is a purely magnetic force. Express the force in terms of the electric current in the conductor and the distance r.

Ans.: (a) $F = \dfrac{Qe}{2\pi\varepsilon_0 r}\left(\dfrac{1}{d_+} - \dfrac{1}{d_-}\right)$, (b) $V = \dfrac{c^2}{\upsilon_-}\left(1 - \dfrac{d_-}{d_+}\right)$, (c) $F' = \dfrac{\mu_0}{2\pi}\dfrac{I'}{r'}QV$

References

1. J.G. King, Phys. Rev. Let. **5**, 562 (1960)
2. J.G. King, in A personal communication to J.D. Jackson, who mentions it in his book *Classical Electrodynamics* 3rd ed. (Wiley, New York, 1999) Section 11.9
3. V.W. Hughes, L.J. Fraser, E.R. Carlson, Z. Phys, D – Atoms. Mol. Clusters **10**, 145 (1988)
4. W.G.V. Rosser, *Classical Electromagnetism Via Relativity* (Butterworths, London, 1968). Appendix D

Chapter 10
Experiments

In this chapter, we will examine the experimental foundation of the Special Theory of Relativity both with experiments performed before the publication of the theory and as well as after. There are an enormous number of such experiments and, obviously, it is not possible for us to examine but a small proportion of them. We have chosen historical experiments, which had an important contribution in the foundation of the theory, as well as experiments performed soon after the publication of the theory and either improved the accuracy of previous important experiments or demonstrated a significant originality in their methodology. The reader who is interested in more experiments or in more details of those described here, is referred to books examining exclusively the experimental foundation and the verification of the Special Theory of Relativity [1].

Many experiments have already been mentioned in some detail in the previous chapters of the book. For these, we simply give here a brief summary and the reference to the sections of the book where they are discussed in more detail. The experiments have been grouped together, according to the physical magnitude or the phenomenon they examine. Within each group, the experiments are arranged in chronological order. In all cases, references are given to the original literature or to appropriate review articles, where the interested reader may study the experiments in more detail.

10.1 The Speed of Light

10.1.1 Historic Measurements of the Speed of Light

1632 Galileo was the first to try to measure the speed of light. He had no success, however, and he concluded that the speed of light was very large and certainly greater than the speed of sound (Sect. 1.7).

1676 The first measurement of the speed of light by Rømer, with observations of the eclipses of Jupiter's satellite Io. It constitutes the first measurement of a universal constant (Sect. 1.3).

© Springer International Publishing Switzerland 2016
C. Christodoulides, *The Special Theory of Relativity*,
Undergraduate Lecture Notes in Physics, DOI 10.1007/978-3-319-25274-2_10

1725 Discovery of the phenomenon of the aberration of light by Bradley [2]. From his measurements and the recognition that the effect is due to the motion of the Earth, a fairly accurate value for the speed of light in vacuum is derived. The observation of the aberration of light will be used later in order to draw the conclusion that the medium in which light was assumed to propagate, the aether, was not dragged by the Earth in its motion through it (Sect. 1.5).

1849–62 The first accurate measurements of the speed of light were performed in the laboratory by Fizeau and Foucault (Sect. 1.7).

1983 By international agreement, the relationship between the unit of length (meter, m) and time (second, s) was defined so that the speed of light in vacuum is, by definition, equal to $c \equiv 299\,792\,458$ m/s (Sect. 1.7).

10.1.2 The Non-dependence of the Speed of Light in Vacuum on the Motion of the Source or of the Observer

1810 Arago [3] tried to detect possible differences in the speed of light coming from stars, as the Earth, moving on its orbit, changes its speed relative to the source. His results were negative. In order to explain these, Fresnel [4] suggested the idea that the aether is partially dragged by the Earth as it moves through it (Sect. 1.6).

1962 *The speed of light and the theory of the re-emission of light in matter.* Fox [5] remarked that, according to the theory of the propagation of light in matter, when an electromagnetic wave impinges on a material, the electromagnetic field causes the free electrons of the material to oscillate. In this way, the wave is reemitted by the oscillating charges and so the transmitted wave is a superposition of the incident wave and the reemitted wave. Taking into account the number density n of the free electrons in the material and the Thomson cross section for the scattering of light by the electrons, σ, it follows that the characteristic length for the reduction by a factor $\mathrm{e} = 2.718\ldots$ of the intensity of the incident electromagnetic wave in matter due to scattering by electrons, the *extinction length*, is of the order of $l \sim 1/(10^5 n\sigma)$. After the light moves inside the material to a depth of a few times the length l, we may assume that it has acquired the speed of light in that material and all information regarding its speed in vacuum is lost. This is the Ewald-Oseen extinction theorem [6] . In materials such as glass, where it is $n \approx 10^{23}\ \mathrm{cm}^{-3}$, and, for visible light, $\sigma \approx 6 \times 10^{-25}\ \mathrm{cm}^2$, this relation gives an extinction length of $l \sim 1\ \mu\mathrm{m}$. In the atmosphere, the extinction length for visible light is $l \sim 1$ mm.

As we have mentioned, Arago tried to measure the difference in the speed of light due to the motion of the Earth towards and away from a star, by measuring the focal length of the lens of his telescope for light from a star. Fox's objection definitely rules out that Arago could measure a different speed for the light from the star, even if such a difference existed. What he would measure would be the speed of light in the material of his telescope's objective lens or, rather, in the air (Sect. 1.6).

The case of the eclipsing binary stars (Sect. 1.6) is more complex. The idea is that one would observe effects which would be due to a different speed by which light would start off from the star and with which it would travel through space. However, even in space, the density of electrons is of the order of $n \approx 0.04\ \text{cm}^{-3}$, as deduced by measurements of the dispersion suffered by short pulses of radio waves from pulsars [7]. This density gives for visible light an extinction length of about 2 light years, which is smaller even than the distance of the star nearest to Earth. Only at distances of this order of magnitude would light retain the speed with which it left the star, while for the rest of the distance to the Earth it would travel with the speed of light in 'vacuum' with $n \approx 0.04\ \text{cm}^{-3}$. There are, however, indications that binary stars are surrounded by a space containing gases which scatter stellar light [8]. Given that these gases will not participate to any significant degree to the motion of the one star relative to the other, any possible difference in the speed of light due to the speed of one star relative to the other would be eliminated and the light which would arrive at the Earth would originate from its reemission by the gases.

The extinction effect is not important in the case of a binary star one member of which emits pulses of X rays. For X rays, the extinction length is thousands of light years in galactic space. For example, for X rays with an energy of 70 keV, the extinction length is equal to $l \approx 3 \times 10^{20}\ \text{m} \approx 20\ \text{kpc} \approx 65\,000$ light years. Brecher [9] used the results of measurements performed with pulses from three systems, one member of which is an X ray pulsar. The sources of X rays which he examined are:

Her X-1, in the constellation of Hercules, at a distance of about $D \approx 6$ kpc $\approx$ 20 000 light years from the Earth, has a radial component of velocity with respect to the Earth equal to $\upsilon_r = 169$ km/s, it emits X ray pulses at a frequency of $f = 0.81\ \text{s}^{-1}$ and has an orbital period of $T = 1.70$ days.

Cen X-3, in the constellation of Centaurus, with $D \approx 8$ kpc $\approx$ 26 000 light years, $\upsilon_r = 415$ km/s, $f = 0.207\ \text{s}^{-1}$ and $T = 2.09$ days.

SMC X-1, in the Small Magellanic Cloud, with $D > 60$ kpc $\approx$ 200 000 light years, $\upsilon_r = 299$ km/s, $f = 1.41\ \text{s}^{-1}$ and $T = 3.89$ days.

The regularity of these pulses sets some limits on the influence that the speed of the source may have on the X rays emitted. Assuming for the speed of the X rays relative to the Earth a relation of the form $c' = c + k\upsilon_r$, where c is the speed of light in vacuum and υ_r is the radial component of the source relative to the Earth, Brecher finds that it is $k < 2 \times 10^{-9}$. This value means that the source does not impart to the X rays even 2 billionths of its own speed. This is a very strict limit, the smaller at present. It is interesting that an idea of de Sitter (see Sect. 1.6) bears fruit a century after its original proposal, applied to X ray pulsars, objects which were unknown at that time.

1963 Sadeh [10] used the annihilation of electron-positron pairs, in order to check whether the speed of the photons produced depends on the speed of the source that emits them. During the incidence of a positron, which moves almost with the speed c of light in vacuum, the center of mass of the electron-positron system will be moving with a speed equal to $c/2$. Relative to this center of mass, two γ rays are emitted, each with energy of at least 511 keV. The times were measured which were needed for the two photons emitted to arrive at two detectors situated at equal distances from the point of the annihilation of the two particles. It was found that the speeds of the two photons, when these were moving on the original line of motion of the positron, were equal within the experimental error of 10 %. This certainly differs from the non-relativistic prediction for speeds $\upsilon_- = 0.5c$ and $\upsilon_+ = 1.5c$ for the γ rays.

1964 Filippas and Fox [11] investigated experimentally the possibility of a dependence of the speed of high-energy photons on the speed of the source which emits them. They bombarded a liquid hydrogen target with π^- particles from the synchrocyclotron of the Carnegie Institute of Technology, in order to produce π^0 particles via the reaction $\pi^- + \mathrm{p} \rightarrow \pi^0 + \mathrm{n}$. The π^0 particles from this reaction move with a speed of $\upsilon = 0.2c$ and decay, with a mean lifetime of $\tau = 8.4 \times 10^{-17}$ s, into two photons, each with an energy of 68 MeV, moving in opposite directions. The photons were detected by two detectors situated symmetrically relative to the liquid hydrogen target. The measurement of the difference between the arrival times of the photons at the detectors would make it possible to evaluate the difference in their speeds. Assuming that the speeds of the two photons, which are emitted in opposite directions by the π^0 particles moving with a speed υ, are equal to $c \pm k\upsilon$, Filippas and Fox established that the coefficient k certainly does not have the value of unity predicted by non-relativistic Mechanics and that its value is $k \leq 0.5$ at a confidence level of 99.9 % and $k \leq 0.4$ at the confidence level of 90 %.

The two researchers examined the possibility that the extinction effect influenced their results, by performing the measurements at two different target-detector distances, 7 inches (18 cm) and 47 inches (120 cm). The distributions of the differences of the arrival times about the value of zero were identical for the two distances, with dispersions of the order of 1 ns about zero. This was proof that the extinction was negligible, something which is a clear advantage relative to previous measurements. In Sadeh's measurements (1963), for example, the distance used for the measurement of the speed of the photons with 0.5 MeV energy, was 60 cm, when the extinction length in air for these photons is estimated to be 19 cm. Alväger et al. [12] measured the speed of 4.4 MeV photons in a distance of 1.7 m, with an extinction length equal to 5 m. For the energy of 68 MeV used by Filippas and Fox, the extinction length is expected to be equal to 25 m, approximately, and so the target-detectors distances used, 18 cm and 1.2 m, were significantly smaller than this.

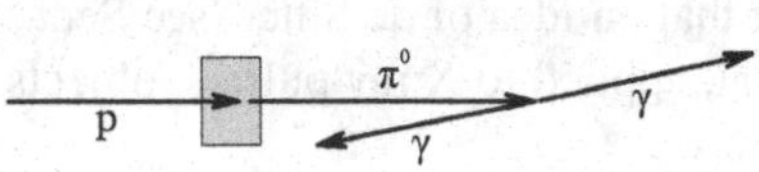

1964 Alväger et al. [13] bombarded a target with high-energy protons, producing π^0 particles, which had a speed of $V = 0.99975c$. These decay, with a mean lifetime of $\tau = 0.9 \times 10^{-16}$ s, by emitting two photons ($\pi^0 \to 2\gamma$). The photons produced, with energies ≥ 6 GeV, which were moving almost in the same direction as the π^0 particles, were found to have a speed

$$\upsilon = (2.9977 \pm 0.0004) \times 10^8 \text{ m/s} = (0.99997 \pm 0.00013)c.$$

We conclude that, within experimental error, even for a source speed which is short of c by only $0.00025c$, the speed of the photons differs from c, if indeed it does differ, by less than $0.00013c$ or 0.01 %.

10.1.3 The Non-dependence of the Speed of Light in Vacuum on Frequency

1964 Astronomical observations, due to the large distances of the objects observed, offer themselves for the comparison of the times of arrival of a pulse of visible light or of other electromagnetic radiation at the Earth, for different frequencies. Lovell et al. [14], first, in 1964, used the arrival times of signals from stellar flares, in the region of visible light and in the region of radio waves, and found that, if there is a difference in the speeds of the waves at these frequencies, it must be $\Delta c/c < 10^{-6}$, where $\Delta c/c$ is defined as the proportional difference of the speeds at the two frequency regions relative to the value c of the speed of visible light in vacuum.

1969 Warner and Nather [15] , in 1969, reported on results of the measurement of the arrival times of pulses from the pulsar PSR B0531+21 in the constellation of Cancer at the wavelengths of 540 nm in the visible spectrum and of 1.2 m for radio waves. This pulsar which is a neutron star, remnant of the supernova SN 1054 which was observed in 1054 A.D. by Chinese astronomers, emits pulses with a period of 33 ms. They reported that the proportional difference in the speeds of the electromagnetic waves at these two frequencies, if it exists, must be $\Delta c/c < 4 \times 10^{-7}$. In parallel, they found that the difference in the arrival times of pulses observed at the two wavelengths of 350 and 550 nm in the region of visible light is smaller than 10 μs. This sets a limit for these two wavelengths equal to $\Delta c/c < 5 \times 10^{-17}$.

1999 Schaefer [16] used the measurements which were performed on pulses from cosmological sources in a wide range of frequencies, from radio waves to very energetic γ rays in the region of TeV, to derive limits for the possible dependence on frequency of the speed of electromagnetic waves in vacuum. He chose objects with a large red shift in their spectrum, which, therefore, lie at great distances, of the order of hundreds of Mpc, with the exception of pulsar PSR B0531+21 in the constellation of Cancer, at a distance of only 2 kpc from the Earth. Most of the other sources, which were sources of γ ray bursts (GRB), are at distances up to 1600 Mpc or 5×10^9 light years, i.e. a considerable fraction of the radius of the visible universe, which is 47×10^9 light years. He also included the supernovas SN 1994G and SN 1997ap and the γ ray burst with energies in the region of TeV, in the near galaxy Mkn 421, which was observed on 15 May 1996. The observations referred to the difference in the times of arrival to the Earth of pulses of electromagnetic radiation at two different frequencies. From this difference, Schaefer evaluated the maximum possible difference in the speeds of the electromagnetic waves at the two frequencies. The results are summarized in Fig. 10.1.

In the figure, the measurements made on each of the objects examined are presented by a straight line, the ends of which correspond to the two frequencies at which the pulses were observed. From the upper limit derived from the measurements for the difference in the arrival times of the pulses at the two frequencies and the distance of the object emitting the pulses, a value $(\Delta c/c)_{\text{lim}}$ is calculated, for the maximum possible value of the reduced difference $\Delta c/c$ of the speeds at the two frequencies, if such a difference exists. This value is given along the vertical axis of

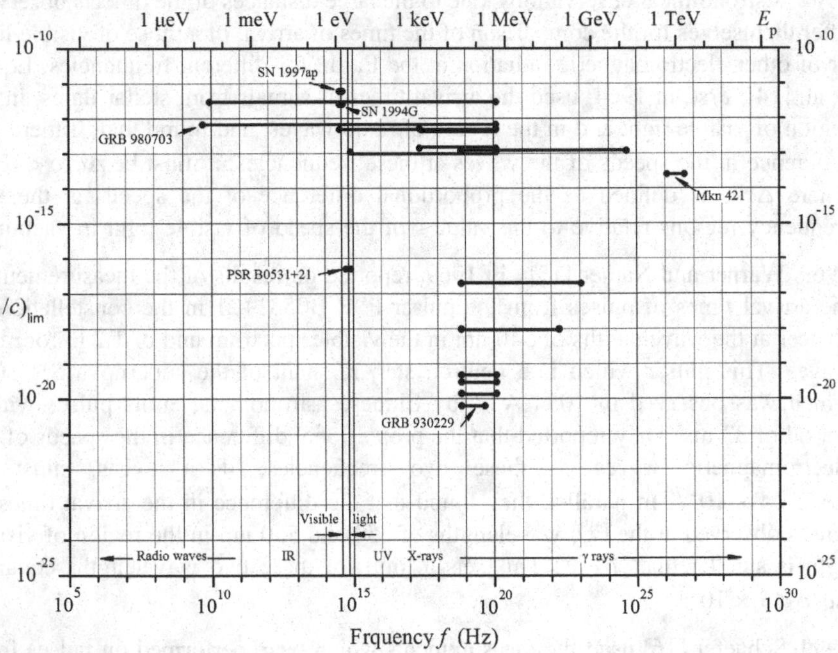

Fig. 10.1 A plot of the measurements used by Schaefer [16] in the investigation of the possibility that the speed of electromagnetic waves in vacuum depends on their frequency, f. Each one of the horizontal lines corresponds to one source. The ends of a line define the two frequencies at which the pulse of a signal was observed. The vertical axis gives the limit $(\Delta c/c)_{\text{lim}}$ in the ratio of the fractional difference $\Delta c/c$ of the speed from the value of c, the speed of visible light in vacuum. Also given is the axis of photon energy E_p corresponding to f

the figure, for each pair of measurements. We notice that the measurements which cover the widest range of frequencies are those of GRB 980703 (5.0×10^9 Hz – 1.2×10^{20} Hz) for which the limit of $(\Delta c/c)_{\text{lim}} = 6.6 \times 10^{-13}$ is derived. The measurements at the highest frequencies are those on the γ rays from galaxy Mkn 421, with energies in the region of TeV, from which the limit of $(\Delta c/c)_{\text{lim}} = 2.5 \times 10^{-14}$ follows. These are the highest energies at which such measurements were made. The strictest limit for the region of visible light is given by the measurements on PSR B0531+21, and is $(\Delta c/c)_{\text{lim}} = 5 \times 10^{-17}$. Finally, the strictest limit for $\Delta c/c$ was set by the γ ray burst GRB 930229 (7.2×10^{18} Hz – 4.8×10^{19} Hz) from which the limit of $(\Delta c/c)_{\text{lim}} = 6.3 \times 10^{-21}$ is found.

We conclude that the limits in $\Delta c/c$, for frequencies between 10^{10} Hz and 10^{27} Hz, range between 10^{-11} and 10^{-20}. The speed of electromagnetic waves can, therefore, be considered to be independent of frequency within very strict limits. Research on the matter will, of course, be continued, not only in order to test the Special Theory of Relativity, but, mainly, because quantum theories of gravity have been proposed which predict a dependence of the speed of electromagnetic waves on their frequencies.

10.2 The Aether

1851 Measurement of the speed of light in moving water by Fizeau [17]. The speed of light in moving water was found to be different than its speed in the same medium when this is stationary. The difference was of the form kv, where v is the speed of the fluid relative to the laboratory and k is a 'drag coefficient' of the aether by the fluid, as proposed by Fresnel, different for each fluid (Sect. 1.8.1).

1868 Hoek [18] used an interferometer in an effort to measure the dragging of aether by a moving optical medium, namely water, if this is moving relative to the aether. His results were in agreement with Fresnel's theory for the dragging of aether (Sect. 1.8.2).

1871 Measurement by Airy [19], of the aberration of light using a telescope filled with water. He found no change in the angle of aberration of 19.8″ in the North-South direction which was measured by Bradley with an ordinary telescope for the star γ *Draconis*.

1881 The first experiment of Michelson [20] was performed, for the detection of the motion of the Earth relative to the aether, by using an interferometer. The results showed no movement of the Earth relative to the aether, within the limits of the sensitivity of the experiment.

1887 Improved experiment of Michelson and Morley [21]. An improved interferometer, using longer light paths, was used for the detection of the motion of the

Earth relative to the aether. The apparatus was 10 times more sensitive than that of the 1881 experiment. The results were again negative, setting an upper limit of 8 km/s on the possible speed of the Earth relative to the aether (Sect. 1.10).

1932 Kennedy and Thorndike [22] repeated the experiment of Michelson and Morley with an interferometer similar to the Michelson interferometer, but having arms which were not mutually perpendicular and had very unequal lengths, with the result that the two light paths had very different lengths. The interferometer was stationary in the laboratory and so changed orientation following the motions of the Earth. The interference fringes were photographed in order to register any possible shift. Measurements were performed during successive and different seasons, thanks to the stabilization of temperature to within 0.001 °C. The results were negative, in agreement with those of the Michelson-Morley experiment. The contraction hypothesis of Lorentz [23] and FitzGerald [24] would not interpret the results of the experiment, although it interprets those of the Michelson-Morley experiment (Sect. 1.11).

1978 Brillet and Hall [25] performed, using modern means, an experiment similar to that of Michelson and Morley, with the intention of detecting any possible motion of the Earth relative to the aether. Their experimental arrangement is shown very schematically in Fig. 10.2.

One part of the apparatus Brillet and Hall used, rested on a granite base which could be rotated with negligible vibrations. Situated on this base was a He–Ne laser, whose frequency was variable, which provided radiation with a wavelength of 3.39 μm. The beam, with suitable reflections on mirrors and transmission through half-silvered glass plates, followed the path ABCDE and entered a Fabry-Perot optical cavity (F-P in the figure). Inside the Fabry-Perot interferometer, the radiation was reflected between two parallel mirrors, whose separation was constant. In this particular case, the laser frequency was adjusted by a feedback mechanism so that a

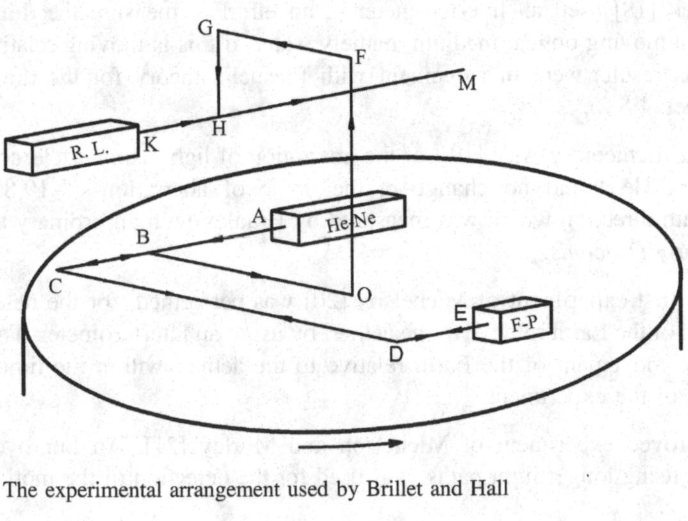

Fig. 10.2 The experimental arrangement used by Brillet and Hall

pattern of standing waves was maintained inside the cavity. Since the geometry of the cavity remained unchanged, the wavelength of the standing waves λ inside the cavity was stabilized. Thus, since it is $c = f\lambda$, if the speed of the electromagnetic radiation c did vary, for the standing waves inside the cavity to remain unchanged, the frequency would have to be appropriately adjusted.

A portion of the radiation emerged from the cavity and, moving along the path EDCBO, reached the axis of revolution of the apparatus, OF. The beam then followed the path OFGHM. In the last portion of the path, HM, the beam was mixed with a beam KHM from a reference laser, whose frequency was maintained constant for comparison. If the frequencies of the two beams were different, their superposition led to the production of beats. Using a suitable optoelectronic arrangement, these beats could be detected at point M.

As the part of the apparatus which was situated on the granite base was rotated about the axis OF, if there was a change in the speed of the electromagnetic radiation with orientation, due to the motion of the Earth relative to the aether (or any other reason), a variation in the frequency of the radiation from the Fabry-Perot interferometer would appear, with a dependence on the azimuthal orientation angle, ϕ, according to the value of $\cos(2\phi)$. The sensitivity of the arrangement of Brillet and Hall was such that the change in the frequency measured was $\Delta f/f = (1.5 \pm 2.5) \times 10^{-15}$. The conclusion was that, within the limits of experimental error, no variation of the speed of light with direction was observed. As with the experiment of Michelson and Morley, given the orbital speed V of the Earth, we would expect a variation of the order of magnitude of the ratio $(V/c)^2$, i.e. 10^{-8}, the limit set by the experiment of Brillet and Hall in the variation $\Delta f/f$ is 7 orders of magnitude smaller than that predicted.

10.3 The Dilation of Time

1941 Rossi [26] and his co-workers studied the decrease with decreasing height, of the surviving μ paricles produced in the atmosphere by cosmic rays. It was found that the muons survived for times which were larger than those expected from measurements performed on slow moving muons, in agreement with the time dilation expected for their speeds (Sect. 4.1).

1963 Frisch and Smith [27] repeated the 1941 experiments of Rossi and his co-workers [26] with much greater accuracy. They detected muons with velocities close to $0.993c$, which had a Lorentz factor of $\gamma = 8.4 \pm 0.2$. Between Mount Washington and Cambridge, Massachusetts, whose elevations differ by 1907 m, the mean counting rate of muons was found to decrease from 563 to 412 per hour. This was translated to a dilation factor for the mean lifetime of the muons equal to 8.8 ± 0.8, which agrees, within experimental error, with the Lorentz factor given above (Sect. 4.1).

1960 The construction of high-energy particle accelerators led to the availability of fast particle beams in which the relativistic focusing effect was observed. The effect is simply explained using time dilation (Sect. 4.2).

1971 Hafele and Keating [28] loaded 4 cesium beam atomic clocks on a commercial airplane which circumnavigated the Earth, first in 41 h moving eastwards and then in 49 h moving westwards, and compared their readings with those of similar clocks that had remained stationary on the surface of the Earth, at the initial starting point. The differences in the indications of the clocks were found to agree, within experimental error, with the predictions of the Special and the General Theories of Relativity (Sect. 4.3).

1976 The influence of gravity on the rate of a clock, predicted by the General Theory of Relativity, was verified in 1976, when a NASA spacecraft, Gravity Probe A, carried a hydrogen maser clock to a height of 10 000 km and back [29]. The clock's readings were continuously compared with those of an identical clock on the surface of the Earth. Monitoring the position and the speed of the spacecraft during the 2 h that the flight lasted, the scientists were able to subtract that part of the difference in the clock's readings which was due to speed and the Special Theory of Relativity and to measure the acceleration in the clock's rate of counting time which was due to the decrease of the gravitational potential. The agreement with theory was exceptional, at the level 70 parts in a million (0.007 %). The experiment was performed in order to test a prediction of the General Theory of Relativity, but to achieve the sensitivity we mentioned, it was necessary to subtract the difference in the times due to the speed, as predicted by the Special Theory of Relativity. Therefore, unless the two theories conspire to cover each other's differences in the experimental measurements (!), we must accept that the particular experiment also verifies the dilation of time due to speed, with the same accuracy.

1977 Bailey [30] and his co-workers at CERN, measured the mean lifetime of $\mu^{\pm}$ muons which were circulating in a storage ring with a diameter of 14 m with a speed which corresponded to $\beta = 0.9994$ and $\gamma = 29.3$. A lengthening of the mean lifetime of the muons was found, by the expected factor of γ, with an accuracy of about 0.1 %. The centripetal acceleration of the muons in this experiment was of the order of 10^{16} m/s^2, a fact that did not influence the validity of the predictions of the theory (Sect. 4.1.1).

10.4 The Relativistic Doppler Effect

1938–41 Ives and Stilwell [31] performed measurements in order to test the relativistic term in the Doppler shift. The Doppler shift was measured in light that was emitted by excited hydrogen atoms returning to their ground state, both in the direction of their motion and in the opposite direction. For beams of fast hydrogen atoms with energies corresponding to values of V/c in the region of 0.005, the shifts measured were in very good agreement with the relativistic predictions (Sect. 5.3.3).

1960 Hay et al. [32] , performed measurements in order to test the relativistic Doppler effect or the dilation of time, using the Mössbauer effect (Rudolf Ludwig Mössbauer 1929–2011). Before examining the experiment, we will give a brief description of this phenomenon.

The Mössbauer effect. In most radioactive disintegrations, the daughter nucleus is left, after the decay, in an excited state. The surplus energy is emitted in the form of photons, until the nucleus ends up in its ground state. These photons have well defined energies, which correspond to the differences between the energy levels of the excited nucleus. As is the case with the electronic transitions in the atom, the nucleus produced may absorb a photon with exactly the right energy and be excited to the corresponding energy state. We refer to this process as *resonance absorption*. The absorption is possible only if the photon has energy in a very narrow range about this resonance value. In some cases, the resonance curve has an extremely small width. For example, the nucleus of the isotope $^{57}_{26}\mathrm{Fe}$ which results from the electron capture by a $^{57}_{27}\mathrm{Co}$ nucleus (Fig. 10.3), reaches its ground state in two different ways: in 10 % of the cases by the emission of a γ with an energy of 136 keV and in 90 % of the cases by the emission of two photons, with energies of 122 and 14.4 keV, respectively.

As we have seen in Example 6.5, during the emission of a photon by a nucleus a fraction of the available energy is given to the nucleus as recoil energy, in order to conserve both momentum and mass-energy. Thus, if E_0 is the available energy, the recoiling nucleus will acquire a kinetic energy R, say, and the remaining energy, $E_0 - R$, will be the energy of the photon emitted. Of course, the photons will not be absolutely monoenergetic, but will have a narrow energy distribution about the mean value of $E_0 - R$. The curve on the left-hand-side of Fig. 10.4a shows one such energy spectrum. The full width at half maximum of the peak, Γ, is a measure of the dispersion of the energies of the emitted photons. For the 14.4 keV photons from $^{57}_{26}\mathrm{Fe}$, a value of $\Gamma = 4.6 \times 10^{-9}$ eV was measured. The recoil energy of the nucleus is $R \approx E_0^2/2Mc^2 = 0.002$ eV, approximately.

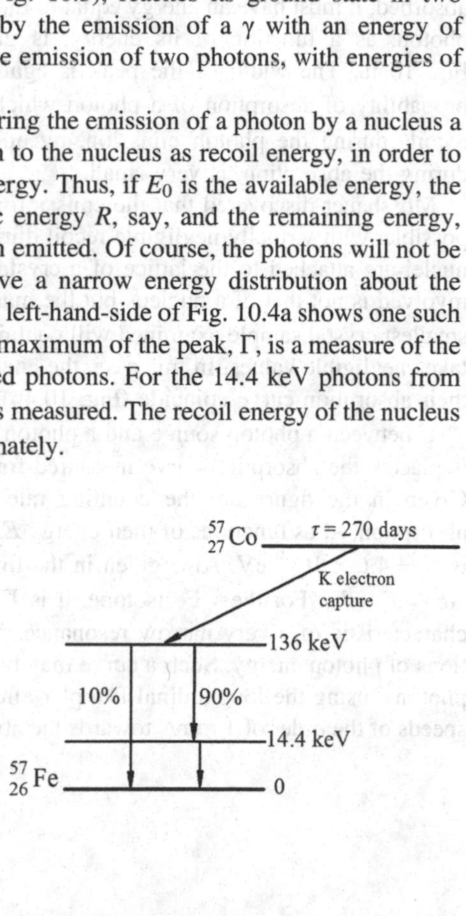

Fig. 10.3 The nuclear energy diagram of isotope $^{57}_{26}\mathrm{Fe}$

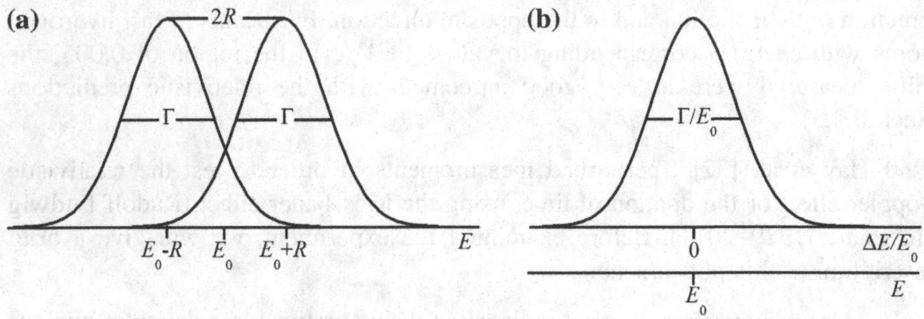

Fig. 10.4 **a** The energy spectrum (*left*) and the absorption curve (*right*) of the photons from $^{57}_{26}$Fe, with nuclear recoil. **b** The energy spectrum and the absorption curve without nuclear recoil. The curve is given on the scale of reduced energy difference from E_0, $\Delta E/E_0$, but the energy scale E is also given

In the inverse phenomenon of photon absorption, the $^{57}_{26}$Fe nucleus would have a maximum probability for absorbing a photon, if the latter had an energy exactly equal to E_0. However, since during the absorption the nucleus will recoil with a kinetic energy of R, again, for the photon to have maximum probability of being absorbed, it must have an energy equal to $E_0 + R$. The probability of absorption of a photon, as a function of its energy, is given by the right-hand-side curve of Fig. 10.4a. The width of the peak is again equal to Γ. Since it is $R \gg \Gamma$, the probability of absorption of a photon which is emitted by a $^{57}_{26}$Fe nucleus, which recoils during the photon emission, by another $^{57}_{26}$Fe nucleus, which will recoil during the absorption, is very small.

Mössbauer discovered that the emission and absorption of photons by nuclei is possible, with virtually negligible recoil during both of these processes [33]. If the nuclei are attached to the lattice of a crystal, then, during the recoil, the mass M involved is not that of a nucleus but the mass of the whole crystal. Since even the smallest crystal sample examined will contain at least 10^{10} atoms, the recoil energy takes negligible values. In this case, the energy spectrum curve of the photons and their absorption curve coincide (Fig. 10.4b). This is the Mössbauer effect.

If between a photon source and a photon counter an absorber consisting of $^{57}_{26}$Fe is placed, the absorption curve measured for the photons will be that of Fig. 10.5. Given in the figure are the counting rate of the photons, R, and their rate of absorption, A, as functions of their energy E. The full width of the absorption curve is $\Gamma = 4.6 \times 10^{-9}$ eV. Also given in the figure is the scale of $\Delta E/E_0$, where it is $\Delta E = E - E_0$. For the $^{57}_{26}$Fe isotope, it is $\Gamma/E_0 = 3.2 \times 10^{-13}$. This magnitude is characteristic of a very narrow resonance curve, which is very sensitive to variations of photon energy. Such a curve may be recorded by varying the energy of the photons, using the longitudinal Doppler effect, by moving the photon source with speeds of the order of 1 mm/s towards the absorber or away from it. Due to the very

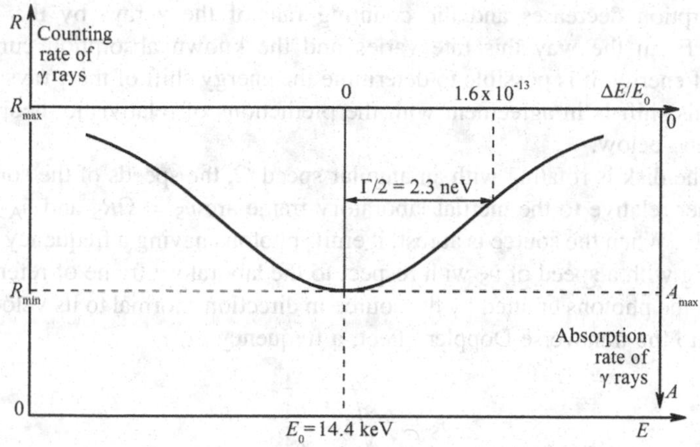

Fig. 10.5 The curve of absorption by a $^{57}_{26}$Fe absorber of the γ rays from $^{57}_{26}$Fe

small speeds involved, even the classical form of the Doppler equation is satisfactory.

Hay, Schiffer, Cranshaw and Egelstaff placed on a rotating disk (Fig. 10.6) a ^{57}Co γ ray source, S, and an iron absorber, A, which was enriched with ^{57}Fe. Source and absorber lied on a diameter of the disk, in opposite directions from the center and at distances equal to R_S and R_A from it, respectively.

When the disk is at rest, the γ rays emitted from the source, with energies of 14.4 keV, suffer resonance absorption in the absorber and the number of γ rays transmitted is at a minimum. As the disk starts rotating, the energy of the γ rays changes, due to the transverse Doppler effect, and the absorption is reduced. When the diameter on which source and absorber are situated coincides with the axis of detection of a γ ray detector, the transmitted γ rays are counted. As the speed of revolution Ω increases, the energy change of the γ rays becomes greater, with the result that the γ rays have energies that differ even more from the resonance energy,

Fig. 10.6 The rotating disk with the photon source S and the absorber A, used by Hay et al. (1960) for the verification of the relativistic Doppler effect

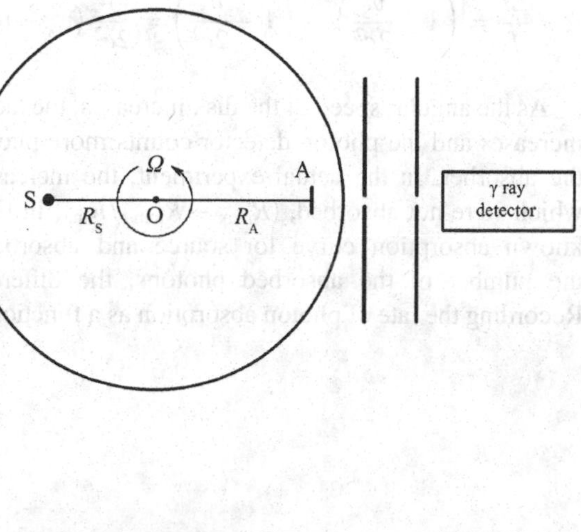

their absorption decreases and the counting rate of the γ rays by the detector increases. From the way this rate varies and the known absorption curve as a function of energy, it is possible to determine the energy shift of the γ rays and test whether this shift is in agreement with the predictions of relativistic theory as we will describe below.

When the disk is rotating with an angular speed Ω, the speeds of the source and the absorber relative to the inertial laboratory frame are $\upsilon_S = \Omega R_S$ and $\upsilon_A = \Omega R_A$, respectively. When the source is at rest, it emits photons having a frequency f. When it is moving with a speed of υ_S with respect to the laboratory frame of reference, in this frame, the photons emitted by the source in directions normal to its velocity will have, due to the transverse Doppler effect, a frequency of

$$f_S = f\sqrt{1-\frac{\upsilon_S^2}{c^2}}. \tag{10.1}$$

The absorber, when at rest, exhibits resonance absorption of the photons with frequency f. When it is moving with a speed υ_A relative to the laboratory frame of reference, the resonance frequency for the photons it absorbs from directions normal to its velocity will be, in the laboratory frame of reference, equal to

$$f_A = f\sqrt{1-\frac{\upsilon_A^2}{c^2}}. \tag{10.2}$$

As the disk rotates, the frequency of emission and the resonance absorption frequency of the photons differ by $\Delta f = f_S - f_A$, for which it is

$$\frac{\Delta f}{f} = \frac{f_S - f_A}{f} = \sqrt{1-\frac{\upsilon_S^2}{c^2}} - \sqrt{1-\frac{\upsilon_A^2}{c^2}}. \tag{10.3}$$

For speeds of the source and the absorber which are small compared to c, we may expand the square roots and keep terms only up to the second powers of the reduced speeds. We derive the approximate, but accurate enough, relation

$$\frac{\Delta f}{f} = \left(1-\frac{\upsilon_S^2}{2c^2}\right) - \left(1-\frac{\upsilon_A^2}{2c^2}\right) = \frac{1}{2c^2}\left(\upsilon_A^2 - \upsilon_S^2\right) = \frac{\Omega^2}{2c^2}\left(R_A^2 - R_S^2\right). \tag{10.4}$$

As the angular speed of the disk increases, the lack of resonance in the absorption increases and the photon detector counts more photons which are not absorbed by the absorber. In the actual experiment, the increase in the proportion of photons which were not absorbed, $(R_{max} - R_{min})/R_{max}$ in Fig. 10.5, reached 6 %. From the known absorption curve for source and absorber at rest and the decrease in the number of the absorbed photons, the difference Δf could be determined. Recording the rate of photon absorption as a function of angular speed of the disk, Ω, the value of the angular speed $\Omega_{1/2}$ at which the absorption rate falls to half its maximum value, will be such that

$$\frac{\Omega_{1/2}^2}{2c^2}\left(R_A^2 - R_S^2\right) = \frac{\Delta f_{1/2}}{f} = \frac{\Gamma/2}{E_0}. \qquad (10.5)$$

This equality will hold if the relativistic prediction for the transverse Doppler effect and, equivalently, the dilation of time, is correct. The agreement found between the two sides of Eq. (10.5) was to within 2 %. It is worth noting that, for $R_S = 5$ cm and $R_A = 10$ cm it must be $\Omega_{1/2} = 1960$ rad/s or 312 revolutions per second. The speed of the absorber, in this case, is 700 km/h or $6.5 \times 10^{-7}c$.

1961 Champeney and Moon [34] repeated the experiment of Hay et al. with the source and the absorber at equal distances from the center of the rotating disk, $R_S = R_A$. With angular speeds which corresponded to speeds $\upsilon_S = \upsilon_A$ between $8 \times 10^{-8}c$ and $5 \times 10^{-7}c$, it was found that the resonance absorption was not affected. This was considered to imply that the Mössbauer absorption was not affected and that this constitutes a confirmation of Eq. (10.4).

1962 Mandelberg and Witten [35], with experiments similar to that of Ives and Stilwell (1938-41), found for the coefficient 1/2 of the relativistic term $(\upsilon/c)^2$ in the Doppler shift, the experimental value of 0.498 ± 0.025 (Sect. 5.3.3).

1973 Olin et al. [36] measured the Doppler shift in photons with an energy of 8.64 MeV, which are emitted when the excited ^{20}Ne nuclei return to their ground states. They found a transverse Doppler shift equal, in energy, to 10.1 ± 0.4 keV, in agreement with theory at the level of 3.5 %, for speeds of the ions which corresponded to $\beta = 0.05$. For the coefficient 1/2 of the relativistic term $(\upsilon/c)^2$, they measured the value of 0.491 ± 0.017 (Sect. 5.3.3).

1979 Hasselkamp et al. [37] measured the Doppler shift in the light emitted by hydrogen atoms in a direction normal to their velocity. The speeds of the atoms corresponded to values of V/c between 0.008 and 0.031. For the numerical coefficient in the equation for the relativistic Doppler shift, they found the value of 0.52 ± 0.03, the expected value being 0.5 (Sect. 5.3.3).

10.5 The Contraction of Length

There are no experiments that measure the contraction of length directly. Indirectly, the consequences of the predictions of the Special Theory of Relativity may be observed in many phenomena. The evaluation of the force exerted by an electric current on a moving charge is an excellent demonstration of the consequences of length contraction. What is actually done is the extraction of the law of Biot-Savart from the law of Coulomb and the phenomenon of the contraction of length (Sect. 9.5).

The ionization caused by a relativistic charged particle. It is known that when a charged particle moves in a material it causes ionization of the material's atoms, due to the forces the particle's charge exerts on their electrons. At high particle speeds, an increased ionization is observed in directions normal to the particle's velocity and this is due to the concentration of the field lines of the electric field produced by the particle in these directions. This is a consequence of the purely relativistic effect of the contraction of length (Sect. 9.4.1).

10.6 The Test of the Predictions of Relativistic Kinematics

Synchrotron radiation. The relativistic beaming or the headlight effect is demonstrated in synchrotron radiation, which is emitted by charged particles moving with high speeds on circular orbits in a synchrotron. The centripetal acceleration of the particles, due to which they move on a circular orbit, leads to the emission of photons with energies of between a fraction of an eV and keV. It is found that, as predicted by theory, these photons move forming very small angles with the velocities of the emitting charges at every point of the accelerator, i.e. they are emitted in directions near the plane of the orbit and, approximately, tangentially to the circular orbit (Sect. 5.4.1).

1926 *Thomas precession.* An electron, as it moves in its orbit around the nucleus, suffers a precession, as predicted on purely kinematical reasons by the Special Theory of Relativity. This effect was proposed by Thomas [38] in order to complete the quantitative explanation of the fine structure of atomic spectra. It was rigorously interpreted theoretically by Wigner [39]. The Wigner rotation and its special case of the Thomas precession are a consequence of the laws of relativistic kinematics, whose success in the interpretation of spectroscopic observations must be considered as experimental verification of the Special Theory of Relativity (Sect. 4.8.1).

10.7 The Sagnac Effect

1897 Michelson [40] tried to detect the rotation of the Earth with respect to the aether. With an interferometer in which two beams of light traversed in opposite directions a closed path, which enclosed a vertical area $S = 900\ \text{m}^2$, oriented in the direction east-west, he tried to measure the shift of the interference fringes as the Earth rotated. The results did not lead to a positive conclusion, as any shift of the fringes was smaller than the limit of sensitivity of the arrangement, which was 1/20 of a fringe.

1911 F Harress, during work for his doctoral thesis, used an interferometer in which two beams of light traversed in opposite directions a closed, almost circular, path, inside ten prisms, so that he could investigate the dispersion of light in various glasses. During the rotation of the apparatus, he noticed a shift in the positions of the fringes, as expected on the basis of the, then unknown, Sagnac effect. The conclusions on the dispersion of light he extracted in this way were not in agreement with the results of other methods of measurement. Harress attributed the shift of the fringes to the dragging of the aether by the rotating glass.

1913 Sagnac [41] was the first to demonstrate that it was possible, with optical measurements, to detect and measure the rotation of a frame of reference with respect to the inertial frames of reference. In the experimental arrangement he used, two beams of light, traveling in opposite directions, traversed a closed path in a rotating interferometer. The interference fringes were photographed when the interferometer was rotating about a vertical axis in one direction, and then in the opposite direction. Sagnac used light with $\lambda_0 = 436$ nm, area enclosed by the beams of light equal to $S = 866$ cm^2 and a rate of revolution of 2 turns per second, i.e. an angular velocity of $\Omega = 6.28$ rad/s. He observed a shift of the interference pattern equal to $2\Delta n = 0.07$ of a fringe, which was easy to measure. The shift of the fringes observed was in agreement with the theoretical prediction, within the limits of experimental error (Sect. 4.2).

1925 Michelson and Gale [42], and Michelson, Gale and Pearson [43], used a very large interferometer, in the shape of a rectangular parallelogram with sides of lengths of 640 and 320 m, in order to observe, utilizing the Sagnac effect, the rotation of the Earth about its axis. The beams propagated in evacuated tubes, so that Fresnel aether dragging effects would be avoided. With partial reflection of one of the beams, a part of it was diverted into a much smaller interferometer, which provided the reference positions of the fringes when no detectable shift was present. They observed a shift of $\Delta n = 0.230 \pm 0.005$ of a fringe, in excellent agreement with the theoretically expected shift of 0.236 of a fringe.

1926 Pogany [44] achieved a significant improvement in the sensitivity of the Sagnac experimental arrangement, using an interferometer with $S = 1178$ cm^2, $\Omega = 157.43$ rad/s and $\lambda_0 = 546$ nm. The agreement with theory was at the level of 2 %, which was improved to 1 % two years later [45]. It is worth mentioning that the apparatus could be used for measurements even up to a rate of rotation equal to 500 revolutions per second!

1993 With the advancement of modern technology, it became possible to construct extremely sensitive experimental arrangements which use the Sagnac effect for a variety of applications. One such arrangement is the interferometer built by the universities of Canterbury, New Zealand, and Oklahoma, USA, in Christchurch, New Zealand [46]. Constructed away from city noises, at a depth of 30 meters under the surface of the ground, it has $S = 0.7547$ m^2, it uses green light from a stabilized He–Ne laser, with a light path length equal to 3.477 m. The angular velocity of the Earth is monitored with a sensitivity of one part in a million ($\Delta\Omega = 10^{-6}\Omega_{\text{Earth}}$). This constitutes an improvement by 12 orders of magnitude over the interferometer of Michelson, Gale and Pearson, of 1925, if it is taken into account that the area of the interferometer is about 300 000 times smaller. Apart from monitoring the rate of rotation of the Earth, the interferometer may be used to measure the rotational component of earthquakes, something that conventional seismometers cannot do. Today (2012), the arrangement, with an area of one square meter and a stabilized laser with a quality factor of $Q = 10^{12}$, has been improved to the extent that it can detect a variation in the Earth's angular velocity, Ω_{Earth}, of the order of $\Omega = 1.8 \times 10^{-12}$ rad/s $= 2.5 \times 10^{-8}\Omega_{\text{Earth}}$, i.e. 25 billionths of Ω_{Earth}. This would be equivalent to a rotation of one thickness of a human hair, 10 μm, per second, as seen from a distance of 560 km. With such sensitivity, the apparatus may be used to measure its rotation due to the revolution of the Earth around the Sun and other such motions. The dramatic improvement of science's arguments ever since Galileo whispered his historic phrase *Eppur si muove,*[1] is impressive.

More information on experimental arrangements and measurements based on the Sagnac effect may be found in the two excellent review articles mentioned in the bibliography [47].

10.8 The Relativistic Mass

1902 Kaufmann [48] studied the motion of relativistic particles in an electric and a magnetic field. He observed a variation of the inertial mass with speed for electrons moving with speeds 0.8-0.95 times the speed of light in vacuum (Sect. 1.12).

1909, 1915 Bucherer [49] and Guye and Lavanchy [50] measured the variation of the inertial mass with speed (Sect. 1.12).

1940 The predictions of the Special Theory of Relativity for the variation of the inertial mass with speed are taken into account in the construction of high-energy particle accelerators (synchrocyclotrons and synchrotrons) and are verified with accuracy up to very high energies (Sect. 7.3).

10.9 The Equivalence of Mass and Energy

1932 Cockcroft and Walton [51] bombarded a lithium target with fast protons, forming excited ${}^{8}_{4}\text{Be}^{*}$ nuclei, which immediately split into two α particles. The kinetic energies of the α particles could be determined from their ranges in air.

---

[1] *And yet it does move.*

The energy of the α particles in excess of the proton's energy agree with the energy equivalent of the mass defect of the reaction (Sect. 7.4.2).

1939 Smith [52] repeated the measurements of Cockcroft and Walton (1932) with greater accuracy. He found for the difference between the kinetic energies of the α particles and the impinging protons the value of $\Delta K = 17.28 \pm 0.03$ MeV, which agreed with theory with an accuracy of better than 1 % (Sect. 7.4.2).

1944 Dushman [53] compared the theoretical prediction for the liberated energy with the experimentally measured one for some then well known nuclear reactions. The agreement of experiment with theory was found to be very good (Sect. 7.4.2).

10.10 The Test of the Predictions of Relativistic Dynamics

1922 *The Compton effect*. Compton [54] applied the laws of relativistic Mechanics in the interpretation of his observations during the scattering of X rays by light elements. His success is due to the fact that he considered the X rays as consisting of photons, with particle properties, which collide elastically with free electrons, according to the relativistic laws (Sect. 7.1).

1932 Champion [55] put under experimental test the relativistic theory for the collision of two identical particles, by observing in a cloud (Wilson) chamber the orbits of the two electrons after the collision of moving electrons (β particles) with the stationary electrons of the gas in the chamber. The experiment provided a verification of the relativistic predictions regarding the variation of the inertial mass and of the momentum of a body with its speed, as well as of the validity of the laws of conservation of momentum and of energy at relativistic energies and speeds (Sect. 6.12.1).

1964 Bertozzi [56] confirmed the relationship between speed and kinetic energy predicted by the Special Theory of Relativity. For this purpose he used short pulses of electrons which had been accelerated by a van de Graaf generator to kinetic energies equal to 0.5, 1, 1.5, 4.5 and 15 MeV. The experimental arrangement used is shown in Fig. 10.7.

The determination of the speed of the electrons at each energy was achieved directly with the measurement of the time of flight of the electrons between two points, A and B in the figure, at a distance of 8.4 m between them. As a pulse passed point A, a part of the total charge fell on a suitably situated metallic conductor, from which a signal emerged, in the form of a charge pulse. The same happened at the end of the path, B. The two signals were viewed on the screen of an oscilloscope, from where the time difference between the two signals was read. The cables used for the transmission of the two signals from A and B had equal lengths so that no additional time delay was introduced in the times of arrival of the pulses at the oscilloscope. The times of flight of the electrons between the points A and B

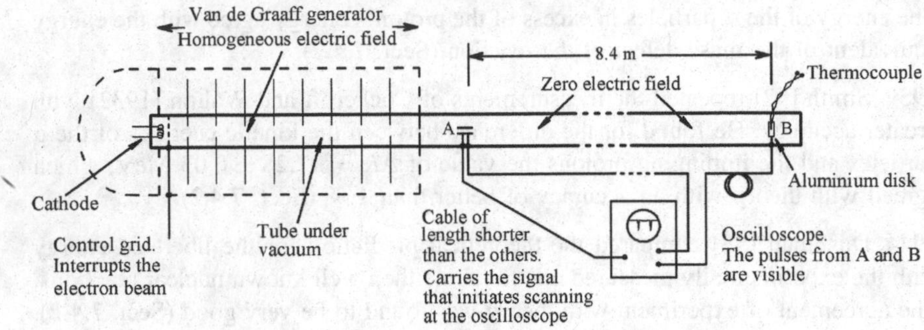

Fig. 10.7 The experimental arrangement used by W. Bertozzi for the determination of the relationship between the speed and the kinetic energy of electrons

ranged between 32 and 28 ns, depending on their energy. The error in the measurement of these time intervals was about 0.7 ns.

The kinetic energy of the electrons was determined by two different methods: From the accelerating potential difference and by a calorimetric method at the end of the path, when the electrons fell on an aluminium disk and were absorbed. The calorimetric method was used in order to avoid objections concerning whether the electric force exerted on an accelerated charge is independent of its speed. By measuring the temperature of the aluminium disk B, the total kinetic energy of the electrons falling on the disk in an interval typically equal to 7 min was determined. The number of electrons falling on the target was determined from the total charge appearing on the disk in the relevant time interval. The ratio of the two quantities gave the kinetic energy per electron.

The equations giving the square of the speed of a body of mass m_0 and kinetic energy K are, according to Classical and Relativistic Mechanics,

$$\left(\frac{\upsilon}{c}\right)^2 = \frac{2K}{m_o c^2} \quad \text{and} \quad \left(\frac{\upsilon}{c}\right)^2 = 1 - \left(\frac{m_0 c^2}{m_0 c^2 + K}\right)^2,$$

respectively. The experimental verification of one of these two equations was the purpose of the experiment.

Shown in Fig. 10.8 are the two curves, as these are predicted by the two equations, and the experimental results. Logarithmic scales were used in the figure, in order to show the differences between the two theories better. It is obvious that the measurements agree with the Special Theory of Relativity, while they are in marked disagreement with Classical Mechanics. The calorimetric measurements confirmed, in addition, that even at kinetic energies as high as 15 MeV, the kinetic energy K acquired by a charge e in an electric field in the direction of its motion, x, is indeed equal to $K = \int eEdx$, independent of the charge's speed. This also constitutes a verification of the invariance of the electric charge with speed.

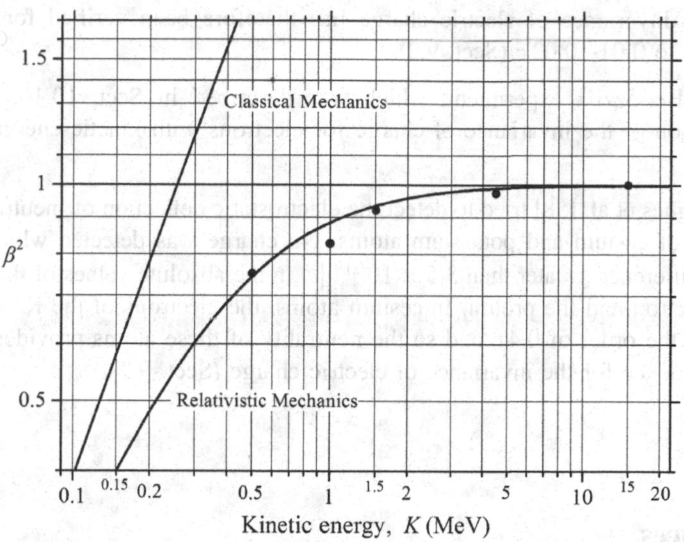

Fig. 10.8 The results of Bertozzi's experiment. The variation of the square of the electrons' reduced speed, $\beta^2 = (\upsilon/c)^2$, with their kinetic energy K. The two lines give the predictions of both classical and relativistic Mechanics. The dots are the experimental results

10.11 The Invariance of Electric Charge

1940 The design and operation of high-energy particle accelerators is based on the equation of motion $\dfrac{d}{dt}\left(\dfrac{m_0\boldsymbol{\upsilon}}{\sqrt{1-\upsilon^2/c^2}}\right) = q\mathbf{E} + q\boldsymbol{\upsilon}\times\mathbf{B}$ for a particle with rest mass m_0 and charge q, moving with a velocity $\boldsymbol{\upsilon}$ in a space where there is an electric field $\mathbf{E}$ and a magnetic field $\mathbf{B}$. In these accelerators, the variation of the inertial mass with speed is taken into account while the charge q is considered constant. Even at energies up to 25 GeV for electrons and 7 TeV for protons, which have speeds of 0.999 999 999 8 c and 0.999 999 991 3 c, respectively, this equation of motion is adequate, with the provision that a correction is made for the emission of electromagnetic radiation by the accelerated charges (Sect. 7.3).

1960 King [57] demonstrated, with great accuracy, that the molecules of hydrogen and the atoms of helium are neutral. He used a vessel, electrically isolated from the surroundings, which contained hydrogen gas. Gas could escape from a little hole. The sensitivity in measuring the charging of the vessel during this procedure, sets an upper limit of $10^{-19}\,|e|$, in absolute value, on the charge that a 'neutral' hydrogen molecule could possibly have. The same experiment was repeated with the same results with helium gas. The electrons of the atoms of hydrogen move with speeds of the order of 0.007 c, while those of the He atoms have speeds about twice as high. The invariance of electric charge has therefore been verified for electron speeds up to $0.01-0.02\,c$ (Sect. 9.2).

1964 W Bertozzi's experiment, which was described in Sect. 10.10, provides confirmation of the invariance of charge for electrons with kinetic energies up to 15 MeV.

1988 Hughes et al. [58] tried to detect the electrostatic deflection of 'neutral' atoms in beams of cesium and potassium atoms. No charge was detected which would imply a difference greater than $3.5 \times 10^{-19}\,|e|$ in the absolute values of the charges of the electron and the proton. In cesium atoms, the electrons of the K shell have speeds of the order of $0.4\,c$ and so the neutrality of these atoms provides considerable evidence for the invariance of electric charge (Sect. 9.2).

References

1. Y.Z. Zhang, *Special Relativity and its Experimental Foundations* (World Scientific, 1997)
2. J. Bradley, Phil. Trans. Roy. Soc. **35**, 637 (1729). See also A. Stewart, The discovery of stellar aberration. Sci. Am. **210**(3), 100 (1964)
3. F. Arago, Compt. Rend. **8**, 326 (1839) and **36**, 38 (1853)
4. A.J. Fresnel, Ann. Chim. Phys. **9**, 57 (1818)
5. J.G. Fox, Experimental evidence for the second postulate of special relativity. Am. J. Phys. **30**, 297 (1962)
6. V.C. Ballenegger, T.A. Weber, The Ewald-Oseen extinction theorem and extinction lengths. Am. J. Phys. **67**, 599 (1999)
7. M. Grewing, M. Walmsley, On the interpretation of the pulsar dispersion measure. Astron. Astrophys. **11**, 65 (1971). J.H. Taylor, J.M. Cordes, Pulsar distances and the galactic distribution of free electrons. Astrophys. J. **411**, 674 (1993)
8. O. Struve, *Stellar Evolution* (Princeton University Press, Princeton, 1950)
9. K. Brecher, Is the speed of light independent of the velocity of the source? Phys. Rev. Lett. **39**, 1051 (1977)
10. D. Sadeh, Experimental evidence for the constancy of the velocity of gamma rays, using annihilation in flight. Phys. Rev. Lett. **10**, 271–273 (1963)
11. T.A. Filippas, J.G. Fox, Velocity of gamma rays from a moving source. Phys. Rev. B **135**, 1071 (1964)
12. T. Alväger, A. Nilsson, J. Kjellman, Nature **197**, 1191 (1963)
13. T. Alväger, F.J.M. Farley, J. Kjellman, I. Wallin, Test of the second postulate of special relativity in the GeV region. Phys. Lett. **12**, 260 (1964)
14. B. Lovell, F. Whipple, L. Solomon, Relative velocity of light and radio waves in space. Nature **202**, 377 (1964)
15. B. Warner, R. Nather, Wavelength independence of the velocity of light in space. Nature **222**, 157 (1969)
16. B.E. Schaefer, Severe limits on variations of the speed of light with frequency. Phys. Rev. Lett. **82**, 4964 (1999)
17. H. Fizeau, C.r. hebd. Seanc. Acad. Sci. Paris **33**, 349 (1851)
18. M. Hoek, Archs. Nèerl. Sci. **3**, 180 (1968)

19. G.B. Airy, Proc. Roy. Soc. A **20**, 35 (1871). G.B. Airy, Phil.Mag. **43**, 310 (1872). G.B. Airy, Proc. Roy. Soc. A **21**, 121 (1873)
20. A.A. Michelson, Am. J. Sci. **22**, 20 (1881)
21. A.A. Michelson, E.W. Morley, Am. J. Sci. **34**, 333 (1887)
22. R.J. Kennedy, E.M. Thorndike, Experimental establishment of the relativity of time. Phys. Rev. **42**, 400–418 (1932)
23. H.A. Lorentz, Verh. K. Akad. Wet. **1**, 74 (1892)
24. G.F. FitzGerald, The aether and the Earth's atmosphere. Science **13**, 328, 390 (1889)
25. A. Brillet, J.L. Hall, Improved laser test of the isotropy of space. Phys. Rev. Lett. **42**, 549–52 (1979)
26. B. Rossi, D.B. Hall, Variation of the rate of decay of mesotrons with momentum. Phys. Rev. **59**, 223 (1941). B. Rossi, K. Greisen, J.C. Stearns, D.K. Froman, P.G. Koontz, Further measurements of the mesotron lifetime. Phys. Rev. **61**, 675 (1942)
27. D.H. Frisch, J.H. Smith, Measurement of the relativistic time dilation using μ-mesons. Am. J. Phys. **31**, 342 (1963)
28. J.C. Hafele, R.E. Keating, Science **177**, 166 (1972), and Science **177**, 168 (1972). Also, J.C. Hafele, Relativistic time for terrestrial circumnavigations. Am. J. Phys. **40**, 81–85 (1971)
29. R.F.C. Vessot, M.W. Levine, E.M. Mattison, E.L. Blomberg, T.E. Hoffman, G.U. Nystrom, B.F. Farrel, R. Decher, P.B. Eby, C.R. Baugher, J.W. Watts, D.L. Teuber, F.D. Wills, Test of relativistic gravitation with a space-borne hydrogen maser. Phys. Rev. Lett. **45**, 2081–2084 (1980)
30. J. Bailey, K. Borer, F. Combley, H. Drumm, F. Krienen, F. Lange, E. Picasso, W. von Ruden, F.J.M. Farley, J.H. Field, W. Flegel, P.M. Hattersley, Measurements of relativistic time dilatation for positive and negative muons in a circular orbit. Nature **268**, 301–5 (1977). J. Bailey, K. Borer, F. Combley, H. Drumm, C. Eck, F.J.M. Farley, J.H. Field, W. Flegel, P.M. Hattersley, F. Krienen, F. Lange, G. Lebée, E. McMillan, G. Petrucci, E. Picasso, O. Rúnolfsson, W. von Rüden, R.W. Williams, S. Wojcicki, Final report on the CERN muon storage ring including the anomalous magnetic moment and the electric dipole moment of the muon, and a direct test of relativistic time dilation. Nucl. Phys. B **150**, 1–75 (1979)
31. H.E. Ives, G.R. Stilwell, J. Opt. Soc. Am. **28**, 215 (1938) and **31**, 369 (1941)
32. H.J. Hay, J.P. Schiffer, T.E. Cranshaw, P.A. Egelstaff, Phys. Rev. Lett. **4**, 165 (1960)
33. R.L. Mössbauer, Z. Phys. **151**, 124 (1958). Naturwissenschaften **45**, 538 (1958). Z. Naturforsch. **14a**, 211 (1959)
34. D.C. Champeney, P.B. Moon, Proc. Phys. Soc. **77**, 350 (1961)
35. H. Mandelberg, L. Witten, J. Opt. Soc. Am. **52**, 529 (1962)
36. A. Olin, T.K. Alexander, O. Häuser, A.B. McDonald, Phys. Rev. D **8**, 1633 (1973)
37. D. Hasselkamp, E. Mondry, A. Scharmann, Direct observation of the transversal Doppler-shift. Z. Phys. A **289**, 151 (1979)
38. L.H. Thomas, Motion of the spinning electron. Nature **117**, 514 (1926)
39. E.P. Wigner, On unitary representations of the inhomogeneous Lorentz group. Ann. Math. **40**, 149–204 (1939)
40. A.A. Michelson, The relative motion of the earth and the ether. Am. J. Sci. **3**, 475–8 (1897), 4th series
41. G. Sagnac, L'éther lumineux démontré par l'effet du vent relatif d'éther dans un interféromètre en rotation uniforme. C. R. Acad. Sci. (Paris) **157**, 708–10 (1913). Sur la preuve de la réalité de l'éther lumineux par l'expérience de l'interférographe tournant, C. R. Acad. Sci. (Paris) **157**, 1410-13 (1913). Also, Effet tourbillonnaire optique. La circulation de l'éther lumineux dans un interférographe tournant. J. Phys. Radium Ser.5 **4**, 177–195 (1914)
42. A.A. Michelson, H.G. Gale, Nature **115**, 566 (1925)
43. A.A. Michelson, H.G. Gale, F. Pearson, The effect of the earth's rotation on the velocity of light. Astrophys. J. **61**, 137–145 (1925)
44. B. Pogany, Ann. Phys. **80**, 217 (1926)
45. B. Pogany, Ann. Phys. **85**, 244 (1928)

46. H.R. Bilger, G.E. Stedman, M.P. Poulton, C.H. Rowe, Z. Li, P.V. Wells, Ring laser for precision measurement of nonreciprocal phenomena. IEEE Trans. Instr. Meas. (special issue for CPEM/92) **42**, 407–11 (1993). G.E. Stedman, H.R. Bilger, Li Ziyuan, M.P. Poulton, C.H. Rowe, I. Vetharaniam, P.V. Wells, Canterbury ring laser and tests for nonreciprocal phenomena. Aust. J. Phys. **46**, 87–101 (1993)
47. E.J. Post, Sagnac effect. Rev. Mod. Phys. **39**, 475 (1967). R. Anderson, H.R. Bilger, G.E. Stedman, 'Sagnac' effect: a century of Earth-rotated interferometers. Am. J. Phys. **63**, 975 (1994)
48. W. Kaufmann, Göttingen Nach. **2**, 143 (1901)
49. A.H. Bucherer, Ann. d. Phys. **28**, 513 (1909)
50. C.E. Guye, C. Lavanchy, *Compt. Rend.* **161**, 52 (1915)
51. J.D. Cockcroft, G.T.S. Walton, Proc. R. Soc. A **137**, 229 (1932)
52. N.M. Smith Jr, Phys. Rev. **56**, 548 (1939)
53. S. Dushman, Gen. Electr. Rev. **47**, 6–13 (1944)
54. A.H. Compton, Bulletin Nat. Res. Council **20**, 16 (1922) and Phys. Rev. **21**, 715 and **22**, 409 (1922)
55. F.C. Champion, Proc. R. Soc. A **136**, 630 (1932)
56. W. Bertozzi, Speed and kinetic energy of relativistic electrons. Am. J. Phys. **32**, 551 (1964)
57. J.G. King, Phys. Rev. Lett. **5**, 562 (1960)
58. V.W. Hughes, L.J. Fraser, E.R. Carlson, Z. Phys, D – Atoms. Mol. Clusters **10**, 145 (1988)

Appendix 1
The Paradox of the Room and the Rod

A1.1 The Room and the Rod–A Relativistic Paradox

There are many 'paradoxes' in the literature of the Theory of Special Relativity [1], which appear to violate the simple rules of logic we acquire from everyday experiences. One of the best known is the paradox of the room and the rod, also known by other names.

Let a room have a rest length of 4 m, along the x-axis and a rod have a rest length of $L_0 = 8$ m, also along the x-axis (Fig. A1.1a). The room has one door at each of its ends (α and β), initially closed. The center of the room, O, is the origin of the inertial frame of reference S (Fig. A1.1b).

The rod moves relative to frame S with a speed $V = 2.903 \times 10^8$ m/s along the x-axis, towards the positive direction. Thus, $\beta = V/c = 0.968$ and $\gamma = 4$. In frame S, the rod has, therefore, length equal to $L = L_0/\gamma = 8$ m $/4 = 2$ m and needs time equal to $\tau = 1$ m$/V = 3.45$ ns in order to move a distance of 1 m. At the moment $t = 0$, the center of the rod coincides with the center of the room. In the frame of reference of the rod, S′, the room moves with a speed $-V$, has a length of 4 m /4 = 1 m and travels a distance of 1 m in time $\tau = 3.45$ ns (Fig. A1.1c).

The frame of reference of the rod, S′, has axes which are parallel to those of S. At the moment $t = 0$, it is $t' = 0$, and the axes of the two frames coincide. Therefore, $x' = \gamma(x - Vt)$ and $t' = \gamma(t - (V/c^2)x)$. We find the x, t values of frame S and the corresponding values x', t' in frame S′, for the following events:

Event			Frame S (room)		Frame S′ (rod)	
			x (m)	t (ns)	x' (m)	t' (ns)
(a)	End A at door α	Door α opens	−2	−10.35	+4	−15.6
(b)	End B at door α	Door α closes	−2	−3.45	−4	+12
(c)	The centers of S and S′ coincide		0	0	0	0
(d)	End A at door β	Door β opens	+2	+3.45	+4	−12
(e)	End B at door β	Door β closes	+2	+10.35	−4	+15.6

© Springer International Publishing Switzerland 2016

C. Christodoulides, *The Special Theory of Relativity*,
Undergraduate Lecture Notes in Physics, DOI 10.1007/978-3-319-25274-2

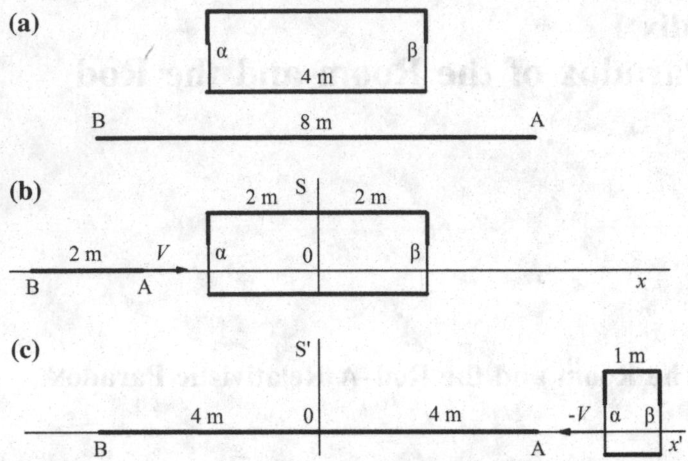

Fig. A1.1 **a** The room and the rod, compared when both are at rest in the same frame of reference. **b** The frame of reference of the room, S, with the rod moving with a velocity $\mathbf{V} = V\hat{\mathbf{x}}$, for which it is $\gamma = 4$. **c** The frame of reference of the rod, S′, in which the room moves with a velocity $-\mathbf{V} = -V\hat{\mathbf{x}}$

We notice that the sequence of events in time is not the same in both frames of reference. The events are drawn in the right temporal orders in the two frames in Fig. A1.2 The first column shows what an observer in frame S of the room would see and the second column what is seen by an observer in the frame S′ of the rod.

According to an observer in S, between $t = -3.45$ ns and $t = +3.45$ ns the rod moves inside the room, with both doors closed. Something like this cannot happen in frame S′, as the length of the room is much smaller than that of the rod. An observer in S′ sees both doors open and a part of the rod inside the room, between $t' = -15.6$ ns and $t' = +15.6$ ns.

The orders of events are not the same in the two frames of reference. In frame S, door α opens and closes and then door β opens and closes. In frame S′, door α opens and then door β opens. Then door α closes and then door β closes. According to the Special Theory of Relativity, there is no paradox, since, as we have already seen, the two observers do not agree as to which events are simultaneous and which are not. However, in neither of the frames of reference is the temporal relationship between cause and effect violated.

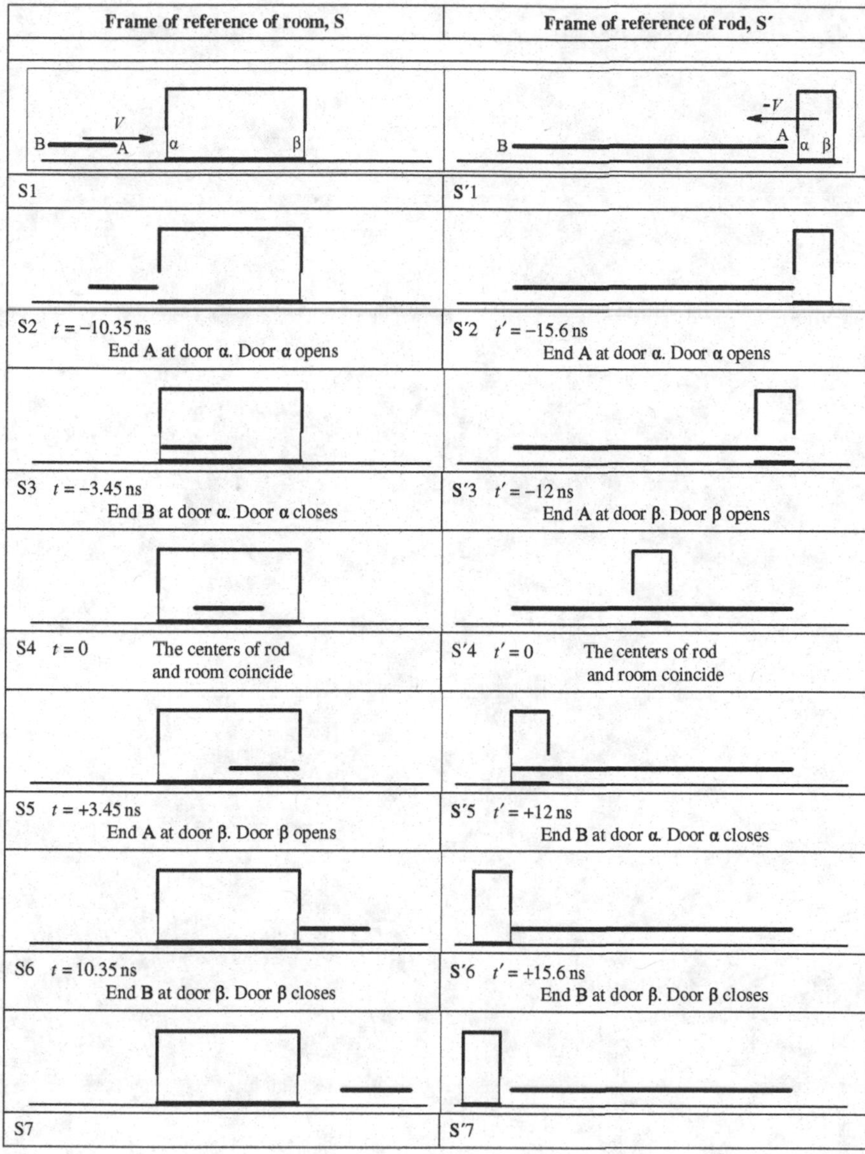

Fig. A1.2 The paradox of the room and the rod. The sequence of events in the two frames of reference

Problem

A1.1 Draw two Minkowski diagrams for the room and rod paradox: one with main frame of reference that of the room and another with main frame of reference that of the rod.

Appendix 2
The Appearance of Moving Bodies

In his historic work of 1905 [2], Einstein states that: 'A solid body which, measured at rest, has the shape of a sphere, has therefore when in motion—if we are looking at it from a stationary frame of reference—the shape of an ellipsoid by revolution, with axes $R(1 - \upsilon^2/c^2)^{1/2}$, R, R. Thus, while the dimensions Y and Z of the sphere (as well as of any other solid body, independently of shape) do not appear to change due to motion, dimension X appears to have suffered contraction by the ratio 1: $(1 - \upsilon^2/c^2)^{1/2}$, i.e. the larger the value of υ, the greater the contraction. For $\upsilon = c$ all the moving objects—if we are looking at them from a 'stationary' frame of reference—shrink to plane shapes'. Obviously, Einstein uses the term 'looking' in the sense of the observation in a frame of reference, as he himself defines the process, with infinite observers at all the points of the frame, each with his synchronized clock. If, however, we take the word 'looking' literally, then this statement is wrong. There is a difference between the shape a moving body *has* and that which it *appears to have*. This was first pointed out by the Austrian physicist Anton Lampa [3], in a brief publication of his in German, which went almost unnoticed. The same fact was pointed out in 1959, independently of each other, by Penrose [4] and Terrel [5]. The problem of the shape a moving body *appears* to have we will examine below.

A2.1 The 'Experimental Arrangement'

We will examine the following problem: Let a body, of known shape when at rest, move with some speed V in the laboratory frame of reference S. If we use, in the laboratory frame (Fig. A2.1), an extremely fast camera and photograph the moving body, which is the picture that will be imprinted? The camera will record a 'snapshot' of the moving body, with light that enters its lens at the same moment of time. We will agree that the camera is situated at a great distance from the moving object, so that the light entering the camera consists of parallel rays. We will take the plane of the camera's lens to be parallel to the xz plane. The photograph is 'taken', therefore, with light which crosses a plane parallel to the xz plane at the same moment.

© Springer International Publishing Switzerland 2016

C. Christodoulides, *The Special Theory of Relativity*,
Undergraduate Lecture Notes in Physics, DOI 10.1007/978-3-319-25274-2

Fig. A2.1 The hypothetical arrangement that might be used in photographing a moving body

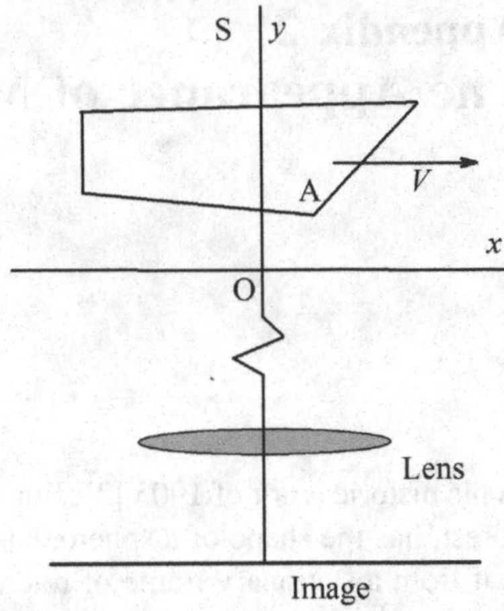

The important fact to remember is that the light crossing the xz plane at a given moment, and which finally forms the image of the object, has travelled different distances from different points of the moving object and, as a consequence, it was emitted at different times from each point and not simultaneously. Thus, the position from which a point such as A in the figure emitted the light that is imprinted on the film, depends on the distance of point A from the camera or, equivalently, from the xz plane.

A2.2 The Appearance of Moving Bodies

We will examine some cases of the appearance of moving bodies. More may be found it the rich literature on the subject [6].

A2.2.1 Square

In a square with sides having rest length L, side OC, in the direction of motion of the square, contracts by a factor γ in the frame of reference S (Fig. A2.2a).

If the light which is imprinted on the photograph left point O at time $t = 0$, that from point A was emitted when this point was at point A_0 at time $-\delta t$, where $\delta t = L/c$. Therefore, $V\delta t = LV/c = \beta L$ and point A_0 is, at the image, at a distance $-\beta L$ from O. The vertices of the square are imprinted on the photograph at the points $A_0(x = -\beta L)$, $O(x = 0)$, $B_0(x = (1/\gamma - \beta)L)$ and $C(x = L/\gamma)$. Angle ϕ in the figure is equal to $\phi = \arctan \beta$. It is worth noting that the same photograph would have resulted from a stationary square, $OA_1B_1C_1$ (Fig. A2.2b), with sides of

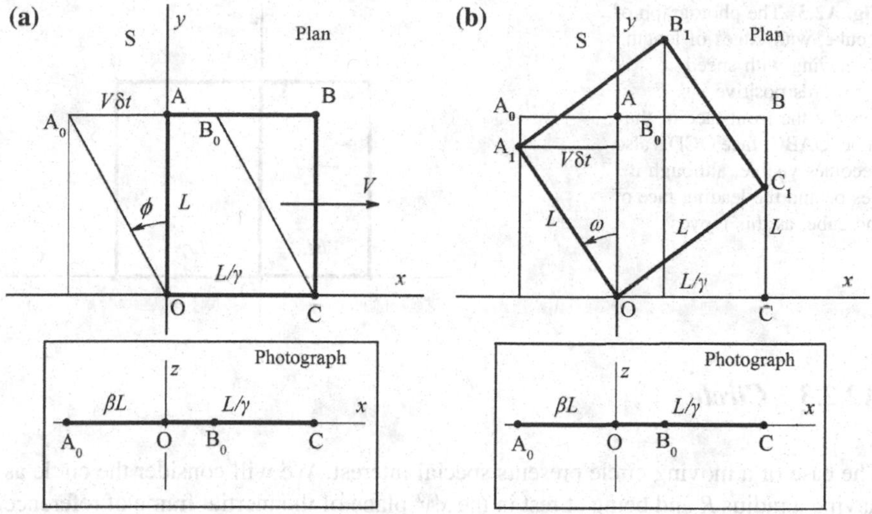

Fig. A2.2 **a** A square with sides of rest length L, moving with a speed V in the direction of positive x's. The photograph of the object is shown in the inset at the bottom of the figure. The image of the object is the straight line A_0OB_0C. **b** The same photograph would result from a stationary square with sides equal to L, which has been rotated by an angle $\omega = \arcsin\beta$, as shown in the top figure

length L, which had been rotated relative to the real square by angle ω, where $\sin\omega = \beta$. The motion of the square has, according to the Special Theory of Relativity, the result that it appears as if it had been rotated by an angle $\omega = \arcsin\beta$. This rotation effect is present in general in moving bodies, as we will see below. It should be noted that, as a result of this rotation, points on the side OA which would not be visible in a square at rest become visible when the cube is moving. This is not an exclusively relativistic effect and is observed in general.

A2.2.2 Cube

Based on what has been said for the case of the square, we can conclude what the appearance of a moving cube will be, when it is photographed. The base of the cube is a rectangle, such as the one shown in Fig. A2.2a, with one side with length equal to L and the other equal to L/γ. The height of the cube remains equal to L. The photograph of the cube will be as shown in Fig. A2.3. Apart from the face OABC, which lies along the x-axis, face OCDE, which lies behind the leading face of the cube as it moves, also becomes visible. Again, the cube appears as if it has been rotated by an angle $\omega = \arcsin\beta$.

Fig. A2.3 The photograph of a cube, with edges of length *L*, moving with speed *V* towards positive *x*'s. Besides the front face of the cube, OABC, face OCDE also becomes visible, although it lies behind the leading face of the cube, as this moves

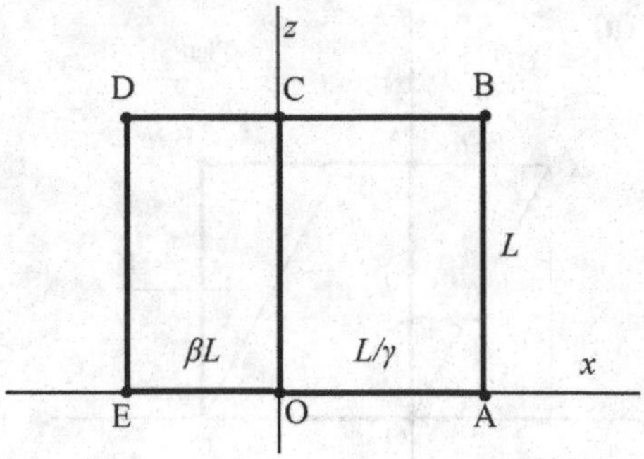

A2.2.3 Circle

The case of a moving circle presents special interest. We will consider the circle as having a radius R and being at rest in the $x'y'$ plane of the inertial frame of reference S′, with its center at the origin of the axes. A point A of the circle is determined by the polar angle θ' (Fig. A2.4). In frame S′, point A is at the position

$$x' = R\cos\theta', \quad y' = R\sin\theta', \quad z' = 0, \tag{A2.1}$$

for every value the time t'. In another inertial frame of reference S, which is moving relative to frame S′ with velocity $\mathbf{V} = V\hat{\mathbf{x}}$, the circle has the shape of an ellipse with its center at O when it is $t = t' = 0$. The circle contracts by a factor γ in the x direction. Point A will be at position A_0 at the moment $t = 0$, with coordinates:

$$x = \frac{R}{\gamma}\cos\theta', \quad y = R\sin\theta', \quad z = 0. \tag{A2.2}$$

At time t, the position of A will be

$$x = \frac{R}{\gamma}\cos\theta' + Vt, \quad y = R\sin\theta', \quad z = 0. \tag{A2.3}$$

Light leaving point A at this moment, will reach the point

$$X = \frac{R}{\gamma}\cos\theta' + Vt, \quad Y = 0, \quad Z = 0 \tag{A2.4}$$

of the plane of the image at time

$$T = t + \frac{l + y}{c} \quad \text{or} \quad T = t + \frac{l + R\sin\theta'}{c}. \tag{A2.5}$$

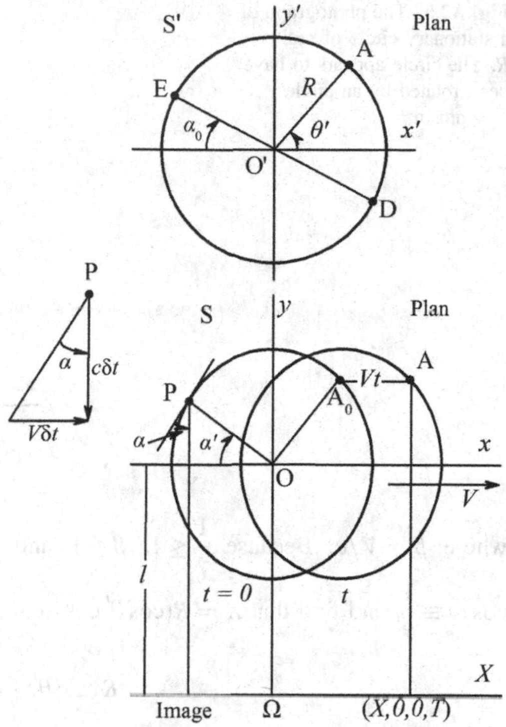

Fig. A2.4 A circle of rest radius R, stationary in the frame of reference S′, moves with a speed V towards positive x's in the frame of reference S

Solving the second of the Eqs. (A2.5) for t and substituting in the first of the Eqs. (A2.4), we find that

$$X = \frac{R}{\gamma}\cos\theta' + VT - \frac{V}{c}(l + R\sin\theta'). \tag{A2.6}$$

If, for mathematical simplicity and without loss of generality, we 'take' the photograph when it is $T = l/c$, i.e. at the moment the center of the circle is at O, then

$$X = R\left(\frac{1}{\gamma}\cos\theta' - \beta\sin\theta'\right), \tag{A2.7}$$

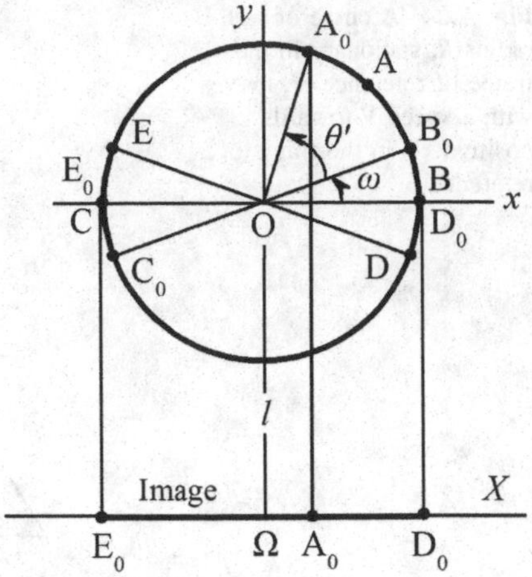

Fig. A2.5 The photograph of a stationary circle of radius *R*. The circle appears to have been rotated by an angle $\omega = \arcsin \beta$

where $\beta = V/c$. Because $\frac{1}{\gamma} \leq 1$, $\beta \leq 1$ and $\frac{1}{\gamma^2} + \beta^2 = 1$, we take $\sin \omega \equiv \beta$, $\cos \omega \equiv \frac{1}{\gamma}$, and find that $X = R(\cos \theta' \cos \omega - \sin \theta' \sin \omega)$ or, finally,

$$X = R\cos(\theta' + \omega). \tag{A2.8}$$

X has a maximum equal to $X_{\max} = R$ when it is $\theta' = -\omega$ and a minimum equal to $X_{\min} = -R$ when it is $\theta' = \pi - \omega$. The points are recorded in the photograph as the points of a stationary circle of radius R which has been rotated, relative to its initial position, by an angle $\omega = \arcsin \beta$ in a counterclockwise direction (Fig. A2.5) for motion of the circle towards positive x's. Thus, point A, which on the stationary circle is at an angle θ', in the photograph of the moving circle will be at position A_0, which corresponds to angle $\theta' + \omega$. We note that points which would have been visible in the photograph when the circle was stationary, will be invisible now that the circle is moving. For example, point B would have been the extreme point visible on the right in the photograph of the stationary circle. This point will now be at point B_0 and invisible. On the other hand, point D will now be the extreme point visible on the right, at D_0. The points between D and B will be invisible in the photograph. In the diametrically opposite region, the extreme point to the left in the image will now be, instead of C, point E_0 which corresponds to point E. The points between C and E become now visible in the moving circle.

The extreme points that become visible may be found with the aid of the triangle in the inset of Fig. A2.4. If such a point of the circle, P, lies in the direction forming an angle α' with the negative x-axis, in the frame of reference S, the condition for this to be just visible is the following: In the time δt which the light needs in order to travel the distance $c\delta t$, the points in the vicinity of point P must 'move aside' by a distance $V\delta t$, if the light from point P is to be able to move towards the camera, in the y direction. From the right-angled triangle, it is $\tan\alpha = (V\delta t/c\delta t) = (V/c)$ and $\alpha = \arctan\beta$. It may be easily shown that angles α' and α in Fig. A2.4 are related by the equation $|\tan\alpha| = (1/\gamma^2)|\tan\alpha'|$. It follows that $\tan\alpha' = \beta\gamma^2$. In the frame of reference of the circle, S, the points which are visible in the photograph are those below the line ED (Fig. A2.4), which forms an angle α_0 with the negative x-axis, where $\tan\alpha_0 = (1/\gamma)\tan\alpha' = \beta\gamma$.

A2.2.4 Sphere

We have seen that a circle with radius R, moving with a speed V in the plane of the circle, appears as a circle of radius R which has been rotated by an angle $\omega = \arcsin\beta$. Now, a sphere may be considered to be build by discs of appropriate radii. When the sphere moves with a speed V in some direction, the discs forming the sphere will appear rotated by the same angle $\omega = \arcsin\beta$ about an axis which is normal to the plane formed by the velocity vector and the point of observation. The overall effect is that the sphere of radius R appears as a sphere of the same radius, which has been rotated by an angle $\omega = \arcsin\beta$ about an axis which is normal to the plane formed by the velocity vector of the sphere and the point of observation. This is shown in Fig. A2.6.

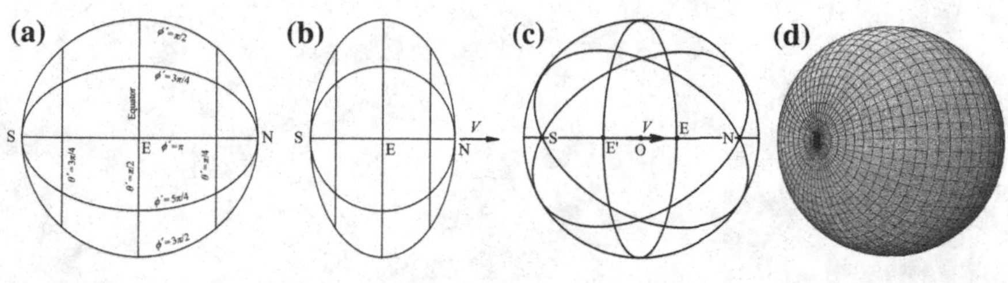

Fig. A2.6 **a** A stationary sphere. The two poles and the circles of constant geographical latitude and longitude are shown. The equator of the sphere corresponds to $\theta' = \pi/2$. **b** Moving sphere. A sphere moving with a speed V in the x-direction, contracts by a factor γ in this direction. **c** Appearance of a moving sphere. When photographed, the moving sphere appears spherical. The sphere also appears to have been rotated by an angle $\omega = \arcsin\beta$ around an axis which is parallel to the plane of the photograph and normal to the direction of the motion of the sphere. **d** The Photograph of the sphere

Appendix 3
The Derivation of the Expression for the Relativistic Mass in the General Case

A3.1 The Transformation for the Total Mass and Momentum of a System of Particles

We will examine the concept of mass and how this is defined in accordance with the Special Theory of Relativity. The so-called *inertial mass* is defined in terms of the acceleration a body acquires when acted upon by a certain force. Classical mass, m_0, which remains constant with speed, is obviously unsuitable for the analysis of dynamics problems in Relativity. When we examine a collision of particles, for example, we find that, if momentum, defined as $\mathbf{p} = m_0\mathbf{\upsilon}$, is conserved in one inertial frame of reference, it will not be conserved when we transform the velocities of the particles in another inertial frame of reference using the Lorentz transformation.

We will investigate the possibility that the mass of a body depends on its speed, υ, through a relationship of the form

$$m = m_0 f(\upsilon), \tag{A3.1}$$

where the function $f(\upsilon)$ will tend to unity for low speeds of the body, in agreement with classical Dynamics. For this reason, m_0 will be called *rest mass* of the body (or even *proper* mass, since it is the mass of the body in its own frame of reference, in which it is at rest). The magnitude m will be called *relativistic mass*. The function $f(\upsilon)$ may also possibly tend to infinity as the speed of the body approaches that of light in vacuum, c, since Relativity forbids the acceleration of a body to speeds higher than c. Momentum will now be defined as

$$\mathbf{p} = m\mathbf{\upsilon} = m_0 f(\upsilon)\mathbf{\upsilon}. \tag{A3.2}$$

We will try to determine that function $f(\upsilon)$ which leads to two laws of conservation which have proved to be very useful in classical Mechanics:

(a) the law of conservation of relativistic mass and
(b) the law of conservation of momentum.

© Springer International Publishing Switzerland 2016

C. Christodoulides, *The Special Theory of Relativity*,
Undergraduate Lecture Notes in Physics, DOI 10.1007/978-3-319-25274-2

The laws of conservation of magnitudes such as the energy and the momentum stem from symmetries in space and time [7]. The law of conservation of energy, for example, is due to the invariance of the physical laws with time, i.e. with displacement in time. The law of conservation of momentum is due to the homogeneity of space, i.e. the invariance of the laws of Physics on displacement in space. For this reason, the laws of conservation cannot be the privilege of one frame of reference, but must hold in all inertial frames of reference. Besides, the principle of Relativity dictates the validity of the same laws in all inertial frames of reference.

We will consider two inertial frames of reference with axes which are correspondingly parallel to each other, S and S′, in relative motion between them. The velocity of frame S′ with respect to frame S is $\mathbf{V}$. The origins and the axes of the two coordinate systems coincide at the moment $t = t' = 0$. We follow the method of McCrea [8] and we examine a system of N particles. The i-th particle has velocities $\boldsymbol{\upsilon}_i$ and $\boldsymbol{\upsilon}'_i$ in the frames S and S′, respectively. To the speeds V, υ_i and υ'_i there correspond the Lorentz factors

$$\gamma = \frac{1}{\sqrt{1 - V^2/c^2}}, \quad \gamma_i = \frac{1}{\sqrt{1 - \upsilon_i^2/c^2}}, \quad \text{and} \quad \gamma'_i = \frac{1}{\sqrt{1 - \upsilon_i'^2/c^2}}. \tag{A3.3}$$

From Eq. (3.49) it is

$$\gamma_i = \gamma\gamma'_i\left(1 + \frac{\boldsymbol{\upsilon}'_i \cdot \mathbf{V}}{c^2}\right), \tag{A3.4}$$

which, combined with Eq. (3.38), gives

$$\gamma_i \boldsymbol{\upsilon}_i = \gamma'_i\left[\boldsymbol{\upsilon}'_i + (\gamma - 1)\frac{(\boldsymbol{\upsilon}'_i \cdot \mathbf{V})\mathbf{V}}{V^2} + \gamma\mathbf{V}\right]. \tag{A3.5}$$

The total mass and the total momentum of the system of particles in frame S is

$$M = \sum_{i=1}^{N} m_i \quad \text{and} \quad \mathbf{P} = \sum_{i=1}^{N} m_i \boldsymbol{\upsilon}_i \tag{A3.6}$$

respectively, where m_i is the inertial mass of the i-th particle in frame S. Using Eq. (A3.5) we find for the momentum

$$\mathbf{P} = \sum_{i=1}^{N} m_i \boldsymbol{\upsilon}_i = \sum_{i=1}^{N} \left\{ m_i \frac{\gamma'_i}{\gamma_i}\left[\boldsymbol{\upsilon}'_i + (\gamma - 1)\frac{(\boldsymbol{\upsilon}'_i \cdot \mathbf{V})\mathbf{V}}{V^2} + \gamma\mathbf{V}\right]\right\} \tag{A3.7}$$

If we define

$$m'_i \equiv m_i \frac{\gamma'_i}{\gamma_i}, \tag{A3.8}$$

$$M' \equiv \sum_{i=1}^{N} m'_i \quad \text{and} \quad \mathbf{P}' \equiv \sum_{i=1}^{N} m'_i \boldsymbol{\upsilon}'_i, \tag{A3.9}$$

Equation (A3.7) may be written as

$$\mathbf{P} = \mathbf{P}' + (\gamma - 1)\frac{(\mathbf{P}' \cdot \mathbf{V})\mathbf{V}}{V^2} + \gamma \mathbf{V} M'. \tag{A3.10}$$

Using Eq. (A3.4), we find for the total mass

$$M = \sum_{i=1}^{N} m_i = \sum_{i=1}^{N} m'_i \frac{\gamma_i}{\gamma'_i} = \gamma \sum_{i=1}^{N} m'_i \left(1 + \frac{\boldsymbol{\upsilon}'_i \cdot \mathbf{V}}{c^2}\right) \tag{A3.11}$$

or

$$M = \gamma \left(M' + \frac{\mathbf{P}' \cdot \mathbf{V}}{c^2}\right). \tag{A3.12}$$

We have thus expressed, with Eqs. (A3.12) and (A3.10), the quantities M and $\mathbf{P}$ as linear functions of M' and $\mathbf{P}'$, with coefficients which depend only on the velocity $\mathbf{V}$ of frame S relative to frame S′. We notice that the quantities $\mathbf{P}$ and M transform from frame S′ to frame S in the same way that $\mathbf{r}$ and t do, respectively [Eq. (3.25)]. The inverse transformations, of M' and $\mathbf{P}'$ in terms of M and $\mathbf{P}$, may easily be stated, by analogy. The necessary and sufficient condition for the quantities M and $\mathbf{P}$ to be conserved during a process, is that M' and $\mathbf{P}'$ be conserved. The quantities M' and $\mathbf{P}'$ play the same role in frame S′ that M and $\mathbf{P}$ do in frame S. They are, therefore, respectively, the total mass and the total momentum of the system of the N particles.

From Eq. (A3.8) we see that the magnitude m_i/γ_i remains constant on transition from one inertial frame of reference to another and, therefore, must be independent of the relative motion of the two frames and a constant of the i-th particle. Let this constant quantity, which has dimensions of mass, be symbolized by m_{0i}. Then, the following relation is true

$$\frac{m_i}{\gamma_i} = \frac{m'_i}{\gamma'_i} = m_{0i} \tag{A3.13}$$

and, omitting the indices i, we have for the mass of a particle which is moving with velocity $\boldsymbol{\upsilon}$ in a frame S, the expression

$$m = \frac{m_0}{\sqrt{1 - \upsilon^2/c^2}}, \tag{A3.14}$$

and for its momentum the expression

$$\mathbf{p} = \frac{m_0 \boldsymbol{\upsilon}}{\sqrt{1 - \upsilon^2/c^2}}. \tag{A3.15}$$

Because it is $m(\upsilon = 0) = m_0$, the mass m_0 is called *rest mass* of the particle.

Appendix 4
The Invariance of the Equations of Maxwell and the Wave Equation Under the Lorentz Transformation

A4.1 The Partial Derivative

As we will use the concept of the partial derivative, we will give a very brief introduction to it. For a better familiarization with the partial derivative, a book on Mathematical Analysis may be consulted. This section may be omitted by readers familiar with these subjects.

The derivative of a function $f(x)$ of one variable is defined as

$$f'(x) = \frac{dy}{dx} = \lim_{\Delta x \to 0} \frac{f(x+\Delta x) - f(x)}{\Delta x}. \tag{A4.1}$$

For functions of two or more variables, corresponding limits are defined with respect to any one of the variables, with the condition that the rest of the variables are kept constant. For example, if the function $f(x,y)$ is a function of the variables x and y, the limit

$$\lim_{\substack{\Delta x \to 0 \\ y = \text{const}}} \frac{f(x+\Delta x,\ y) - f(x,y)}{\Delta x} \tag{A4.2}$$

is called the *partial derivative of* $f(x,y)$ *with respect to* x and is denoted by $\frac{\partial f}{\partial x}$ or f_x. The function $f(x,y)$ has two partial derivatives of the first order, namely

$$\frac{\partial f}{\partial x} = \lim_{\substack{\Delta x \to 0 \\ y = \text{const}}} \frac{f(x+\Delta x,\ y) - f(x,y)}{\Delta x} \quad \text{and} \quad \frac{\partial f}{\partial y} = \lim_{\substack{\Delta y \to 0 \\ x = \text{const}}} \frac{f(x,\ y+\Delta y) - f(x,y)}{\Delta y}. \tag{A4.3}$$

Differentiating the function $\frac{\partial f}{\partial x}$ with respect to x, we get the second partial derivative $\frac{\partial^2 f}{\partial x^2}$ of f with respect to x. If we differentiate $f(x,y)$ first with respect to

© Springer International Publishing Switzerland 2016

C. Christodoulides, *The Special Theory of Relativity*,
Undergraduate Lecture Notes in Physics, DOI 10.1007/978-3-319-25274-2

x and then with respect to y, we get $\frac{\partial^2 f}{\partial y \partial x}$. This procedure may be continued with other combinations of the independent variables.

A4.1.1 Differentials

If $f(x, y, z)$ is a function of the variables $x,\ y,\ z$, the expression

$$df = \frac{\partial f}{\partial x} dx + \frac{\partial f}{\partial y} dy + \frac{\partial f}{\partial z} dz \tag{A4.4}$$

is called the *total differential* or, simply, the *differential* of the function $f(x, y, z)$. The quantities dx, dy and dz are the differentials of $x,\ y,\ z$, respectively, which may be thought of as being infinitesimal changes in the three variables.

The variables $x,\ y,\ z$ are possibly functions of other variables. For 'very small' variations dx, dy and dz in the variables $x,\ y,\ z$, respectively, the differential df may be considered to be the resulting variation in $f(x, y, z)$. The smaller the dx, dy and dz are, the better approximation is the differential to the variation of $f(x, y, z)$. If $f(x, y, z) = \text{constant}$, then $df = 0$.

When a number of small variations, which are independent between them, happen simultaneously in a system, the total result is the sum of the partial results, as seen in Eq. (A4.4). From a physical point of view, this is equivalent to the principle of superposition.

A4.1.1.1 Differentiation of Composite Functions

If $f(x, y, z)$ is a function of the variables $x,\ y,\ z$, and these are functions of another variable, say t, the differential of Eq. (A4.4),

$$df = \frac{\partial f}{\partial x} dx + \frac{\partial f}{\partial y} dy + \frac{\partial f}{\partial z} dz$$

we obtain the rate of variation of $f(x, y, z)$ with respect to t as

$$\frac{df}{dt} = \frac{\partial f}{\partial x}\frac{dx}{dt} + \frac{\partial f}{\partial y}\frac{dy}{dt} + \frac{\partial f}{\partial z}\frac{dz}{dt}. \tag{A4.5}$$

$\frac{df}{dt}$ is called the *total derivative* of $f(x, y, z)$ with respect to t. The physical interpretation of this relation is the following: The rate of change of $f(x, y, z)$ is the superposition of three terms, each of which gives the rate of change of $f(x, y, z)$ which is due to the variation, separately, of each one of the variables $x,\ y,\ z$. In more detail, $\frac{\partial f}{\partial x}$ is the rate of change of $f(x, y, z)$ with respect to x when y and z are kept constant, and

$\frac{dx}{dt}$ is the rate of change of x with respect to t. Therefore, $\frac{\partial f}{\partial x}\frac{dx}{dt}$ is the rate of change of $f(x,y,z)$ with t, due to the variation of x alone. The other terms have similar interpretations. The sum gives the total rate of variation of $f(x,y,z)$ with t. Of course, by the substitution of $x(t)$, $y(t)$ and $z(t)$ in $f(x,y,z)$, we find the function $f[x(t),\ y(t),\ z(t)] = f(t)$, whose derivative $\frac{df}{dt}$ may be evaluated directly.

A4.1.2 Differentiation Operators

In Eq. (A4.4), in the position of $f(x,y,z)$ there could be any other function of x, y and z or some of them. This can be expressed by defining the *differentiation operator*

$$d = dx\frac{\partial}{\partial x} + dy\frac{\partial}{\partial y} + dz\frac{\partial}{\partial z}. \tag{A4.6}$$

If we place any function $f(x,y,z)$ to the right of the operator, the operator will act on the function, giving its total differential, as in Eq. (A4.4).

Similarly, the operator for the total derivative with respect to t is found from Eq. (A4.5) to be:

$$\frac{d}{dt} = \frac{dx}{dt}\frac{\partial}{\partial x} + \frac{dy}{dt}\frac{\partial}{\partial y} + \frac{dz}{dt}\frac{\partial}{\partial z}. \tag{A4.7}$$

If this operator acts on the function $f(x,y,z)$, it will give the total derivative with respect to t, as this is expressed by Eq. (A4.5).

From vector analysis the nabla operator, ∇, is known, which in Cartesian coordinates is written as:

$$\nabla \equiv \hat{\mathbf{x}}\frac{\partial}{\partial x} + \hat{\mathbf{y}}\frac{\partial}{\partial y} + \hat{\mathbf{z}}\frac{\partial}{\partial z}. \tag{A4.8}$$

Using ∇, we may define magnitudes such as

the *gradient* of a scalar function:

$$grad f = \nabla f = \hat{\mathbf{x}}\frac{\partial f}{\partial x} + \hat{\mathbf{y}}\frac{\partial f}{\partial y} + \hat{\mathbf{z}}\frac{\partial f}{\partial z}, \tag{A4.9}$$

the *divergence* of a vector function:

$$div\mathbf{E} = \nabla \cdot \mathbf{E} = \frac{\partial E_x}{\partial x} + \frac{\partial E_y}{\partial y} + \frac{\partial E_z}{\partial z}, \tag{A4.10}$$

the *curl* or *rotation* of a vector function:

$$\nabla \times \mathbf{E} \equiv \mathbf{curl}\,\mathbf{E} \equiv \mathbf{rot}\,\mathbf{E} \equiv \left(\hat{\mathbf{x}}\frac{\partial}{\partial x} + \hat{\mathbf{y}}\frac{\partial}{\partial y} + \hat{\mathbf{z}}\frac{\partial}{\partial z}\right) \times (E_x\,\hat{\mathbf{x}} + E_y\,\hat{\mathbf{y}} + E_z\,\hat{\mathbf{z}}). \tag{A4.11}$$

It is
$$\nabla \times \mathbf{E} = \begin{vmatrix} \hat{\mathbf{x}} & \hat{\mathbf{y}} & \hat{\mathbf{z}} \\ \frac{\partial}{\partial x} & \frac{\partial}{\partial y} & \frac{\partial}{\partial z} \\ E_x & E_y & E_z \end{vmatrix} = \left(\frac{\partial E_z}{\partial y} - \frac{\partial E_y}{\partial z}\right)\hat{\mathbf{x}} + \left(\frac{\partial E_x}{\partial z} - \frac{\partial E_z}{\partial x}\right)\hat{\mathbf{y}} + \left(\frac{\partial E_y}{\partial x} - \frac{\partial E_x}{\partial y}\right)\hat{\mathbf{z}}. \tag{A4.12}$$

The gradient ∇f of a scalar quantity f is a vector on 3 dimensions, which points in the direction in which f increases with the maximum rate with distance and whose magnitude gives this rate of change. In a sense, it is a slope generalized in 3 dimensions.

The divergence $\nabla \cdot \mathbf{E}$ of a vector quantity $\mathbf{E}$ expresses the flux of this quantity through an infinitesimal closed surface, per unit volume enclosed by the surface. The flux of a fluid from a *source* is expressed by a positive divergence of its velocity. A *sink* is expressed by a negative divergence.

The curl $\nabla \times \mathbf{E}$ of a vector quantity $\mathbf{E}$ expresses the degree by which the field differs from being conservative. A conservative field has a curl equal to zero. In this case, the vector field may be expressed as the gradient of a scalar function. This is the case for the electrostatic field, for which it is $\nabla \times \mathbf{E} = 0$ and a scalar function ϕ (the potential) may be found which gives $\mathbf{E} = -\nabla\phi$.

The *Laplace operator* $\nabla^2 \equiv \nabla \cdot \nabla \equiv$ *div grad* is also defined. It is

$$\nabla^2 \equiv \nabla \cdot \nabla \equiv \left(\hat{\mathbf{x}}\frac{\partial}{\partial x} + \hat{\mathbf{y}}\frac{\partial}{\partial y} + \hat{\mathbf{z}}\frac{\partial}{\partial z}\right) \cdot \left(\hat{\mathbf{x}}\frac{\partial}{\partial x} + \hat{\mathbf{y}}\frac{\partial}{\partial y} + \hat{\mathbf{z}}\frac{\partial}{\partial z}\right) = \left(\frac{\partial^2}{\partial x^2} + \frac{\partial^2}{\partial y^2} + \frac{\partial^2}{\partial z^2}\right). \tag{A4.13}$$

The *Laplacian* of a scalar function ψ is defined as

$$\nabla^2\psi \equiv \frac{\partial^2\psi}{\partial x^2} + \frac{\partial^2\psi}{\partial y^2} + \frac{\partial^2\psi}{\partial z^2}. \tag{A4.14}$$

The *Laplacian* of a vector function $\mathbf{E}$ is also defined

$$\nabla^2\mathbf{E} \equiv \frac{\partial^2\mathbf{E}}{\partial x^2} + \frac{\partial^2\mathbf{E}}{\partial y^2} + \frac{\partial^2\mathbf{E}}{\partial z^2}. \tag{A4.15}$$

These operators are widely used in electromagnetic theory.

A4.2 The Differentiation Operators of the Transformations of Galileo and Lorentz

A4.2.1 The Differentiation Operators of the Galilean Transformation

The Galilean transformation from the frame of reference $S : (x, y, z, t)$ to the frame of reference $S' : (x', y', z', t')$, which is moving relative to S with speed V in the direction of positive values of x, is performed using the relations

$$x' = x - Vt, \qquad y' = y, \qquad z' = z, \qquad t' = t. \tag{A4.16}$$

We also need to transform the derivatives with respect to x, y, z and t into derivatives with respect to x', y', z' and t'. For example, the variable x is a function of the variables x' and t', and, therefore, according to the chain rule, the partial derivative with respect to x of the function $\psi(x, y, z, t)$ is:

$$\frac{\partial \psi}{\partial x} = \frac{\partial \psi}{\partial x'}\frac{\partial x'}{\partial x} + \frac{\partial \psi}{\partial t'}\frac{\partial t'}{\partial x} \tag{A4.17}$$

Symbolically, we may write the operator of partial differentiation with respect to x, i.e. $\partial/\partial x$, as

$$\frac{\partial}{\partial x} = \frac{\partial x'}{\partial x}\frac{\partial}{\partial x'} + \frac{\partial t'}{\partial x}\frac{\partial}{\partial t'}. \tag{A4.18}$$

From the first of Eq. (A4.16) we find that it is $\dfrac{\partial x'}{\partial x} = 1$ and from the fourth that it is $\dfrac{\partial t'}{\partial x} = 0$. Substituting in Eq. (A4.18), it follows that

$$\frac{\partial}{\partial x} = \frac{\partial}{\partial x'}. \tag{A4.19}$$

In other words, the partial derivative with respect to x is equal to the partial derivative with respect to x'. Repeating the procedure, we find

$$\frac{\partial^2}{\partial x^2} = \frac{\partial^2}{\partial x'^2}. \tag{A4.20}$$

As it is $y' = y$ and $z' = z$, it also follows that

$$\frac{\partial}{\partial y} = \frac{\partial}{\partial y'} \quad \text{and} \quad \frac{\partial}{\partial z} = \frac{\partial}{\partial z'}. \tag{A4.21}$$

and that

$$\frac{\partial^2}{\partial y^2} = \frac{\partial^2}{\partial y'^2} \quad \text{and} \quad \frac{\partial^2}{\partial z^2} = \frac{\partial^2}{\partial z'^2}. \tag{A4.22}$$

For the derivatives with respect to t, we have

$$\frac{\partial}{\partial t} = \frac{\partial t'}{\partial t}\frac{\partial}{\partial t'} + \frac{\partial x'}{\partial t}\frac{\partial}{\partial x'} \tag{A4.23}$$

and as from the fourth of Eqs. (A4.16) it is $\frac{\partial t'}{\partial t} = 1$ and from the fourth it is $\frac{\partial x'}{\partial t} = -V$, it follows that

$$\frac{\partial}{\partial t} = \frac{\partial}{\partial t'} - V\frac{\partial}{\partial x'}. \tag{A4.24}$$

Repeating the partial differentiation with respect to t, we have

$$\frac{\partial^2}{\partial t^2} = \left(\frac{\partial}{\partial t'} - V\frac{\partial}{\partial x'}\right)\left(\frac{\partial}{\partial t'} - V\frac{\partial}{\partial x'}\right) = \frac{\partial^2}{\partial t'^2} + V^2\frac{\partial^2}{\partial x'^2} - 2V\frac{\partial}{\partial t'}\frac{\partial}{\partial x'} \tag{A4.25}$$

where we have used the fact that, except in rare occasions, it is $\frac{\partial}{\partial t'}\frac{\partial}{\partial x'} = \frac{\partial}{\partial x'}\frac{\partial}{\partial t'}$.

A4.2.2 The Differentiation Operators of the Lorentz Transformation

The Lorentz transformation from the frame of reference S : $(x,\ y,\ z,\ t)$ to the frame of reference S′ : $(x',\ y',\ z',\ t')$, which is moving relative to S with speed V in the positive x direction, is done by using the relations

$$x' = \gamma\,(x - Vt) \quad y' = y \quad z' = z \quad t' = \gamma\left(t - \frac{V}{c^2}x\right). \tag{A4.26}$$

From the fact that x is a function of x' and t', we have the relation

$$\frac{\partial}{\partial x} = \frac{\partial x'}{\partial x}\frac{\partial}{\partial x'} + \frac{\partial t'}{\partial x}\frac{\partial}{\partial t'}. \tag{A4.27}$$

From the first of the Eqs. (A4.26) we find $\frac{\partial x'}{\partial x} = \gamma$ and from the fourth that it is $\frac{\partial t'}{\partial x} = -\gamma \frac{V}{c^2}$. Substituting in Eq. (A4.27), it follows that

$$\frac{\partial}{\partial x} = \gamma \frac{\partial}{\partial x'} - \gamma \frac{V}{c^2} \frac{\partial}{\partial t'} = \gamma \left(\frac{\partial}{\partial x'} - \frac{V}{c^2} \frac{\partial}{\partial t'} \right). \qquad \text{(A4.28)}$$

Repeating the procedure, we find

$$\frac{\partial^2}{\partial x^2} = \gamma \left(\frac{\partial}{\partial x'} - \frac{V}{c^2} \frac{\partial}{\partial t'} \right) \gamma \left(\frac{\partial}{\partial x'} - \frac{V}{c^2} \frac{\partial}{\partial t'} \right) \quad \text{or}$$

$$\frac{\partial^2}{\partial x^2} = \gamma^2 \left(\frac{\partial^2}{\partial x'^2} - 2 \frac{V}{c^2} \frac{\partial^2}{\partial x' \partial t'} + \frac{V^2}{c^4} \frac{\partial^2}{\partial t'^2} \right). \qquad \text{(A4.29)}$$

Also, since it is $y' = y$ and $z' = z$, we have

$$\frac{\partial}{\partial y} = \frac{\partial}{\partial y'} \quad \text{and} \quad \frac{\partial}{\partial z} = \frac{\partial}{\partial z'} \qquad \text{(A4.30)}$$

as well as

$$\frac{\partial^2}{\partial y^2} = \frac{\partial^2}{\partial y'^2} \quad \text{and} \quad \frac{\partial^2}{\partial z^2} = \frac{\partial^2}{\partial z'^2}. \qquad \text{(A4.31)}$$

From the fact that t is a function of x' and t', it follows that

$$\frac{\partial}{\partial t} = \frac{\partial t'}{\partial t} \frac{\partial}{\partial t'} + \frac{\partial x'}{\partial t} \frac{\partial}{\partial x'}. \qquad \text{(A4.32)}$$

Since from the fourth of the Eqs. (A4.26) it is $\frac{\partial t'}{\partial t} = \gamma$ and from the first it is $\frac{\partial x'}{\partial t} = -V\gamma$, it follows that

$$\frac{\partial}{\partial t} = \gamma \left(\frac{\partial}{\partial t'} - V \frac{\partial}{\partial x'} \right). \qquad \text{(A4.33)}$$

Repeating the partial differentiation with respect to t, we have

$$\frac{\partial^2}{\partial t^2} = \gamma \left(\frac{\partial}{\partial t'} - V \frac{\partial}{\partial x'} \right) \gamma \left(\frac{\partial}{\partial t'} - V \frac{\partial}{\partial x'} \right) \quad \text{or}$$

$$\frac{\partial^2}{\partial t^2} = \gamma^2 \left(\frac{\partial^2}{\partial t'^2} - 2V \frac{\partial^2}{\partial t' \partial x'} + V^2 \frac{\partial^2}{\partial x'^2} \right). \qquad \text{(A4.34)}$$

A4.3 Maxwell's Equations

The mathematical description of the laws of electromagnetism is given by Maxwell's equations, which are:

$$\nabla \cdot \mathbf{E} = \frac{\rho}{\varepsilon_0}, \qquad \nabla \cdot \mathbf{B} = 0, \qquad \nabla \times \mathbf{E} = -\frac{\partial \mathbf{B}}{\partial t}, \qquad \nabla \times \mathbf{B} = \varepsilon_0 \mu_0 \frac{\partial \mathbf{E}}{\partial t} + \mu_0 \mathbf{J}. \tag{A4.35}$$

In empty space, without free charges ($\rho = 0$) or currents ($\mathbf{J} = 0$), the equations simplify to

$$\nabla \cdot \mathbf{E} = 0, \qquad \nabla \cdot \mathbf{B} = 0, \qquad \nabla \times \mathbf{E} = -\frac{\partial \mathbf{B}}{\partial t}, \qquad \nabla \times \mathbf{B} = \varepsilon_0 \mu_0 \frac{\partial \mathbf{E}}{\partial t}, \tag{A4.36}$$

which may be combined to give one equation for the electric field **E** and one equation for the magnetic field **B**:

$$\nabla^2 \mathbf{E} = \frac{1}{c^2} \frac{\partial^2 \mathbf{E}}{\partial t^2}, \qquad \nabla^2 \mathbf{B} = \frac{1}{c^2} \frac{\partial^2 \mathbf{B}}{\partial t^2}. \tag{A4.37}$$

These equations describe an electromagnetic wave which moves with speed

$$c = \frac{1}{\sqrt{\varepsilon_0 \mu_0}} = 299\,792\,458 \text{ m/s}. \tag{A4.38}$$

Equations (A4.36) and (A4.37) may be written in Cartesian coordinates. The equation $\nabla \cdot \mathbf{E} = 0$ is written as:

$$\frac{\partial E_x}{\partial x} + \frac{\partial E_y}{\partial y} + \frac{\partial E_z}{\partial z} = 0 \tag{A4.39}$$

Equation $\nabla \cdot \mathbf{B} = 0$ is also written as:

$$\frac{\partial B_x}{\partial x} + \frac{\partial B_y}{\partial y} + \frac{\partial B_z}{\partial z} = 0 \tag{A4.40}$$

Equation $\nabla \times \mathbf{E} = -\dfrac{\partial \mathbf{B}}{\partial t}$ is also written as:

$$\frac{\partial E_z}{\partial y} - \frac{\partial E_y}{\partial z} = -\frac{\partial B_x}{\partial t}, \qquad \frac{\partial E_x}{\partial z} - \frac{\partial E_z}{\partial x} = -\frac{\partial B_y}{\partial t}, \qquad \frac{\partial E_y}{\partial x} - \frac{\partial E_x}{\partial y} = -\frac{\partial B_z}{\partial t}. \tag{A4.41 – 43}$$

Equation $\nabla \times \mathbf{B} = \dfrac{1}{c^2}\dfrac{\partial \mathbf{E}}{\partial t}$ is also written as:

$$\frac{\partial B_z}{\partial y} - \frac{\partial B_y}{\partial z} = \frac{1}{c^2}\frac{\partial E_x}{\partial t}, \qquad \frac{\partial B_x}{\partial z} - \frac{\partial B_z}{\partial x} = \frac{1}{c^2}\frac{\partial E_y}{\partial t}, \qquad \frac{\partial B_y}{\partial x} - \frac{\partial B_x}{\partial y} = \frac{1}{c^2}\frac{\partial E_z}{\partial t}. \tag{A4.44 – 46}$$

Equation $\nabla^2 \mathbf{E} = \dfrac{1}{c^2}\dfrac{\partial^2 \mathbf{E}}{\partial t^2}$ is also written as:

$$\frac{\partial^2 E_x}{\partial x^2} + \frac{\partial^2 E_x}{\partial y^2} + \frac{\partial^2 E_x}{\partial z^2} = \frac{1}{c^2}\frac{\partial^2 E_x}{\partial t^2}, \tag{A4.47}$$

$$\frac{\partial^2 E_y}{\partial x^2} + \frac{\partial^2 E_y}{\partial y^2} + \frac{\partial^2 E_y}{\partial z^2} = \frac{1}{c^2}\frac{\partial^2 E_y}{\partial t^2}, \tag{A4.48}$$

$$\frac{\partial^2 E_z}{\partial x^2} + \frac{\partial^2 E_z}{\partial y^2} + \frac{\partial^2 E_z}{\partial z^2} = \frac{1}{c^2}\frac{\partial^2 E_z}{\partial t^2}. \tag{A4.49}$$

Equation $\nabla^2 \mathbf{B} = \dfrac{1}{c^2}\dfrac{\partial^2 \mathbf{B}}{\partial t^2}$ is also written as:

$$\frac{\partial^2 B_x}{\partial x^2} + \frac{\partial^2 B_x}{\partial y^2} + \frac{\partial^2 B_x}{\partial z^2} = \frac{1}{c^2}\frac{\partial^2 B_x}{\partial t^2}, \tag{A4.50}$$

$$\frac{\partial^2 B_y}{\partial x^2} + \frac{\partial^2 B_y}{\partial y^2} + \frac{\partial^2 B_y}{\partial z^2} = \frac{1}{c^2}\frac{\partial^2 B_y}{\partial t^2}, \tag{A4.51}$$

$$\frac{\partial^2 B_z}{\partial x^2} + \frac{\partial^2 B_z}{\partial y^2} + \frac{\partial^2 B_z}{\partial z^2} = \frac{1}{c^2}\frac{\partial^2 B_z}{\partial t^2}. \tag{A4.52}$$

The advantages of vector notation are obvious.

These equations describe the behavior of the fields **E** and **B**, as this was observed in the laboratory. Unless there is a reason to assume that the Earth is a privileged frame of reference, the equations must have the same form for all observers.

A4.4 The Galilean Transformation and the Wave Equation

The wave equation of a scalar magnitude ψ has, in three dimensions, the form

$$\nabla^2\psi - \frac{1}{c^2}\frac{\partial^2 \psi}{\partial t^2} = 0 \quad \text{or} \quad \frac{\partial^2 \psi}{\partial x^2} + \frac{\partial^2 \psi}{\partial y^2} + \frac{\partial^2 \psi}{\partial z^2} - \frac{1}{c^2}\frac{\partial^2 \psi}{\partial t^2} = 0. \tag{A4.53}$$

In place of ψ we may substitute the components $E_x,\ E_{y,},\ E_z$ of the electric field, the components $B_x,\ B_{y,},\ B_z$ of the magnetic field or even the vectors **E** and **B**. We will transform the wave equation from frame S : (x, y, z, t) to frame S' : (x', y', z', t') through the Galileo transformation

$$x' = x - Vt, \qquad y' = y, \qquad z' = z, \qquad t' = t, \tag{A4.54}$$

in order to find the form of the wave equation for an observer moving with speed V in the positive x direction relative to an observer for which Maxwell's equations have the form given above.

Using the operators of differentiation of Eqs. (A4.20), (A4.22) and (A4.25) in the left hand side of the wave equation, Eq. (A4.53), we have

$$\frac{\partial^2\psi}{\partial x^2} + \frac{\partial^2\psi}{\partial y^2} + \frac{\partial^2\psi}{\partial z^2} - \frac{1}{c^2}\frac{\partial^2\psi}{\partial t^2} = \\ = \frac{\partial^2\psi}{\partial x'^2} + \frac{\partial^2\psi}{\partial y'^2} + \frac{\partial^2\psi}{\partial z'^2} - \frac{1}{c^2}\frac{\partial^2\psi}{\partial t'^2} - \frac{V^2}{c^2}\frac{\partial^2\psi}{\partial x'^2} + 2\frac{V}{c^2}\frac{\partial^2\psi}{\partial t'\partial x'} = 0$$

and, therefore,

$$\frac{\partial^2\psi}{\partial x'^2} + \frac{\partial^2\psi}{\partial y'^2} + \frac{\partial^2\psi}{\partial z'^2} - \frac{1}{c^2}\frac{\partial^2\psi}{\partial t'^2} = \frac{V^2}{c^2}\frac{\partial^2\psi}{\partial x'^2} - 2\frac{V}{c^2}\frac{\partial^2\psi}{\partial t'\partial x'}. \tag{A4.55}$$

For the electric field vector, for example, if we start from the wave equation in the inertial frame of reference S,

$$\nabla^2\mathbf{E} = \frac{1}{c^2}\frac{\partial^2\mathbf{E}}{\partial t^2}, \tag{A4.56}$$

in order for this equation to remain invariant under the transformation to another inertial frame of reference S′, the equation that should be produced is

$$\nabla'^2\mathbf{E}' = \frac{1}{c^2}\frac{\partial^2\mathbf{E}'}{\partial t'^2}. \tag{A4.57}$$

with a suitable transformation giving $\mathbf{E}'$ in terms of $\mathbf{E}$. If we apply the Galilean transformation to the wave equation, we find, according to Eq. (A4.55),

$$\nabla'^2\mathbf{E} = \frac{1}{c^2}\frac{\partial^2\mathbf{E}}{\partial t'^2} + \frac{1}{c^2}\left(V^2\frac{\partial^2\mathbf{E}}{\partial x'^2} - 2V\frac{\partial^2\mathbf{E}}{\partial x'\partial t'}\right) \tag{A4.58}$$

and no transformation of $\mathbf{E}$ is able to reduce this equation to Eq. (A4.57).

Obviously the wave equation is not invariant but it changes form when it is transformed according to the transformation of Galileo[1].

A4.5 The Lorentz Transformation and the Wave Equation

Historically, Maxwell's equations were known before the advent of the Theory of Relativity. The fact that the Galilean transformation does not leave the equations invariant had already been noted. Lorentz determined that transformation, which has to replace that of Galileo for two inertial frames of reference in relative motion and which leaves invariant the wave equation and Maxwell's equations in general. The transformation Lorentz found, which satisfies these conditions, is the one known today by his name:

$$x' = \gamma(x - Vt) \quad y' = y \quad z' = z \quad t' = \gamma\left(t - \frac{V}{c^2}x\right). \tag{A4.59}$$

Following a procedure which is opposite to that of Lorentz, we will show that this transformation does leave the wave equation invariant. We will now transform the wave equation from the frame of reference S : (x, y, z, t) to the frame of reference S' : (x', y', z', t') using the Lorentz transformation.

Replacing in the wave equation (A4.53) the partial derivatives with respect to the unprimed variables in terms of those with respect to the primed variables, Eqs. (A4.29), (A4.31) and (A4.34), we find

$$\gamma^2\left(\frac{\partial^2\psi}{\partial x'^2} - 2\cancel{\frac{V}{c^2}\frac{\partial^2\psi}{\partial x'\partial t'}} + \frac{V^2}{c^4}\frac{\partial^2\psi}{\partial t'^2}\right) + \frac{\partial^2\psi}{\partial y'^2} + \frac{\partial^2\psi}{\partial z'^2} - \frac{\gamma^2}{c^2}\left(\frac{\partial^2\psi}{\partial t'^2} - \cancel{2V\frac{\partial^2\psi}{\partial t'\partial x'}} + V^2\frac{\partial^2\psi}{\partial x'^2}\right) = 0 \tag{A4.60}$$

$$\gamma^2\left(1 - \frac{V^2}{c^2}\right)\frac{\partial^2\psi}{\partial x'^2} + \frac{\partial^2\psi}{\partial y'^2} + \frac{\partial^2\psi}{\partial z'^2} - \frac{\gamma^2}{c^2}\left(1 - \frac{V^2}{c^2}\right)\frac{\partial^2\psi}{\partial t'^2} = 0 \tag{A4.61}$$

or, finally,

---

[1]It is worth noting that Schrödinger's equation in Quantum Mechanics, $-(\hbar^2/2m)\nabla'^2\psi' + U\psi' = (i\hbar)\,\partial\psi'/\partial t'$, is invariant under the Galilean transformation, i.e. it is $-(\hbar^2/2m)\nabla^2\psi + U\psi = (i\hbar)\,\partial\psi/\partial t$, with the provision that the potential energy function U is invariant under the Galilean transformation and the wave function ψ' transforms into ψ on the basis of $\psi = \psi' \exp[i(m/\hbar)\mathbf{V}\cdot\mathbf{r} - i(mV^2/2\hbar)t]$. Schrödinger's equation is not, of course, relativistic.

$$\frac{\partial^2\psi}{\partial x'^2}+\frac{\partial^2\psi}{\partial y'^2}+\frac{\partial^2\psi}{\partial z'^2}-\frac{1}{c^2}\frac{\partial^2\psi}{\partial t'^2}=0. \tag{A4.62}$$

We see that the Lorentz transformation does leave the wave equation invariant. Of course it may do so but not leave Maxwell's equations invariant. It is proved that it does also leave Maxwell's equations invariant, provided the fields **E** and **B** and the charge density ρ and current density **J** transform in a certain way. This is the subject we will examine in the next two sections.

A4.6 The Lorentz Transformations and Maxwell's Equations

We first examine Eq. (A4.43), $\frac{\partial E_x}{\partial z}-\frac{\partial E_z}{\partial x}=-\frac{\partial B_y}{\partial t}$. Substituting in this the differentiation operators from Eqs. (A4.28), (A4.30) and (A4.33), we find:

$$\frac{\partial E_x}{\partial z'}-\gamma\left(\frac{\partial E_z}{\partial x'}-\frac{V}{c^2}\frac{\partial E_z}{\partial t'}\right)=-\gamma\left(\frac{\partial B_y}{\partial t'}-V\frac{\partial B_y}{\partial x'}\right), \tag{A4.63}$$

which we may rewrite as

$$\frac{\partial E_x}{\partial z'}-\frac{\partial}{\partial x'}\gamma(E_z+VB_y)=-\frac{\partial}{\partial t'}\gamma\left(B_y+\frac{V}{c^2}E_z\right). \tag{A4.64}$$

This equation has the form $\frac{\partial E'_x}{\partial z'}-\frac{\partial E'_z}{\partial x'}=-\frac{\partial B'_y}{\partial t'}$ if we accept that

$$E'_x=E_x,\qquad E'_z=\gamma(E_z+VB_y),\qquad B'_y=\gamma\left(B_y+\frac{V}{c^2}E_z\right). \tag{A4.65 – 67}$$

Similarly, from Eq. (A4.43), $\frac{\partial E_y}{\partial x}-\frac{\partial E_x}{\partial y}=-\frac{\partial B_z}{\partial t}$, we end up with the relation

$$\frac{\partial}{\partial x'}\gamma(E_y-VB_z)-\frac{\partial E_x}{\partial y'}=-\frac{\partial}{\partial t'}\gamma\left(B_z-\frac{V}{c^2}E_y\right). \tag{A4.68}$$

This is identical to $\frac{\partial E'_y}{\partial x'}-\frac{\partial E'_x}{\partial y'}=-\frac{\partial B'_z}{\partial t'}$, if we accept that:

$$E'_y=\gamma(E_y-VB_z),\qquad E'_x=E_x,\qquad B'_z=\gamma\left(B_z-\frac{V}{c^2}E_y\right). \tag{A4.69 – 71}$$

To find the transformation for the component B_x we use Eq. (A4.40) $\frac{\partial B_x}{\partial x}+\frac{\partial B_y}{\partial y}+\frac{\partial B_z}{\partial z}=0$. Substituting the differentiation operators from Eqs. (A4.28) and (A4.30) in Eq. (A4.40), we find

$$\gamma\left(\frac{\partial B_x}{\partial x'} - \frac{V}{c^2}\frac{\partial B_x}{\partial t'}\right) + \frac{\partial B_y}{\partial y'} + \frac{\partial B_z}{\partial z'} = 0. \tag{A4.72}$$

From Eq. (A4.67), $B'_y = \gamma\left(B_y + \frac{V}{c^2}E_z\right)$, and Eq. (A4.71), $B'_z = \gamma\left(B_z - \frac{V}{c^2}E_y\right)$, we may find the inverse transformations for B_y and B_z, by interchanging primed symbols with unprimed and changing the sign of V:

$$B_y = \gamma\left(B'_y - \frac{V}{c^2}E'_z\right) \quad \text{and} \quad B_z = \gamma\left(B'_z + \frac{V}{c^2}E'_y\right). \tag{A4.73 – 74}$$

Substituting in Eq. (A4.72),

$$\gamma\left(\frac{\partial B_x}{\partial x'} - \frac{V}{c^2}\frac{\partial B_x}{\partial t'}\right) + \gamma\frac{\partial}{\partial y'}\left(B'_y - \frac{V}{c^2}E'_z\right) + \gamma\frac{\partial}{\partial z'}\left(B'_z + \frac{V}{c^2}E'_y\right) = 0, \tag{A4.75}$$

which, on cancelling γ and rearranging terms gives:

$$\frac{\partial B_x}{\partial x'} + \frac{\partial B'_y}{\partial y'} + \frac{\partial B'_z}{\partial z'} = \frac{V}{c^2}\left(\frac{\partial E'_z}{\partial y'} - \frac{\partial E'_y}{\partial z'} + \frac{\partial B_x}{\partial t'}\right). \tag{A4.76}$$

Similarly, substituting in Eq. (A4.43), $\frac{\partial E_z}{\partial y} - \frac{\partial E_y}{\partial z} = -\frac{\partial B_x}{\partial t}$, from Eqs. (A4.30) and (A4.32), we find

$$\frac{\partial E_z}{\partial y'} - \frac{\partial E_y}{\partial z'} = -\gamma\left(\frac{\partial B_x}{\partial t'} - V\frac{\partial B_x}{\partial x'}\right). \tag{A4.77}$$

The components of E_z and E_y are acquired by inverting Eqs. (A4.66) and (A4.69):

$$E_z = \gamma\left(E'_z - VB'_y\right), \quad E_y = \gamma\left(E'_y + VB'_z\right). \tag{A4.78 – 79}$$

Substituting in Eq. (A4.77), we find

$$\gamma\left(\frac{\partial E'}{\partial y'} - V\frac{\partial B'_y}{\partial y'}\right) - \gamma\left(\frac{\partial E'}{\partial z'} + V\frac{\partial B'}{\partial z'}\right) = -\gamma\left(\frac{\partial B_x}{\partial t'} - V\frac{\partial B_x}{\partial x'}\right), \tag{A4.80}$$

which, on cancelling γ and rearranging terms gives:

$$-V\left(\frac{\partial B_x}{\partial x'} + \frac{\partial B'_y}{\partial y'} + \frac{\partial B'_z}{\partial z'}\right) + \frac{\partial E'_z}{\partial y'} - \frac{\partial E'_y}{\partial z'} + \frac{\partial B_x}{\partial t'} = 0 \tag{A4.81}$$

or

$$\frac{\partial B_x}{\partial x'} + \frac{\partial B'_y}{\partial y'} + \frac{\partial B'_z}{\partial z'} = \frac{1}{V}\left(\frac{\partial E'_z}{\partial y'} - \frac{\partial E'_y}{\partial z'} + \frac{\partial B_x}{\partial t'}\right). \tag{A4.82}$$

Given that it is $V \neq c$, Eqs. (A4.76) and (A4.82) are compatible only if it is

$$\frac{\partial B_x}{\partial x'} + \frac{\partial B'_y}{\partial y'} + \frac{\partial B'_z}{\partial z'} = 0 \tag{A4.83}$$

and

$$\frac{\partial E'_z}{\partial y'} - \frac{\partial E'_y}{\partial z'} + \frac{\partial B_x}{\partial t'} = 0. \tag{A4.84}$$

If Maxwell's equations are the same in frames S and S′, then, in S′ Eq. (A4.40) is written as

$$\frac{\partial B'_x}{\partial x'} + \frac{\partial B'_y}{\partial y'} + \frac{\partial B'_z}{\partial z'} = 0 \tag{A4.85}$$

and Eq. (A4.41) as

$$\frac{\partial E'_z}{\partial y'} - \frac{\partial E'_y}{\partial z'} = -\frac{\partial B'_x}{\partial t'}. \tag{A4.86}$$

Comparing Eq. (A4.83) with Eq. (A4.85) or Eq. (A4.84) with Eq. (A4.86), it follows that it is:

$$B'_x = B_x \tag{A4.87}$$

Summarizing, we found that those of Maxwell's equations which we have used, remain invariant under the Lorentz transformation, if the components of the fields transform as follows:

$$\begin{aligned} E'_x &= E_x & E_x &= E'_x \\ E'_y &= \gamma(E_y - VB_z) & E_y &= \gamma\left(E'_y + VB'_z\right) \\ E'_z &= \gamma(E_z + VB_y) & E_z &= \gamma\left(E'_z - VB'_y\right) \end{aligned} \tag{A4.88}$$

$$\begin{aligned} B'_x &= B_x & B_x &= B'_x \\ B'_y &= \gamma(B_y + VE_z/c^2) & B_y &= \gamma\left(B'_y - VE'_z/c^2\right) \\ B'_z &= \gamma(B_z - VE_y/c^2) & B_z &= \gamma\left(B'_z + VE'_y/c^2\right) \end{aligned} \tag{A4.89}$$

Using the same procedure, we may verify that these transformations, together with the Lorentz transformation for x, y, z and t, leave all of Maxwell's equations, as well as the wave equation, invariant.

A4.7 The Transformation of the Charge Density ρ and the Current Density J

The equation of continuity

$$\nabla \cdot \mathbf{J} + \frac{\partial \rho}{\partial t} = 0 \tag{A4.90}$$

expresses the conservation of charge. According to the first postulate of the Special Theory of Relativity, if it is true in one inertial frame of reference S, it must be true in all the others as well. In another frame of reference, S′, the equation will have the form

$$\nabla' \cdot \mathbf{J}' + \frac{\partial \rho'}{\partial t'} = 0. \tag{A4.91}$$

We will determine how the quantities ρ and **J** must transform if the equation of continuity is to remain invariant under the Lorentz transformation. We rewrite Eq. (A4.90) as

$$\frac{\partial J_x}{\partial x} + \frac{\partial J_y}{\partial y} + \frac{\partial J_z}{\partial z} + \frac{\partial \rho}{\partial t} = 0. \tag{A4.92}$$

The differentiation operators $\frac{\partial}{\partial x} = \gamma\left(\frac{\partial}{\partial x'} - \frac{V}{c^2}\frac{\partial}{\partial t'}\right)$, $\frac{\partial}{\partial y} = \frac{\partial}{\partial y'}$, $\frac{\partial}{\partial z} = \frac{\partial}{\partial z'}$ and $\frac{\partial}{\partial t} = \gamma\left(\frac{\partial}{\partial t'} - V\frac{\partial}{\partial x'}\right)$, applied to Eq. (A4.92), give

$$\gamma\left(\frac{\partial J_x}{\partial x'} - \frac{V}{c^2}\frac{\partial J_x}{\partial t'}\right) + \frac{\partial J_y}{\partial y'} + \frac{\partial J_z}{\partial z'} + \gamma\left(\frac{\partial \rho}{\partial t'} - V\frac{\partial \rho}{\partial x'}\right) = 0 \tag{A4.93}$$

or

$$\frac{\partial}{\partial x'}[\gamma(J_x - V\rho)] + \frac{\partial J_y}{\partial y'} + \frac{\partial J_z}{\partial z'} + \frac{\partial}{\partial t'}\left[\gamma\left(\rho - \frac{VJ_x}{c^2}\right)\right] = 0 \tag{A4.94}$$

Equation (A4.91) is written as

$$\frac{\partial J'_x}{\partial x'} + \frac{\partial J'_y}{\partial y'} + \frac{\partial J'_z}{\partial z'} + \frac{\partial \rho'}{\partial t'} = 0. \tag{A4.95}$$

Equations (A4.94) and (A4.95) are identical if

$$J'_x = \gamma(J_x - V\rho), \quad J'_y = J_y, \quad J'_z = J_z, \quad \rho' = \gamma\left(\rho - \frac{V}{c^2}J_x\right). \tag{A4.96}$$

This transformation of ρ and $\mathbf{J}$ leaves the equation of continuity invariant under the Lorentz transformation from one inertial frame of reference to another.

A4.8 The Invariance of Electric Charge as a Consequence of the Transformation of ρ and J

We will consider an element of volume dv containing a charge $dQ = \rho\, dv$, at rest in the inertial frame of reference S. Since the charge is at rest in S, in this frame of reference and at that point it will be $\mathbf{J} = 0$. The transformation of ρ and $\mathbf{J}$ gives for these quantities in another inertial frame of reference, S′, which is moving with velocity $\mathbf{V} = V\hat{\mathbf{x}}$ relative to S:

$$J'_x = \gamma(J_x - V\rho) = -\gamma V\rho, \quad J'_y = J_y = 0, \quad J'_z = J_z = 0,$$
$$\rho' = \gamma\left(\rho - \frac{V}{c^2}J_x\right) = \gamma\rho. \tag{A4.97}$$

The important result is the last one, $\rho' = \gamma\rho$. Since, due to the contraction of length in the x direction, the volume dv will occupy a volume $dv' = dv/\gamma$ in frame S′, the total charge in the element dv' will be equal to $dQ' = \rho'\, dv' = \gamma\rho\, dv/\gamma = \rho\, dv = dQ$. The charge enclosed in the element of volume remains, therefore, *invariant* under the Lorentz transformation from S to S′.

If the charge enclosed in a certain volume v consists of N particles, each of which has a charge e, say, and given that the number of particles remains invariant during the transformation from one inertial frame of reference S to another S′ (conservation of the number of particles), the invariance of the total charge $Q = Ne$ also implies the invariance of the charge quantum e. This has been verified experimentally to a very high accuracy (Sect. 9.2). The theoretical proof given is based on Maxwell's equations. They, however, resulted from experimental studies of electromagnetism. We must not forget that Physics is an experimental science and that experiment is the final judge of the validity of its theories. It is obvious that electromagnetic theory, as expressed by Maxwell's equations, is a relativistic theory, whose equations needed no modification in order to become compatible with the Theory of Relativity, at least as these apply to the vacuum.

Problem

A4.1 A region of a space is characterized by a charge density and a current density which are the functions of position, $\rho(\mathbf{r})$ and $\mathbf{J}(\mathbf{r})$, respectively, as measured in a frame of reference S. Consider an element of volume $d\tau$, centered around a point at $\mathbf{r}$. At this point, let the values of the charge density and the current density be ρ and $\mathbf{J}$, respectively, and that the choice of the axes of frame S is such that $\mathbf{J} = J\hat{\mathbf{x}}$. The volume element and the charge it contains move, momentarily, with a velocity equal

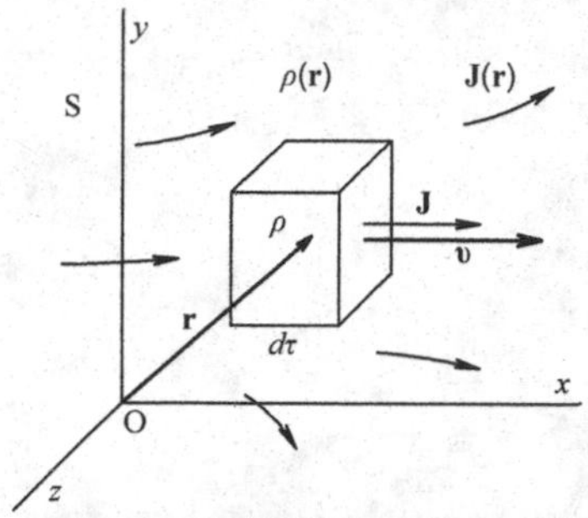

to $\mathbf{v} = \upsilon\hat{\mathbf{x}}$ in frame S. If the number density of the charge carriers in the element of volume is n and each charge carrier has a charge e, it is $\rho = ne$, $\mathbf{J} = ne\,\mathbf{v} = \rho\mathbf{v} = \rho\upsilon\hat{\mathbf{x}}$ and $J = \rho\upsilon$. Now view the element of volume from another frame of reference, S′, which is moving with a velocity $\mathbf{v} = \upsilon\hat{\mathbf{x}}$ with respect to frame S.

(a) In the frame of reference S′, find the volume $d\tau'$ of the element, and the charge density ρ' and the current density $\mathbf{J}'$ at the point considered.

(b) Prove, from first principles, that the magnitude $c^2\rho^2 - J^2$ is invariant under the Lorentz transformation.

(c) Use the transformation formulas for the magnitudes ρ and $\mathbf{J}$ in order to show that $c^2\rho^2 - J^2$ is invariant under the Lorentz transformation.

Ans.: (a) $d\tau' = \gamma d\tau$, $\rho' = \rho_0 = \rho/\gamma$ and $\mathbf{J}' = 0$.

(b) and (c) $c^2\rho^2 - J^2 = c^2\rho'^2 - J'^2 = c^2\rho_0^2$

Appendix 5
Tachyons

A5.1 The Tachyons and Some of Their Properties

From what we have said concerning the transformation of velocities and the nature of the relativistic mass and energy, it must be clear that the speed of light in vacuum, c, constitutes a kind of a barrier for the speeds of material bodies. The Special Theory of Relativity forbids the acceleration of a body of non-zero rest mass from low speeds to speeds which are equal or higher than c. On the other hand, the theory says nothing about the existence of bodies which move with a speed higher than c. It simply forbids the passage through the barrier at c. The photons are examples of entities that move with a speed equal to c, but have not acquired their speed through their being accelerated from lower speeds. They are simply created having this speed. The question, therefore, arises as to whether it is possible for bodies to exist, which move with speeds higher than c, simply because they have these speeds for some reason, at present unknown, since the moment of their creation. These bodies are called *tachyons*, from the Greek word *ταχύς* meaning fast [9]. It must be made perfectly clear that we have at present no experimental evidence for the existence of such bodies. They are simply theoretical creations, whose existence is considered possible on the basis of the empirical 'law' which states that '*Anything not expressly forbidden by Nature, very possibly exists*[2].'

We may examine some of the properties of these hypothetical tachyons. From the arguments presented in Sect. 3.2.4, if a body is a tachyon in one inertial frame of reference, it will be a tachyon in all inertial frames of reference which move, relative to this frame of reference, with speeds $V < c$. A tachyon cannot, therefore, by a transition in another frame of reference change its *tachyonic* nature. Also, photons are photons in all the frames of reference.

In Fig. A5.1, a Minkowski diagram was drawn for the frame of reference S of an observer and for another frame of reference, S′, which moves relative to the S with a

[2]Some people go as far as to state the 'law' in its stronger form, i.e. that '*Anything not expressly forbidden to exist by Nature, does exist*' or '*Anything not expressly forbidden to happen by Nature, does happen*'.

© Springer International Publishing Switzerland 2016

C. Christodoulides, *The Special Theory of Relativity*,
Undergraduate Lecture Notes in Physics, DOI 10.1007/978-3-319-25274-2

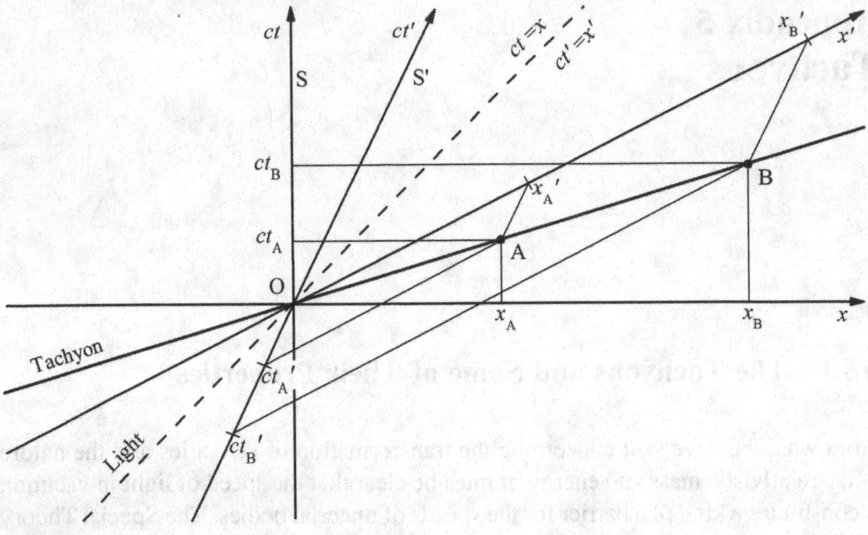

Fig. A5.1 The world line of a tachyon (line OAB), in the Minkowski diagram for an observer in the inertial frame of reference S and in another, S′, which moves with respect to S at a constant speed V along the x-axis

speed V (smaller than c) along the x-axis. Also drawn in the diagram is the world line of a tachyon which is moving relative to the frame S with a speed $\upsilon > c$ along the x-axis. Let us examine two points on the world line of the tachyon, A and B. In frame S, these two events happen at times t_{A} and t_{B}, respectively, where it is $t_{\mathrm{B}} > t_{\mathrm{A}}$. The spatial positions of the two events are x_{A} and x_{B}, respectively, where it is $x_{\mathrm{B}} > x_{\mathrm{A}}$. To find the corresponding times in frame S′, we draw from these two points straight lines parallel to the x'-axis of frame S′. These lines intersect the ct'-axis at two points which correspond to time values t'_{A} and t'_{B}. We notice that it is $t'_{\mathrm{B}} < t'_{\mathrm{A}}$. The temporal sequence of the two events has been reversed in frame S′. While in one frame of reference event A, which lies on the world line of the tachyon, happens *before* event B, there exist other frames of reference in which event A happens *after* event B. From the figure, it appears that this reversal will not occur if the straight line representing the world line of the tachyon lies between the straight line $ct' = x'$ of the light wave front and the axis of x'.

The same effect is demonstrated in Fig. A5.2. Let a tachyon be created in the frame of reference S at point O and that is disappears at point A, at a moment $t_{\mathrm{A}} > 0$. In frame S′, the point of disappearance corresponds to time $t'_{\mathrm{A}} < 0$. In frame S′, therefore, the disappearance of the tachyon occurs before its creation.

Another strange effect appears if we examine a tachyon which emits pulses at regular intervals. Figure A5.3 shows, in the frame of reference S, the world line of a tachyon which is moving along the x-axis, emitting light pulses. For simplicity in the figure, we will study the pulses which are emitted at regular intervals at the positions we have numbered as (−5), (−4),…, (0), (1), (2), (3),… To find the

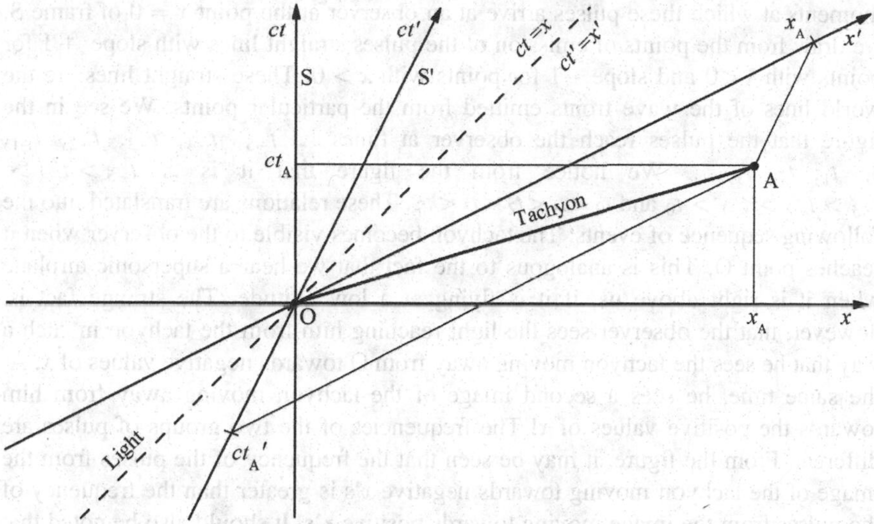

Fig. A5.2 The creation (O) and the absorption (A) of a tachyon, as seen in two inertial frames of reference, S and S′. The temporal sequence at which these two events happen is opposite in the two frames. In both frames, the creation of the tachyon happens at O ($t_O = t'_O = 0$, $x_O = x'_O = 0$). In frame S, the tachyon is created before it is absorbed and $t_A > 0$. In frame S′, it is $t_A' < 0$ and, therefore, the tachyon is absorbed before its creation

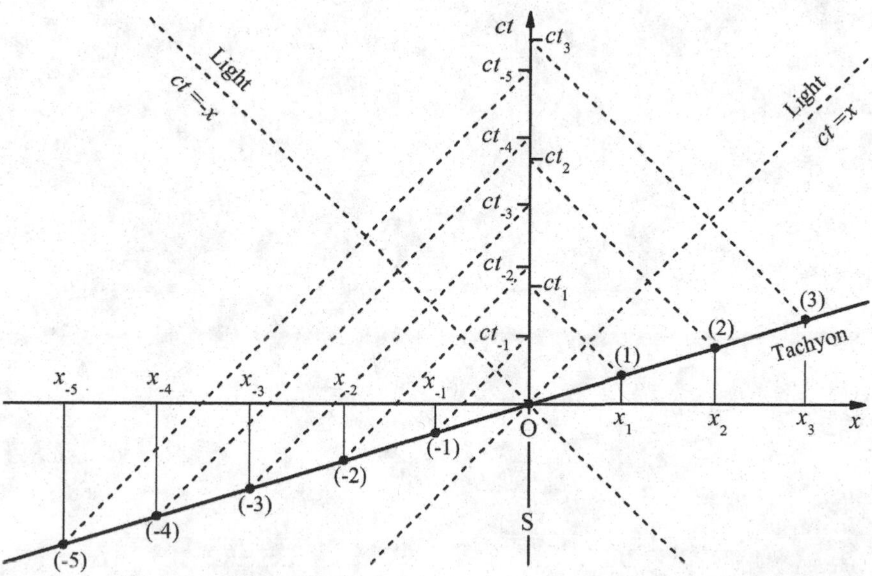

Fig. A5.3 The observation, in frame of reference S, of a tachyon moving with constant velocity along the *x*-axis, and emitting light pulses at a constant rate

moments at which these pulses arrive at an observer at the point $x = 0$ of frame S, we draw from the points of emission of the pulses straight lines with slope $+1$ for points with $x < 0$ and slope -1 for points with $x > 0$. These straight lines are the world lines of the wave fronts emitted from the particular points. We see in the figure that the pulses reach the observer at times $\ldots t_{-5},\ t_{-4},\ t_{-3},\ t_{-2},\ t_{-1},\ t_0,\ t_1,\ t_2,\ t_3, \ldots$. We notice from the figure that it is $\ldots t_{-5} > t_{-4} > t_{-3} > t_{-2} > t_{-1} > t_0$ and $t_0 < t_1 < t_2 < t_3 < \ldots$ These relations are translated into the following sequence of events: The tachyon becomes visible to the observer when it reaches point O. This is analogous to the fact that we hear a supersonic airplane when it is right above us, if it is flying at a low altitude. The strange fact is, however, that the observer sees the light reaching him from the tachyon in such a way that he sees the tachyon moving away from O towards negative values of x. At the same time, he sees a second image of the tachyon moving away from him towards the positive values of x! The frequencies of the two groups of pulses are different. From the figure, it may be seen that the frequency of the pulses from the image of the tachyon moving towards negative x's is greater than the frequency of the pulses from the image moving towards positive x's. It should also be noted that the speeds with which the two tachyon images move away from the observer are different.

These, and many other paradoxes, make the tachyons a very interesting theoretical concept.

Appendix 6
The Lorentz Transformation in Matrix Form

We review very briefly the topic of matrices for the needs of this book and their use in the formulation of the Lorentz transformation. The interested reader may find more information in a book on advanced Mathematics.

A6.1 Matrices

A matrix A or (a_{ij}), with dimensions $n \times m$, is an arrangement of mn quantities, which are denoted by a_{ij} with $i = 1, 2, 3, \ldots n$ and $j = 1, 2, 3, \ldots m$, placed in a rectangular arrangement with n lines and m columns:

$$\mathsf{A} = (a_{ij}) = \begin{pmatrix} a_{11} & a_{12} & \ldots & a_{1m} \\ a_{21} & a_{22} & \ldots & a_{2m} \\ \ldots & \ldots & \ldots & \ldots \\ a_{n1} & a_{n2} & \ldots & a_{nm} \end{pmatrix}. \tag{A6.1}$$

Bold sans serif letters are used to denote a matrix.

If $m = n$, the matrix is called a *square matrix* of order n. If $m = 1$, the matrix is called column-matrix or a vector in n dimensions:

$$\mathsf{x} = (x_i) = \begin{pmatrix} x_1 \\ x_2 \\ \ldots \\ x_n \end{pmatrix}. \tag{A6.2}$$

Two $n \times m$ matrices A and B are equal if, and only if, it is $a_{ij} = b_{ij}$ for every i and j. The sum of two $n \times m$ matrices $\mathsf{A} = (a_{ij})$ and $\mathsf{B} = (b_{ij})$ is equal to the matrix $\mathsf{C} = (c_{ij})$, where $c_{ij} = a_{ij} + b_{ij}$. The difference of two matrices is defined in a similar manner. Multiplication of a matrix by a number k means that $k\mathsf{A} = k(a_{ij}) = (ka_{ij})$.

The product of an $n \times m$ matrix A and an $m \times p$ matrix B, is defined as the $n \times p$ matrix

© Springer International Publishing Switzerland 2016

C. Christodoulides, *The Special Theory of Relativity*,
Undergraduate Lecture Notes in Physics, DOI 10.1007/978-3-319-25274-2

$$\mathbf{C} = (c_{ij}) = \mathbf{AB} = \left(\sum_k a_{ik}a_{kj}\right). \tag{A6.3}$$

Example:

$$\mathbf{C} = \mathbf{AB} = \begin{pmatrix} 1 & 0 & 2 \\ 2 & -1 & 3 \\ 0 & 4 & 1 \end{pmatrix} \begin{pmatrix} 1 & 3 \\ 0 & 2 \\ 2 & 4 \end{pmatrix}$$

$$= \begin{pmatrix} (1)(1)+(0)(0)+(2)(2) & (1)(3)+(0)(2)+(2)(4) \\ (2)(1)+(-1)(0)+(3)(2) & (2)(3)+(-1)(2)+(3)(4) \\ (0)(1)+(4)(0)+(1)(2) & (0)(3)+(4)(2)+(1)(4) \end{pmatrix} = \begin{pmatrix} 5 & 11 \\ 8 & 14 \\ 2 & 12 \end{pmatrix}.$$

We write for the general term of matrix $\mathbf{C}$,

$$c_{ij} = a_{ik}b_{kj} \tag{A6.4}$$

and imply the summation with regard to the repeated index k.

A6.2 Square Matrices

A matrix $\mathbf{A}$ with dimensions $n \times n$

$$\mathbf{A} = (a_{ij}) = \begin{pmatrix} a_{11} & a_{12} & \dots & a_{1n} \\ a_{21} & a_{22} & \dots & a_{2n} \\ \dots & \dots & \dots & \dots \\ a_{n1} & a_{n2} & \dots & a_{nn} \end{pmatrix}, \tag{A6.5}$$

is called a *square matrix* of order n. The determinant of the matrix $\mathbf{A}$ is

$$a = |a_{ij}| = \begin{vmatrix} a_{11} & a_{12} & \dots & a_{1n} \\ a_{21} & a_{22} & \dots & a_{2n} \\ \dots & \dots & \dots & \dots \\ a_{n1} & a_{n2} & \dots & a_{nn} \end{vmatrix}. \tag{A6.6}$$

If $\mathbf{A} = (a_{ij})$ and $\mathbf{B} = (b_{ij})$ are two $n \times n$ square matrices, in general it is $\mathbf{AB} \neq \mathbf{BA}$.

A square matrix of dimensions $n \times n$

$$\mathbf{I} = \begin{pmatrix} 1 & 0 & \dots & 0 \\ 0 & 1 & \dots & 0 \\ \dots & \dots & \dots & \dots \\ 0 & 0 & \dots & 1 \end{pmatrix} \tag{A6.7}$$

is called *identity* or *unit* matrix and it satisfies the relations

$$\mathbf{IA} = \mathbf{AI} = \mathbf{A}. \tag{A6.8}$$

The matrix which is the inverse to matrix $\mathbf{A}$ is defined as

$$\mathbf{A}^{-1} = \begin{pmatrix} \frac{A_{11}}{a} & \frac{A_{12}}{a} & \cdots & \frac{A_{1n}}{a} \\ \frac{A_{11}}{a} & \frac{A_{22}}{a} & \cdots & \frac{A_{2n}}{a} \\ \cdots & \cdots & \cdots & \cdots \\ \frac{A_{11}}{a} & \frac{A_{n2}}{a} & \cdots & \frac{A_{nn}}{a} \end{pmatrix}, \tag{A6.9}$$

where a is the determinant of the matrix $\mathbf{A}$ and A_{ij} is the coefficient of the element a_{ij} in the determinant $a = |a_{ij}|$, i.e. $A_{ij} = (-1)^{i+j}M_{ij}$, where M_{ij} is the determinant that results by deleting in the determinant a the column and the line which include the element a_{ij}. It is known from the theory of determinants that $a = |a_{ij}| = \sum_{\substack{i=1 \\ j=const.}}^{i=n} a_{ij}A_{ij} = \sum_{\substack{j=1 \\ i=const.}}^{j=n} a_{ij}A_{ij}$, where in the first sum we expand the determinant with respect to the j-th column, while in the second sum we expand the determinant with respect to the i-th line.

It may be shown that $\mathbf{AA}^{-1} = \mathbf{A}^{-1}\mathbf{A} = \mathbf{I}$.

A6.3 Linear Transformations

The totality of the linear relationships

$$\mathbf{Y} = \mathbf{AX} \quad \text{or} \quad y_i = a_{ij}x_j, \quad i,j = 1,\ 2,\ 3, \ldots n, \tag{A6.10}$$

where summation over all j is implied and the a_{ij} are constants, define a *linear transformation*. These relations may also be written in a compact manner as $\mathbf{Y} = \mathbf{AX}$, where

$$\mathbf{X} = (x_i) = \begin{pmatrix} x_1 \\ x_2 \\ \cdots \\ x_n \end{pmatrix}, \quad \mathbf{Y} = (y_i) = \begin{pmatrix} y_1 \\ y_2 \\ \cdots \\ y_n \end{pmatrix},$$
$$\mathbf{A} = (a_{ij}) = \begin{pmatrix} a_{11} & a_{12} & \cdots & a_{1n} \\ a_{21} & a_{22} & \cdots & a_{2n} \\ \cdots & \cdots & \cdots & \cdots \\ a_{n1} & a_{n2} & \cdots & a_{nn} \end{pmatrix} \tag{A6.11}$$

The transformation which is the inverse to the transformation of Eq. (A6.10) is

$$\mathbf{X} = \mathbf{A}^{-1}\mathbf{Y} \quad \text{or} \quad x_i = \frac{A_{ij}}{a} y_j, \tag{A6.12}$$

with the repeated index implying summation for all the values $j = 1, 2, 3, \dots n$.

A6.4 Rotation of the Axes

Consider two frames of reference (x, y, z) and (x', y', z'), the origins of which coincide. If $\hat{\mathbf{x}}, \hat{\mathbf{y}}, \hat{\mathbf{z}}$ and $\hat{\mathbf{x}}', \hat{\mathbf{y}}', \hat{\mathbf{z}}'$ are the unit vectors along the axes of the two frames, respectively, then the angles formed by the corresponding axes have cosines which are given by:

$$\begin{array}{lll} \cos\theta_{11} = \hat{\mathbf{x}}\cdot\hat{\mathbf{x}}' & \cos\theta_{12} = \hat{\mathbf{y}}\cdot\hat{\mathbf{x}}' & \cos\theta_{13} = \hat{\mathbf{z}}\cdot\hat{\mathbf{x}}' \\ \cos\theta_{21} = \hat{\mathbf{x}}\cdot\hat{\mathbf{y}}' & \cos\theta_{22} = \hat{\mathbf{y}}\cdot\hat{\mathbf{y}}' & \cos\theta_{23} = \hat{\mathbf{z}}\cdot\hat{\mathbf{y}}' \\ \cos\theta_{31} = \hat{\mathbf{x}}\cdot\hat{\mathbf{z}}' & \cos\theta_{32} = \hat{\mathbf{y}}\cdot\hat{\mathbf{z}}' & \cos\theta_{33} = \hat{\mathbf{z}}\cdot\hat{\mathbf{z}}' \end{array} \tag{A6.13}$$

The transformation of the coordinates of a point $\mathbf{r} = x\hat{\mathbf{x}} + y\hat{\mathbf{y}} + z\hat{\mathbf{z}} = x'\hat{\mathbf{x}}' + y'\hat{\mathbf{y}}' + z'\hat{\mathbf{z}}'$ from the one frame to the other is:

$$\begin{aligned} x' &= \cos\theta_{11}\, x + \cos\theta_{12}\, y + \cos\theta_{13}\, z \\ y' &= \cos\theta_{21}\, x + \cos\theta_{22}\, y + \cos\theta_{23}\, z \\ z' &= \cos\theta_{31}\, x + \cos\theta_{32}\, y + \cos\theta_{33}\, z \end{aligned} \tag{A6.14}$$

Using matrices, these relations are expressed as $\mathbf{x}' = \mathbf{R}\mathbf{x}$, where it is

$$\mathbf{x} = \begin{pmatrix} x \\ y \\ z \end{pmatrix}, \quad \mathbf{x}' = \begin{pmatrix} x' \\ y' \\ z' \end{pmatrix} \quad \text{and} \quad \mathbf{R} = \begin{pmatrix} \cos\theta_{11} & \cos\theta_{12} & \cos\theta_{13} \\ \cos\theta_{21} & \cos\theta_{22} & \cos\theta_{23} \\ \cos\theta_{31} & \cos\theta_{32} & \cos\theta_{33} \end{pmatrix}. \tag{A6.15}$$

The angles θ_{ij} are not, of course, independent of each other but may be expressed in terms of three independent angles.

For a rotation by an angle ϕ only about the z-axis, which coincides with the z'-axis, it is

$$\begin{aligned} x' &= x\cos\phi + y\sin\phi \\ y' &= -x\sin\phi + y\cos\phi \\ z' &= z \end{aligned} \tag{A6.16}$$

or $\mathbf{x}' = \mathbf{R}\mathbf{x}$, where it is

$$\mathbf{x} = \begin{pmatrix} x \\ y \\ z \end{pmatrix}, \quad \mathbf{x}' = \begin{pmatrix} x' \\ y' \\ z' \end{pmatrix} \quad \text{and} \quad \mathbf{R} = \begin{pmatrix} \cos\phi & \sin\phi & 0 \\ -\sin\phi & \cos\phi & 0 \\ 0 & 0 & 1 \end{pmatrix}. \tag{A6.17}$$

A6.5 The Lorentz Transformation in Matrix Form

Let the axes of the inertial frame of reference S coincide, at the moment $t = t' = 0$, with those of the inertial frame of reference S′, which moves with a constant velocity $\mathbf{V} = V\hat{\mathbf{x}}$ relative to S. The special Lorentz transformation from S to S′ is

$$x' = \gamma(x - \beta ct), \qquad y' = y, \qquad z' = z, \qquad t' = \gamma(t - (\beta/c)x).$$

Defining the matrices

$$\mathbf{x} = (x_i) = \begin{pmatrix} ct \\ x \\ y \\ z \end{pmatrix}, \quad \mathbf{x}' = (x'_i) = \begin{pmatrix} ct' \\ x' \\ y' \\ z' \end{pmatrix} \quad \text{and}$$

$$\mathbf{L} = (L_{ij}) = \begin{pmatrix} \gamma & -\gamma\beta & 0 & 0 \\ -\gamma\beta & \gamma & 0 & 0 \\ 0 & 0 & 1 & 0 \\ 0 & 0 & 0 & 1 \end{pmatrix}, \tag{A6.18}$$

the transformation may be written as $\mathbf{x}' = \mathbf{L}\mathbf{x}$.
The inverse transformation is $\mathbf{x} = \mathbf{L}^{-1}\mathbf{x}'$, where

$$\mathbf{L}^{-1} = \begin{pmatrix} \gamma & \gamma\beta & 0 & 0 \\ \gamma\beta & \gamma & 0 & 0 \\ 0 & 0 & 1 & 0 \\ 0 & 0 & 0 & 1 \end{pmatrix}. \tag{A6.19}$$

The Lorentz transformation in the general case when the velocity $\mathbf{V} = c\boldsymbol{\beta}$ has any direction, is

$$\mathbf{r}' = \mathbf{r} + (\gamma - 1)\frac{\mathbf{r}\cdot\boldsymbol{\beta}}{\beta^2}\boldsymbol{\beta} - \gamma\boldsymbol{\beta}ct, \qquad ct' = \gamma(ct - \mathbf{r}\cdot\boldsymbol{\beta})$$

where $\mathbf{r} = (x, y, z)$ and $\mathbf{r}' = (x', y', z')$. In this case the transformation is $\mathbf{x}' = \mathbf{L}\mathbf{x}$, where $\mathbf{x} = (ct, \mathbf{r})$ and

$$\mathbf{L} = (L_{ij}) = \begin{pmatrix} \gamma & -\gamma\beta_x & -\gamma\beta_y & -\gamma\beta_z \\ -\gamma\beta_x & 1+(\gamma-1)\dfrac{\beta_x^2}{\beta^2} & (\gamma-1)\dfrac{\beta_x\beta_y}{\beta^2} & (\gamma-1)\dfrac{\beta_x\beta_z}{\beta^2} \\ -\gamma\beta_y & (\gamma-1)\dfrac{\beta_x\beta_y}{\beta^2} & 1+(\gamma-1)\dfrac{\beta_y^2}{\beta^2} & (\gamma-1)\dfrac{\beta_y\beta_z}{\beta^2} \\ -\gamma\beta_z & (\gamma-1)\dfrac{\beta_x\beta_z}{\beta^2} & (\gamma-1)\dfrac{\beta_z\beta_y}{\beta^2} & 1+(\gamma-1)\dfrac{\beta_z^2}{\beta^2} \end{pmatrix} \quad \text{(A6.20)}$$

For $\mathbf{V} = V\hat{\mathbf{x}}$, this matrix assumes the form of the matrix in Eq. (A6.18). When during the transformation there is no rotation of the axes, the transformation matrix is symmetric with respect to the main diagonal, like the matrices of Eqs. (A6.18) and (A6.20). In this case the transformation is called a *Lorentz boost*. A transformation consisting of a Lorentz boost and one or more rotations of the axes has a non-symmetric matrix. In the case when the axes of the frames of reference are mutually parallel but do not coincide at any moment, we have the *inhomogeneous Lorentz transformation* or the *Poincaré transformation*, $\mathbf{x}' = \mathbf{L}\mathbf{x} + \mathbf{a}$.

A6.6 Two Successive Lorentz Transformations

We will investigate the transformation that results from two successive applications of the Lorentz transformation. Initially, we will study the simplest example of two transformations in the same direction, followed by the study of two transformations in two mutually perpendicular directions. This will lead us to the presentation of the phenomenon of Wigner rotation and its special case of the Thomas precession (see also Sect. 4.8). We will adopt the method of Ferraro and Thibeault [10] for the purposes of this section. For mathematical simplicity we will use matrices and the rules of linear Algebra in the presentation of the theory for the linear transformations (Lorentz) involved in this study.

A6.6.1 Two Successive Lorentz Transformations in the Same Direction

Consider three inertial frame of reference S_1, S_2 and S_3, whose axes coincided when the time in them was $t_1 = t_2 = t_3 = 0$, respectively. Frame S_2 moves with a velocity $\mathbf{V}_1 = V_1\hat{\mathbf{x}}$ relative to frame S_1 and frame S_3 moves with a velocity $\mathbf{V}_2 = V_2\hat{\mathbf{x}}$ relative to frame S_2. The Lorentz transformation from frame S_1 to frame S_3 is found by transforming first from S_1 to S_2 and then from S_2 to S_3. Defining $\beta_1 = V_1/c$, $\beta_2 = V_2/c$, $\gamma_1 = 1/\sqrt{1-\beta_1^2}$ and $\gamma_2 = 1/\sqrt{1-\beta_2^2}$, the matrix of the combination of the two transformations is,

$$\mathsf{L}_{1\to 3} = \mathsf{L}_{2\to 3}\mathsf{L}_{1\to 2} = \begin{pmatrix} \gamma_2 & -\gamma_2\beta_2 & 0 & 0 \\ -\gamma_2\beta_2 & \gamma_2 & 0 & 0 \\ 0 & 0 & 1 & 0 \\ 0 & 0 & 0 & 1 \end{pmatrix}\begin{pmatrix} \gamma_1 & -\gamma_1\beta_1 & 0 & 0 \\ -\gamma_1\beta_1 & \gamma_1 & 0 & 0 \\ 0 & 0 & 1 & 0 \\ 0 & 0 & 0 & 1 \end{pmatrix} \tag{A6.21}$$

$$\mathsf{L}_{1\to 3} = \mathsf{L}_{2\to 3}\mathsf{L}_{1\to 2} = \begin{pmatrix} \gamma_1\gamma_2(1+\beta_1\beta_2) & -\gamma_1\gamma_2(\beta_1+\beta_2) & 0 & 0 \\ -\gamma_1\gamma_2(\beta_1+\beta_2) & \gamma_1\gamma_2(1+\beta_1\beta_2) & 0 & 0 \\ 0 & 0 & 1 & 0 \\ 0 & 0 & 0 & 1 \end{pmatrix} \tag{A6.22}$$

Defining the quantities

$$\beta_{1,3} \equiv \frac{\beta_1+\beta_2}{1+\beta_1\beta_2} \quad \text{and} \quad \gamma_{1,3} \equiv \gamma_1\gamma_2(1+\beta_1\beta_2), \tag{A6.23}$$

the matrix of the transformation may also be written as

$$\mathsf{L}_{1\to 2\to 3} = \mathsf{L}_{2\to 3}\mathsf{L}_{1\to 2} = \begin{pmatrix} \gamma_{1,3} & -\gamma_{1,3}\beta_{1,3} & 0 & 0 \\ -\gamma_{1,3}\beta_{1,3} & \gamma_{1,3} & 0 & 0 \\ 0 & 0 & 1 & 0 \\ 0 & 0 & 0 & 1 \end{pmatrix} \tag{A6.24}$$

which is symmetric with respect to its main diagonal. We conclude that two successive Lorentz boosts (transformations without rotation of the axes) in the same direction, are equivalent to a single Lorentz boost (without rotation of the axes). The transformation corresponds to the velocity $\mathbf{V}_{1,3} = V_{1,3}\,\hat{\mathbf{x}} = c\beta_{1,3}\hat{\mathbf{x}}$, which must be considered as the velocity of frame S_3 relative to frame S_1. The law of addition follows for two velocities which are parallel, as well as the transformation of the Lorentz factor [Eq. (A6.23)].

A6.6.2 Two Successive Lorentz Transformations in Directions Normal to Each Other

Consider three inertial frames of reference S_1, S_2 and S_3, whose axes coincided when the time in them was $t_1 = t_2 = t_3 = 0$, respectively. Frame S_2 moves with a velocity $\mathbf{V}_1 = V_1\,\hat{\mathbf{x}}$ relative to frame S_1, and frame S_3 moves with a velocity $\mathbf{V}_2 = V_2\,\hat{\mathbf{y}}$ relative to S_2. The Lorentz transformation from frame S_1 to frame S_2 is given by the matrix

$$\mathbf{L}_{1\to2} = \begin{pmatrix} \gamma_1 & -\gamma_1\beta_1 & 0 & 0 \\ -\gamma_1\beta_1 & \gamma_1 & 0 & 0 \\ 0 & 0 & 1 & 0 \\ 0 & 0 & 0 & 1 \end{pmatrix} \tag{A6.25}$$

and that from frame S_2 to S_3 by the matrix

$$\mathbf{L}_{2\to3} = \begin{pmatrix} \gamma_2 & 0 & -\gamma_2\beta_2 & 0 \\ 0 & 1 & 0 & 0 \\ -\gamma_2\beta_2 & 0 & \gamma_2 & 0 \\ 0 & 0 & 0 & 1 \end{pmatrix}. \tag{A6.26}$$

The combination of the two is

$$\mathbf{L}_{1\to2\to3} = \mathbf{L}_{2\to3}\mathbf{L}_{1\to2} = \begin{pmatrix} \gamma_2 & 0 & -\gamma_2\beta_2 & 0 \\ 0 & 1 & 0 & 0 \\ -\gamma_2\beta_2 & 0 & \gamma_2 & 0 \\ 0 & 0 & 0 & 1 \end{pmatrix} \begin{pmatrix} \gamma_1 & -\gamma_1\beta_1 & 0 & 0 \\ -\gamma_1\beta_1 & \gamma_1 & 0 & 0 \\ 0 & 0 & 1 & 0 \\ 0 & 0 & 0 & 1 \end{pmatrix} \tag{A6.27}$$

$$\mathbf{L}_{1\to2\to3} = \begin{pmatrix} \gamma_1\gamma_2 & -\gamma_1\gamma_2\beta_1 & -\gamma_2\beta_2 & 0 \\ -\gamma_1\beta_1 & \gamma_1 & 0 & 0 \\ -\gamma_1\gamma_2\beta_2 & \gamma_1\gamma_2\beta_1\beta_2 & \gamma_2 & 0 \\ 0 & 0 & 0 & 1 \end{pmatrix}. \tag{A6.28}$$

This matrix is not symmetrical with respect to its main diagonal. We will assume that the transformation may be expressed as a Lorentz transformation without rotation of the axes (a Lorentz boost), which is expressed by an unknown matrix $\mathbf{L}_{1\to3}$, and a rotation of the axes. As the elements of the last line are (0, 0, 0, 1), the rotation, say by an angle θ, will take place about the z-axis, and is expressed by the matrix

$$\mathbf{R}_z = \begin{pmatrix} 1 & 0 & 0 & 0 \\ 0 & \cos\theta & \sin\theta & 0 \\ 0 & -\sin\theta & \cos\theta & 0 \\ 0 & 0 & 0 & 1 \end{pmatrix}. \tag{A6.29}$$

Therefore

$$\mathbf{L}_{1\to2\to3} = \mathbf{R}_z\mathbf{L}_{1\to3}. \tag{A6.30}$$

In order to find the matrix $\mathbf{L}_{1\to 3}$, we multiply the two sides of Eq. (A6.30) by the inverse matrix of matrix $\mathbf{R}_z$,

$$\mathbf{R}_z^{-1} = \begin{pmatrix} 1 & 0 & 0 & 0 \\ 0 & \cos\theta & -\sin\theta & 0 \\ 0 & \sin\theta & \cos\theta & 0 \\ 0 & 0 & 0 & 1 \end{pmatrix} \tag{A6.31}$$

in which case we have the equation

$$\mathbf{L}_{1\to 3} = \mathbf{R}_z^{-1}\mathbf{L}_{1\to 2\to 3} = \begin{pmatrix} 1 & 0 & 0 & 0 \\ 0 & \cos\theta & -\sin\theta & 0 \\ 0 & \sin\theta & \cos\theta & 0 \\ 0 & 0 & 0 & 1 \end{pmatrix} \begin{pmatrix} \gamma_1\gamma_2 & -\gamma_1\gamma_2\beta_1 & -\gamma_2\beta_2 & 0 \\ -\gamma_1\beta_1 & \gamma_1 & 0 & 0 \\ -\gamma_1\gamma_2\beta_2 & \gamma_1\gamma_2\beta_1\beta_2 & \gamma_2 & 0 \\ 0 & 0 & 0 & 1 \end{pmatrix} \tag{A6.32}$$

$$\mathbf{L}_{1\to 3} = \begin{pmatrix} \gamma_1\gamma_2 & -\gamma_1\gamma_2\beta_1 & -\gamma_2\beta_2 & 0 \\ -\gamma_1\beta_1\cos\theta + \gamma_1\gamma_2\beta_2\sin\theta & \gamma_1\cos\theta - \gamma_1\gamma_2\beta_1\beta_2\sin\theta & -\gamma_2\sin\theta & 0 \\ -\gamma_1\beta_1\sin\theta - \gamma_1\gamma_2\beta_2\cos\theta & \gamma_1\sin\theta + \gamma_1\gamma_2\beta_1\beta_2\cos\theta & \gamma_2\cos\theta & 0 \\ 0 & 0 & 0 & 1 \end{pmatrix}. \tag{A6.33}$$

A necessary condition for this matrix to be symmetrical with respect to its main diagonal is that it should be

$$\gamma_1\sin\theta + \gamma_1\gamma_2\beta_1\beta_2\cos\theta = -\gamma_2\sin\theta \tag{A6.34}$$

or
$$\tan\theta = -\frac{\gamma_1\gamma_2\beta_1\beta_2}{\gamma_1+\gamma_2}, \quad \sin\theta = -\frac{\gamma_1\gamma_2\beta_1\beta_2}{1+\gamma_1\gamma_2}, \quad \cos\theta = \frac{\gamma_1+\gamma_2}{1+\gamma_1\gamma_2}. \tag{A6.35}$$

Apart from being necessary, the condition of Eq. (A6.34) is also sufficient, since, by substituting the relations of Eqs. (A6.35) in Eq. (A6.33) we have the symmetric matrix

$$\mathbf{L}_{1\to 3} = \begin{pmatrix} \gamma_1\gamma_2 & -\gamma_1\gamma_2\beta_1 & -\gamma_2\beta_2 & 0 \\ -\gamma_1\gamma_2\beta_1 & 1+\dfrac{\gamma_1^2\gamma_2^2\beta_1^2}{1+\gamma_1\gamma_2} & \dfrac{\gamma_1\gamma_2^2\beta_1\beta_2}{1+\gamma_1\gamma_2} & 0 \\ -\gamma_2\beta_2 & \dfrac{\gamma_1\gamma_2^2\beta_1\beta_2}{1+\gamma_1\gamma_2} & \dfrac{\gamma_1(\gamma_1+\gamma_2)}{1+\gamma_1\gamma_2} & 0 \\ 0 & 0 & 0 & 1 \end{pmatrix}. \tag{A6.36}$$

If we compare the elements of this matrix with those of the matrix of Eq. (A6.20), which represents a Lorentz boost in the xy plane, we may find the velocity $\mathbf{V}_{1\to3} = c\boldsymbol{\beta}_{1\to3}$ to which the transformation corresponds. From the elements of the first line we have the relations

$$\gamma_{1\to3} = \gamma_1\gamma_2, \quad -\gamma_{1\to3}\beta_{x,1\to3} = -\gamma_1\gamma_2\beta_1, \quad -\gamma_{1\to3}\beta_{y,1\to3} = -\gamma_2\beta_2, \quad -\gamma_{1\to3}\beta_{z,1\to3} = 0. \tag{A6.37}$$

Substituting the first of these equations in the others, we find

$$\beta_{x,1\to3} = \beta_1, \qquad \beta_{y,1\to3} = \frac{\beta_2}{\gamma_1}, \qquad \beta_{z,1\to3} = 0, \tag{A6.38}$$

which also satisfy the first equation. Substituting for β_1, β_2, γ_1 and γ_2 in terms of $\beta_{x,1\to3}$, $\beta_{y,1\to3}$, $\beta_{z,1\to3}$ and $\gamma_{1\to3}$ in Eq. (A6.36), we find that the matrix in Eq. (A6.36) is exactly of the form of the matrix of Eq. (A6.20), i.e. it is a Lorentz transformation without rotation of the axes. This transformation corresponds to the velocity $\mathbf{V}_{1\to3} = c\boldsymbol{\beta}_{1\to3}$, i.e.

$$\mathbf{V}_{1\to3} = V_1\hat{\mathbf{x}} + \frac{V_2}{\gamma_1}\hat{\mathbf{y}}. \tag{A6.39}$$

It is easily verified that this velocity is equal to the relativistic composition of the velocities $V_1\hat{\mathbf{x}}$ and $V_2\hat{\mathbf{y}}$.

Summarizing, we found that

a Lorentz transformation, without rotation of the axes, from the inertial frame of reference S_1 to the inertial frame of reference S_2, which moves with a velocity $\mathbf{V}_1 = V_1\hat{\mathbf{x}}$ relative to S_1, followed by

a Lorentz transformation, without rotation of the axes, from the inertial frame of reference S_2 to the inertial frame of reference S_3, which moves with a velocity $\mathbf{V}_2 = V_2\hat{\mathbf{y}}$ relative to S_3, is equivalent to

a Lorentz transformation, without rotation of the axes, from the inertial frame of reference S_1 to another inertial frame of reference which moves with a velocity $\mathbf{V}_{1\to3} = V_1\hat{\mathbf{x}} + \frac{V_2}{\gamma_1}\hat{\mathbf{y}}$ relative to S_1, followed by

a rotation about the z-axis by an angle equal to

$$\theta = -\arctan\left(\frac{\gamma_1\gamma_2\beta_1\beta_2}{\gamma_1+\gamma_2}\right). \tag{A6.40}$$

This is the Wigner rotation angle.

Appendix 7
Table of Some Functions of the Speed

$\beta = \upsilon/c$, reduced speed $\quad \gamma = \dfrac{1}{\sqrt{1-\beta^2}}$, Lorentz factor

$D = \sqrt{\dfrac{1-\beta}{1+\beta}}, \quad \dfrac{1}{D} = \sqrt{\dfrac{1+\beta}{1-\beta}}$, Doppler factors

$\gamma - 1$ = (Kinetic energy)/(Rest energy) $\quad \dfrac{\gamma-1}{\gamma}$ = (Kinetic energy)/(Total energy)

$\beta\gamma$ = (Momentum)/(Rest mass×c)

In Fig. A7.1 some of these functions are plotted for values of β between 0 and 0.99.

β		$\sqrt{\frac{1-\beta}{1+\beta}}$	$\sqrt{\frac{1+\beta}{1-\beta}}$	$\frac{1}{\beta}$	γ	$\gamma-1$	$\frac{\gamma-1}{\gamma}$	$\beta\gamma$
0.01	1/100	0.99005	1.01005	100	1.00005	0.00005	0.00005	0.01000
0.02	1/50	0.98020	1.02020	50	1.00020	0.00020	0.00020	0.02000
0.05	1/20	0.95119	1.05131	20	1.00125	0.00125	0.00125	0.05006
0.1	1/10	0.90453	1.10554	10	1.00504	0.00504	0.00501	0.10050
0.15		0.85973	1.16316	6.66667	1.01144	0.01144	0.01131	0.15172
0.2	1/5	0.81650	1.22474	5	1.02062	0.02062	0.02020	0.20412
0.25	1/4	0.77460	1.29099	4	1.03280	0.03280	0.03175	0.25820
0.3		0.73380	1.36277	3.33333	1.04828	0.04828	0.04606	0.31449
0.333333	1/3	0.70711	1.41421	3	1.06066	0.06066	0.05719	0.35355
0.4	2/5	0.65465	1.52753	2.5	1.09109	0.09109	0.08348	0.43644
0.5	1/2	0.57735	1.73205	2	1.15470	0.15470	0.13397	0.57735
0.6	3/5	0.5	2	1.66667	1.25000	0.25	0.2	0.75
0.666667	2/3	0.44721	2.23607	1.5	1.34164	0,34164	0.25464	0.89443
0.7		0.42008	2.38048	1.42857	1.40028	0.40028	0.28586	0.98020
0.745356		0.38197	2.61803	1.34164	1.5	0.5	0.33333	1.11803
0.75	3/4	0.37796	2.64575	1.33333	1.51186	0.51186	0.33856	1.13389
0.8	4/5	0.33333	3	1.25	1.66667	0.66667	0.4	1.33333
0.85		0.28475	3.51188	1.17647	1.89832	0.89832	0.47322	1.61357
0.866025		0.26795	3.73204	1.15470	2	1	0.5	1.73205
0.9		0.22942	4.35890	1.11111	2.29416	1.29416	0.56411	2.06474
0.91		0.21707	4.60676	1.09890	2.41192	1.41192	0.58539	2.19484
0.92		0.20412	4.89898	1.08696	2.55155	1.55155	0.60808	2.34743
0.93		0.19045	5.25085	1.07527	2.72065	1.72065	0.63244	2.53020
0.94		0.17586	5.68624	1.06383	2.93105	1.93105	0.65883	2.75519

(continued)

© Springer International Publishing Switzerland 2016

C. Christodoulides, *The Special Theory of Relativity*,
Undergraduate Lecture Notes in Physics, DOI 10.1007/978-3-319-25274-2

(continued)

β	$\sqrt{\frac{1-\beta}{1+\beta}}$	$\sqrt{\frac{1+\beta}{1-\beta}}$	$\frac{1}{\beta}$	γ	$\gamma-1$	$\frac{\gamma-1}{\gamma}$	$\beta\gamma$
0.942809	0.17157	5.82842	1.06066	3	2	0.66667	2.82843
0.95	0.16013	6.24500	1.05263	3.20256	2.20256	0.68775	3.04243
0.96	0.14286	7	1.04167	3.57143	2.57143	0.72	3.42857
0.968246	0.12702	7.87300	1.03280	4	3	0.75	3.87299
0.97	0.12340	8.10350	1.03093	4.11345	3.11345	0.75690	3.99005
0.975	0.11251	8.88819	1.02564	4.50035	3.50035	0.77780	4.38784
0.979796	0.10102	9.89900	1.02062	5	4	0.8	4.89899
0.98	0.10050	9.94987	1.02041	5.02519	4.02519	0.80100	4.92469
0.985	0.08693	11.5036	1.01523	5.79528	4.79528	0.82745	5.70835
0.99	0.07089	14.1067	1.01010	7.08881	6.08881	0.85893	7.01792
0.991	0.06723	14.8735	1.00908	7.47039	6.47039	0.86614	7.40315
0.992	0.06337	15.7797	1.00806	7.92155	6.92155	0.87376	7.85818
0.993	0.05926	16.8735	1.00705	8.46637	7.46637	0.88189	8.40711
0.994	0.05485	18.2300	1.00604	9.14243	8.14243	0.89062	9.08758
0.994987	0.05013	19.9490	1.00504	10	9	0.9	9.94987
0.995	0.05006	19.9750	1.00503	10.01252	9.01252	0.90013	9.962460
0.996	0.04477	22.3383	1.00402	11.19154	10.19154	0.91065	11.14677
0.997	0.03876	25.8005	1.00301	12.91964	11.91964	0.92260	12.88088
0.998	0.03164	31.6070	1.00200	15.81930	14.81930	0.93679	15.78766
0.998749	0.02502	39.9715	1.00125	20	19	0.95	19.97498
0.999	0.02237	44.7102	1.00100	22.36627	21.36627	0.95529	22.34391
0.9991	0.02122	47.1298	1.0009	23.57553	22.57553	0.95758	23.55431
0.9992	0.02	49.9900	1.0008	25.00500	24.00500	0.96001	24.98500
0.9993	0.01871	53.4429	1.0007	26.73080	25.73080	0.96259	26.71209
0.9994	0.01732	57.7264	1.0006	28.87184	27.87184	0.96536	28.85452
0.9995	0.01581	63.2377	1.0005	31.62673	30.62673	0.96838	31.61092
0.9996	0.01414	70.7036	1.0004	35.35888	34.35888	0.97172	35.34473
0.9997	0.01225	81.6435	1.0003	40.82789	39.82789	0.97551	40.81564
0.9998	0.01	99.9950	1.0002	50	49	0.98	49.99250
0.9999	0.00707	141.4178	1.0001	70.71245	69.71245	0.98586	70.70537
0.99995	0.00500	199.9975	1.00005	100	99	0.99	99.99625
0.99999	0.00224	447.2125	1.00001	223.6074	222.6074	0.99553	223.6051
0.999995	0.00158	632.4547	1.000005	316.2282	315.2282	0.99684	316.2266
0.999999	0.00071	1414.210	1.000001	707.1070	706.1070	0.99859	707.1063

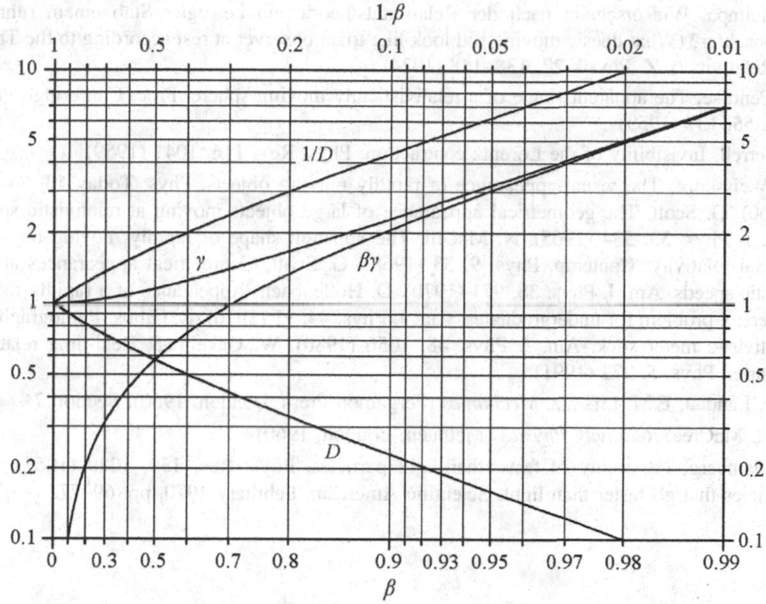

Fig. A7.1 The variation with β, of the quantities $\gamma = \dfrac{1}{\sqrt{1-\beta^2}}$, $\beta\gamma$, $D = \sqrt{\dfrac{1-\beta}{1+\beta}}$ and $\dfrac{1}{D} = \sqrt{\dfrac{1+\beta}{1-\beta}}$

References

1. V.A. Ugarov, *Special Theory of Relativity* (Mir Publishers, Moscow, 1979), ch. 8: "On certain paradoxes of the special theory of relativity". C. Møller, *The Theory of Relativity*, 2nd edn. (Clarendon Press, Oxford, 1972). §8.17, "The clock paradox". P. Langevin, L'évolution de l'espace et du temps. Scientia **10**, 31–54 (1911). W. Rindler, Length contraction paradox. Am. J. Phys. **29**, 365–366 (1961). J.W. Buttler, The Lewis-Tolman lever paradox. Am. J. Phys. **38**, 360–368 (1970). F.W. Sears, Another relativistic paradox. Am. J. Phys. **40**, 771–773 (1972). J. C. Nickerson, Right-angle lever paradox. Am. J. Phys. **43**, 615–621 (1975). D. Gareth Jensen, The paradox of the L-shaped object. Am. J. Phys. **57**, 553–555 (1989). C. Iyer, G.M. Prahbu, Differing observations on the landing of the rod into the slot. Am. J. Phys. **74**, 998–1001 (2006). E. Pierce, The lock and key paradox and the limits of rigidity in special relativity. Am. J. Phys. **75**, 610–614 (2007). D.V. Redžić, Note on Dewan-Beran-Bell's spaceship problem. Eur. J. Phys. **29**, N11–N19 (2008). J. Franklin, Lorentz contraction, Bell's spaceships and rigid body motion in special relativity. Eur. J. Phys. **31**, 291–298 (2010). Y.-Q. Gu, Some paradoxes in special relativity and their resolutions. Adv. Appl. Clifford Algebras **21**, 103–119 (2010)
2. A. Einstein, Ann. Physik **17**, 891 (1905). Translated into English in A. Einstein, H.A. Lorentz, H. Weyl, H. Minkowski, *The Principle of Relativity* (Dover Publications, New York, 1952)

3. A. Lampa, Wie erscheint nach der Relativitatstheorie ein bewegter Stab einem ruhrnden Beobachter? (What does a moving rod look like to an observer at rest according to the Theory of Relativity?). Z. Physik **27**, 138–148 (1924)
4. R. Penrose, The apparent shape of a relativistically moving sphere. Proc. Cambridge Philos. Soc. **55**, 137 (1959)
5. J. Terrell, Invisibility of the Lorentz contraction. Phys. Rev. **116**, 1041 (1959)
6. V. Weisskopf, The visual appearance of rapidly moving objects. Phys. Today **13**, 24 Sept. (1960). G. Scott, The geometrical appearance of large objects moving at relativistic speeds. Am. J. Phys. **33**, 534 (1965). N. McGill, The apparent shape of rapidly moving objects in special relativity. Contemp. Phys. **9**, 33 (1968). G. Scott, Geometrical appearances at relativistic speeds. Am. J. Phys. **38**, 971 (1970). D. Hollenbach, Appearance of a rapidly moving sphere: a problem for undergraduates. Am. J. Phys. **44**, 91 (1976). R. Gibbs, Photographing a relativistic meter stick. Am. J. Phys. **48**, 1056 (1980). W. Gekelman, Real-time relativity. Comput. Phys. **5**, 372 (1991)
7. L.D. Landau, E.M. Lifshitz, *Mechanics* (Pergamon Press, London, 1960). Section 7
8. W.H. McCrea, *Relativity Physics* (Methuen, London, 1960)
9. G. Feinberg, Possibility of faster-than-light particles. Phys. Rev. **159**, 1089 (1967). Also, Particles that go faster than light. Scientific American, February 1970, pp. 69–77.

Solutions to the Problems

Chapter 3
Relativistic Kinematics

3.1 Show that for an event (x, y, z, t), the quantity $s^2 = c^2t^2 - x^2 - y^2 - z^2$ is invariant under the special Lorentz transformation. Also show that s^2 is not invariant under the Galilean transformation.

Solution

In the frame of reference S, it is

$$s^2 = c^2t^2 - x^2 - y^2 - z^2. \tag{1}$$

In the frame of reference S′, it is

$$s'^2 = c^2t'^2 - x'^2 - y'^2 - z'^2. \tag{2}$$

The special Lorentz transformation gives: $x' = \gamma(x - Vt)$, $y' = y$, $z' = z$, $t' = \gamma\left(t - \frac{V}{c^2}x\right)$.

Substituting in Eq. (2) we have:

$$s'^2 = c^2\gamma^2\left(t - \frac{V}{c^2}x\right)^2 - \gamma^2(x - Vt)^2 - y^2 - z^2 =$$

$$= \gamma^2\left(c^2t^2 - 2Vxt + \frac{V^2}{c^2}x^2 - x^2 + 2Vxt - V^2t^2\right) - y^2 - z^2$$

$$s'^2 = \gamma^2(c^2 - V^2)t^2 - \gamma^2\left(1 - \frac{V^2}{c^2}\right)x^2 - y^2 - z^2 = c^2t^2 - x^2 - y^2 - z^2 = s^2.$$

It has been shown that $s'^2 = s^2$ and that the magnitude $s^2 = c^2t^2 - x^2 - y^2 - z^2$ is invariant under the special Lorentz transformation.

© Springer International Publishing Switzerland 2016

C. Christodoulides, *The Special Theory of Relativity*,
Undergraduate Lecture Notes in Physics, DOI 10.1007/978-3-319-25274-2

The Galilean transformation is $x' = x - Vt, \quad y' = y, \quad z' = z, \quad t' = t$. Substituting in Eq. (2) we have

$$\begin{aligned} s'^2 &= t^2 - (x - Vt)^2 - y^2 - z^2 = t^2 - x^2 - 2Vtx + V^2t^2 - y^2 - z^2 \\ &= s^2 - 2Vtx + V^2t^2, \end{aligned}$$

which shows that $s^2 = c^2t^2 - x^2 - y^2 - z^2$ is not invariant under the Galilean transformation.

3.2 Show that if in an inertial frame of reference S it is $x = ct$ (light pulse), then in every other inertial frame of reference S′ it will be $x' = ct'$.

Solution

From the special Lorentz transformation,

$$x' = \frac{x - Vt}{\sqrt{1-\beta^2}} = \frac{ct - Vt}{\sqrt{1-\beta^2}} = ct\frac{1-\beta}{\sqrt{1-\beta^2}} = ct\sqrt{\frac{1-\beta}{1+\beta}}$$

and

$$t' = \frac{t - (V/c^2)x}{\sqrt{1-\beta^2}} = \frac{t - (V/c^2)ct}{\sqrt{1-\beta^2}} = t\frac{1-\beta}{\sqrt{1-\beta^2}} = t\sqrt{\frac{1-\beta}{1+\beta}}.$$

Therefore, $x' = ct'$.

3.3 In an inertial frame of reference S, two events are separated by a distance of $\Delta x = 600$ m and an interval of time $\Delta t = 0.8$ μs. There exists a frame of reference S′, which moves with a constant velocity $\mathbf{V} = V\hat{\mathbf{x}}$ with respect to S, in which these two events happen at the same moment. Find the value of V. What is the spatial distance of the two events in S′?

Solution

The two events, 1 and 2, are observed in the frame of reference S′ to happen at times

$$t_1' = \gamma[t_1 - (\beta/c)x_1] \quad \text{and} \quad t_2' = \gamma[t_2 - (\beta/c)x_2].$$

Subtracting

$$t_2' - t_1' = \gamma[(t_2 - t_1) - (\beta/c)(x_2 - x_1)].$$

Substituting in this $t_2' - t_1' = 0$, $t_2 - t_1 = 8 \times 10^{-7}$ s and $(x_2 - x_1)/c = 600 \text{ m}/3 \times 10^8 \text{ m/s} = 2 \times 10^{-6}$ s, we have $\beta = \dfrac{(t_2 - t_1)c}{x_2 - x_1} = \dfrac{8 \times 10^{-7}}{2 \times 10^{-6}} = \dfrac{2}{5} = 0.4$. The Lorentz factor for this speed is $\gamma = \frac{5}{\sqrt{21}}$.

Because $x'_1 = \gamma(x_1 - \beta c t_1)$ and $x'_2 = \gamma(x_2 - \beta c t_2)$, it follows that

$$x'_2 - x'_1 = \gamma[(x_2 - x_1) - \beta c(t_2 - t_1)] = \frac{5}{\sqrt{21}}\left(600 - \frac{2}{5} \times 3 \times 10^8 \times 8 \times 10^{-7}\right)$$
$$= 550 \text{ m}.$$

3.4 Two events happen at the same point in an inertial frame of reference S. Show that the temporal sequence in which the two events happen is the same in all inertial frames of reference. Also show that the time interval between them is minimum in frame S.

Solution

Let the coordinates of the two events in the frame of reference S be:

$$\text{Event 1}: \quad x_1 = x, \quad y_1 = y, \quad z_1 = z, \quad t_1.$$
$$\text{Event 2}: \quad x_2 = x, \quad y_2 = y, \quad z_2 = z, \quad t_2.$$

Without loss of generality, we will assume that $t_2 > t_1$, i.e. that event 1 happens before event 2 in frame S. The corresponding coordinates in another frame of reference S′ are:

$$\text{Event 1}: \quad x'_1 = \gamma(x_1 - Vt_1) = \gamma(x - Vt_1), \quad t'_1 = \gamma\left(t_1 - \frac{V}{c^2}x_1\right) = \gamma\left(t_1 - \frac{V}{c^2}x\right).$$

$$\text{Event 2}: \quad x'_2 = \gamma(x_2 - Vt_2) = \gamma(x - Vt_2), \quad t'_2 = \gamma\left(t_2 - \frac{V}{c^2}x_2\right) = \gamma\left(t_2 - \frac{V}{c^2}x\right).$$

The difference between the two time values for the two events is $t'_2 - t'_1 = \gamma(t_2 - t_1)$.

Because γ is always positive, if it is $t_2 - t_1 > 0$, then $t'_2 - t'_1 > 0$ and the temporal sequence in which the two events happen is the same in all inertial frames of reference.

Also, it is $\gamma \geq 1$ and therefore $|t'_2 - t'_1| \geq |t_2 - t_1|$. This means that the time interval between the two events is minimum for $\gamma = 1$, i.e. in frame S, in which the two events happen at the same point.

3.5 Define the parameter ϕ, usually called *rapidity*, according to the relation $\tanh\phi = \beta$. Show that

$$\sinh\phi = \beta\gamma, \qquad \cosh\phi = \gamma, \qquad e^{\phi} = \sqrt{\frac{1+\beta}{1-\beta}}$$

and that the special Lorentz transformation for x and t may be written as

$$x' = x\cosh\phi - ct\sinh\phi, \qquad ct' = -x\sinh\phi + ct\cosh\phi.$$

Also show that

$$ct' + x' = e^{-\phi}(ct + x) \quad \text{and} \quad ct' - x' = e^{\phi}(ct - x).$$

Solution

It is $\sinh^2\phi = \cosh^2\phi - 1 = \dfrac{\sinh^2\phi}{\tanh^2\phi} - 1$ and $\sinh^2\phi\left(\dfrac{1}{\tanh^2\phi} - 1\right) = 1$.

Substituting, $\sinh\phi = \left(\dfrac{1}{\tanh^2\phi} - 1\right)^{-1/2} = \left(\dfrac{1}{\beta^2} - 1\right)^{-1/2} = \dfrac{\beta}{\sqrt{1-\beta^2}}$ or $\sinh\phi = \beta\gamma$.

It follows that $\cosh\phi = \sqrt{1 + \sinh^2\phi} = \sqrt{1 + \beta^2\gamma^2} = \sqrt{1 + \dfrac{\beta^2}{1-\beta^2}} = \dfrac{1}{\sqrt{1-\beta^2}}$

or $\cosh\phi = \gamma$.

The special Lorentz transformation is

$$x' = \gamma\,(x - \beta ct) = \gamma x - \beta\gamma ct \quad \text{and} \quad ct' = \gamma\,(ct - \beta x) = \gamma ct - \beta\gamma x.$$

Putting $\gamma = \cosh\phi$, $\beta\gamma = \sinh\phi$, we have

$$x' = x\cosh\phi - ct\sinh\phi \;\; ct' = -x\sinh\phi + ct\cosh\phi.$$

From the relation $\tanh\phi = \beta$, it follows that $\dfrac{e^{\phi} - e^{-\phi}}{e^{\phi} + e^{-\phi}} = \beta$, from which we get $\dfrac{e^{2\phi} - 1}{e^{2\phi} + 1} = \beta$.

Therefore, $e^{2\phi} = \dfrac{1+\beta}{1-\beta}$ and $e^{\phi} = \sqrt{\dfrac{1+\beta}{1-\beta}}$. Using these, we have

$$\begin{aligned} ct' + x' &= \gamma ct - \beta\gamma x + \gamma x - \beta\gamma ct = \gamma(1-\beta)(ct + x) = \sqrt{\frac{1-\beta}{1+\beta}}(ct + x) \\ &= e^{-\phi}(ct + x) \end{aligned}$$

and, similarly,

$$\begin{aligned} ct' - x' &= \gamma ct - \beta\gamma x - \gamma x + \beta\gamma ct = \gamma(1+\beta)(ct - x) = \sqrt{\frac{1+\beta}{1-\beta}}(ct - x) \\ &= e^{\phi}(ct - x). \end{aligned}$$

3.6 Show that, as the speed υ approaches that of light in vacuum, c, the Lorentz factor corresponding to υ can be written, in S.I. units, as $\gamma \approx 12\,243/\sqrt{c - \upsilon}$, approximately.

Solution

From $\gamma = \dfrac{1}{\sqrt{1-\beta^2}} = \dfrac{c}{\sqrt{(c+\upsilon)(c-\upsilon)}}$, it is for $\upsilon \approx c$

$\gamma \approx \dfrac{c}{\sqrt{2c(c-\upsilon)}} = \dfrac{\sqrt{c/2}}{\sqrt{c-\upsilon}} = \dfrac{12\,243}{\sqrt{c-\upsilon}}$ in S.I. units.

This equation can be solved to give $c - \upsilon \approx \left(\frac{12243}{\gamma}\right)^2$ m/s $= \frac{150\,000}{\gamma^2}$ km/s.

As an example, for $\gamma = 100$, υ is smaller than c by only 15 km/s.

3.7 Show that in the frames of reference S and S′, in relative motion with respect to each other, there exists a plane on which the clocks of the two frames show the same time. Find the positions of the plane in the two frames.

Solution

Putting $t' = t$ in $t' = \gamma\,[t - (\beta/c)x]$, we have $t = \gamma\,[t - (\beta/c)x]$ or $x = \frac{c}{\beta}\left(1 - \frac{1}{\gamma}\right)t$.

This equation gives the position of the plane in frame S as a function of time. It is seen that the plane moves towards positive x's with a speed of $\frac{c}{\beta}\left(1 - \frac{1}{\gamma}\right)$.

Starting from $t = \gamma\,[t' + (\beta/c)x']$ and putting $t = t'$ we get $x' = -\frac{c}{\beta}\left(1 - \frac{1}{\gamma}\right)t'$,

which is a plane moving with speed $-\frac{c}{\beta}\left(1 - \frac{1}{\gamma}\right)$ towards negative x's.

The equal and opposite speeds are expected due to the symmetry of the two frames of reference.

3.8 A rectilinear rod is parallel to the x'-axis in the inertial frame of reference S′, which moves with a velocity $\mathbf{V} = V\hat{\mathbf{x}}$ relative to another frame of reference, S. In S′, the rod moves in the direction of positive y''s, with constant speed υ. The rest length of the rod is L_0 and its center remains on the y'-axis. Find the position of the rod in frame S, as a function of time.

Solution

Let the rod coincide with the x'-axis in the inertial frame of reference S′ at the moment $t' = 0$. To find the position of the rod as a function of time in frame S, we will examine the position of one of its points, A, which in frame S′ has x' coordinate equal to

$$x'_A = \alpha L_0/2 \quad (-1 \le \alpha \le 1).$$

Its position in S′ at time t' is (see figure below): $x'_A = \alpha L_0/2$, $y'_A = \upsilon t'$, $z'_A = 0$.

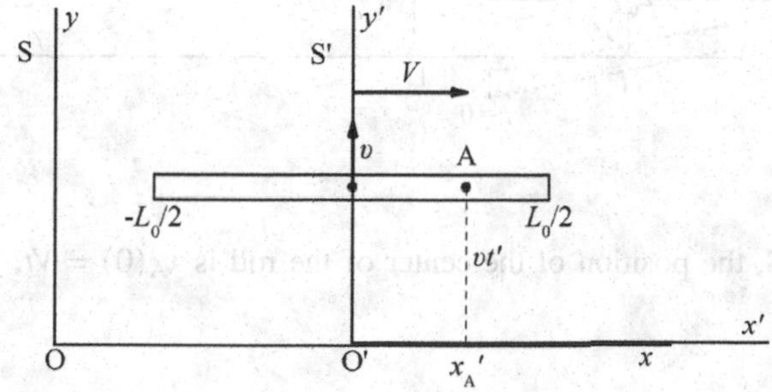

In frame S, it is

$$x_{\rm A} = \gamma(x'_{\rm A} + Vt') = \gamma\left[\frac{1}{2}\alpha L_0 + V\gamma\left(t - \frac{V}{c^2}x_{\rm A}\right)\right] = \frac{1}{2}\gamma\alpha L_0 + V\gamma^2 t - \frac{V^2}{c^2}\gamma^2 x_{\rm A},$$

which finally gives $x_{\rm A}(\alpha) = \dfrac{\alpha L_0}{2\gamma} + Vt$.

Also,

$$y_{\rm A} = y'_{\rm A} = \upsilon t' = \upsilon\gamma\left(t - \frac{V}{c^2}x_{\rm A}\right) = \upsilon\gamma\left[t - \frac{V}{c^2}\left(\frac{\alpha L_0}{2\gamma} + Vt\right)\right]$$

$$= \upsilon\gamma t - \frac{\upsilon V}{2c^2}\alpha L_0 - \upsilon\gamma\frac{V^2}{c^2}t$$

from which it follows that $y_{\rm A}(\alpha) = -\dfrac{\upsilon V}{2c^2}\alpha L_0 + \dfrac{\upsilon t}{\gamma}$.

The relationship between $y_{\rm A}$ and $x_{\rm A}$ has already been found. Without indices,

$$y = \upsilon\gamma t - \frac{\upsilon V}{c^2}\gamma x.$$

For given t, this is the equation of a straight line with a slope $\tan\theta = -\left(\dfrac{\upsilon V}{c^2}\gamma\right)$.

The rod coincides with this straight line in frame S and, therefore, forms with the x-axis an angle

$$\theta = -\arctan\left(\frac{\upsilon V}{c^2}\gamma\right),$$

as shown in the figure below.

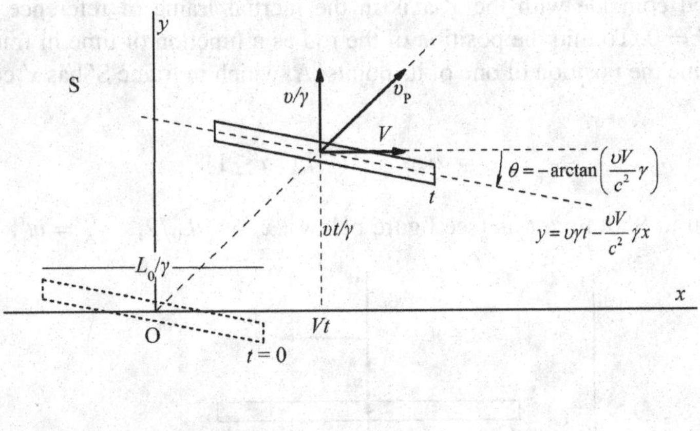

In frame S, the position of the center of the rod is $x_{\rm A}(0) = Vt$, $y_{\rm A}(0) = \dfrac{\upsilon t}{\gamma}$.

One end of the rod (for $\alpha = -1$) is found, as a function of time, at the position

$$x_A(-1) = -\frac{L_0}{2\gamma} + Vt, \qquad y_A(-1) = \frac{\upsilon V}{2c^2}L_0 + \frac{\upsilon t}{\gamma}.$$

The other end (for $\alpha = 1$) is found at the position

$$x_A(1) = \frac{L_0}{2\gamma} + Vt, \qquad y_A(1) = -\frac{\upsilon V}{2c^2}L_0 + \frac{\upsilon t}{\gamma}.$$

As expected due to the contraction of length in the x direction, it is $x_A(1) - x_A(-1) = L_0/\gamma$. The total length of the rod in the frame of reference S is equal to

$$L = \sqrt{[x_A(1) - x_A(-1)]^2 + [y_A(1) - y_A(-1)]^2} = \sqrt{(L_0/\gamma)^2 + (-\upsilon V L_0/c^2)^2}$$
$$= L_0\sqrt{1 - \frac{V^2}{c^2}\left(1 - \frac{\upsilon^2}{c^2}\right)}.$$

We note that it is: $L_0/\gamma \le L \le L_0$.

The velocity of any point of the rod may be found using the equations $x_A = \frac{\alpha L_0}{2\gamma} + Vt$ and $y_A = -\frac{\upsilon V}{2c^2}\alpha L_0 + \frac{\upsilon t}{\gamma}$. Its components are $\upsilon_{Ax} = \frac{dx_A}{dt} = V$ and $\upsilon_{Ay} = \frac{dy_A}{dt} = \frac{\upsilon}{\gamma}$. The speed of any point of the rod is

$$\upsilon_A = \sqrt{\upsilon_{Ax}^2 + \upsilon_{Ay}^2} = \sqrt{V^2 + \frac{\upsilon^2}{\gamma^2}} = \sqrt{V^2 + \upsilon^2\left(1 - \frac{V^2}{c^2}\right)} = \sqrt{V^2 + \upsilon^2 - \frac{\upsilon^2 V^2}{c^2}}.$$

3.9 An airplane has rest length equal to $L_0 = 10$ m. It moves, relative to an observer on the ground, with a speed equal to that of sound in air, $V = 343$ m/s. Calculate the contraction in the length of the airplane that the observer will measure.

Solution

For $V = 343$ m/s, it is $\beta = 343/(3 \times 10^8) = 1.14 \times 10^{-6}$ and

$$\gamma = \frac{1}{\sqrt{1 - (1.14 \times 10^{-6})^2}} \approx 1 + \frac{1}{2}(1.14 \times 10^{-6})^2 = 1 + 6.5 \times 10^{-13}$$
$$= 1.000\,000\,000\,000\,65.$$

The observer will find that the airplane has a length $L = L_0/\gamma$. The contraction he will measure in the length of the airplane will, therefore, be

$$\Delta L = L_0 - L = L_0(1 - 1/\gamma) = 10\left[1 - 1/\left(1 + 6.5 \times 10^{-13}\right)\right]$$
$$\approx 10\left[1 - \left(1 - 6.5 \times 10^{-13}\right)\right] = 6.5 \times 10^{-12}\ \text{m}.$$

This means that the contraction is of the order of 1/100 of the atomic diameters!

3.10 At what speed does the length of a body contract to 99 % of its rest length?

Solution

It is $L/L_0 = \sqrt{1 - (\upsilon/c)^2} = 0.99$. Therefore, $1 - (\upsilon/c)^2 = (1 - 0.01)^2$ or $(\upsilon/c)^2 = 0.02 - 0.0001$, $\upsilon = 0.141c \approx c/7$.

3.11 An observer S is stationary at the middle of a straight line AB, which has a length that the observer measures to be equal to $2L$. Another observer, S′, moves along the line with a speed of $V = \frac{3}{5}c$ relative to S. Both observers are at the origin of their respective frame of reference and coincide when their clocks show $t = t' = 0$.

(a) What is the length of AB as measured by S′?
(b) In frame S, at the moment $t = 0$ two pulses are emitted simultaneously from the points A and B. Find, in the frame of reference of S′, the positions at which the pulses are emitted as well as the time at which each pulse is emitted.
(c) At which values of time, T_A and T_B, will the pulses arrive at S, and at which (T'_A and T'_B) will they arrive at S′?

Solution

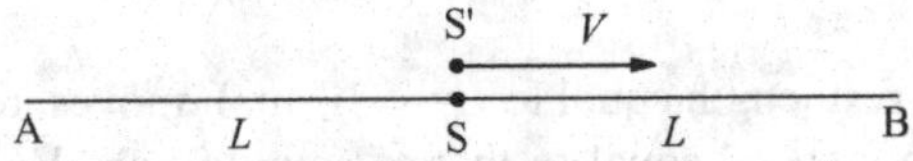

(a) Observer S′ sees the line AB moving with a speed of $V = \frac{3}{5}c$, to which there corresponds a Lorentz factor of $\gamma = 5/4$. He will, therefore, measure the length of the line as equal to $2L/\gamma = 8L/5 = 1.6L$.
(b) The two pulses emitted in frame S are defined as the following two events:

Event A: Emission of pulse A, with coordinates $x_A = -L$, $t_A = 0$.
Event B: Emission of pulse B, with coordinates $x_B = L$, $t_B = 0$.

The coordinates of the two events in frame S′ are:

Event A: Emission of pulse A, with coordinates

$$x'_{\rm A} = \gamma\,(x_{\rm A} - V t_{\rm A}) = \frac{5}{4}\left(-L - \frac{3}{5}c \times 0\right) = -\frac{5}{4}L \quad \text{and}$$

$$t'_{\rm A} = \gamma\left(t_{\rm A} - \frac{V}{c^2}x_{\rm A}\right) = \frac{5}{4}\left(0 - \frac{3}{5c}(-L)\right) = \frac{3}{4}\frac{L}{c}.$$

Event B: Emission of pulse B, with coordinates

$$x'_{\rm B} = \gamma\,(x_{\rm B} - V t_{\rm B}) = \frac{5}{4}\left(L - \frac{3}{5}c \times 0\right) = \frac{5}{4}L \quad \text{and}$$

$$t'_{\rm B} = \gamma\left(t_{\rm B} - \frac{V}{c^2}x_{\rm B}\right) = \frac{5}{4}\left(0 - \frac{3}{5c}(L)\right) = -\frac{3}{4}\frac{L}{c}.$$

(c) The time, T_i, at which a pulse reaches an observer is found by adding the time at which the emission of the pulse occurred in the observer's frame of reference to the time it takes for the pulse to travel from its point of emission to the observer: $T_i = t_i + |x_i|/c$ and $T'_i = t'_i + |x'_i|/c$. It follows that

$$T_{\rm A} = t_{\rm A} + \frac{|x_{\rm A}|}{c} = 0 + \frac{|-L|}{c} = \frac{L}{c}, \quad T_{\rm B} = t_{\rm B} + \frac{|x_{\rm B}|}{c} = 0 + \frac{|L|}{c} = \frac{L}{c},$$

$$T'_{\rm A} = t'_{\rm A} + \frac{|x'_{\rm A}|}{c} = \frac{3}{4}\frac{L}{c} + \frac{5}{4}\frac{L}{c} = 2\frac{L}{c}, \quad T'_{\rm B} = t'_{\rm B} + \frac{|x'_{\rm B}|}{c} = -\frac{3}{4}\frac{L}{c} + \frac{5}{4}\frac{L}{c} = \frac{1}{2}\frac{L}{c}.$$

For observer S′, pulse A was emitted after pulse B. Compare with the case of Einstein's train, Sects. 2.3 and 4.6.

3.12 A rod, having a rest length of L_0, lies along the x'-axis of the inertial frame of reference S′ and moves with a constant velocity $\mathbf{V} = V\hat{\mathbf{x}}$ with respect to another frame of reference, S. An observer in frame S notes the times $t_{\rm A}$ and $t_{\rm B}$ at which the two ends of the rod pass in front of him and estimates the length of the rod as $(t_{\rm B} - t_{\rm A})V$. Show that the phenomenon of length contraction is exhibited in his result.

Solution

Let the observer be at point $x = 0$ in the frame of reference S. He observes the coordinates of place and time for the two events:

Event A: the leading end of the rod passes the point $x = 0$: $x_{\rm A} = 0$, $t_{\rm A}$.
Event B: the back end of the rod passes the point $x = 0$: $x_{\rm B} = 0$, $t_{\rm B}$.

He concludes that the length of the rod is equal to $L = (t_{\rm B} - t_{\rm A})V$.
Transforming, we find the times at which the two events are observed in S′:

$$\text{Event A:} \quad t'_{\rm A} = \gamma\left(t_{\rm A} + \frac{V}{c^2}x_{\rm A}\right) = \gamma t_{\rm A}$$

$$\text{Event B:} \quad t'_{\rm B} = \gamma\left(t_{\rm B} + \frac{V}{c^2}x_{\rm B}\right) = \gamma t_{\rm B}$$

The observer in the frame of reference S′, in which the rod is at rest, sees the other observer moving with a speed $-V$ and passing near the one end of the rod at time $t'_A = \gamma t_A$ and the other at time $t'_B = \gamma t_B$. This observer will estimate the (proper) length of the rod to be

$$L_0 = V(t'_B - t'_A) = \gamma V(t_B - t_A) = \gamma L.$$

Therefore, $L = L_0/\gamma$ and observer S finds a contraction in the length of the rod.

3.13 *The transformation of an angle*. A straight line lies in the xy plane and forms an angle θ with the x-axis, in the inertial frame of reference S. What is the angle θ' formed by the line with the x'-axis, in the frame of reference S′ which moves with a constant velocity $\mathbf{V} = V\hat{\mathbf{x}}$ relative to the frame S′?

Solution

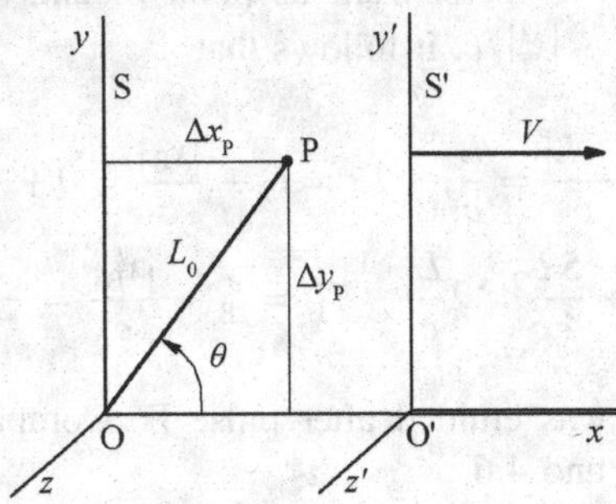

The projections of the straight line OP on the two axes of the frame of reference S have lengths equal to

$$\Delta x_P = L_0 \cos\theta \;\text{ and }\; \Delta y_P = L_0 \sin\theta.$$

In the frame of reference S′, these lengths will be

$$\Delta x'_P = \Delta x_P/\gamma = (L_0 \cos\theta)/\gamma$$

and

$$\Delta y'_P = \Delta y_P = L_0 \sin\theta.$$

Therefore, the slope of the straight line in the frame of reference S′ is given by

$$\tan\theta' = \frac{\Delta y'_P}{\Delta x'_P} = \frac{L_0 \sin\theta}{(L_0 \cos\theta)/\gamma}.$$

It follows that $\tan\theta' = \gamma \tan\theta$ or $\theta' = \arctan(\gamma \tan\theta)$.

3.14 A straight rod moves in an inertial frame of reference S with a constant velocity $\mathbf{v} = v\hat{\mathbf{x}}$, and has a slope λ with respect to the x-axis. Another frame of reference, S′, moves with a constant velocity $\mathbf{V} = V\hat{\mathbf{x}}$ in S. Show, by two methods, that, in frame S′, the slope of the rod with respect to the x'-axis is $\lambda\gamma(1 - Vv/c^2)$.

Solution

First Method

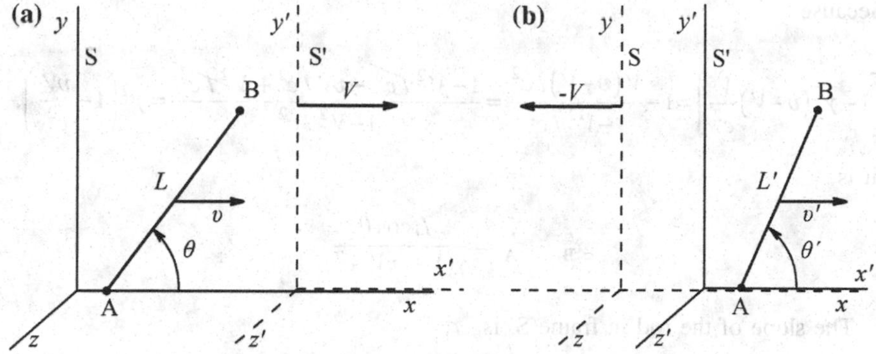

We assume that at time $t = t' = 0$ the end A of the rod is situated at the origins of the axes of the two frames of reference, S and S′, which coincide. In general, the coordinates of the two ends of the rod, in frame S, are:

$$x_A = vt_A, \qquad y_A = 0, \qquad z_A = 0, \qquad t_A$$
$$x_B = vt_B + L\cos\theta, \qquad y_B = L\sin\theta, \qquad z_B = 0, \qquad t_B.$$

The corresponding values in frame S′ are

$$x'_A = \gamma(x_A - Vt_A), \quad y'_A = y_A = 0 \quad \text{and}$$

$$x'_B = \gamma(x_B - Vt_B), \quad y'_B = y_B = L\sin\theta.$$

However, it is $t_A = \gamma\left(t'_A + \frac{V}{c^2}x'_A\right)$ and $t_B = \gamma\left(t'_B + \frac{V}{c^2}x'_B\right)$ and, hence,
$x'_B - x'_A = \gamma(x_B - x_A) - \gamma V(t_B - t_A) = \gamma[v(t_B - t_A) + L\cos\theta] - \gamma V(t_B - t_A)$

or,

$$\begin{aligned} x'_B - x'_A &= \gamma(v - V)(t_B - t_A) + \gamma L\cos\theta \\ &= \gamma(v - V)\left[\gamma(t'_B - t'_A) + \gamma\frac{V}{c^2}(x'_B - x'_A)\right] + \gamma L\cos\theta. \end{aligned}$$

We are interested in the positions of the two ends of the rod at the same time in frame S′, and so it is $t'_B = t'_A$. Thus,

$$x'_B - x'_A = \gamma^2 (\upsilon - V)\frac{V}{c^2}(x'_B - x'_A) + \gamma L\cos\theta$$

$$(x'_B - x'_A)\left[1 - \gamma^2 (\upsilon - V)\frac{V}{c^2}\right] = \gamma L\cos\theta.$$

Because

$$\left[1-\gamma^2(\upsilon-V)\frac{V}{c^2}\right]=1-\frac{V(\upsilon-V)/c^2}{1-V^2/c^2}=\frac{1-\cancel{V^2/c^2}-\upsilon V/c^2+\cancel{V^2/c^2}}{1-V^2/c^2}=\gamma^2\left(1-\frac{\upsilon V}{c^2}\right),$$

it is

$$x'_B - x'_A = \frac{L\cos\theta}{\gamma(1-\upsilon V/c^2)}.$$

The slope of the rod in frame S′ is

$$\tan\theta' = \frac{y'_B - y'_A}{x'_B - x'_A} = \frac{\gamma(1-\upsilon V/c^2)L\sin\theta}{L\cos\theta}$$

and

$$\tan\theta' = \gamma(1-\upsilon V/c^2)\tan\theta \qquad \text{or} \qquad \lambda' = \lambda\gamma(1-\upsilon V/c^2).$$

Second Method

In this method, we first transform the positions of the ends of the rod into the rod's frame of reference, S_0 [see figures (c) and (d), below]. Since the rod moves with a velocity $\boldsymbol{\upsilon} = \upsilon\hat{\mathbf{x}}$ with respect to frame of reference S, this will be the velocity of S_0 with respect to S. We define $\gamma = \sqrt{1-(V/c)^2}$ and $\gamma_0 = \sqrt{1-(\upsilon/c)^2}$. Then,

$$x_{A0} = \gamma_0 (x_A - \upsilon t_A) = 0, \quad x_{B0} = \gamma_0 (x_B - \upsilon t_B) = \gamma_0 L\cos\theta = L_0\cos\theta_0.$$

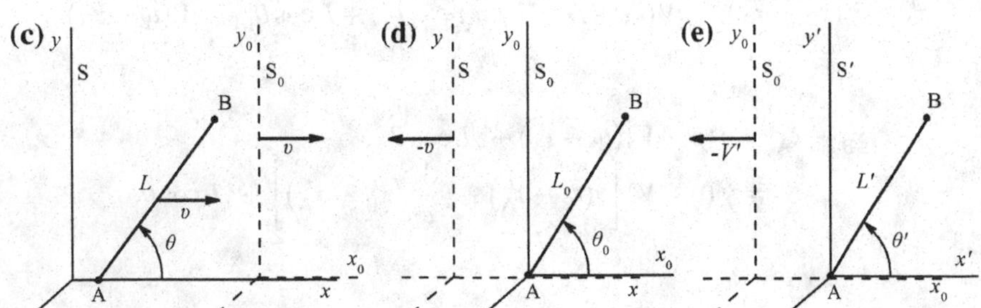

In frame S, the speed of frame S′ is V. In frame S_0, the speed of S′ is

$$V' = \frac{V - \upsilon}{1 - \upsilon V/c^2}.$$

To this there corresponds a Lorentz factor

$$\gamma' = \frac{1}{\sqrt{1 - \dfrac{(V-\upsilon)^2}{(1-\upsilon V/c^2)^2}}} = \frac{1 - \dfrac{\upsilon V}{c^2}}{\sqrt{1 + \dfrac{\upsilon^2 V^2}{c^4} - 2\dfrac{\upsilon V}{c^2} - \dfrac{\upsilon^2}{c^2} + 2\dfrac{\upsilon V}{c^2} - \dfrac{V^2}{c^2}}}.$$

This gives $\gamma' = \dfrac{1 - \dfrac{\upsilon V}{c^2}}{\sqrt{\left(1 - \dfrac{\upsilon^2}{c^2}\right) - \dfrac{V^2}{c^2}\left(1 - \dfrac{\upsilon^2}{c^2}\right)}}$ and, finally, $\gamma' = \gamma\gamma_0\left(1 - \dfrac{\upsilon V}{c^2}\right)$.

Transforming from frame S_0 to frame S′, we have

$$x'_A = \gamma'(x_{A0} - V' t_{A0}), \quad x'_B = \gamma'(x_{B0} - V' t_{B0}), \quad y'_A = y_{A0} = 0, \quad y'_B = y_{B0} = L \sin\theta$$
$$t'_A = \gamma'\left(t_{A0} - (V'/c^2)x_{A0}\right), \quad t'_B = \gamma'\left(t_{B0} - (V'/c^2)x_{B0}\right).$$

We may now find

$$\tan\theta' = \frac{y'_B - y'_A}{x'_B - x'_A} = \frac{L\sin\theta}{\gamma'[(x_{B0} - x_{A0}) - V'(t_{B0} - t_{A0})]}$$

The measurements of the positions of A and B must be performed simultaneously in frame S′ and so it must be $t'_A = t'_B$ and, therefore,

$$t_{A0} - \frac{V'}{c^2}x_{A0} = t_{B0} - \frac{V'}{c^2}x_{B0} \quad \text{or} \quad t_{B0} - t_{A0} = \frac{V'}{c^2}(x_{B0} - x_{A0}).$$

and

$$\tan\theta' = \frac{L\sin\theta}{\gamma'(x_{B0} - x_{A0})(1 - V'^2/c^2)} = \gamma' \frac{L\sin\theta}{\gamma_0 L\cos\theta} = \frac{\gamma'}{\gamma_0}\tan\theta.$$

Using $\gamma' = \gamma\gamma_0(1 - \upsilon V/c^2)$, we finally find $\tan\theta' = \gamma(1 - \upsilon V/c^2)\tan\theta$ or $\lambda' = \lambda\gamma(1 - \upsilon V/c^2)$, as above.

3.15 With what constant speed must a spaceship be moving, relative to the Earth, if it is to travel across the Galaxy in 40 years, as this time is measured inside the spaceship? In the Earth's frame of reference, which is considered to be inertial, the diameter of the Galaxy is $D = 10^5$ light years. What will the diameter of the Galaxy appear to be to an observer inside the spaceship? How much time, t_a, would be required, in the Earth's frame of reference, for the spaceship to reach this speed, if it could move at a constant acceleration of $g = 9.81\ \text{m/s}^2$?

Solution

Since D is the diameter of the Galaxy, then, according to an observer on Earth, the time needed for the spaceship to cross the Galaxy will be: $\Delta t = D/V$.

This time will be equal to $\Delta t' = \Delta t/\gamma$ for the spaceship, where γ is the Lorentz factor corresponding to the speed of the spaceship relative to the Earth. Therefore,

$$\gamma = \frac{D}{V\Delta t'} \quad \Rightarrow \quad \beta\gamma = \frac{D/c}{\Delta t'} \quad \Rightarrow \quad \frac{\beta^2}{1-\beta^2} = \left(\frac{D/c}{\Delta t'}\right)^2 \quad \Rightarrow \quad \beta = \sqrt{\frac{\left(\frac{D/c}{\Delta t'}\right)^2}{1+\left(\frac{D/c}{\Delta t'}\right)^2}}$$

However, $\dfrac{D/c}{\Delta t'} = \dfrac{100\,000}{40} = 2500$ and it follows that

$$\beta = \sqrt{\frac{2500^2}{1+2500^2}} \approx 1 - \frac{1}{2 \times 2500^2} = 1 - 8 \times 10^{-8} \quad \text{and} \quad \gamma = 2501.$$

The diameter of the Galaxy for an observer inside the spaceship will be

$$D' = \frac{D}{\gamma} = V\Delta t' = \frac{V}{c}(c\Delta t') = \beta(c\Delta t') \approx 1 \times (c\Delta t') \approx 40 \text{ light years.}$$

The speed that must be reached by the spaceship differs very little from the speed of light. Therefore, the time needed for the spaceship to achieve this speed is

$$t_a = \frac{V}{g} \approx \frac{c}{g} = \frac{3 \times 10^8}{9.81} = 3.06 \times 10^7\ \text{s} \approx 1 \text{ year.}$$

Of course, there are very serious reasons why this cannot happen!

3.16 Two events happen at the same point in the inertial frame of reference S′, which is moving with a constant velocity $\mathbf{V} = V\hat{\mathbf{x}}$ with respect to another frame of reference S. An observer at rest in frame S notes the positions x_A and x_B at which these two events happen in his own frame of reference and estimates the time interval between the events as $(x_B - x_A)/V$. Show that the phenomenon of time dilation is exhibited in the observer's result.

Solution

In the frame of reference S′, the coordinates of the two events are:

$$\text{Event A}: \quad x'_A, \quad t'_A \qquad\qquad \text{Event B}: \quad x'_B, t'_B.$$

For the observer in frame S′, the time interval between the two events is $\Delta t' = t'_B - t'_A$.
The coordinates of the two events in frame S are:

$$\text{Event A}: \quad x_A = \gamma(x'_A + Vt'_A), \qquad t_A = \gamma(t'_A + (V/c^2)x'_A).$$
$$\text{Event B}: \quad x_B = \gamma(x'_B + Vt'_B), \qquad t_B = \gamma(t'_B + (V/c^2)x'_B).$$

The observer in the frame of reference S sees the point at which these two events happen in S′ to shift, with speed V, by a distance $x_B - x_A$. He concludes that the time interval between the two events is

$$\Delta t = (x_B - x_A)/V.$$

Therefore,

$$\Delta t = \gamma(x'_B + Vt'_B - x'_A - Vt'_A)/V = \gamma(x'_B - x'_A)/V + \gamma(t'_B - t'_A).$$

However, in frame S′ the two events happen at the same point and hence $x'_B = x'_A$. Then, it is

$$\Delta t = \gamma(x'_B - x'_A)/V = \gamma \Delta t'.$$

We have found that $\Delta t = \gamma \Delta t'$ and a dilation of time is observed in frame S.

3.17 An observer A remains on Earth and watches his twin sister B move away from him with a speed of 5000 km/s for 6 months, and then return to him with the same speed. How much younger than A will B be when they meet again?

Solution

To the speed of 5000 km/s there correspond a $\beta = 5 \times 10^6/3 \times 10^8 = 0.0167$ and a Lorentz factor of $\gamma = 1\big/\sqrt{1 - 0.0167^2} = 1.000\,139$. If for observer A time t passes, for observer B the time elapsed will be smaller and equal to $t' = t/\gamma$. The difference is

$$\Delta t = t - t' = t(1 - 1/\gamma) = 365 \times 24 \times 60 \times 60 \times 0.000\,139 = 4384\ \text{s}$$
$$= 1.22\ \text{hour}.$$

3.18 Frame of reference S′ moves with velocity $\mathbf{V} = V\hat{\mathbf{x}}$ relative to frame S. In S′, $2N + 1$ events occur (N being a positive integer). The events are equally separated in space and in time. Event k $(-N \le k \le N)$ occurs at the point $x'_k = ka$ when $t'_k = k\tau$, with a and τ constants. In frame S′, the events are spread over a time interval $T' = t'_N - t'_{-N} = 2N\tau$ and a length $L' = x'_N - x'_{-N} = 2Na$. What are the corresponding T and L in frame S? What is the condition for all the events to be simultaneous in frame S? What is then the value L_0 of L?

Solution

In frame S′, the events happen at $x_k' = ka$ and $t_k' = k\tau$, $(-N \le k \le N)$.

In frame S, the events occur at $x_k = \gamma(x_k' + Vt_k')$ and $t_k = \gamma\left(t_k' + \dfrac{V}{c^2}x_k'\right)$.

Substituting, we have: $x_k = \gamma k(a + V\tau)$ and $t_k = \gamma k\left(\tau + \dfrac{V}{c^2}a\right)$.

The spreads of the events in space and time in frame S are

$$L = x_N - x_{-N} = 2\gamma N(a + V\tau) \quad \text{and} \quad T = t_N - t_{-N} = 2\gamma N\left(\tau + \frac{V}{c^2}a\right).$$

All the events are simultaneous in S if $T = 0$ or $t_k = 0$ for all k.

This happens if $\tau + \dfrac{V}{c^2}a = 0$ or $V = -\dfrac{\tau}{a}c^2$.

For this value of V, it is $\gamma = \dfrac{1}{\sqrt{1 - V^2/c^2}} = \dfrac{1}{\sqrt{1 - c^2\tau^2/a^2}}$ and

$L_0 = 2N\sqrt{a^2 - c^2\tau^2} = 2Na/\gamma = L'/\gamma$.

We see that it is possible to have $L_0 = 0$, but then it must be $a = \pm c\tau$, in which case it is $V = \pm c$.

This problem is examined in another way in Example 3.4.

3.19 Three galaxies, A, B and C, lie on a straight line. Relative to A, which lies between B and C, the other two galaxies are moving away with speeds $0.7c$. The rate at which the distance between B and C is changing is $1.4c$, as measured by A. Does this violate the conclusion derived in the Special Theory of Relativity that nothing can travel with a speed higher than c? What is the speed of B as measured by C?

Solution

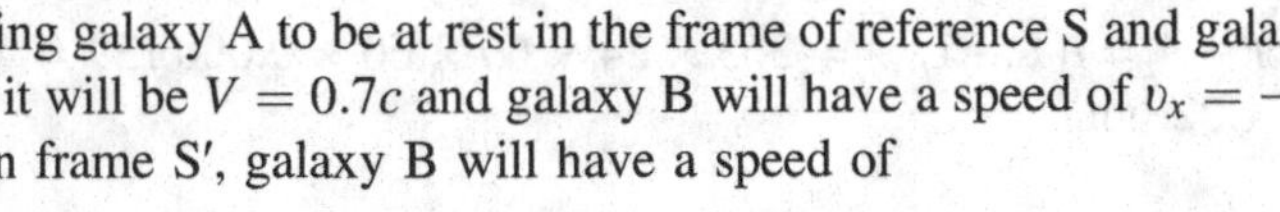

The fact that the rate at which the distance between B and C is changing is $1.4c$, as measured by A, does not violate Special Relativity since no information or mass moves at this speed.

Considering galaxy A to be at rest in the frame of reference S and galaxy C to be at rest in S′, it will be $V = 0.7c$ and galaxy B will have a speed of $\upsilon_x = -0.7c$ in S. Therefore, in frame S′, galaxy B will have a speed of

$$\upsilon_x' = \frac{\upsilon_x - V}{1 - \dfrac{\upsilon_x V}{c^2}} = \frac{-0.7c - 0.7c}{1 - \dfrac{(-0.7c) \times 0.7c}{c^2}} = \frac{-1.4c}{1.49} = -0.94c.$$

3.20 Relative to the Earth, spaceship A moves in one direction with speed $0.8c$ and another spaceship, B, moves in the opposite direction with speed $0.6c$. What is the speed of spaceship A as measured by B?

Solution

Let S be the frame of the Earth and S′ the frame of reference of B. Then, $V = 0.6c$ and $\upsilon_x = -0.8c$ and the speed of spaceship A as measured by B is

$$\upsilon'_x = \frac{\upsilon_x - V}{1 - V\upsilon_x/c^2} = \frac{-0.8c - (0.6c)}{1 - \dfrac{(0.6c)(-0.8c)}{c^2}} = \frac{-1.4c}{1.48} = -0.946c.$$

3.21 A spaceship A departs from Earth for a trip to *α Centauri*, at constant speed. The distance of the star from the Earth is $D = 4$ light years, approximately. Consider the Earth as frame of reference S and the spaceship as frame of reference S′, which moves with speed V away from the Earth.

(a) What must the value of V be, if the trip is to last for $\Delta t' = 4$ years for the passengers of the spaceship?
(b) What is the duration of the trip, Δt, for an observer on Earth?
(c) A second spaceship B is returning from *α Centauri* with a speed of $\upsilon_{Bx} = -c/\sqrt{2}$ relative to the Earth. What is the speed of spaceship B as measured by an observer in spaceship A?
(d) If the rest length of spaceship B is $l_{B0} = 48$ m, what is its length, l'_B, as measured by an observer in spaceship A?

Solution

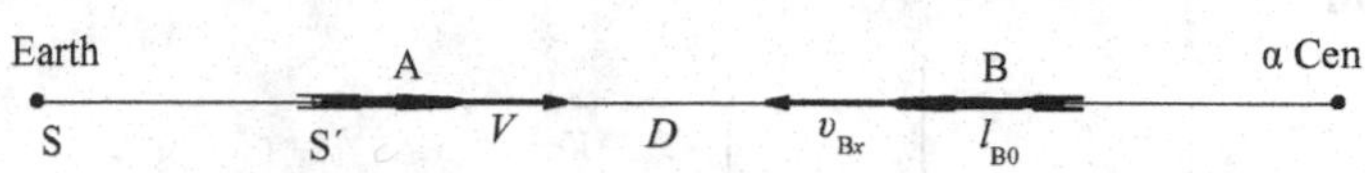

(a) The duration of the journey for an observer inside the spaceship is: $\Delta t' = 4$ years.
For an observer on the Earth, the journey lasts for a time: $\Delta t = D/V$.
The relationship between the two times is $\Delta t' = \Delta t/\gamma$.
Therefore,

$$\Delta t' = \frac{\Delta t}{\gamma} = \frac{D/V}{\gamma} \quad \Rightarrow \quad V\gamma = \frac{D}{\Delta t'}, \quad \beta\gamma = \frac{D/c}{\Delta t'}.$$

However, $\Delta t' = 4$ years and $D/c = 4$ years, and, therefore, $\beta\gamma = 1$.
This gives:

$$(\beta\gamma)^2 = \frac{\beta^2}{1 - \beta^2} = 1 \quad \Rightarrow \quad \beta = \frac{V}{c} = \frac{\sqrt{2}}{2} \quad \text{and} \quad \gamma = \sqrt{2}.$$

(b) The duration of the journey for the observer on the Earth will be
$\Delta t = \gamma\,\Delta t' = \sqrt{2} \times 4 = 5.7$ years.

(c) In the frame of reference S, the speed of spaceship B is $\upsilon_{\mathrm{B}x} = -c/\sqrt{2}$. In the frame of reference S′ it will, therefore, be

$$\upsilon'_{\mathrm{B}x} = \frac{\upsilon_{\mathrm{B}x} - V}{1 - \dfrac{\upsilon_{\mathrm{B}x} V}{c^2}} = \frac{-c/\sqrt{2} - c/\sqrt{2}}{1 + c^2/2c^2} = \frac{-\sqrt{2}c}{3/2} = -\frac{2\sqrt{2}}{3}c.$$

(d) To the speed $\upsilon'_{\mathrm{B}x} = -(2\sqrt{2}/3)c$ measured by A for B, there corresponds a Lorentz factor

$$\gamma'_{\mathrm{B}} = \frac{1}{\sqrt{1 - (\upsilon'_{\mathrm{B}x}/c)^2}} = \frac{1}{\sqrt{1 - (2\sqrt{2}/3)^2}} = \frac{1}{\sqrt{1 - 8/9}} = 3.$$

Therefore,

$$l'_{\mathrm{B}} = \frac{l_{\mathrm{B}0}}{\gamma'_{\mathrm{B}}} = \frac{48}{3} = 16 \text{ m}.$$

3.22 A point moves with constant speed υ' in a direction which forms an angle θ' with the x'-axis of the frame of reference S′. Frame S′ moves with a velocity $\mathbf{V} = V\hat{\mathbf{x}}$ relative to another frame of reference S. What is the angle θ formed by the direction of motion of the point with the x-axis of S? What is the relationship between the two angles as $\upsilon' \to c$?

Solution

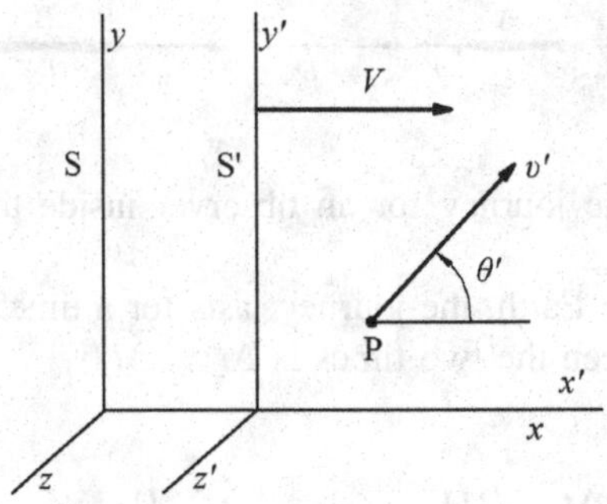

The components of the velocity in the frame of reference S′ are

$$\upsilon'_x = \upsilon' \cos\theta', \qquad \upsilon'_y = \upsilon' \sin\theta'.$$

Transforming, we find the corresponding components in the frame of reference S as

$$\upsilon_x = \frac{\upsilon'_x + V}{1 + \upsilon'_x V/c^2} = \frac{\upsilon' \cos\theta' + V}{1 + (\upsilon' V/c^2)\cos\theta'}$$

$$\upsilon_y = \frac{\upsilon'_y}{\gamma(1 + \upsilon'_x V/c^2)} = \frac{\upsilon' \sin\theta'}{\gamma(1 + (\upsilon' V/c^2)\cos\theta')}.$$

The slope of the velocity vector $\boldsymbol{v}$ in frame S is

$$\tan\theta = \frac{v_y}{v_x} = \frac{v'\sin\theta'}{\gamma(V + v'\cos\theta')} \quad \text{or} \quad \tan\theta = \frac{\beta'\sin\theta'}{\gamma(\beta + \beta'\cos\theta')}.$$

In the limit $v' \to c$ we have $\beta' \to 1$ and $\tan\theta = \dfrac{\sin\theta'}{\gamma(\beta + \cos\theta')}$.

3.23 The inertial frame of reference S′ moves with a constant velocity $V\hat{\mathbf{x}}$ relative to another frame of reference S. In S′, a photon has velocity components $v'_x = c\,\cos\theta'$ and $v'_y = c\,\sin\theta'$. Find the values of the velocity components in S and show that the speed of the photon in S is c. What angle is formed with the x-axis by the direction of motion of the photon in S?

Solution

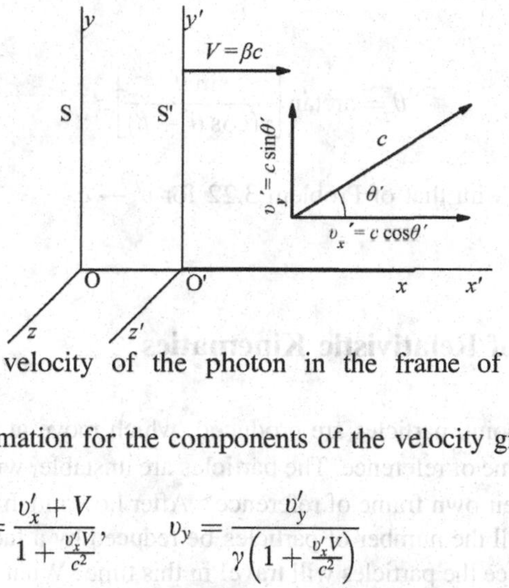

The square of the velocity of the photon in the frame of reference S′ is: $v'^2_x + v'^2_y = c^2$.

The Lorentz transformation for the components of the velocity give

$$v_x = \frac{v'_x + V}{1 + \frac{v'_x V}{c^2}}, \qquad v_y = \frac{v'_y}{\gamma\left(1 + \frac{v'_x V}{c^2}\right)}$$

or

$$v_x = c\frac{\cos\theta' + \beta}{1 + \beta\cos\theta'}, \qquad v_y = c\frac{\sin\theta'}{\gamma(1 + \beta\cos\theta')}$$

where $\beta = V/c$ and $\gamma = 1\Big/\sqrt{1-\beta^2}$.

From these equations it follows that $v_x^2 + v_y^2 = c^2\dfrac{(\cos\theta' + \beta)^2 + \sin^2\theta'(1-\beta^2)}{(1 + \beta\cos\theta')^2}$.

Expanding,

$$\upsilon_x^2+\upsilon_y^2 = c^2\frac{\cos^2\theta'+2\beta\cos\theta'+\beta^2+\sin^2\theta'-\beta^2\sin^2\theta'}{(1+\beta\cos\theta')^2} =$$
$$= c^2\frac{1+2\beta\cos\theta'+\beta^2\cos^2\theta'}{(1+\beta\cos\theta')^2} = c^2$$

As expected, a velocity of magnitude c in frame S′ also has a magnitude equal to c in frame S.

If θ is the angle formed with the x-axis by the direction of propagation of the photon in frame S, it will be

$$\tan\theta = \frac{\upsilon_y}{\upsilon_x} = \frac{\upsilon_y'}{\gamma(\upsilon_x'+V)} = \frac{c\sin\theta'}{\gamma(c\cos\theta'+V)} \quad \Rightarrow \tan\theta = \frac{\sin\theta'}{\gamma(\cos\theta'+\beta)}$$

or

$$\theta = \arctan\left[\frac{\sin\theta'}{\gamma(\cos\theta'+\beta)}\right].$$

The result agrees with that of Problem 3.22 for $\upsilon'\to c$.

Chapter 4
Applications of Relativistic Kinematics

4.1 At a certain point, particles are produced, which move at a speed of $\upsilon = \frac{4}{5}c$ in the laboratory frame of reference. The particles are unstable, with a mean lifetime of $\tau = 10^{-8}$ s (in their own frame of reference). After how much time, as measured in the laboratory, will the number of particles be reduced by a factor e = 2.71828...? What is the distance the particles will travel in this time? What is this distance in the frame of reference of the particles?

Solution

In the frame of reference of the particles, according to the law $N/N_0 = \mathrm{e}^{-t/\tau}$, the number of the surviving particles decreases by a factor e in a time $\Delta t' = \tau = 10^{-8}$ s. In the laboratory frame of reference, this time is $\Delta t = \gamma\Delta t'$, where $\gamma = 1/\sqrt{1-(\upsilon/c)^2} = 1/\sqrt{1-(4/5)^2} = 5/3$.

Therefore,

$$\Delta t = (5/3)\times 10^{-8}\ \text{s} = 1.67\times 10^{-8}\ \text{s}.$$

In this time, the particles will travel, in the laboratory frame of reference, a distance of

$$l = \beta c \Delta t = \frac{4}{5} \times 3 \times 10^8 \times (5/3) \times 10^{-8} = 4 \text{ m}.$$

In the frame of reference of the particles, this distance appears equal to

$$l' = l/\gamma = (4 \text{ m})/(5/3) = 12/5 = 2.4 \text{ m}.$$

Naturally, in their own frame of reference, the particles are at rest. An observer in the frame of reference of the particles will see the laboratory moving by a distance of 2.4 m in the time needed for the number of the particles to decrease by a factor of e.

4.2 The mean lifetime of muons at rest is $\tau = 2.2 \times 10^{-6}$ s. The muons in a beam are observed in the laboratory to have a mean lifetime of $\tau_L = 1.5 \times 10^{-5}$ s. What is the speed of the muons in the laboratory frame of reference?

Solution

The mean lifetimes of the muons in their frame of reference and in the laboratory frame of reference are related by the equation $\tau_L = \gamma\tau$, where γ is the Lorentz factor corresponding to the speed of the muons. Therefore,

$$\frac{1}{\gamma^2} = 1 - \beta^2 = \left(\frac{\tau}{\tau_L}\right)^2, \quad |\beta| = \sqrt{1 - \left(\frac{\tau}{\tau_L}\right)^2},$$

$$|\beta| = \sqrt{1 - \left(\frac{2.2 \times 10^{-6}}{1.5 \times 10^{-5}}\right)^2} = 0.989$$

and the speed of the muons is $V = 0.989c$.

4.3 The particles in a beam move in the laboratory with a common speed of $V = \frac{4}{5}c$. It is observed that the number of the particles is reduced by a factor of e = 2.71828...after they travel a distance of 300 m in the laboratory frame of reference. What is the mean lifetime τ of the particles in their own frame of reference?

Solution

To travel a distance of 300 m in the laboratory frame of reference, the particles need a time

$$t_L = \frac{300 \text{ m}}{V} = \frac{300 \text{ m}}{\frac{4}{5} \times 3 \times 10^8 \text{ m/s}} = 1.25 \times 10^{-6} \text{ s}.$$

For the particles, the time elapsed is equal to τ. Due to the dilation of time, it is $t_L = \gamma\tau$.

Therefore, $\tau = \dfrac{t_L}{\gamma}$, where $\gamma = \dfrac{1}{\sqrt{1-(4/5)^2}} = \dfrac{5}{3}$.

Substituting, we find

$$\tau = \frac{1.25}{5/3} \times 10^{-6}\ \text{s} = 0.75\ \mu\text{s}$$

for the mean lifetime of the particles in their own frame of reference.

4.4 The mean lifetime of π^+ mesons, in their own frame of reference, is $\tau_0 = 2.8 \times 10^{-8}$ s. A pulse of 10^4 π^+ mesons travels a distance of 59.4 m in the laboratory, with a speed of $\upsilon = 0.99c$.

(a) Approximately how many mesons will survive till the end of the trip?

(b) How many mesons would have survived after the same interval of time, had they been at rest?

Solution

(a) In the laboratory frame of reference, the mesons travel a distance of 59.4 m with a speed of $\upsilon = 0.99c$. The time needed for this, in the laboratory frame of reference, is

$$t' = \frac{s}{\upsilon} = \frac{59.4}{0.99 \times 3 \times 10^8} = 20 \times 10^{-8}\ \text{s}.$$

In the frame of reference of the particles, this time will appear smaller by a factor of γ, where $\gamma = 1\Big/\sqrt{1-0.99^2} = 7.09$. Hence, it is

$$t = \frac{t'}{\gamma} = \frac{20 \times 10^{-8}}{7.09} = 2.82 \times 10^{-8}\ \text{s}.$$

Thus, at the end of the trip, the number of surviving particles will be

$$N = N_0 \mathrm{e}^{-t/\tau_0} = 10^4 \exp(-2.82 \times 10^{-8}/2.8 \times 10^{-8}) = 10^4 \mathrm{e}^{-1},$$
$$N \approx 3700\ \text{mesons}.$$

(b) After time 20×10^{-8} s, in the frame of reference of the mesons, the number of surviving particles will be approximately equal to

$$N = N_0 \mathrm{e}^{-t/\tau_0} = 10^4 \exp(-20 \times 10^{-8}/2.8 \times 10^{-8}) = 10^4 \mathrm{e}^{-7}, \qquad N \approx 9\ \text{mesons}.$$

The numbers are, of course, only approximate, due to the statistical nature of the phenomenon, but they show the difference between the behavior of the particles in motion as compared to that of the particles at rest.

4.5 A beam of μ particles is produced at some height in the atmosphere. The particles move with a speed of $\upsilon_\mu = 0.99c$ vertically downwards. μ particles disintegrate into electrons and neutrinos ($\mu^- \to e^- + \bar{\nu}_e + \nu_\mu$) with a mean lifetime of $\tau_\mu = 2.2\ \mu s$ in their own frame of reference.

(a) Find the height at which the particles are produced, if a fraction of 1% of them survive for long enough to reach the surface of the Earth.

(b) What is this distance as seen by the particles?

Solution

(a) In the frame of reference of the Earth, S′, the fraction of surviving particles after a time t' is

$$N/N_0 = e^{-t'/\tau'_\mu},$$

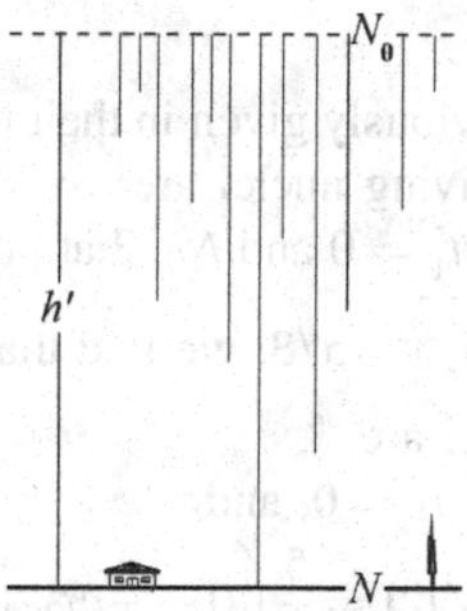

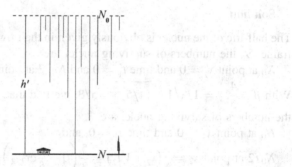

where $\tau'_\mu = \gamma_\mu \tau_\mu$ is the mean lifetime of the muons in frame S′. For $\beta_\mu = 0.99$ it is $\gamma_\mu = 7.09$ and $\tau'_\mu = 7.09 \times 2.2 \times 10^{-6} = 1.56 \times 10^{-5}$ s. The time t' for which it will be $N/N_0 = 0.01$, is therefore given by the equation $0.01 = e^{-t'/1.56\times10^{-5}\,\text{s}}$, from which we find $\ln 0.01 = -\ln 100 = -4.605 = -\dfrac{t'}{1.56 \times 10^{-5}\ \text{s}}$ and finally, $t' = 7.2 \times 10^{-5}$ s.

During this time, in frame S′, the muons travel a distance equal to

$$h' = \beta_\mu c t' = 21\ \text{km}$$

and this is the altitude above the surface of the Earth at which the muons are produced.

(b) In the muons' frame of reference, S, it is $N/N_0 = e^{-t/\tau_\mu}$.

Now it is
$$\ln 0.01 = -\ln 100 = -4.605 = -\frac{t'}{2.2 \times 10^{-6}\ \text{s}}$$

and
$$t = 1.02 \times 10^{-5}\ \text{s}.$$

[Alternatively, $t' = \gamma_\mu t$, and $t = t'/\gamma_\mu = (7.2 \times 10^{-5}\ \text{s})/7.09 = 1.02 \times 10^{-5}$ s.]

This is the time that the particles measure as the duration of their journey in the atmosphere. During this time, the muons observe the atmosphere to move, in their own frame of reference S, by a distance of

$$h = \beta_\mu ct = 3 \text{ km}.$$

This is the thickness of the atmosphere as seen by the muons.

4.6 A spaceship, which will serve as frame of reference S′, has a rest length $L_0 = 10$ m. It moves with a speed $\frac{4}{5}c$ with respect to another frame of reference, S. Inside the spaceship, and at the point $x = x' = 0$, $y = y' = 0$, $z = z' = 0$, there is a quantity of radioactive material, which at the moment $t = t' = 0$ consists of N_0 nuclei. The half-life of the nuclei is $\tau_{1/2} = 2\ \mu\text{s}$. When and at what position will the number of the surviving nuclei be $N_0/2$?

Solution

The half-life of the nuclei is obviously given in their own frame of reference. In this frame, S′, the numbers of surviving nuclei are

N_0 at point $x_1' = 0$ and time $t_1' = 0$ and $N_0/2$ at point $x_2' = 0$ and time $t_2' = \tau_{1/2}$.

With $\beta = \frac{4}{5}$, $\gamma = 1/\sqrt{1-(4/5)^2} = 5/3$, we find that, in the frame of reference S, the numbers of surviving nuclei are

N_0 at point $x_1 = 0$ and time $t_1 = 0$, and

$N_0/2$ at point $x_2 = \gamma\left(x_2' + Vt_2'\right) = \frac{5}{3}\left(0 + \frac{4}{5}c\tau_{1/2}\right) = \frac{4}{3}c\tau_{1/2} = 800$ m

and time $t_2 = \gamma\left(t_2' + \frac{V}{c^2}x_2'\right) = \frac{5}{3}(\tau_{1/2} + 0) = \frac{5}{3}\tau_{1/2} = 3.33\ \mu\text{s}$.

4.7 A spaceship will travel to *Proxima Centauri*, at a distance of $L = 4.4$ light years from the Earth, and return. The spaceship will cover the first half of the distance accelerating at a constant proper acceleration of $g = 9.81\ \text{m/s}^2$ and the other half at a constant proper deceleration of $g = -9.81\ \text{m/s}^2$, and following the reverse procedure on the return trip. How much time will be needed for the whole trip, in the frame of reference of the Earth and in that of the spaceship?

Solution

Due to the symmetry of the journey, the time that will be needed will be 4 times the time needed for the spaceship to travel a distance equal to $L/2$ with a constant proper acceleration g.

The total time needed, as this will be measured in the frame of reference of the Earth, is given by the equation

$$t_{\text{tot}} = 4\sqrt{\frac{x}{c}\left(\frac{x}{c} + 2\tau_0\right)}$$

with $x = L/2$ and, here, $\tau_0 = 0.97$ year. Putting $x/c = 2.2$ years, we find $t_{\text{tot}} = 12$ years.

In the frame of reference of the spaceship it is

$$\tau_{tot} = 4\tau_0 \ln\left[1 + \frac{x}{c\tau_0} + \sqrt{\frac{x}{c\tau_0}\left(\frac{x}{c\tau_0} + 2\right)}\right]$$

with $x = L/2$. Substituting $x/c\tau_0 = 2.2/0.97 = 2.27$, we find $\tau_{tot} = 7.2$ years.

Let us note that for a journey to the center of our galaxy, which is at a distance of approximately $L = 50\,000$ light years from the Earth, the corresponding times would be $t_{tot} = 100\,000$ and $\tau_{tot} = 42$ years.

Chapter 5
Optics

5.1 A certain line in the spectrum of the light from a nebula has a wavelength of 656 nm instead of the 434 nm measured in the laboratory. If the nebula is moving radially, what is its speed relative to the Earth?

Solution

If β is considered positive when the source is moving away from the observer, the relationship between the wavelengths is

$$\lambda = \lambda_0 \sqrt{\frac{1+\beta}{1-\beta}}$$

where, here, $\lambda_0 = 434$ nm is the wavelength in the frame of reference of the source and $\lambda = 656$ nm is the wavelength in the frame of reference of the observer. Solving for β, we find

$$\beta = \frac{\lambda^2 - \lambda_0^2}{\lambda^2 + \lambda_0^2} = \frac{656^2 - 434^2}{656^2 + 434^2} = 0.391.$$

Because β is positive, the nebula is moving away from the Earth with a speed of $V = 0.391c$.

5.2 Show that the light from a source which is moving away from us with a speed of $0.6c$ has twice the wavelength it has in the frame of reference of the source.

Solution

According to the Doppler effect, the relationship between the wavelengths is $\lambda = \lambda_0 \sqrt{\frac{1+\beta}{1-\beta}}$, where β is considered to be positive when the source is moving away from us. For $\beta = 0.6$, it is

$$\lambda = \lambda_0\sqrt{\frac{1+0.6}{1-0.6}} = \lambda_0\sqrt{\frac{1.6}{0.4}} = \lambda_0\sqrt{4} = 2\lambda_0.$$

5.3 A spaceship moves with speed V relative to an observer O, on a straight line which passes very near the observer. A source on the spaceship emits light of wavelength $\lambda_O = 500$ nm in the frame of reference of the spaceship. For what range of the spaceship's speed will the light be visible to the observer? The wavelengths of visible light stretches from $\lambda_A = 400$ nm to $\lambda_B = 700$ nm, approximately.

Solution

For motion of the source with speed β, it is $\lambda = \lambda_O\sqrt{\frac{1+\beta}{1-\beta}}$ with β positive when the source is moving away from the observer. Solving, $\beta = \frac{\lambda^2 - \lambda_O^2}{\lambda^2 + \lambda_O^2}$.

For $\lambda = \lambda_A = 400$ nm,

$$\beta_A = \frac{4^2 - 5^2}{4^2 + 5^2} = -\frac{25 - 16}{16 + 25} = -\frac{9}{41}$$

and if the source is approaching the observer with speed $-\frac{9}{41} < \beta$ (i.e. $|\beta| < 9/41$), the light from the source will be visible.

For $\lambda = \lambda_B = 700$ nm,

$$\beta_B = \frac{7^2 - 5^2}{7^2 + 5^2} = \frac{45 - 16}{45 + 25} = \frac{24}{74} = \frac{12}{37}$$

and if the source is moving away from observer with speed $\beta < \frac{12}{37}$, the light from the source will be visible.

It was found that the light from the source on the spaceship will be visible to the observer for source speeds in the range $-\frac{9}{41} < \beta < \frac{12}{37}$ or $-0.22 < \beta < 0.33$.

5.4 A light source moves in a circular orbit with a speed of $0.5c$. What is the displacement, due to the Doppler effect, of the sodium yellow line, as observed at the center of the circle? The line has a wavelength of 589 nm in the laboratory.

Solution

It is $\beta = 0.5$ and, therefore, $\gamma = 2/\sqrt{3} = 1.1547$. The shift in wavelength is exclusively due to the transverse Doppler effect and, therefore, from the relationship $\lambda = \lambda_0\gamma(1 + \beta\cos\theta)$ with $\theta = 90^\circ$, it is $\lambda = \lambda_0\gamma$, where $\lambda_0 = 589$ nm is the wavelength in the source's frame of reference and λ the wavelength in the frame of reference of the observer. The shift of the spectral line will be:

$$\Delta\lambda = \lambda - \lambda_0 = \lambda_0(\gamma - 1) = 589 \times (1.1547 - 1) = 91 \text{ nm.}$$

5.5 The Sun has a radius of 7.0×10^8 m, approximately, and a period of rotation about its axis equal to 24.7 days. What is the Doppler shift of a spectral line with laboratory wavelength 500 nm, in the light emitted (a) from the center of the Sun and (b) from the edges of the Sun's disk at its equator?

Solution

The speed of a point at the Sun's equator is:

$$\upsilon = \frac{2\pi R}{T} = \frac{2\pi \times 7 \times 10^8}{24.7 \times 24 \times 60 \times 60} = 2061 \text{ m/s.}$$

Because $\beta = \dfrac{2061}{3 \times 10^8} = 6.87 \times 10^{-6}$, it is

$$\gamma = \frac{1}{\sqrt{1 - (6.87 \times 10^{-6})^2}} \approx 1 + \frac{1}{2}\left(6.87 \times 10^{-6}\right)^2 = 1 + 2.36 \times 10^{-11}.$$

(a) The displacement of a spectral line in the light from the center of the Sun is due exclusively to the transverse Doppler effect and, therefore, it is $\lambda = \lambda_0\gamma$, where $\lambda_0 = 500$ nm is the light's wavelength in the source's frame of reference and λ is the light's wavelength in the observer's frame of reference. The displacement of the spectral line will be:

$$\Delta\lambda = \lambda - \lambda_0 = \lambda_0(\gamma - 1) = 500 \times 2.36 \times 10^{-11} = 1.18 \times 10^{-8} \text{ nm.}$$

(b) For light from a point at the edge of the Sun's disk at its equator, which is moving away from the Earth, the relationship between the wavelengths is

$$\lambda = \lambda_0\sqrt{\frac{1+\beta}{1-\beta}}$$

where β is positive. Substituting,

$$\lambda = 500\sqrt{\frac{1 + 6.87 \times 10^{-6}}{1 - 6.87 \times 10^{-6}}} \approx 500 \times \left(1 + 6.87 \times 10^{-6}\right) \text{ nm}$$

and the displacement of the spectral line is $\Delta\lambda = 500 \times 6.87 \times 10^{-6} = 3.44 \times 10^{-3}$ nm.

For a point at the Sun's edge on its equator, which is moving towards the Earth, the displacement of the spectral line is: $\Delta\lambda = -3.44 \times 10^{-3}$ nm. Therefore, $\Delta\lambda = \pm 0.00344$ nm.

Chapter 6
Relativistic Dynamics

6.1 What is the speed of an electron whose kinetic energy is 2 MeV? What is the ratio of its relativistic mass to its rest mass? The rest mass of the electron is $m_0 = 0.511\ \text{MeV}/c^2$.

Solution

The kinetic energy of the electron is given by the relation $K = \dfrac{m_0c^2}{\sqrt{1-\upsilon^2/c^2}} - m_0c^2$, where υ is the speed of the electron. For $m_0c^2 = 0.511$ MeV, this becomes $2\ \text{MeV} = (0.511\ \text{MeV}) \times \gamma - 0.511\ \text{MeV}$, from which it follows that $\gamma = 5$ and $\upsilon = 0.98c$.

The ratio of the relativistic mass of an electron with a kinetic energy of 2 MeV to its rest mass is:

$$\frac{m}{m_0} = \gamma = 5.$$

6.2 The extremely rare event of the indirect observation, in cosmic radiation, of a particle with an energy of the order of 10^{20} eV (16 J!) occurred a few years ago [see J. Linsey, *Phys. Rev. Lett.* **10**, 146 (1963)]. Assuming the particle was a proton, which has a rest energy of 1 GeV, approximately, evaluate its speed relative to the Earth. How much time would this proton need, in its own frame of reference, in order to cross our galaxy, whose diameter is 10^5 light years? What is the diameter of the Galaxy as seen by the proton?

Solution

From the equation $E = \gamma m_0c^2$, with $E = 10^{20}$ eV and $m_0c^2 = 10^9$ eV, we find that $\gamma = 10^{11}$ for the proton. Naturally, its speed differs very little from c. The time we would see that the particle needs in order to cross the Galaxy would be equal to $T = 10^5$ years. For the particle, this journey would last for $T_0 = 10^5/\gamma = 10^5/10^{11} = 10^{-6}$ years or 32 s. For the particle, the diameter of the Galaxy would be equal to 32 'light seconds'. i.e. about 1/16 of the distance of the Earth from the Sun or 10^{10} m approximately.

For $\gamma = 10^{11}$, it is $\beta = 1 - 5 \times 10^{-23}$, which corresponds to a speed of the proton equal to $\upsilon = 0.999\,999\,999\,999\,999\,999\,999\,95\ c$. This speed is smaller than the speed of light by an amount equal to $c - \upsilon = 1.5 \times 10^{-14}$ m/s, i.e. one nuclear diameter per second! If this particle and light raced across the Galaxy, at the end of the race the light would lead by only 5 cm.

By the way, for the benefit of those people who were recently (2012) very concerned that during the experiments with the LHC at CERN black holes would be created, that would 'swallow' the Earth, it might be worth pointing out that this proton had an energy which was 10 million times higher than the $2 \times 7 = 14$ TeV available to the protons in the LHC.

6.3 The kinetic energy and the momentum of a particle were measured and found equal to 250 MeV and 368 MeV/c, respectively. Find the rest mass of the particle in MeV/c^2.

Solution

If E, p, K and m_0 are, respectively, the total energy, the momentum, the kinetic energy and the rest mass of the particle, the relations $E = m_0c^2 + K$ and $E^2 = (m_0c^2)^2 + (pc)^2$ hold true.

Therefore, it is $(m_0c^2 + K)^2 = (m_0c^2)^2 + (pc)^2$, from which we have

$\cancel{(m_0c^2)^2} + 2m_0c^2K + K^2 = \cancel{(m_0c^2)^2} + (pc)^2$ and, finally, $m_0c^2 = \frac{(pc)^2 - K^2}{2K}$.

Substituting, we get $m_0c^2 = \frac{368^2 - 250^2}{2 \times 250} = 270$ MeV

and, therefore, the rest mass of the particle is $m_0 = 270$ MeV/c^2.

6.4 At what value of the speed of a particle is its kinetic energy equal to (a) its rest energy and (b) 10 times its rest energy?

Solution

If m_0c^2 is the rest energy of the particle, its kinetic energy is $K = m_0c^2(\gamma - 1)$.

Therefore, $\gamma = 1 + K/m_0c^2$ and

$$\beta = \sqrt{1 - \frac{1}{\gamma^2}} = \sqrt{1 - \frac{1}{(1 + K/m_0c^2)^2}} = \frac{\sqrt{(K/m_0c^2)^2 + 2(K/m_0c^2)}}{1 + K/m_0c^2}$$
$$= \frac{\sqrt{1 + 2m_0c^2/K}}{1 + m_0c^2/K}.$$

(a) For $m_0c^2/K = 1$, it is $\beta = \frac{\sqrt{3}}{2} = 0.866$.

(b) For $m_0c^2/K = 1/10$, it is $\beta = \frac{\sqrt{1 + 2/10}}{1 + 1/10} = \frac{\sqrt{12 \times 10}}{11} = \frac{2\sqrt{30}}{11} = 0.996$.

6.5 An electron with an energy of 100 MeV moves along a tube which is 5 m long. What is the length of the tube in the frame of reference of the electron? The rest energy of the electron is $E_0 = m_0c^2 = 0.511$ MeV, where m_0 is its rest mass.

Solution

If γ is the Lorentz factor that corresponds to the speed of the electron, its total energy is $E = m_0c^2\gamma = \gamma E_0$ and, therefore, $\gamma = \frac{E}{E_0} = \frac{100}{0.511} = 196$.

Since $l = 5$ m is the length of the tube in the laboratory frame of reference, its length in the frame of reference of the electron will be $l_0 = l/\gamma = 5\ \text{m}/196 = 0.026\ \text{m} = 26$ mm.

6.6 A beam of identical particles with the same speed is produced by an accelerator. The particles of the beam travel, inside a tube, its full length of $l = 2400$ m in a time $\Delta t = 10\ \mu\text{s}$, as measured by an observer in the laboratory frame of reference.

(a) Find β and γ for the particles and the duration $\Delta t'$ of the trip as measured in their own frame of reference.

(b) If the particles are unstable, with a mean lifetime of $\tau = 10^{-6}$ s, what proportion of the particles is statistically expected to reach the end of the tube? (Use: $e^3 \approx 20$).

(c) If the rest energy of each particle is $m_0 c^2 = 3$ GeV, find their kinetic energy.

(d) Determine the quantity pc for a particle of the beam, where p is its momentum, and express p in units of GeV$/c$.

Solution

(a) The speed of the particles in the laboratory frame of reference is $V = \dfrac{l}{\Delta t} = \dfrac{2400}{10^{-5}} = 2.4 \times 10^8$ m/s, to which there correspond $\beta = 4/5$ and $\gamma = 5/3$. The journey lasts, in the frame of reference of the particles, for a time $\Delta t' = \dfrac{\Delta t}{\gamma} = \dfrac{10\ \mu\text{s}}{5/3} = 6\ \mu\text{s}$.

(b) The proportion of the particles reaching the end of the tube $N/N_0 = \mathrm{e}^{-t'/\tau}$. In the frame of reference of the particles, it is $\tau = 10^{-6}$ s and $t' = \Delta t' = 6 \times 10^{-6}$ s and, therefore,

$$N/N_0 = \mathrm{e}^{-6} = 1/(e^3)^2 \approx 1/(20)^2 \approx 1/400 \approx 2.5 \times 10^{-3}.$$

(c) The kinetic energy of a particle is $K = (\gamma - 1)m_0 c^2$. With $m_0 c^2 = 3$ GeV and $\gamma = 5/3$, it is $K = (5/3 - 1) \times 3\ \text{GeV} = 2$ GeV.

(d) The momentum of each particle is given by $p = m_0 \gamma \upsilon$. Therefore, $pc = \beta \gamma m_0 c^2$. Substituting in this last equation we have

$$pc = \beta\gamma\left(m_0 c^2\right) = \frac{4}{5} \times \frac{5}{3} \times 3\ \text{GeV} = 4\ \text{GeV} \quad \text{and} \quad p = 4\ \text{GeV}/c.$$

6.7 Two particles, A and B, each having a rest mass of $m_0 = 1\ \text{GeV}/c^2$, move, in the frame of reference of an accelerator, on the x-axis and in opposite directions, approaching each other. In this frame, particle A moves with a velocity of $\upsilon_{\text{A}x} = -0.6c$ and particle B with a velocity of $\upsilon_{\text{B}x} = 0.6c$. In the frame of reference of particle A,

(a) what is the speed of particle B?

(b) what is the energy (in GeV) and the momentum (in GeV$/c$) of particle B?

Solution

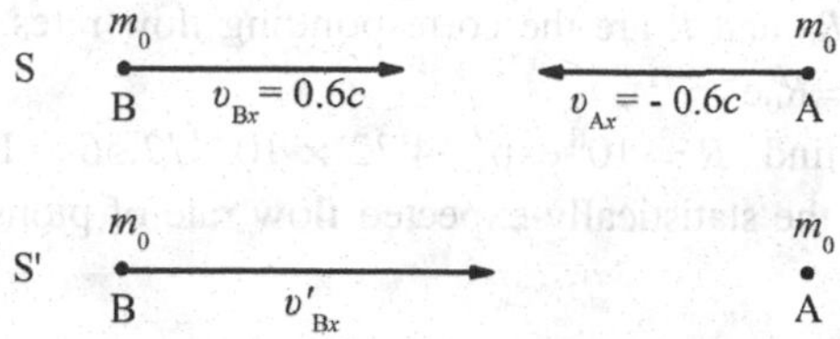

(a) The frame of reference S′ of particle A moves in the laboratory frame of reference, S, with a speed of $V = \upsilon_{Ax} = -0.6c$. In the same frame of reference, particle B moves with a speed of $\upsilon_{Bx} = 0.6c$.

The transformation of the components of velocity give for the speed of B in frame S′:

$$\upsilon'_{Bx} = \frac{\upsilon_{Bx} - V}{1 - \dfrac{\upsilon_{Bx} V}{c^2}} \qquad \upsilon'_{Bx} = \frac{0.6c - (-0.6c)}{1 - \dfrac{0.6c \times (-0.6c)}{c^2}} = \frac{1.2}{1.36}c = 0.88c.$$

(b) In the frame of reference S′, for $\upsilon'_{Bx} = 0.88c$, we have $\beta_B = 0.88$ and $\gamma_B = 2.11$. The energy of B is, therefore,

$$E'_B = m_0 c^2 \gamma_B = (1\ \text{GeV}/c^2) \times c^2 \times 2.11 = 2.11\ \text{GeV}$$

and its momentum is

$$p'_{Bx} = m_0 c \gamma_B \beta_B = (1\ \text{GeV}/c^2) \times c \times 2.11 \times 0.88 = 1.86\ \text{GeV}/c.$$

6.8 A beam of π^+ particles, each with an energy of 1 GeV, has a total flow rate of 10^6 particles/s at the start of a trip that has a length of 10 m in the laboratory frame of reference. What is the flow rate of particles at the end of the trip? π^+ has a rest mass of $m_\pi = 140\ \text{MeV}/c^2$ and a mean lifetime of $\tau_\pi = 2.56 \times 10^{-8}$ s.

Solution

The energy of a pion is $E = m_\pi c^2 \gamma$ and, therefore, $\gamma = E/m_\pi c^2 = 1000/140 = 7.14$. The reduced speed of the pions is $\beta = \sqrt{1 - 1/\gamma^2} = \sqrt{1 - 1/7.14^2} = 0.990$. The time needed, in the laboratory frame of reference, for the particles to travel the distance of 10 m is, therefore, $\Delta t = D/c\beta = 10/(3 \times 10^8 \times 0.990) = 3.37 \times 10^{-8}$ s.

In the frame of reference of the particles, this time is equal to

$$\Delta t' = \Delta t/\gamma = 3.37 \times 10^{-8}/7.14 = 4.72 \times 10^{-9}\ \text{s}.$$

During this time, the number of surviving particles is reduced from N_0 to $N = N_0 e^{-\Delta t'/\tau_\pi}$. If R_0 and R are the corresponding flow rates, in particles per unit time, it will be $R = R_0 e^{-\Delta t'/\tau_\pi}$.
Substituting, we find $R = 10^6 \exp(-4.72 \times 10^{-9}/2.56 \times 10^{-8}) = 10^6 e^{-0.184} = 0.83 \times 10^6\ \text{s}^{-1}$ for the statistically expected flow rate of pions per unit time at the end of the trip.

6.9 Show that, for a body of rest mass m_0, which moves with a speed υ and has momentum p and kinetic energy K, it is $\dfrac{p\upsilon}{K} = 1 + \dfrac{1}{1 + K/m_0c^2}$.

Solution

The kinetic energy of the body is

$$K = m_0c^2(\gamma - 1) \tag{1}$$

and its momentum is

$$p = m_0\gamma\upsilon. \tag{2}$$

Equation (1) gives $\gamma = 1 + \dfrac{K}{m_0c^2}$, $\dfrac{1}{\gamma^2} = 1 - \beta^2 = \dfrac{1}{(1 + K/m_0c^2)^2}$, $\beta^2 = 1 - \dfrac{1}{(1 + K/m_0c^2)^2}$.

From Eqs. (1) and (2), it follows that $\dfrac{p\upsilon}{K} = \dfrac{m_0\gamma\upsilon^2}{m_0c^2(\gamma - 1)} = \dfrac{\beta^2}{1 - 1/\gamma}$.
Substituting for β and γ, we get

$$\frac{p\upsilon}{K} = \left(1 - \frac{1}{(1 + K/m_0c^2)^2}\right)\left(\frac{1}{1 - \dfrac{1}{1 + K/m_0c^2}}\right)$$
$$= \left(\frac{2K/m_0c^2 + (K/m_0c^2)^2}{(1 + K/m_0c^2)^2}\right)\left(\frac{1 + K/m_0c^2}{K/m_0c^2}\right)$$

and, finally,

$$\frac{p\upsilon}{K} = \frac{2 + K/m_0c^2}{1 + K/m_0c^2} = 1 + \frac{1}{1 + K/m_0c^2}.$$

6.10 What is the mass m_ϕ which corresponds to the energy E_ϕ of a photon of wavelength 500 nm?

Solution

It is $m_\phi c^2 = E_\phi = hf = hc/\lambda$ or $m_\phi = h/c\lambda$

and, therefore,

$$m_\phi = \frac{6.63 \times 10^{-34}\ \text{J} \cdot \text{s}}{(3 \times 10^8\ \text{m/s}) \times (5 \times 10^{-7}\ \text{m})} = 4.42 \times 10^{-36}\ \text{kg}.$$

6.11 The wavelength of the photons from a laser is 633 nm. What is the momentum of one such photon?

Solution

The momentum of a photon is given by $p = \frac{E}{c} = \frac{hf}{c} = \frac{h}{\lambda}$, where E is the energy of the photon and $h = 4.136 \times 10^{-15}$ eV · s is Planck's constant. Therefore,

$$pc = E = \frac{hc}{\lambda} = \frac{(4.136 \times 10^{-15}) \times (3 \times 10^8)}{633 \times 10^{-9}} = \frac{4.136 \times 3}{6.33} = 1.96\ \text{eV}$$

and the momentum of the photon is $p = 1.96$ eV/c.

6.12 A spring has a constant equal to $k = 2 \times 10^4$ N/m. What is the increase in the mass of the spring when it is compressed by $x = 5$ cm from its natural length?

Solution

The increase in the spring's potential energy will be:

$$\Delta U = \tfrac{1}{2}kx^2 = \tfrac{1}{2} \times 2 \times 10^4 \times 0.05^2 = 25\ \text{J}.$$

The corresponding change in the mass of the spring will be

$$\Delta m = \Delta U/c^2 = (25\ \text{J})\Big/\left(3 \times 10^8\ \text{m/s}\right)^2 = 2.8 \times 10^{-16}\ \text{kg},$$

which is impossible to measure using today's means.

6.13 A particle with rest mass m, is stationary in the laboratory. The particle splits into two others: one with rest mass m_1 which moves with a speed $V_1 = \frac{3}{5}c$, and another with rest mass m_2 which moves with a speed $V_2 = \frac{4}{5}c$.

(a) Find m_1 and m_2 as fractions of m.

(b) What are the kinetic energies, K_1 and K_2, of the two particles, in terms of m?

Solution

(a) For momentum to be conserved, the two particles produced will move in opposite directions in the laboratory frame of reference.

Defining $|\beta_1| = \dfrac{V_1}{c} = \dfrac{3}{5}$, $|\beta_2| = \dfrac{V_2}{c} = \dfrac{4}{5}$

and $\gamma_1 = \dfrac{1}{\sqrt{1-(3/5)^2}} = \dfrac{5}{4}$, $\gamma_2 = \dfrac{1}{\sqrt{1-(4/5)^2}} = \dfrac{5}{3}$,

the conservation of energy gives

$$mc^2 = m_1\gamma_1 c^2 + m_2\gamma_2 c^2 \qquad (1)$$

and the conservation of momentum gives

$$m_1\gamma_1 V_1 = m_2\gamma_2 V_2. \qquad (2)$$

From Eq. (2) $m_1\dfrac{5}{4}\times\dfrac{3}{5} = m_2\dfrac{5}{3}\times\dfrac{4}{5}$ and, therefore, $\dfrac{m_2}{m_1} = \left(\dfrac{3}{4}\right)^2 = \dfrac{9}{16}$. Substituting in Eq. (1),

$$m = m_1\frac{5}{4} + \frac{9}{16}m_1\frac{5}{3} = \frac{35}{16}m_1$$

and, finally,

$$m_1 = \frac{16}{35}m, \qquad m_2 = \frac{9}{35}m.$$

We note that $m_1 + m_2 = \dfrac{5}{7}m < m$, which is as expected since part of the energy is given as kinetic energy to m_1 and m_2.

(b) The kinetic energies of the particles are

$$K_1 = (\gamma_1 - 1)m_1c^2 = \left(\frac{5}{4} - 1\right)\frac{16}{35}mc^2 = \frac{4}{35}mc^2$$

and

$$K_2 = (\gamma_2 - 1)m_1c^2 = \left(\frac{5}{3} - 1\right)\frac{9}{35}mc^2 = \frac{6}{35}mc^2.$$

As expected, it is

$$m_1c^2 + K_1 + m_2c^2 + K_2 = mc^2.$$

6.14 A particle with rest mass m moves in the laboratory with speed $V = \frac{3}{5}c$. The particle disintegrates into two others: one with rest mass m_1 which remains stationary and another with rest mass m_2 which moves with a speed of $\upsilon = \frac{4}{5}c$. Find the masses m_1 and m_2 in terms of m.

Solution

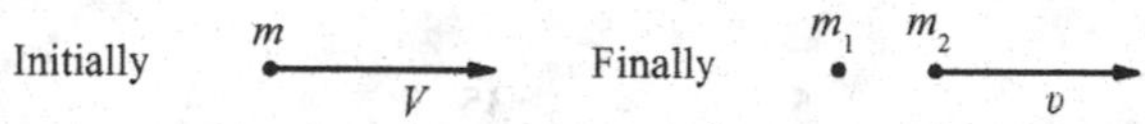

We define the quantities:

$$\beta_m = V/c = \frac{3}{5} \qquad \gamma_m = 1/\sqrt{1-\beta_m^2} = \frac{5}{4}$$

and

$$\beta_2 = \upsilon/c = \frac{4}{5} \qquad \gamma_2 = 1/\sqrt{1-\beta_2^2} = \frac{5}{3}$$

For conservation of momentum,

$$m\gamma_m\beta_m = 0 + m_2\gamma_2\beta_2. \tag{1}$$

For conservation of energy,

$$mc^2\gamma_m = m_1c^2 + m_2c^2\gamma_2. \tag{2}$$

Equation (1) gives

$$m_2 = \frac{\gamma_m\beta_m}{\gamma_2\beta_2}m = \frac{\frac{5}{4}\times\frac{3}{5}}{\frac{5}{3}\times\frac{4}{5}}m = \frac{9}{16}m = 0.56m$$

and Eq. (2) gives, then,

$$m_1 = m\gamma_M - m_2\gamma_2 = m\frac{5}{4} - \frac{9}{16}m\frac{5}{3} = \left(\frac{5}{4} - \frac{15}{16}\right)m = \frac{5}{16}m = 0.31m.$$

6.15 A particle of rest mass $m_1 = 1\ \text{GeV}/c^2$ moves with a speed of $\upsilon_1 = \frac{4}{5}c$ and collides with another particle with rest mass $m_2 = 10\ \text{GeV}/c^2$ which is stationary. After the collision, the two particles form a single body with rest mass M. Find:

(a) The total energy of the system, in GeV.
(b) The total momentum of the system, in GeV/c.
(c) The mass M, in GeV/c^2.

Solution

(a) The total energy of the system is $E_{tot} = E_1 + E_2$, where

$$E_1 = \gamma m_1 c^2, \qquad E_2 = m_2 c^2 \quad \text{and} \quad \beta = \frac{4}{5} \qquad \gamma = \frac{1}{\sqrt{1-(4/5)^2}} = \frac{5}{3}.$$

Therefore,

$$E_{tot} = \frac{5}{3} \times 1 + 10 = \frac{35}{3} = 11.67 \text{ GeV}.$$

(b) The total momentum of the system is $\mathbf{p}_{tot} = \mathbf{p}_1 + \mathbf{p}_2$, where

$$\mathbf{p}_1 = (\beta c)(\gamma m_1)\hat{\mathbf{x}}, \qquad \mathbf{p}_2 = 0.$$

Therefore,

$$p_{tot} = \beta\gamma m_1 c, \qquad cp_{tot} = \beta\gamma m_1 c^2 = \frac{4}{5} \times \frac{5}{3} \times 1 = \frac{4}{3} \text{ GeV}, \qquad p_{tot} = 1.33 \text{ GeV/c}.$$

(c) After the collision, a body of rest mass M is formed. The laws of momentum and energy conservation are valid. Therefore, after the collision it will be $E = E_{tot}$ and $p = p_{tot}$.

From equation $E_{tot}^2 = (p_{tot}c)^2 + (Mc^2)^2$, there follows that

$$Mc^2 = \sqrt{E_{tot}^2 - (p_{tot}c)^2} = \sqrt{\left(\tfrac{35}{3}\right)^2 - \left(\tfrac{4}{3}\right)^2} = \frac{\sqrt{1209}}{3} = 11.59 \text{ GeV and thus}$$

$$M = 11.6 \text{ GeV/c}^2.$$

6.16 Show that, if a photon could disintegrate into two photons, both the photons produced would have to be moving in the same direction as the original photon.

Solution

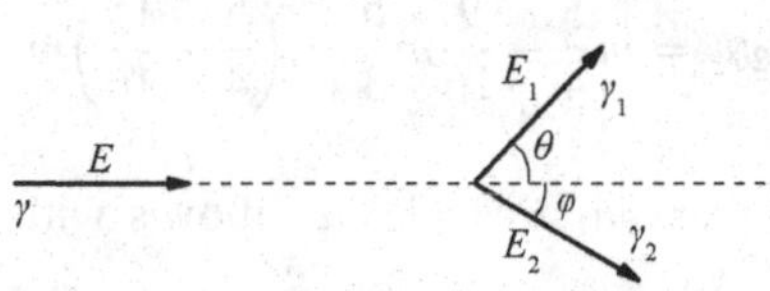

Let the original photon have an energy of E and to disintegrate into two photons, one with energy E_1 moving in a direction forming an angle θ with the direction of propagation of the original photon, and another with energy E_2 moving in a direction forming an angle φ with the direction of propagation of the original photon.

Conservation of energy gives

$$E = E_1 + E_2.$$

Conservation of momentum in the direction of motion of the original photon gives

$$\frac{E}{c} = \frac{E_1}{c}\cos\theta + \frac{E_2}{c}\cos\varphi.$$

Given that $\cos\theta \leq 1$ and $\cos\varphi \leq 1$, and that E, E_1 and E_2 are all positive, the only way both equations can be satisfied is to have $\cos\theta = 1$ and $\cos\varphi = 1$ and, therefore, $\theta = 0$ and $\varphi = 0$. It follows that both the photons produced move in the same direction as the original photon.

6.17 A neutral pion disintegrates into two photons, which move on the same straight line. The energy of the one photon is twice that of the other. Show that the speed of the pion was $c/3$.

Solution

We will assume that the speed of the pion was υ and its total energy E_0 (Figure a). Given that the two photons move on the same straight line and have different energies (E and $2E$) and momenta (E/c και $2E/c$), if momentum is to be conserved in the direction normal to the direction of motion of the pion, the transverse momenta of the photons must be equal to zero. Thus, the two photons must move on the line of motion of the pion.

There are two possibilities:

1. The two photons move in the same direction (Figure b) and
2. The two photons move in opposite directions (Figure c).

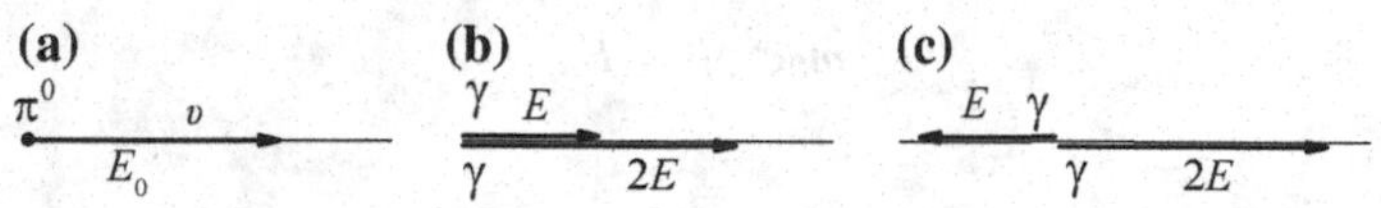

(a) Examining the first possibility, with reference to Figures a and b, the laws of conservation give:

Conservation of energy:

$$E_0 = E + 2E = 3E. \tag{1}$$

Conservation of momentum:

$$m_0\gamma\upsilon = \frac{E}{c} + \frac{2E}{c} = \frac{3E}{c}. \tag{2}$$

Therefore, from Eq. (2),

$$m_0 c^2 \gamma \beta = 3E. \tag{3}$$

However,

$$3E = E_0 = m_0 c^2 \gamma. \tag{4}$$

Equations (3) and (4) give $\beta = 1$ which is impossible for a pion. We therefore conclude that the photons cannot be moving in the same direction. By the way, the solution $\beta = 1$ refers to the (non realistic) case of the photon splitting into two other photons, which we examined in the previous problem.

(b) For the final momentum to be in the same direction as the initial direction of motion of the pion, the photon with the greater energy, $2E$, which also has the greater momentum, must be moving in the direction of motion of the pion. The photons must, therefore, move in the directions shown in Figure c. With reference to Figures a and c, the laws of conservation give:

Conservation of energy:

$$E_0 = E + 2E = 3E. \tag{5}$$

Conservation of momentum:

$$m_0 \gamma \upsilon = \frac{2E}{c} - \frac{E}{c} = \frac{E}{c}. \tag{6}$$

From Eq. (6), therefore,

$$m_0 c^2 \gamma \beta = E. \tag{7}$$

However,

$$E = \frac{1}{3} E_0 = \frac{1}{3} m_0 c^2 \gamma \tag{8}$$

and Eqs. (7) and (8) give $\beta = \frac{1}{3}$.

6.18 A π particle, which is at rest in the laboratory frame of reference, disintegrates into a muon and a neutrino: $\pi \to \mu + \nu$. Show that the energy of μ is $E_\mu = \frac{c^2}{2m_\pi}\left(m_\pi^2 + m_\mu^2 - m_\nu^2\right)$, where m_π, m_μ and m_ν are the rest masses of the three particles, respectively. What is the energy of ν?

Solution

From momentum conservation:

$$\mathbf{p}_{\mu} + \mathbf{p}_{\nu} = 0. \tag{1}$$

Let the two particles, μ and ν, have momenta $\mathbf{p}_{\nu} = p\hat{\mathbf{x}}$ and $\mathbf{p}_{\mu} = -p\hat{\mathbf{x}}$, respectively.

Conservation of energy gives:

$$E_{\mu} + E_{\nu} = m_{\pi}c^2 \tag{2}$$

We also have the relations $E_{\mu}^2 = m_{\mu}^2 c^4 + p^2c^2$ and $E_{\nu}^2 = m_{\nu}^2 c^4 + p^2c^2$.
Taking the difference of the last two equations, we have

$$E_{\mu}^2 - E_{\nu}^2 = \left(m_{\mu}^2 - m_{\nu}^2\right)c^4. \tag{3}$$

Equations (2) and (3) give

$$(E_{\mu} - E_{\nu})(E_{\mu} + E_{\nu}) = (m_{\mu}^2 - m_{\nu}^2)c^4 \qquad \text{or} \qquad (E_{\mu} - E_{\nu})m_{\pi}c^2 = (m_{\mu}^2 - m_{\nu}^2)c^4.$$

Therefore,

$$E_{\mu} = E_{\nu} + \frac{m_{\mu}^2 - m_{\nu}^2}{m_{\pi}}c^2 = m_{\pi}c^2 - E_{\mu} + \frac{m_{\mu}^2 - m_{\nu}^2}{m_{\pi}}c^2$$

and

$$E_{\mu} = \frac{c^2}{2m_{\pi}}\left(m_{\pi}^2 + m_{\mu}^2 - m_{\nu}^2\right).$$

By analogy,

$$E_{\nu} = \frac{c^2}{2m_{\pi}}\left(m_{\pi}^2 + m_{\nu}^2 - m_{\mu}^2\right).$$

As expected, it is $E_{\mu} + E_{\nu} = m_{\pi}c^2$.
The rest mass of the neutrino is either zero or negligible. In this case, the results become:

$$E_{\mu} = \frac{c^2}{2m_{\pi}}\left(m_{\pi}^2 + m_{\mu}^2\right) \qquad \text{and} \qquad E_{\nu} = \frac{c^2}{2m_{\pi}}\left(m_{\pi}^2 - m_{\mu}^2\right).$$

6.19 A stationary particle of rest mass M, disintegrates into a new particle of rest mass m and a photon. Find the energies of the new particle and the photon.

Solution

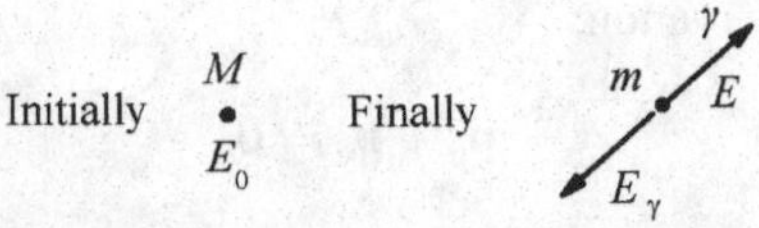

Since originally the momentum is zero in the laboratory frame of reference, after the disintegration the particle produced must have a momentum equal and opposite to that of the photon. The laws of conservation give:

Conservation of energy: $$Mc^2 = mc^2\gamma + E_\gamma \quad (1)$$

Conservation of momentum: $$\frac{E_\gamma}{c} = mc\beta\gamma \quad (2)$$

Substituting for E_γ from Eq. (2) into Eq. (1), we have

$$M = m\gamma + m\beta\gamma = m\gamma(1+\beta) = m\sqrt{\frac{1+\beta}{1-\beta}}.$$

This equation gives

$$\frac{1+\beta}{1-\beta} = \frac{M^2}{m^2}, \qquad \beta = \frac{M^2 - m^2}{M^2 + m^2}$$

and

$$\gamma = \frac{1}{\sqrt{1 - \left(\frac{M^2 - m^2}{M^2 + m^2}\right)^2}} = \frac{M^2 + m^2}{\sqrt{4m^2M^2}} = \frac{M^2 + m^2}{2mM}.$$

Substituting in Eq. (1),

$$E_\gamma = (M - m\gamma)c^2 = \left(M - m\frac{M^2 + m^2}{2mM}\right)c^2 = \left(M - \frac{M^2 + m^2}{2M}\right)c^2.$$

The energy of the photon is, therefore,

$$E_\gamma = \frac{M^2 - m^2}{2M}c^2.$$

The energy of the particle is given by $E = Mc^2 - E_\gamma$, from which it follows that

$$E = \frac{M^2 + m^2}{2M} c^2.$$

6.20 A particle decays, producing a π^+ and a π^-. Both pions have momenta equal to 530 MeV/c and move in directions which are normal to each other. Find the rest mass of the original particle. The rest mass of $\pi^\pm$ is $m_\pi = 140$ MeV/c^2.

Solution

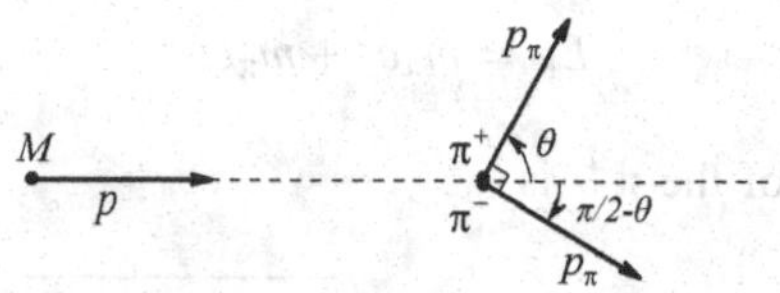

Let the original particle have rest mass M, energy E and momentum p. The two pions produced, π^+ and π^-, have momenta that are normal to each other and have magnitudes equal to p_π. Since the two particles have the same rest mass, they will also have the same energy, E_π.

The laws of conservation give:

Energy: $$E = E_\pi + E_\pi = 2E_\pi \quad (1)$$

y-momentum: $$p_\pi \sin\theta = p_\pi \sin(\pi/2 - \theta) = p_\pi \cos\theta \quad (2)$$

x-momentum: $$p = p_\pi(\cos\theta + \sin\theta) \quad (3)$$

It is given that $p_\pi = 530$ MeV/c and $m_\pi = 140$ MeV/c^2.

Therefore, $E_\pi = \sqrt{(m_\pi c^2)^2 + (p_\pi c)^2} = \sqrt{140^2 + 530^2} = 548$ MeV and $E = 2E_\pi = 1096$ MeV.

Equation (2) gives $\sin\theta = \cos\theta$ and, therefore, $\theta = 45^\circ$ and $\sin\theta = \cos\theta = \sqrt{2}/2$.

Substituting in Eq. (3), $p = 2\frac{\sqrt{2}}{2} p_\pi = 750$ MeV/c.

Finally, $Mc^2 = \sqrt{E^2 - (pc)^2} = \sqrt{1096^2 - 750^2} = 799$ MeV and $M = 799$ MeV/c^2.

6.21 A neutral kaon decays into two pions: $K^0 \to \pi^+ + \pi^-$. If the negative pion produced is at rest, what is the energy of the positive pion? What was the energy of the kaon? Given are the rest masses of K^0, $m_K = 498$ MeV/c^2 and the $\pi^\pm$, $m_\pi = 140$ MeV/c^2.

Solution

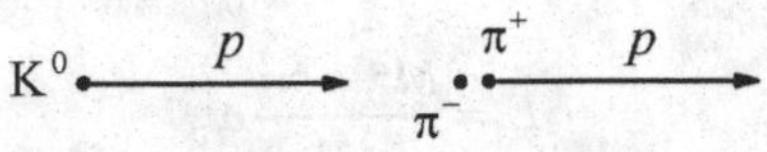

Initially Finally

The decay is shown in the figure. Since π^+ is the only moving particle after the disintegration, conservation of momentum dictates that its momentum is equal to the momentum **p** of K^0. If E_K is the energy of the K^0, conservation of energy gives us

$$E_K = m_\pi c^2 + m_\pi c^2 \gamma. \quad (1)$$

However, the energy of the π^+ is

$$E_\pi = m_\pi c^2 \gamma = \sqrt{(m_\pi c^2)^2 + (pc)^2} \quad (2)$$

and of the kaon

$$E_K = \sqrt{(m_K c^2)^2 + (pc)^2}. \quad (3)$$

Substituting Eqs. (2) and (3) in (1), we get

$$\sqrt{(m_K c^2)^2 + (pc)^2} = m_\pi c^2 + \sqrt{(m_\pi c^2)^2 + (pc)^2}.$$

Squaring,

$$\left(m_K c^2\right)^2 + \cancel{(pc)^2} = \left(m_\pi c^2\right)^2 + \left(m_\pi c^2\right)^2 + \cancel{(pc)^2} + 2m_\pi c^2 \sqrt{\left(m_\pi c^2\right)^2 + (pc)^2}$$

$$(m_K c^2)^2 - 2\left(m_\pi c^2\right)^2 = 2m_\pi c^2 E_\pi$$

and, therefore,

$$E_\pi = m_\pi c^2 \left(\frac{m_K^2}{2m_\pi^2} - 1\right).$$

Substituting in Eq. (1), $E_K = m_\pi c^2 + E_\pi = m_\pi c^2 \left(\frac{m_K^2}{2m_\pi^2}\right)$ or $E_K = \frac{m_K^2 c^2}{2m_\pi}$.

The energy of the positive pion is:

$$E_\pi = 140 \times \left(\frac{498^2}{2 \times 140^2} - 1\right) = 746 \text{ MeV}.$$

For the kaon we find:

$$E_\text{K} = \frac{498^2}{2 \times 140} = 886 \text{ MeV}.$$

6.22 A photon which has an energy equal to E, collides with an electron which is moving in the opposite direction to that of the photon. After the collision, the photon still has an energy E and reverses direction of motion. Show that, for this to happen, the electron must initially have a momentum of magnitude E/c. Also show that the final speed of the electron is $\upsilon = c\Big/\sqrt{1 + (m_0c^2/E)^2}$, where m_0c^2 is its rest energy.

Solution

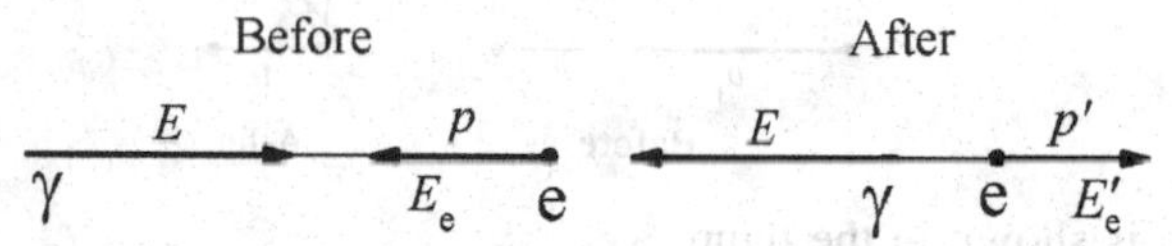

Let E be the energy of the photon and p and p' the momenta and E_e and E'_e the energies of the electron before and after the collision, as shown in the figure. The laws of conservation give:

Energy: $$E + E_\text{e} = E + E'_\text{e} \tag{1}$$

Momentum: $$\frac{E}{c} - |p| = -\frac{E}{c} + |p'| \tag{2}$$

where p and p' have the directions shown in the figure. From Eq. (1) it follows that $E_\text{e} = E'_\text{e}$ and, consequently, also $|p| = |p'|$. Equation (2) then gives

$$\frac{E}{c} - |p| = -\frac{E}{c} + |p| \quad \text{or} \quad \frac{E}{c} - |p| = 0$$

and the total momentum of the system is equal to zero. As a result, Eq. (2) gives

$$p = \frac{E}{c} \quad \text{or} \quad m_0c^2\beta\gamma = E \quad \text{and} \quad \left(\frac{m_0c^2}{E}\right)^2 = \frac{1}{\beta^2\gamma^2} = \frac{1}{\beta^2}(1-\beta^2)$$

$$1 + \left(\frac{m_0c^2}{E}\right)^2 = \frac{1}{\beta^2}, \qquad \beta = \frac{1}{\sqrt{1+(m_0c^2/E)^2}}, \qquad \text{or}$$

$$\upsilon = \frac{c}{\sqrt{1+(m_0c^2/E)^2}}.$$

6.23 A particle which has rest mass m_1 and speed υ_1, collides with a particle at rest, which has a rest mass equal to m_2. The two particles are united into one body which has a rest mass M and moves with speed υ. Show that $\upsilon = (m_1\gamma_1\upsilon_1)/(m_1\gamma_1 + m_2)$ and $M^2 = m_1^2 + m_2^2 + 2\gamma_1 m_1 m_2$, where $\gamma_1 = 1\big/\sqrt{1-\upsilon_1^2/c^2}$.

Solution

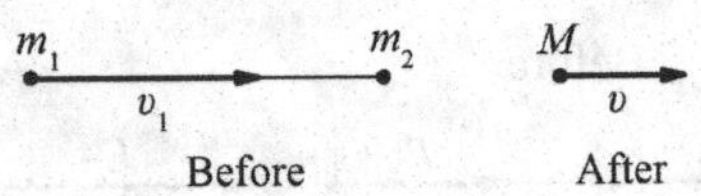

The collision is shown in the figure.
The laws of conservation give

$$\text{Energy:} \qquad m_1c^2\gamma_1 + m_2c^2 = Mc^2\gamma \tag{1}$$

$$\text{Momentum:} \qquad m_1\gamma_1\upsilon_1 = M\gamma\upsilon \tag{2}$$

where $\gamma = 1\big/\sqrt{1-\upsilon^2/c^2}$. Eliminating the product $M\gamma$ between Eqs. (1) and (2), we have

$$m_1\gamma_1\upsilon_1 = m_1\gamma_1\upsilon + m_2\upsilon, \qquad \text{which gives} \qquad \upsilon = \frac{m_1\gamma_1\upsilon_1}{m_1\gamma_1 + m_2}.$$

For this value, it follows that

$$\frac{1}{\gamma^2} = 1 - \beta^2 = 1 - \left(\frac{m_1\gamma_1\beta_1}{m_1\gamma_1 + m_2}\right)^2.$$

Squaring Eq. (1) and substituting for $1/\gamma^2$, we have

$$M^2 = \frac{1}{\gamma^2}(m_1\gamma_1 + m_2)^2 = (m_1\gamma_1 + m_2)^2 - m_1^2\gamma_1^2\beta_1^2,$$

$$M^2 = m_1^2\gamma_1^2 + m_2^2 + 2m_1m_2\gamma_1 - m_1^2\gamma_1^2\beta_1^2, \qquad M^2 = m_1^2\gamma_1^2(1-\beta_1^2) + m_2^2 + 2\gamma_1 m_1 m_2$$

and, finally,

$$M^2 = m_1^2 + m_2^2 + 2\gamma_1 m_1 m_2.$$

6.24 Two identical particles with rest mass m_0 each move towards each other, in the laboratory frame of reference, with the same speed βc. Find the energy of the one particle in the frame of reference of the other.

Solution

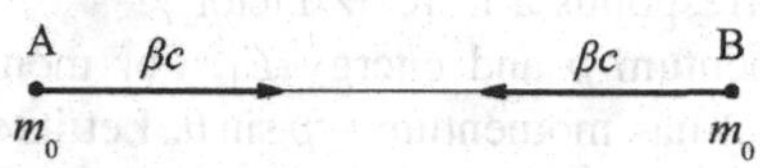

Particles A and B are shown in the figure. We will consider S to be the laboratory frame of reference and S′ to be the frame of reference of particle A. Frame S′ is moving relative to S with velocity $\mathbf{V} = \beta c\hat{\mathbf{x}}$. Particle B, which has a velocity of $-\beta c\hat{\mathbf{x}}$ in S, will have in S′ the speed

$$\upsilon_{\text{BA}} = \upsilon'_{\text{B}} = \frac{\upsilon_{\text{B}} - V}{1 - \frac{\upsilon_{\text{B}} V}{c^2}} = \frac{-\beta c - \beta c}{1 - \frac{(-\beta c)\beta c}{c^2}} = -c\frac{2\beta}{1+\beta^2}.$$

The Lorentz factor corresponding to this speed is

$$\gamma_{\text{BA}} = \frac{1}{\sqrt{1 - \left(\frac{2\beta}{1+\beta^2}\right)^2}} = \frac{1+\beta^2}{\sqrt{1 + 2\beta^2 + \beta^4 - 4\beta^2}} = \frac{1+\beta^2}{1-\beta^2}.$$

The energy of particle B in the frame of reference S′ is, therefore,

$$E_{\text{BA}} = m_0c^2\gamma_{\text{BA}} = m_0c^2\frac{1+\beta^2}{1-\beta^2}.$$

This is the energy of each particle in the frame of reference of the other.

6.25 A particle with rest mass M, moves with velocity $\mathbf{v}$. The particle disintegrates into two others, 1 and 2, with rest masses m_1 and m_2 respectively. If particle 2 moves in a direction perpendicular to $\mathbf{v}$, find the angle θ that the direction of motion of particle 1 makes with $\mathbf{v}$.

Solution

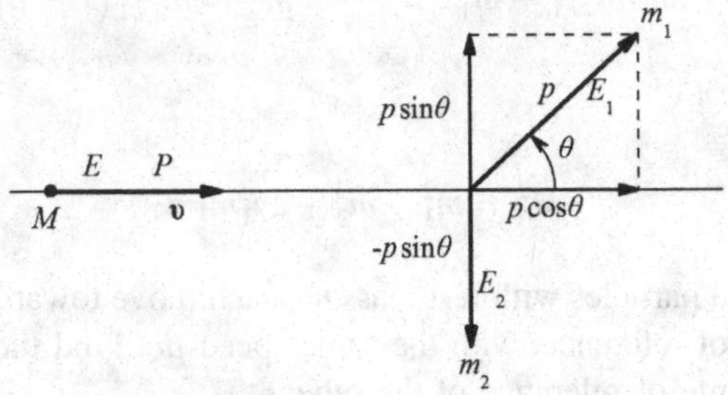

Let the momentum of the original particle be P and its energy E. Its reduced speed is β, to which there corresponds a Lorentz factor γ.

Particle 1 has momentum p and energy E_1. For momentum to be conserved normally to $\mathbf{v}$, particle 2 has momentum $-p\sin\theta$. Let its energy be E_2.

The laws of conservation give:

Energy: $$E = E_1 + E_2 \tag{1}$$

x-momentum: $$M\gamma v = p\cos\theta \tag{2}$$

Because it is $E = Mc^2\gamma$, Eq. (2) can be written as

$$E\beta = cp\cos\theta. \tag{3}$$

It is also true that

$$E_1^2 = \left(m_1c^2\right)^2 + c^2p^2 \tag{4}$$

and

$$E_2^2 = \left(m_2c^2\right)^2 + c^2p^2\sin^2\theta. \tag{5}$$

Subtracting Eq. (4) from Eq. (5),

$$E_2^2 - E_1^2 = \left(m_2c^2\right)^2 - \left(m_1c^2\right)^2 - c^2p^2\cos^2\theta.$$

Making use of Eq. (3), this may be written as

$$E_2^2 - E_1^2 = \left(m_2c^2\right)^2 - \left(m_1c^2\right)^2 - \beta^2E^2.$$

Dividing the left hand side of this equation by $E_1 + E_2$ and the right hand side by E, quantities that are equal according to Eq. (1), we find

$$E_2 - E_1 = \frac{(m_2c^2)^2 - (m_1c^2)^2 - \beta^2E^2}{E}.$$

Taking half the sum of this equation with Eq. (1), we have

$$E_2 = \frac{1}{2}\left[\frac{(m_2c^2)^2 - (m_1c^2)^2 - \beta^2E^2}{E} + E\right] = \frac{(m_2c^2)^2 - (m_1c^2)^2 + (E/\gamma)^2}{2E}$$
$$= \frac{m_2^2 - m_1^2 + M^2}{2\gamma M}c^2,$$

where we have used the fact that $E = \gamma Mc^2$.
Because it is

$$\tan^2\theta = \frac{c^2p^2\sin^2\theta}{c^2p^2\cos^2\theta} = \frac{E_2^2 - m_2^2c^4}{E^2 - M^2c^4} = \frac{(E_2/c^2)^2 - m_2^2}{M^2(\gamma^2 - 1)},$$

with substitution, we find

$$\tan^2\theta = \frac{1}{M^2(\gamma^2 - 1)}\left[\left(\frac{m_2^2 - m_1^2 + M^2}{2\gamma M}\right)^2 - m_2^2\right]$$

and, finally,

$$\tan\theta = \frac{1}{2\beta\gamma^2}\sqrt{\left[1 - \left(\frac{m_1}{M}\right)^2 + \left(\frac{m_2}{M}\right)^2\right]^2 - \left(2\gamma\frac{m_2}{M}\right)^2},$$

which gives the angle θ formed by the direction of motion of particle 1 with the vector $\mathbf{v}$.

The reader must have noticed by now that solving problems in Special Relativity is not a difficult task, once one gets through the masses of algebra involved!

6.26 A Σ particle disintegrates, in motion, into three charged pions, A, B and C. The rest mass of each pion is 140 MeV/c^2. Their kinetic energies are, respectively, $K_A = 190$ MeV, $K_B = 321$ MeV and $K_C = 58$ MeV. The velocities of the pions form with the x-axis angles equal to $\theta_A = 22.4°$, $\theta_B = 0°$ and $\theta_C = -12.25°$, respectively. Evaluate the rest mass of the original particle and the direction of its motion.

Solution

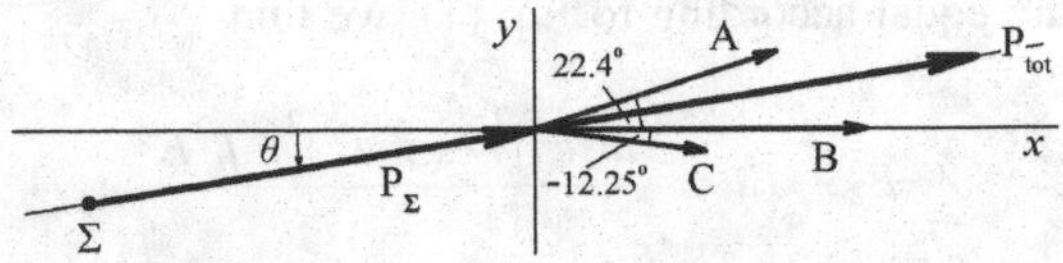

The kinetic and the total energies of the particles are:

A	$K_A = 190$ MeV	$E_A = 190 + 140 = 330$ MeV
B	$K_B = 321$ MeV	$E_B = 321 + 140 = 461$ MeV
C	$K_C = 58$ MeV	$E_C = 58 + 140 = 198$ MeV

Using the relationship $E^2 = m_0^2c^4 + p^2c^2$, we find for the momenta of the particles:

A	$p_Ac = \sqrt{E_A^2 - m_0^2c^4} = \sqrt{330^2 - 140^2} = 299$ MeV	$p_A = 299$ MeV/c
B	$p_Bc = \sqrt{E_B^2 - m_0^2c^4} = \sqrt{461^2 - 140^2} = 439$ MeV	$p_B = 439$ MeV/c
C	$p_Cc = \sqrt{E_C^2 - m_0^2c^4} = \sqrt{198^2 - 140^2} = 140$ MeV	$p_C = 140$ MeV/c

The components of the total momentum of the three particles are found as

$$P_{\text{tot},x}c = p_Ac\cos 22.4^\circ + p_Bc + p_Cc\cos 12.4^\circ = 852 \text{ MeV} \quad P_{\text{tot},x} = 852 \text{ MeV}/c$$
$$P_{\text{tot},y}c = p_Ac\sin 22.4^\circ - p_Cc\sin 12.25^\circ = 84 \text{ MeV} \quad P_{\text{tot},y} = 84 \text{ MeV}/c$$

The magnitude of the momentum of particle Σ is

$$P_\Sigma = P_{\text{tot}} = \sqrt{P_{\text{tot},x}^2 + P_{\text{tot},y}^2} = \sqrt{852^2 + 84^2} = 856 \text{ MeV}/c.$$

The energy of particle Σ is $E_\Sigma = E_A + E_B + E_C = 989$ MeV.
The rest mass of the particle is found, using the relationship $E_\Sigma^2 = M_\Sigma^2c^4 + P_\Sigma^2c^2$, to be

$$M_\Sigma c^2 = \sqrt{E_\Sigma^2 - P_\Sigma^2c^2} = \sqrt{989^2 - 856^2} = 495 \text{ MeV} \qquad \text{or}$$

$$M_\Sigma = 495 \text{ MeV}/c^2.$$

The angle formed by the direction of the Σ particle's momentum with the x-axis is θ, where

$$\tan\theta = \frac{P_{\text{tot},y}}{P_{\text{tot},x}} = \frac{84}{852} = 0.0986 \qquad \text{or} \qquad \theta = 5.6^\circ.$$

6.27 A particle with rest mass M, disintegrates, in motion, into two other particles, 1 and 2. Particle 1 has rest mass m_1, momentum p_1 and energy E_1, while particle 2 has rest mass m_2, momentum p_2 and energy E_2. The angle formed by their directions of motion is θ. Show that $E_1E_2 - p_1p_2c^2\cos\theta = \frac{1}{2}\left(M^2 - m_1^2 - m_2^2\right)c^4$, a quantity that is an invariant.

Solution

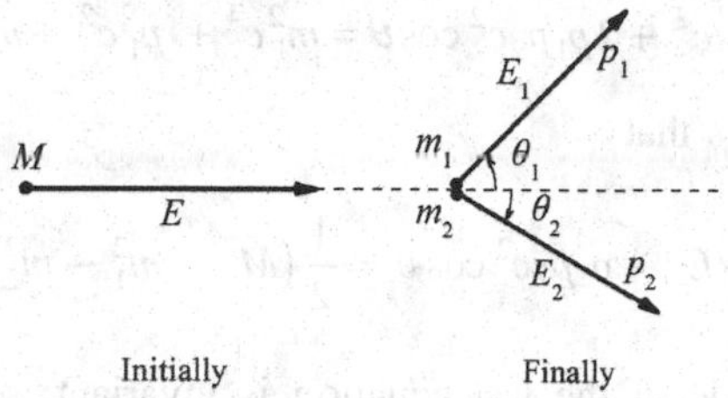

The laws of conservation give:

Energy:

$$E = E_1 + E_2 \tag{1}$$

Component x of momentum:

$$p = p_1\cos\theta_1 + p_2\cos\theta_2 \tag{2}$$

Component y of momentum:

$$0 = p_1\sin\theta_1 - p_2\sin\theta_2 \tag{3}$$

Squaring Eqs. (2) and (3),

$$p^2 = p_1^2\cos^2\theta_1 + p_2^2\cos^2\theta_2 + 2p_1p_2\cos\theta_1\cos\theta_2 \tag{4}$$

$$0 = p_1^2\sin^2\theta_1 + p_2^2\sin^2\theta_2 - 2p_1p_2\sin\theta_1\sin\theta_2 \tag{5}$$

which, when added together, give

$$p^2 = p_1^2 + p_2^2 + 2p_1p_2\cos\theta, \qquad \text{where} \qquad \theta = \theta_1 + \theta_2. \tag{6}$$

From Eq. (1) we have

$$E^2 = E_1^2 + E_2^2 + 2E_1E_2,$$

which, on using the relationship $E^2 = m_0^2 c^4 + p^2 c^2$ for each of the particles, gives

$$\left[\left(Mc^2\right)^2 + (pc)^2\right] = \left[\left(m_1 c^2\right)^2 + (p_1 c)^2\right] + \left[\left(m_2 c^2\right)^2 + (p_2 c)^2\right] + 2E_1 E_2.$$

Using Eq. (6), this leads to

$$M^2 c^4 + \cancel{p_1^2 c^2} + \cancel{p_2^2 c^2} + 2 p_1 p_2 c^2 \cos\theta = m_1^2 c^4 + \cancel{p_1^2 c^2} + m_2^2 c^4 + \cancel{p_2^2 c^2} + 2E_1 E_2$$

from which it follows that

$$E_1 E_2 - p_1 p_2 c^2 \cos\theta = \frac{1}{2}\left(M^2 - m_1^2 - m_2^2\right) c^4.$$

The right-hand-side of the last equation is invariant, so the quantity $E_1 E_2 - p_1 p_2 c^2 \cos\theta$ is also an invariant quantity.

6.28 A neutral pion with energy E disintegrates into two photons, $\pi^0 \to 2\gamma$, with energies E_1 and E_2. Express the angle between the directions of motion of the two photons, θ, in terms of E, the rest mass m_π of the pion and the variable $\varepsilon = (E/2) - E_1 = E_2 - (E/2) = (E_2 - E_1)/2$. What is the minimum value of this angle for $E = 10$ GeV? It is given that $m_\pi c^2 = 0.140$ GeV.

Solution

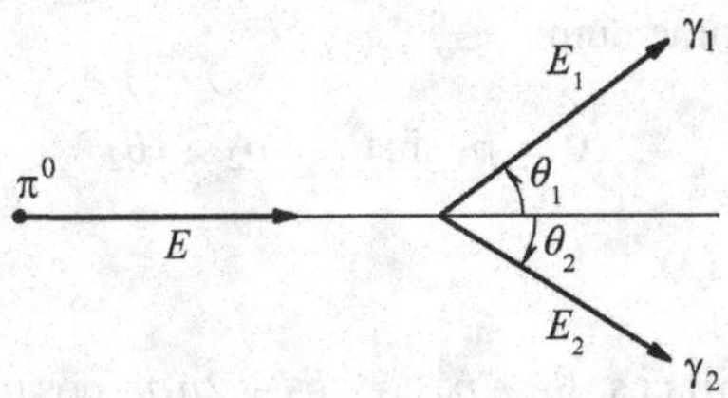

Let the two photons, γ_1 and γ_2 in the figure, move in directions forming angles θ_1 and θ_2, respectively, with the direction of motion of the pion. The laws of conservation give:

Energy:

$$E = E_1 + E_2 \tag{1}$$

x-momentum:

$$m_\pi V \gamma = \frac{E_1}{c}\cos\theta_1 + \frac{E_2}{c}\cos\theta_2 \tag{2}$$

y-momentum:

$$0 = \frac{E_1}{c}\sin\theta_1 - \frac{E_2}{c}\sin\theta_2 \tag{3}$$

Equations (2) and (3) may be rewritten, using $m_\pi c^2\gamma = E$ and $\beta = V/c$, as

$$E\beta = E_1\cos\theta_1 + E_2\cos\theta_2 \tag{4}$$

and

$$0 = E_1\sin\theta_1 - E_2\sin\theta_2. \tag{5}$$

Squaring these two equations and adding, we get

$$E^2\beta^2 = E_1^2 + E_2^2 + 2E_1E_2\cos\theta, \quad \text{where } \theta = \theta_1 + \theta_2. \tag{6}$$

Subtracting Eq. (6) from the square of Eq. (1) it follows that

$$E^2(1-\beta^2) = 2E_1E_2(1-\cos\theta) \qquad \text{or} \qquad (m_\pi c^2)^2 = 4E_1E_2\sin^2(\theta/2) \tag{7}$$

and, therefore,

$$\sin\frac{\theta}{2} = \frac{m_\pi c^2}{2\sqrt{E_1E_2}}. \tag{8}$$

If $E_1 = (E/2) - \varepsilon$ and $E_2 = (E/2) + \varepsilon$, Eq. (8) is written as

$$\sin\frac{\theta}{2} = \frac{m_\pi c^2}{\sqrt{E^2 - 4\varepsilon^2}}. \tag{9}$$

The sine has its minimum value when the denominator in Eq. (9) has its maximum value, i.e. for $\varepsilon = 0$. The minimum value of θ is, therefore, given by

$$\sin\frac{\theta_{\min}}{2} = \frac{m_\pi c^2}{E} = \frac{1}{\gamma}. \tag{10}$$

For $\varepsilon = 0$, it is $E_1 = E_2 = E/2$.

With $E = 10$ GeV and $m_\pi c^2 = 0.140$ GeV, it is $\sin\frac{\theta_{\min}}{2} = \frac{0.140}{10} = 0.0140$, $\theta_{\min} = 1.6^\circ$ and $E_1 = E_2 = 5$ GeV.

6.29 A Λ hyperon disintegrates in motion into a proton and a pion ($\Lambda \to \mathrm{p} + \pi$). These have, in the laboratory frame of reference, momenta p_p and p_π, and energies E_p and E_π, respectively. The angle between p_p and p_π is equal to θ. Show that the energy released in this disintegration is $Q = \sqrt{m_p^2c^4 + m_\pi^2c^4 + 2E_pE_\pi - 2p_pp_\pi c^2 \cos\theta} - (m_p + m_\pi)c^2$, where m_p and m_π are the rest masses of the proton and the pion, respectively.

Solution

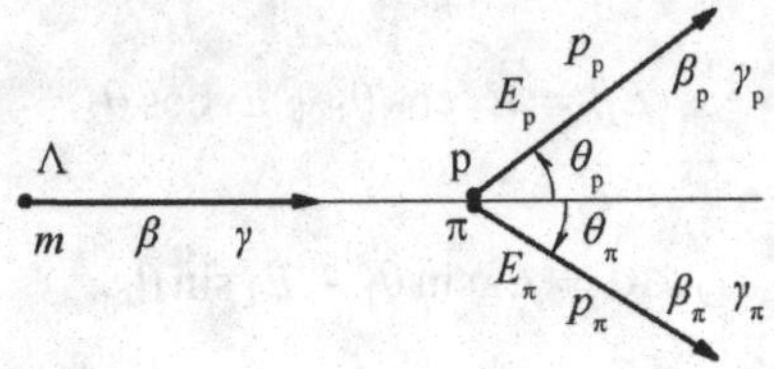

The equation describing the disintegration is

$$\Lambda \to p + \pi + Q$$

where Q is the energy released. The angles formed with the direction of motion of the Λ by the momentum vectors of the two particles, after the disintegration, are θ_p and θ_π. The laws of conservation give:

Energy:

$$mc^2\gamma = E_p + E_\pi \tag{1}$$

x-momentum:

$$mc\beta\gamma = p_p \cos\theta_p + p_\pi \cos\theta_\pi \tag{2}$$

y-momentum:

$$0 = p_p \sin\theta_p - p_\pi \sin\theta_\pi \tag{3}$$

Squaring Eqs. (2) and (3) and adding, we have

$$m^2c^4\beta^2\gamma^2 = p_p^2c^2 + p_\pi^2c^2 + 2p_pp_\pi c^2 \cos\theta \tag{4}$$

where $\theta = \theta_p + \theta_\pi$. The square of Eq. (1) gives

$$m^2c^4\gamma^2 = E_p^2 + E_\pi^2 + 2E_pE_\pi. \tag{5}$$

Subtracting this from Eq. (4), it follows that

$$m^2c^4(1-\beta^2)\gamma^2 = E_p^2 - p_p^2c^2 + E_\pi^2c^2 - p_\pi^2c^2 + 2E_pE_\pi - 2p_pp_\pi c^2 \cos\theta \tag{6}$$

which can also be written as

$$m^2c^4 = m_p^2c^4 + m_\pi^2c^4 + 2E_pE_\pi - 2p_pp_\pi c^2 \cos\theta. \tag{7}$$

If K, K_p and K_π are the kinetic energies of the Λ particle, the proton and the pion, respectively, the energy released during the disintegration is

$$Q = K_p + K_\pi - K = (E_p - m_pc^2) + (E_\pi - m_\pi c^2) - (E - mc^2)$$
$$= mc^2 - (m_p + m_\pi)c^2.$$

This and Eq. (7) give

$$Q = \sqrt{m_p^2c^4 + m_\pi^2c^4 + 2E_pE_\pi - 2p_pp_\pi c^2\cos\theta} - (m_p + m_\pi)c^2.$$

6.30 In the disintegration $K^0 \to \pi^+ + \pi^-$, both the momenta of the particles produced have magnitude equal to $p_\pi = 360\ \text{MeV}/c$ and form an angle of 70° between them. What is the rest mass m_K of K^0 in MeV/c^2? For the pions, it is $m_\pi c^2 = 140$ MeV.

Solution

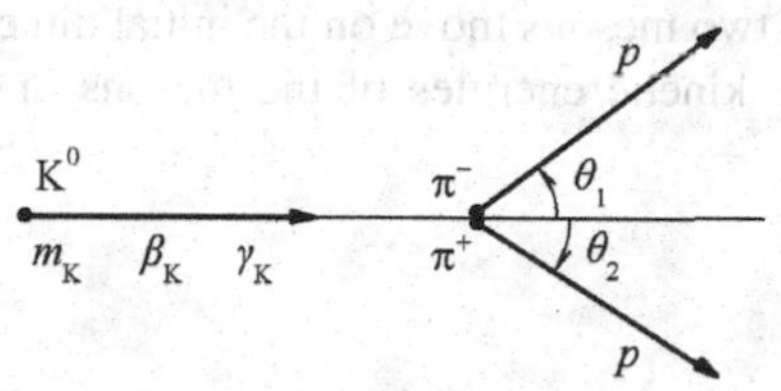

The geometry of the disintegration is shown in the figure. Because the momenta of the two particles produced are equal in magnitude, the particles will have the same energy, E_π say. Also, in order that the transverse momentum should be conserved (equal to zero) it must be

$$\theta_1 = \theta_2 = 70°/2 = 35° \tag{1}$$

The laws of conservation give:

Energy:

$$m_Kc^2\gamma_K = 2E_\pi \tag{2}$$

Longitudinal momentum:

$$m_Kc\beta_K\gamma_K = 2p_\pi\cos\theta_1 \tag{3}$$

Squaring these two equations, after multiplying Eq. (3) by c, and subtracting, we find

$$m_K^2c^4(1-\beta_K^2)\gamma_K^2 = 4E_\pi^2 - 4p_\pi^2c^2\cos^2\theta_1 = 4(m_\pi^2c^4 + p_\pi^2c^2 - p_\pi^2c^2\cos^2\theta_1)$$
$$= 4(m_\pi^2c^4 + p_\pi^2c^2\sin^2\theta_1).$$

As it is $(1-\beta_K^2)\gamma_K^2 = 1$, the last equation can also be written as

$$m_K^2c^4 = 4(m_\pi^2c^4 + p_\pi^2c^2\sin^2\theta_1) \qquad \text{or} \qquad m_Kc^2 = 2\sqrt{m_\pi^2c^4 + p_\pi^2c^2\sin^2\theta_1}.$$

Substituting, we have

$$m_Kc^2 = 2\sqrt{140^2 + 360^2\sin^2 35^\text{o}} = 499\ \text{MeV} \quad \text{and} \quad m_K = 499\ \text{MeV}/c^2.$$

6.31 A K^0 particle, having a rest energy $m_K c^2 = 498$ MeV, disintegrates into two mesons, π^+ and π^-, which have rest masses equal to m_π. In the frame of reference of K^0 both the mesons move with speed $0.83c$.

(a) Find the ratio m_π/m_K of the rest masses and the rest mass m_π of the π particles.

(b) Let the K^0 particle move with a speed of $0.83c$ in the laboratory frame of reference and the two mesons move on the initial direction of motion of the K^0 particle. Find the kinetic energies of the mesons in the laboratory frame of reference.

Solution

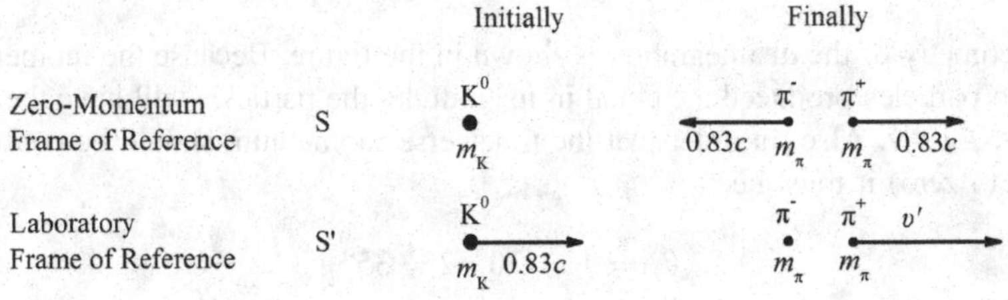

(a) The rest frame of K^0 is the zero-momentum frame of reference, or center-of-mass frame of reference, of the particles π^+ and π^-. In this frame, the two particles move with equal and opposite velocities, with magnitude $0.83c$. For $\beta_\pi = \pm 0.83$, it is $\gamma_\pi = 1.793$. Conservation of energy in the zero-momentum frame of reference gives

$$m_Kc^2 = 2m_\pi\gamma_\pi c^2.$$

Therefore,

$$\frac{m_\pi}{m_K} = \frac{1}{2\gamma_\pi} = \frac{1}{3.59} = 0.279 \quad \text{and} \quad m_\pi c^2 = 139 \text{ MeV}.$$

(b) If we assume that in the laboratory frame of reference, S′, the K^0 particle moves (to the right) with a reduced velocity $\beta'_K = 0.83$ and the π^- particle moves with $\beta = -0.83$ (to the left) in the zero-momentum frame of reference, S, then, in the laboratory frame of reference, particle π^- will be at rest. The speed of π^+ in the frame of reference S′ is

$$\upsilon' = \frac{0.83c - (-0.83c)}{1 - (0.83c)(-0.83c)/c^2} = \frac{1.66c}{1 + 0.83^2} = 0.983c,$$

to which there corresponds a Lorentz factor $\gamma'_\pi = 5.43$.
The kinetic energy of the π^- particle is obviously equal to zero, while that of the π^+ particle is:

$$K'_\pi = m_\pi c^2 (\gamma'_\pi - 1), \qquad K'_\pi = 139 \times 4.43 = 616 \text{ MeV}.$$

6.32 *Symmetric Elastic Collision.* A particle with rest mass m and kinetic energy K collides elastically with a particle at rest that has the same rest mass. After the collision, the two particles move in directions which form equal and opposite angles, $\pm\theta$, with the direction of motion of the initially moving particle. Find angle θ in terms of m and K. *Note:* In the Special Theory of Relativity, the term *elastic collision* implies that the reacting particles remain the same, with the same rest masses, respectively, before and after the collision.

Solution

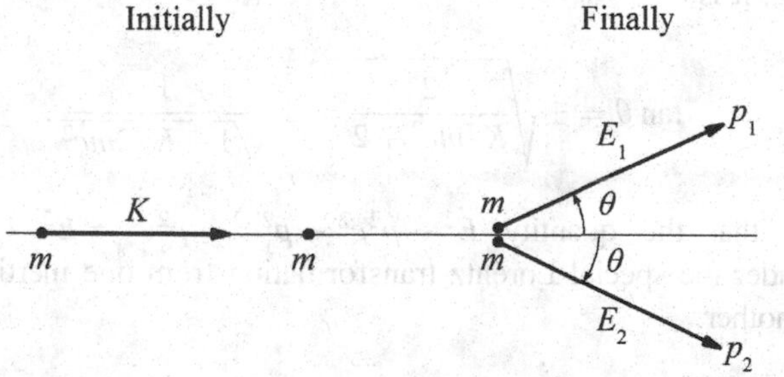

Let E_1, E_2 and p_1, p_2 be the energies and the momenta of the two particles, respectively, after the collision. For transverse momentum to be equal to zero after the collision, it must be $p_1 = p_2$. Since the two particles have equal rest masses, it will also be $E_1 = E_2$.

The laws of conservation of energy and momentum give

$$mc^2\gamma + mc^2 = 2E_1 \quad \text{and} \quad mc\beta\gamma = 2p_1 \cos\theta.$$

Squaring these two equations, we have

$$m^2c^4\gamma^2 + 2m^2c^4\gamma + m^2c^4 = 4E_1^2 \quad \text{and} \quad m^2c^4\beta^2\gamma^2 = 4\left(E_1^2 - m^2c^4\right)\cos^2\theta.$$

Eliminating E_1^2 between them, gives

$$m^2c^4\beta^2\gamma^2 = \left(m^2c^4\gamma^2 + 2m^2c^4\gamma + m^2c^4 - 4m^2c^4\right)\cos^2\theta \quad \text{or}$$
$$\beta^2\gamma^2 = \left(\gamma^2 + 2\gamma + 1 - 4\right)\cos^2\theta.$$

Thus,

$$\cos^2\theta = \frac{\beta^2\gamma^2}{(\gamma - 1)(\gamma + 3)}.$$

Therefore,

$$\sin^2\theta = 1 - \frac{\beta^2\gamma^2}{(\gamma - 1)(\gamma + 3)} = \frac{\gamma^2 + 2\gamma - 3 - \beta^2\gamma^2}{(\gamma - 1)(\gamma + 3)} = \frac{2}{\gamma + 3}$$

and

$$\tan^2\theta = \frac{2(\gamma - 1)}{\beta^2\gamma^2} = \frac{2(\gamma - 1)}{\gamma^2 - 1} = \frac{2}{\gamma + 1}, \qquad \tan\theta = \pm\sqrt{\frac{2}{\gamma + 1}}.$$

This result can also be derived from Eq. (6.68) for $\theta_1 = \theta_2 = \theta$.

The kinetic energy of the initially moving particle is $K = mc^2(\gamma - 1)$, from which $\gamma = 1 + K/mc^2$.

Therefore, it is

$$\tan\theta = \pm\sqrt{\frac{2}{K/mc^2 + 2}} = \pm\frac{1}{\sqrt{1 + K/2mc^2}}.$$

6.33 Show that the quantity $E^2 - p_x^2c^2 - p_y^2c^2 - p_z^2c^2 = E^2 - p^2c^2 = m_0^2c^4$ is invariant under the special Lorentz transformation from one inertial frame of reference to another.

Solution

From the momentum-energy transformation

$$p'_x = \gamma\left(p_x - \frac{\beta}{c}E\right), \qquad p'_y = p_y, \qquad p'_z = p_z, \qquad E' = \gamma(E - c\beta p_x),$$

we have

$$\begin{aligned} E'^2 - p'^2c^2 &= E'^2 - p'^2_x c^2 - p'^2_y c^2 - p'^2_z c^2 \\ &= \gamma^2 (E - c\beta p_x)^2 - \gamma^2\left(p_x - \frac{\beta}{c}E\right)^2 c^2 - p_y^2c^2 - p_z^2c^2 \\ &= E^2\gamma^2(1-\beta^2) - \cancel{2\gamma^2 Ec\beta p_x} - p_x^2c^2\gamma^2(1-\beta^2) + \cancel{2\gamma^2 Ec\beta p_x} - p_y^2c^2 - p_z^2c^2 \\ &= E^2 - p_x^2c^2 - p_y^2c^2 - p_z^2c^2 = E^2 - p^2c^2 = m_0^2c^4 \end{aligned}$$

i.e. $E^2 - p^2c^2 = E'^2 - p'^2c^2 = m_0^2c^4$, an invariant.

6.34 A particle S, which has rest mass M, is stationary in the laboratory frame of reference. The particle disintegrates into a particle with rest mass $M/2$ and a photon. Find:

(a) The speed V of the particle produced, in the laboratory frame of reference.
(b) The energy of the photon, in terms of M,
   (i) in the laboratory frame of reference, E_γ, and
   (ii) in the frame of reference of the particle produced, E'_γ.

Solution

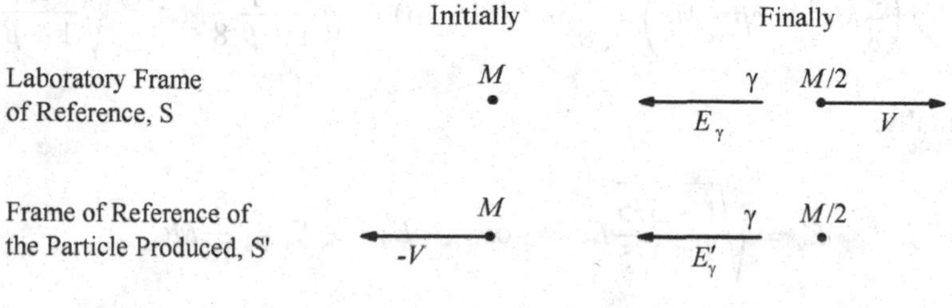

(a) In the laboratory frame of reference, S, if $\gamma = 1/\sqrt{1 - (V/c)^2}$, conservation of energy:

$$Mc^2 = E_\gamma + \frac{M}{2}c^2\gamma, \tag{1}$$

conservation of momentum:

$$0 = -\frac{E_\gamma}{c} + \frac{M}{2}V\gamma. \tag{2}$$

Substituting for E_γ from Eq. (2) into Eq. (1)

$$Mc^2 = \frac{M}{2}Vc\gamma + \frac{M}{2}c^2\gamma \quad \text{or} \quad 1 = \frac{1}{2}\beta\gamma + \gamma\frac{1}{2}, \qquad \gamma(1+\beta) = 2, \quad \sqrt{\frac{1+\beta}{1-\beta}} = 2,$$

from which it follows that $\beta = \frac{3}{5}, \quad \gamma = \frac{5}{4}.$

Therefore,

$$V = \frac{3}{5}c.$$

(b) (i) From Eq. (2)

$$E_\gamma = \frac{\beta\gamma}{2}Mc^2, \qquad E_\gamma = \frac{1}{2}\times\frac{3}{5}\times\frac{5}{4}Mc^2, \qquad E_\gamma = \frac{3}{8}Mc^2.$$

(ii) In S, the photon has energy $E_\gamma = \frac{3}{8}Mc^2$ and momentum $p_{\gamma x} = \dfrac{E_\gamma}{c} = \dfrac{3}{8}Mc$. The frame of reference S′ moves with a speed $V = \beta c$ relative to S. From the transformation of momentum-energy into frame S′, $E' = \gamma\,(E - c\beta\, p_x)$, we find $E'_\gamma = \gamma\left(E_\gamma - c\beta\, p_{\gamma x}\right)$, from which we have

$$E'_\gamma = \gamma\left(\frac{3}{8}Mc^2 - c\beta\,\frac{3}{8}Mc\right) = \frac{3}{8}Mc^2\gamma\,(1-\beta) = \sqrt{\frac{1-\beta}{1+\beta}}\,\frac{3}{8}Mc^2 = \sqrt{\frac{1-\beta}{1+\beta}}\,E_\gamma$$

and, finally,

$$E'_\gamma = \sqrt{\frac{1-3/5}{1+3/5}}\,E_\gamma \qquad \text{or} \qquad E'_\gamma = \frac{1}{2}E_\gamma = \frac{3}{16}Mc^2.$$

6.35 Explain why the following cannot happen:

(a) A photon collides with an electron at rest and imparts all its energy to it.

(b) An isolated photon is transformed into an electron-positron pair. (The positron is the antiparticle of the electron.)

(c) A moving positron collides with an electron at rest and both annihilate into a single photon.

Solution

(a) *A photon collides with an electron at rest and imparts all its energy to it.*

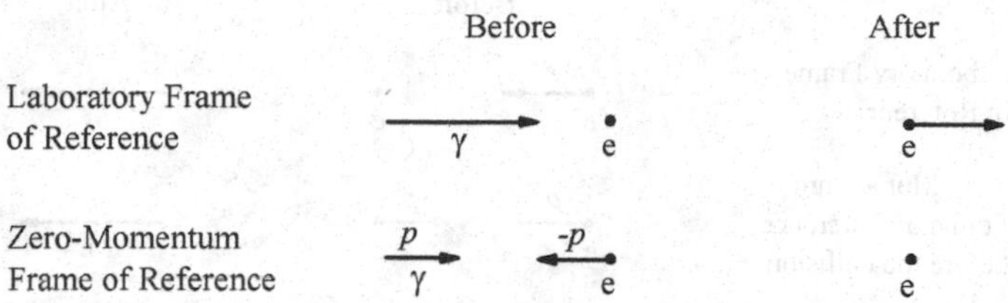

In the zero-momentum frame of reference, for momentum conservation, the electron must be at rest after the collision. However, before the collision, apart from the photon, there was present in this frame of reference a *moving* electron. Energy is obviously not conserved since an electron at rest has less energy than a moving electron plus a photon. This process can therefore not occur.

(b) *An isolated photon is transformed into an electron-positron pair.*

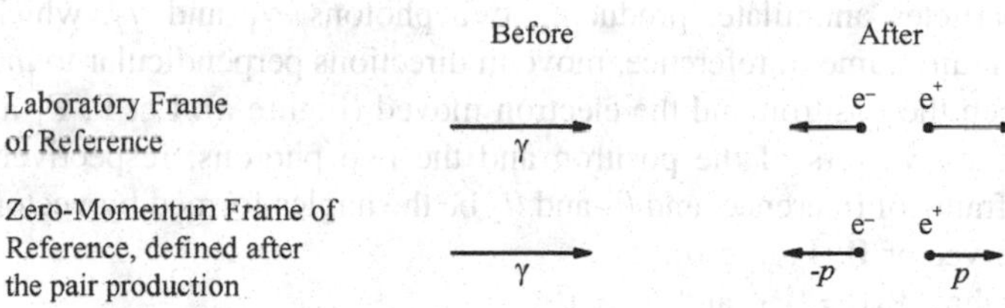

In the zero-momentum frame of reference, defined after the pair production, the electron and the positron have equal and opposite momenta. In the same frame of reference, however, before the pair production, there was a single photon, which obviously had some momentum. If this process was possible, there would be a violation of the conservation of momentum.

Another way of dealing with this problem is to assume that an observer is moving in the direction of the photon with a certain speed V. Naturally, as V increases, the photon appears to this observer to have lower and lower energy due to the Doppler effect. There is a threshold in V, above which the photon will appear to this observer to have an energy smaller than the 1.022 MeV needed for the production of the electron-positron pair. We would then have the unacceptable situation in which the pair may be produced in some frames of reference but not in others. The answer of course is that another body is needed, with which the photon will interact producing the electron-positron pair. As V increases, the kinetic energy of this body increases, making up for any energy loss by the photon and providing enough energy for the pair production to occur.

(c) *A moving positron collides with an electron at rest and both annihilate into a single photon.*

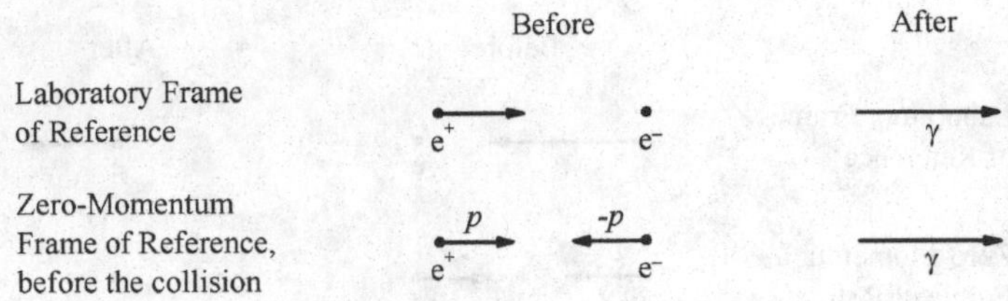

Before the annihilation, in the zero-momentum frame of reference, the electron and the positron have equal and opposite momenta. However, in the same frame of reference, there will be present, after the annihilation, a single photon, which must have some momentum. If this could happen, the law of conservation of momentum would be violated.

6.36 A positron, e^+, with rest mass m and speed v in the laboratory frame of reference, collides with a stationary electron, e^- (which has the same rest mass m). The two particles annihilate, producing two photons, γ_1 and γ_2, which, in the zero-momentum frame of reference, move in directions perpendicular to the straight line on which the positron and the electron moved (Figure a). Let $\mathbf{P}$, $\mathbf{P}_1$ and $\mathbf{P}_2$ be the momentum vectors of the positron and the two photons, respectively, in the laboratory frame of reference, and θ_1 and θ_2 be the angles formed by vectors $\mathbf{P}_1$ and $\mathbf{P}_2$ with the vector $\mathbf{P}$ (Figure b).
(a) Show that $|\mathbf{P}_1| = |\mathbf{P}_2|$ and $\theta_1 = \theta_2$.
(b) Find the angle θ between vectors $\mathbf{P}_1$ and $\mathbf{P}_2$ in terms of v.

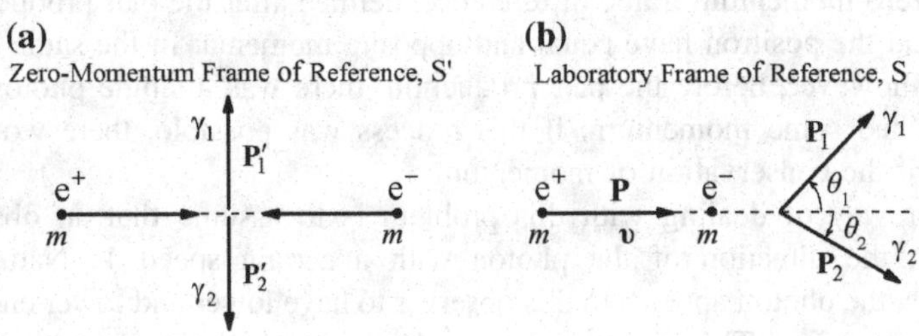

Solution

(a) In the zero-momentum frame of reference $\mathbf{P}'_1 + \mathbf{P}'_2 = 0 \Rightarrow P'_1 = P'_2 \equiv P'$.
Therefore, it is also $E'_1 = E'_2 = cP'$.
If we take momentum $\mathbf{P}'_1$ along the y-axis, we will have:

$$\mathbf{P}'_1 = P'\hat{\mathbf{y}} \qquad \mathbf{P}'_2 = -P'\hat{\mathbf{y}}.$$

The momentum-energy transformation gives

$$P_{1x} = \gamma\left(P'_{1x} + \frac{\beta}{c}E'_1\right) = \beta\gamma P', \qquad P_{1y} = P'_{1y} = P', \qquad P_{1z} = 0$$

$$P_{2x} = \gamma\left(P'_{2x} + \frac{\beta}{c}E'_2\right) = \beta\gamma P', \qquad P_{2y} = P'_{2y} = -P', \qquad P_{2z} = 0$$

and, therefore, $|\mathbf{P}_1| = \sqrt{P_{1x}^2 + P_{1y}^2} = P'\sqrt{\gamma^2\beta^2 + 1} = \gamma P'$,
$|\mathbf{P}_2| = \sqrt{P_{2x}^2 + P_{2y}^2} = P'\sqrt{\gamma^2\beta^2 + 1} = \gamma P'$

$$\tan\theta_1 = \frac{P_{1y}}{P_{1x}} = \frac{P'}{\gamma\beta P'} = \frac{1}{\gamma\beta}, \qquad \tan\theta_2 = \frac{|P_{2y}|}{P_{2x}} = \frac{P'}{\gamma\beta P'} = \frac{1}{\gamma\beta}.$$

Hence, it is $|\mathbf{P}_1| = |\mathbf{P}_2|$ and $\theta_1 = \theta_2$.

(b) Using the values of θ_1 and θ_2, we find that

$$\tan\theta = \tan(\theta_1 + \theta_2) = \frac{\tan\theta_1 + \tan\theta_2}{1 - \tan\theta_1\tan\theta_2} = \frac{2\tan\theta_1}{1 - \tan^2\theta_1} = \frac{2/\gamma\beta}{1 - 1/\gamma^2\beta^2} = \frac{2\gamma\beta}{\gamma^2\beta^2 - 1}$$

$$= \frac{\beta\sqrt{1-\beta^2}}{\beta^2 - 1/2}.$$

Chapter 7
Applications of Relativistic Dynamics

7.1 A photon of energy E_γ collides with an electron at rest, which has a rest mass of m_0. After the collision, the photon moves in a direction opposite to its initial direction of motion. Find, from first principles:

(a) The fraction of the energy E_γ that is given to the electron as kinetic energy.
(b) The change in the wavelength λ of the photon.

Solution

(a) After the collision, the electron moves with speed υ, to which there corresponds a Lorentz factor γ. We will use the laws of conservation of energy and of momentum.

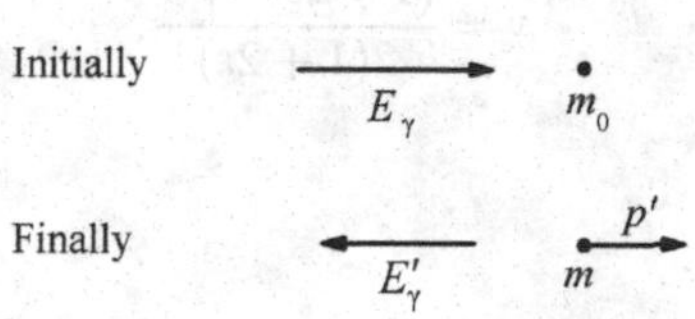

Conservation of energy gives:

$$E_\gamma + m_0c^2 = E'_\gamma + mc^2, \qquad \text{where} \quad m = \gamma m_0.$$

Conservation of momentum gives:

$$\frac{E_\gamma}{c} = -\frac{E'_\gamma}{c} + p', \qquad \text{where} \quad p' = m\upsilon = \beta\gamma c m_0.$$

From these equations we get

$$E_\gamma + m_0c^2 = E'_\gamma + \gamma m_0c^2 \quad \text{and} \quad E_\gamma = -E'_\gamma + \beta\gamma m_0c^2.$$

If we define the variables

$$\varepsilon \equiv \frac{E_\gamma}{m_0c^2}, \qquad \varepsilon' \equiv \frac{E'_\gamma}{m_0c^2},$$

we have $\varepsilon + 1 = \varepsilon' + \gamma$ and $\varepsilon = -\varepsilon' + \beta\gamma$.
Adding,

$$1 + 2\varepsilon = \gamma(1+\beta).$$

Squaring,

$$(1+2\varepsilon)^2 = \gamma^2(1+\beta)^2 = \frac{(1+\beta)^2}{1-\beta^2} = \frac{1+\beta}{1-\beta} \quad \text{and}$$

$$(1+2\varepsilon)^2 - (1+2\varepsilon)^2\beta = 1+\beta$$

from which it follows that

$$\beta = \frac{(1+2\varepsilon)^2 - 1}{(1+2\varepsilon)^2 + 1}$$

and

$$\gamma = \frac{1+2\varepsilon}{1+\beta} = \frac{1+2\varepsilon}{1+\dfrac{(1+2\varepsilon)^2-1}{(1+2\varepsilon)^2+1}} = \frac{(1+2\varepsilon)[(1+2\varepsilon)^2+1]}{2(1+2\varepsilon)^2}.$$

Finally,

$$\gamma = \frac{(1+2\varepsilon)^2+1}{2(1+2\varepsilon)}.$$

The kinetic energy of the electron is $K = m_0c^2(\gamma - 1)$ and, therefore,

$$K = m_0c^2\left[\frac{(1+2\varepsilon)^2+1}{2(1+2\varepsilon)}-1\right] = \frac{m_0c^2}{2}\left(\frac{1+\cancel{4\varepsilon}+4\varepsilon^2+1-2-\cancel{4\varepsilon}}{1+2\varepsilon}\right),$$

$$K = 2E_\gamma\frac{\varepsilon}{1+2\varepsilon}.$$

Thus $\dfrac{K}{E_\gamma} = \dfrac{2\varepsilon}{1+2\varepsilon}$, and, because it is $\varepsilon = \dfrac{E_\gamma}{m_0c^2}$, it follows that

$$\frac{K}{E_\gamma} = \frac{2E_\gamma}{m_0c^2+2E_\gamma}, \quad \text{or} \quad \frac{K}{E_\gamma} = \frac{1}{1+m_0c^2/2E_\gamma}.$$

(b) The kinetic energy of the electron is equal to the change in the energy of the photon:

$$E_\gamma - E'_\gamma = K.$$

From this, it follows that

$$\frac{hc}{\lambda} - \frac{hc}{\lambda'} = \frac{hc}{\lambda}\frac{2\varepsilon}{1+2\varepsilon}, \qquad \frac{1}{\lambda} - \frac{1}{\lambda'} = \frac{1}{\lambda}\frac{2\varepsilon}{1+2\varepsilon}$$

or $\dfrac{\lambda}{\lambda'} = 1 - \dfrac{2\varepsilon}{1+2\varepsilon} = \dfrac{1}{1+2\varepsilon}$, and, finally, $\lambda' = \lambda(1+2\varepsilon)$

$$\lambda' = \lambda\left(1+\frac{2E_\gamma}{m_0c^2}\right) = \lambda\left(1+\frac{2hc}{\lambda m_0c^2}\right) \quad \text{and} \quad \lambda' = \lambda + \frac{2h}{m_0c}.$$

Therefore,

$$\lambda' - \lambda = \frac{2h}{m_0c}.$$

7.2 Show that, during Compton scattering, the scattering angles of the photon and the electron satisfy the relation $\cot\phi = \left(1+\dfrac{E}{m_0c^2}\right)\tan\dfrac{\theta}{2}$.

Solution

Equations (7.2) and (7.3) give: $m\upsilon\cos\phi = \dfrac{E}{c} - \dfrac{E'}{c}\cos\theta$ and $m\upsilon\sin\phi = \dfrac{E'}{c}\sin\theta$.
Taking the ratio of the two equations, we get, using Eq. (7.10) for the ratio E/E',

$$\cot\phi = \frac{\dfrac{E}{c} - \dfrac{E'}{c}\cos\theta}{\dfrac{E'}{c}\sin\theta} = \frac{\dfrac{E}{E'} - \cos\theta}{\sin\theta} = \frac{1 + \dfrac{E}{m_0c^2}(1-\cos\theta) - \cos\theta}{\sin\theta}$$

$$= \left(1 + \frac{E}{m_0c^2}\right)\frac{1-\cos\theta}{\sin\theta}$$

which gives

$$\cot\phi = \left(1 + \frac{E}{m_0c^2}\right)\tan\frac{\theta}{2}.$$

7.3 *The binding energy of an α particle.* The α particle is a helium nucleus, consisting of two protons and two neutrons. Find the binding energy of the α particle and its binding energy per nucleon.

Solution

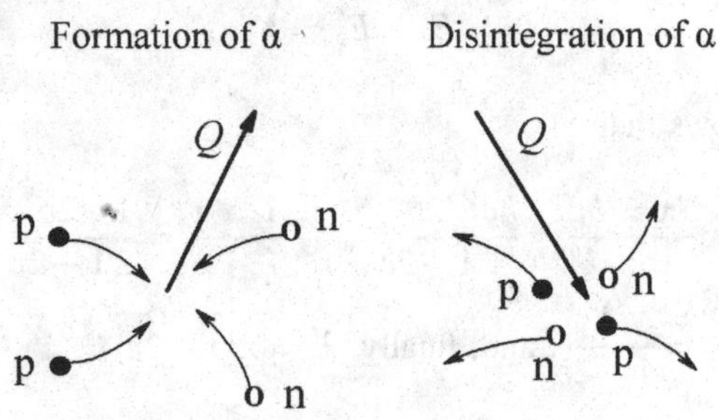

The 'formation' reaction of an α particle is: $2\mathrm{p} + 2\mathrm{n} \to \alpha + Q$.
Q, which must be positive for the α particle to exist, is the energy released during its formation and is its binding energy.
Conservation of mass-energy gives: $2m_\mathrm{p}c^2 + 2m_\mathrm{n}c^2 = m_\alpha c^2 + Q$

$$Q = (2m_\mathrm{p} + 2m_\mathrm{n} - m_\alpha)c^2 = \Delta mc^2$$

where Δm is the mass defect $\Delta m = 2m_\mathrm{p} + 2m_\mathrm{n} - m_\alpha$.
The mass of an α particle may be found from that of a helium-4 atom by subtracting the mass of two electrons and ignoring their binding energy to the nucleus, which is small compared to the other energies:

$$m_\alpha = m_{\mathrm{He}} - 2m_\mathrm{e}.$$

Therefore,

$$\Delta m = (2 \times 1.007\,276 + 2 \times 1.008\,665)\ \mathrm{u} - (4.002\,604 - 2 \times 0.000\,549)\ \mathrm{u}$$

$$\Delta m = 4.031\,882\ \mathrm{u} - 4.001\,506\ \mathrm{u} \quad \Delta m = 0.030\,376\ \mathrm{u}.$$

From the equivalence

$$1\ \mathrm{u} \equiv 931.478\ \mathrm{MeV}$$

it follows that

$$Q = 0.0304 \times 931 = 28.3\ \mathrm{MeV}.$$

The binding energy of the α particle is 28.3 MeV and its binding energy per nucleon is $28.3/4 = 7.07$ MeV/nucleon.

7.4 *Neutron decay.* Free neutrons are unstable and decay with a mean lifetime of $\tau = 898\ \mathrm{s} \approx 15$ minutes, according to $\mathrm{n} \to \mathrm{p} + \mathrm{e}^- + \bar{\nu}_\mathrm{e} + \Delta E$. Find the energy released during the decay.

Solution

The energy released during the decay of the neutron is

$$\Delta E = (m_\mathrm{n} - m_\mathrm{p} - m_\mathrm{e} - m_\nu)c^2\,.$$

The mass of the neutrino, if not zero, is relatively very small and may be ignored in the calculations. Also,

$$m_\mathrm{n}c^2 = 939.67\ \mathrm{MeV} \quad m_\mathrm{p}c^2 = 938.37\ \mathrm{MeV} \quad m_\mathrm{e}c^2 = 0.511\ \mathrm{MeV}$$

and, therefore,

$$\Delta E = 0.79\ \mathrm{MeV}.$$

7.5 *Photofission of uranium-235.* A photon with energy $E_\phi = 6$ MeV collides with a $^{235}_{92}\mathrm{U}$ nucleus at rest and causes the fission $^{235}_{92}\mathrm{U} + \gamma \to {}^{90}_{36}\mathrm{Kr} + {}^{142}_{56}\mathrm{Ba} + 3^1_0\mathrm{n}$. What is the total kinetic energy K of the products of this fission? The isotopic masses are given as $M(^{235}_{92}\mathrm{U}) = 235.043\,930$ u, $M(^{90}_{36}\mathrm{Kr}) = 89.919\,72$ u, $M(^{142}_{56}\mathrm{Ba}) = 141.916\,35$ u and $m_\mathrm{n} = 1.008\,665$ u.

Solution

Conservation of mass-energy gives:

$$M(^{235}_{92}\mathrm{U})c^2 + E_\phi = \left[M(^{90}_{36}\mathrm{Kr}) + M(^{142}_{56}\mathrm{Ba}) + 3m_\mathrm{n}\right]c^2 + K$$

and

$$K = [235.043\,930\text{u} - (89.919\,72 + 141.916\,35 + 3 \times 1.008\,665)\text{u}] \\ \times (931.5\ \text{MeV/u}) + 6\ \text{MeV}$$

$$K = 175\ \text{MeV}.$$

7.6 (a) How much energy is released during the fusion of two deuterium nuclei to form a helium nucleus? The nuclear rest mass of deuterium is $m_D = 2.0136$ u and of helium $m_{He} = 4.0015$ u.

(b) How much energy is released in the fusion of a mass of deuterium equal to 1 kg for the formation of helium? Express your answer in J and in kWh.

Solution

(a) The mass defect during the fusion of two deuterium nuclei for the formation of a helium nucleus is

$$\Delta m = 2 \times m_D - m_{He} = 2 \times 2.0136 - 4.0015 = 0.0257\ \text{u}.$$

The energy equivalent of this mass is

$$\Delta E = \Delta mc^2 = (0.0257\ \text{u}) \times (931.49\ \text{MeV/u}) = 23.9\ \text{MeV}.$$

(b) Because $1\ \text{u} = 1.6605 \times 10^{-27}$ kg, it is $1\ \text{kg} = 1/(1.6605 \times 10^{-27}) = 6.02 \times 10^{26}$ u and this contains $N = 6.02 \times 10^{26}/2.0136 = 3 \times 10^{26}$ deuterium nuclei. During their fusion, 1.5×10^{26} helium nuclei are produced. In the formation of each one of these, the energy released is $23.9\ \text{MeV} = (2.39 \times 10^7\ \text{eV}) \times (1.6 \times 10^{-19}\ \text{J/eV}) = 3.82 \times 10^{-12}$ J.

For $N/2 = 1.5 \times 10^{26}$ fusions, the total energy released is $W = \left(1.5 \times 10^{26}\right) \times (3.82 \times 10^{-12}\ \text{J}) = 5.7 \times 10^{14}$ J.

Because $1\ \text{kWh} = 10^3 \times 60 \times 60 = 3.6 \times 10^6$ J, the energy released is equal to $W = (5.7 \times 10^{14}\ \text{J})/(3.6 \times 10^6\ \text{J/kWh}) = 1.6 \times 10^8\ \text{kWh} = 1.6 \times 10^5\ \text{GWh} =$ 160 TWh.

7.7 The radium isotope $^{226}_{88}\text{Ra}$ decays into radon $^{222}_{86}\text{Rn}$ emitting an α particle, $^{226}_{88}\text{Ra} \rightarrow {}^{222}_{86}\text{Rn} + {}^{4}_{2}\text{He}$. How much energy is released in this decay? The atomic masses $M(^{226}_{88}\text{Ra}) = 226.02541$ u, $M(^{222}_{86}\text{Rn}) = 222.01758$ u and $M(^{4}_{2}\text{He}) =$ 4.00260 u are given.

Solution

Given that the α particles has two positive charges and the atom of $^{222}_{86}\text{Rn}$ produced has two more electrons than it needs for neutrality, the decay may be written as $^{226}_{88}\text{Ra} \rightarrow {}^{222}_{86}\text{Rn}^{2-} + {}^{4}_{2}\text{He}^{2+}$. The variation of the mass during the reaction is, therefore,

$$\delta M = M(^{226}_{88}\text{Ra}) - M(^{222}_{86}\text{Rn}^{2-}) - M(^{4}_{2}\text{He}^{2+}) = M(^{226}_{88}\text{Ra}) - M(^{222}_{86}\text{Rn}) - M(^{4}_{2}\text{He}) =$$
$$= 226.02541 - 222.01758 - 4.00260 = 0.00523\ \text{u}$$

The energy released is:

$$\Delta E = \delta M\, c^2 = (0.00523\ \text{u}) \times (931.494\ \text{MeV/u}) = 4.87\ \text{MeV}.$$

7.8 The β decay of $^{55}_{24}\text{Cr}$ takes place according to $^{55}_{24}\text{Cr} \to {}^{55}_{25}\text{Mn}^{+} + \text{e}^{-} + \bar{\nu}_{\text{e}}$. What kinetic energy is given to the electron and the neutrino if the two nuclei are considered stationary? The atomic masses are given as $M_{\text{Cr-55}} = 54.940\,840$ u and $M_{\text{Mn-55}} = 54.938\,045$ u. By comparison, the mass of the neutrino is negligible, if not zero.

Solution

The variation of the mass during the reaction is:

$$\delta M = M_{\text{Cr-55}} - (M_{\text{Mn-55}} - m_{\text{e}}) - m_{\text{e}} = M_{\text{Cr-55}} - M_{\text{Mn-55}}$$
$$= 54.940\,840 - 54.938\,045 = 0.00280\ \text{u}$$

The energy released is

$$\Delta E = \delta M\, c^2 = (0.00280\ \text{u}) \times (931.494\ \text{MeV/u}) = 2.6\ \text{MeV}$$

and this will be given to the electron and the neutrino as kinetic energy, since the nuclei are considered stationary.

7.9 (a) What is the binding energy per nucleon of the isotope $^{12}_{6}\text{C}$? (b) A $^{12}_{6}\text{C}$ nucleus is to be split into three $^{4}_{2}\text{He}$ nuclei. How much energy is needed for this? It is given that the mass of the $^{12}_{6}\text{C}$ atom is $M_{\text{C}} = 12$ u exactly, of the atom of $^{4}_{2}\text{He}$ $M_{\text{He}} = 4.002603$ u, of the atom of hydrogen $M_{\text{H}} = 1.007825$ u and of the neutron $M_{\text{n}} = 1.008665$ u.

Solution

(a) The mass defect of the nucleus of the isotope $^{12}_{6}\text{C}$ is

$$\Delta M = 6M_{\text{H}} + 6m_{\text{n}} - M_{\text{C}} = 6 \times (1.007825 + 1.008665) - 12 = 0.09894\ \text{u}$$

The binding energy of the nucleus is

$$\Delta E = \Delta M\, c^2 = (0.09894\ \text{u}) \times (931.494\ \text{MeV/u}) = 92.162\ \text{MeV},$$

and its binding energy per nucleon is

$$\frac{\Delta E}{A} = \frac{92.162}{12} = 7.68 \text{ MeV/nucleon.}$$

(b) We want to find ΔE in the reaction ${}^{12}_{6}\text{C} + \Delta E \rightarrow 3{}^{4}_{2}\text{He}$. Using the atomic masses, $M_{\text{C}}c^2 + \Delta E \rightarrow 3M_{\text{He}}c^2$. Therefore, $\Delta E = (3M_{\text{He}} - M_{\text{C}})c^2$ where $3M_{\text{He}} - M_{\text{C}} = 3 \times 4.002603 - 12 = 0.00781$ u.
It follows that

$$\Delta E = (0.00781 \text{ u}) \times (931.494 \text{ MeV/u}) = 7.27 \text{ MeV}.$$

is the energy needed.

7.10 What mass of ${}^{235}_{92}\text{U}$ must undergo fission for the production of thermal energy of 1 GW for a whole day? During the fission of a ${}^{235}_{92}\text{U}$ nucleus, an energy of about 220 MeV is released.

Solution

Power of 1 GW for a day corresponds to an energy of

$$W = 24 \times 60 \times 60 \times 10^9 = 8,64 \times 10^{13} \text{ J}$$
$$= \left(8.64 \times 10^{13} \text{ J}\right)/\left(1.602 \times 10^{-19} \text{ J/eV}\right) = 5.4 \times 10^{32} \text{ eV}.$$

The required number of fissions for this amount of energy to be released is

$$N = \left(5.4 \times 10^{32} \text{ eV}\right)/\left(2.2 \times 10^8 \text{ eV/fission}\right) = 2.5 \times 10^{24} \text{ fissions.}$$

An atom of ${}^{235}_{92}\text{U}$ has a mass equal to

$$m\left({}^{235}_{92}\text{U}\right) = (235.044 \text{ u}) \times \left(1.661 \times 10^{-27} \text{ kg/u}\right) = 3.9 \times 10^{-25} \text{ kg}.$$

The total mass of ${}^{235}_{92}\text{U}$ that must undergo fission is, therefore,

$$M = N \times m\left({}^{235}_{92}\text{U}\right) = \left(2.5 \times 10^{24}\right) \times \left(3.9 \times 10^{-25}\right) = 0.976 \text{ kg}$$

i.e. approximately 1 kg. For comparison, it is worth mentioning that a conventional power station using coal as fuel, would need, for the same amount of energy, 10 000 tons of coal.

7.11 In the *carbon cycle* or the *Bethe cycle*, energy is produced in the stars by the following reactions:

$$^{12}_{6}\mathrm{C}+\mathrm{p}\rightarrow {}^{13}_{7}\mathrm{N}\rightarrow {}^{13}_{6}\mathrm{C}+\mathrm{e}^{+}+\nu_{\mathrm{e}}$$
$$^{13}_{6}\mathrm{C}+\mathrm{p}\rightarrow {}^{14}_{7}\mathrm{N}$$
$$^{14}_{7}\mathrm{N}+\mathrm{p}\rightarrow {}^{15}_{8}\mathrm{O}\rightarrow {}^{15}_{7}\mathrm{N}+\mathrm{e}^{+}+\nu_{e}$$
$$^{15}_{7}\mathrm{N}+\mathrm{p}\rightarrow {}^{12}_{6}\mathrm{C}+{}^{4}_{2}\mathrm{He}$$

Find the total energy released in the cycle. Assume that no orbital electrons are present.

Solution

Adding the four equations together and canceling common terms, we have the net result of the cycle:

$$4\mathrm{p}\rightarrow {}^{4}_{2}\mathrm{He}+2\mathrm{e}^{+}+2\nu_{\mathrm{e}}$$

We will write the mass energy balance. Since we are going to use atomic masses from Table 7.1, we must subtract the masses of the electrons. The mass of the neutrino is negligible. Thus

$$4(M_{\mathrm{H}}-m_{\mathrm{e}})c^{2}=(M_{\mathrm{He}}-2m_{\mathrm{e}})c^{2}+2m_{\mathrm{e}}c^{2}+Q,$$

from which we have

$$Q=(4M_{\mathrm{H}}-M_{\mathrm{He}}-4m_{\mathrm{e}})c^{2}$$
$$=(4\times 1.007\,825\ \mathrm{u}-4.002\,603\ \mathrm{u}-4\times 0.000\,549\ \mathrm{u})\times(931.5\ \mathrm{MeV/u})=24.7\ \mathrm{MeV}.$$

7.12 In the laboratory frame of reference, a moving X particle (with rest mass m), collides with another X particle, at rest, and transforms it to a Y particle (with rest mass $M = 3m$), according to the reaction X + X → X + Y. In the laboratory frame of reference, how much is the threshold energy (kinetic energy) of the moving X particle for this to happen?

Solution

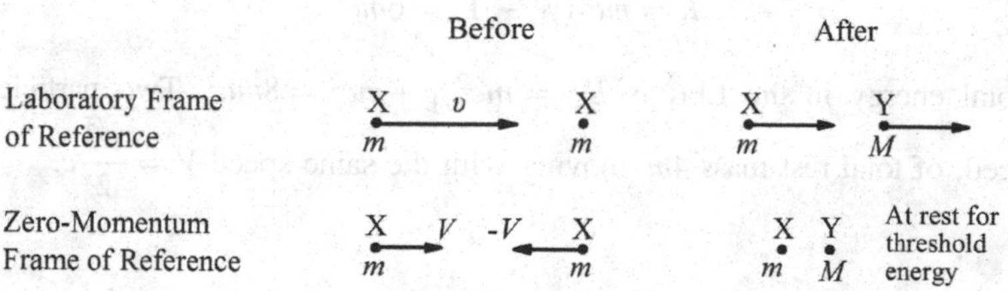

We will initially examine the reaction in the Zero-Momentum Frame of Reference (ZMFR). In this frame of reference, the two original X particles move with equal and opposite velocities, $\pm V$, say. The maximum energy will be available for the creation of the Y particle if the X and Y particles produced are stationary in this frame of reference. In this way the momentum remains equal to zero and the energy available is maximum, since the two particles produced have no kinetic energies. The law of energy conservation gives

$$2mc^2\gamma = (m+M)c^2, \quad \text{where} \quad \gamma = 1/\sqrt{1-(V/c)^2}.$$

Therefore,

$$\gamma = \frac{m+M}{2m} = \frac{m+3m}{2m} = 2 \quad \text{and} \quad V = c\sqrt{1-\frac{1}{\gamma^2}} = \frac{\sqrt{3}}{2}c.$$

Since the particle that is initially at rest in the Laboratory Frame of Reference (LFR) has speed $-V$ in ZMFR, it follows that the ZMFR moves with a speed V relative to the LFR. The speed of the initially moving X particle in the LFR is

$$\upsilon = \frac{V+V}{1+V^2/c^2} = c\frac{2(V/c)}{1+(V/c)^2} = c\frac{\sqrt{3}}{1+3/4}, \qquad \upsilon = \frac{4\sqrt{3}}{7}c.$$

This corresponds to

$$\beta_{\mathrm{L}} = \frac{4\sqrt{3}}{7} \quad \text{and} \quad \gamma_{\mathrm{L}} = 7.$$

Summarizing, we have found that, the X particle initially moves with speed $\upsilon = \frac{4\sqrt{3}}{7}c$ in the LFR, while the produced X and Y particles both move with the same speed $V = \frac{\sqrt{3}}{2}c$.

The threshold energy for the initially moving X particle is its kinetic energy which corresponds to the value $\gamma_{\mathrm{L}} = 7$ found, i.e.

$$K = mc^2(\gamma_{\mathrm{L}} - 1) = 6mc^2.$$

The total energy in the LFR is $E_{\mathrm{tot}} = mc^2\gamma_{\mathrm{L}} + mc^2 = 8mc^2$. Two particles are produced, of total rest mass $4m$, moving with the same speed $V = \frac{\sqrt{3}}{2}c$.

7.13 What is the threshold energy for the production of an electron-positron pair (e^-, e^+) in the collision of a photon, γ, with a stationary electron, according to the reaction $\gamma + e^- \to e^- + e^- + e^+$? For the e^- and e^+, it is given that $m_e c^2 = 0.511$ MeV.

Solution

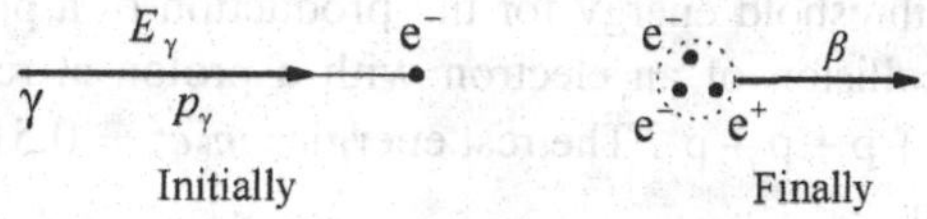

For the available energy to be equal to the threshold energy, the particles produced will be stationary in the zero-momentum frame of reference. In the laboratory frame of reference, therefore, all the particles produced will move with the same velocity, of magnitude βc, say, as shown in the figure.

From the laws of conservation, we have:

Energy:

$$E_\gamma + m_e c^2 = 3 m_e c^2 \gamma \tag{1}$$

Momentum:

$$\frac{E_\gamma}{c} = 3 m_e c \gamma \beta \tag{2}$$

Substituting for E_γ from Eq. (2) in Eq. (1), we have

$$3 m_e c^2 \beta \gamma + m_e c^2 = 3 m_e c^2 \gamma$$

from which there follows that

$$3\beta\gamma + 1 = 3\gamma, \qquad 3(1-\beta)\gamma = 1, \qquad \sqrt{\frac{1-\beta}{1+\beta}} = \frac{1}{3}, \qquad 9(1-\beta) = 1+\beta.$$

Therefore,

$$\beta = \frac{4}{5} \quad \text{and} \quad \gamma = \frac{1}{\sqrt{1-\beta^2}} = \frac{5}{3}.$$

Equation (2) gives

$$E_\gamma = 3 m_e c^2 \beta\gamma = 3 m_e c^2 \frac{4}{5} \times \frac{5}{3} \quad \text{or} \quad E_\gamma = 4 m_e c^2.$$

Putting $m_e c^2 = 0.511$ MeV in $E_\gamma = 4m_e c^2$, we find for the threshold energy of the photon the value of

$$E_\gamma = 2.044 \text{ MeV}.$$

7.14 Which is the threshold energy for the production of a proton-antiproton pair (p, $\bar{\text{p}}$) during the collision of an electron with a proton at rest, according to the reaction $\text{e} + \text{p} \rightarrow \text{e} + \text{p} + \text{p} + \bar{\text{p}}$? The rest energies $m_e c^2 = 0.511$ MeV and $m_p c^2 =$ 938 MeV are given.

Solution

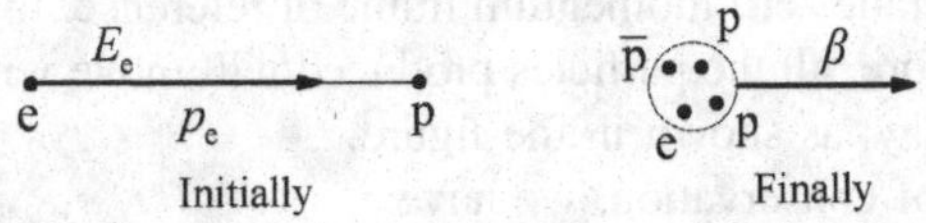

For the available energy to be equal to the threshold energy, the particles produced will be stationary in the zero-momentum frame of reference. In the laboratory frame of reference, therefore, all the particles produced will move with the same velocity, of magnitude βc, say, as shown in the figure.

We have,

conservation of energy

$$E_e + m_p c^2 = (3m_p + m_e)c^2\gamma \quad (1)$$

conservation of momentum

$$p_e c = \sqrt{E_e^2 - (m_e c^2)^2} = (3m_p + m_e)c^2\beta\gamma \quad (2)$$

where E_e and p_e are the energy and momentum, respectively, of the incident electron.

Equation (1) gives

$$E_e^2 = (3m_p + m_e)^2 c^4 \gamma^2 - 2m_p c^2 (3m_p + m_e)c^2\gamma + (m_p c^2)^2$$

and Eq. (2)

$$E_e^2 = (3m_p + m_e)^2 c^4 \beta^2 \gamma^2 + (m_e c^2)^2.$$

Equating and dividing by c^4 we get

$$(3m_p + m_e)^2 \gamma^2 (1 - \beta^2) - 2m_p (3m_p + m_e)\gamma + m_p^2 - m_e^2 = 0.$$

However, $\gamma^2(1-\beta^2)=1$ and so

$$9m_p^{\ 2}+6m_em_p+\cancel{m_e^2}-2m_p(3m_p+m_e)\gamma+m_p^{\ 2}-\cancel{m_e^2}=0$$

$$\gamma=\frac{10m_p^2+6m_em_p}{2m_p(3m_p+m_e)}=\frac{5m_p+3m_e}{3m_p+m_e}.$$

Therefore,

$$E_e=\cancel{(3m_p+m_e)}\frac{5m_p+3m_e}{\cancel{3m_p+m_e}}c^2-m_pc^2=(4m_p+3m_e)c^2.$$

The kinetic energy of the electron is:

$$K_e=E_e-m_ec^2=2(2m_p+m_e)c^2.$$

Substituting, we find the threshold energy $K_e=2(2\times 938+0.511)=$ 3753 MeV = 3.75 GeV.

7.15 Show that, in the laboratory frame of reference, the threshold (kinetic) energy for the production of n pions in the collision of protons with a hydrogen target, $\mathrm{p}+\mathrm{p}\to\mathrm{p}+\mathrm{p}+n\pi$, is $K=2nm_\pi c^2(1+nm_\pi/4m_p)$, where m_π and m_p are, respectively, the rest masses of the pion and the proton. Which is the threshold energy for the reactions $\mathrm{p}+\mathrm{p}\to\mathrm{p}+\mathrm{p}+\pi$ and $\mathrm{p}+\mathrm{p}\to\mathrm{p}+\mathrm{p}+2\pi$, if $m_pc^2=$ 938 MeV and $m_\pi c^2=140$ MeV?

Solution

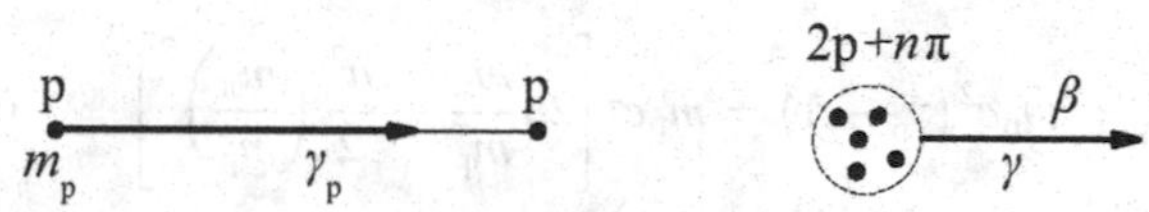

Initially Finally

For the available energy to be equal to the threshold energy, the particles produced will be stationary in the zero-momentum frame of reference. In the laboratory frame of reference, therefore, all the particles produced will move with the same velocity, of magnitude βc, say, as shown in the figure.

We have,

conservation of energy:

$$m_pc^2\gamma_p+m_pc^2=(2m_p+nm_\pi)c^2\gamma$$

conservation of momentum:

$$m_{\mathrm{p}} c\beta_{\mathrm{p}}\gamma_{\mathrm{p}} = (2m_{\mathrm{p}} + nm_{\pi})c\beta\gamma$$

which simplify to

$$m_{\mathrm{p}}\gamma_{\mathrm{p}} + m_{\mathrm{p}} = (2m_{\mathrm{p}} + nm_{\pi})\gamma \qquad \text{and} \qquad m_{\mathrm{p}}\beta_{\mathrm{p}}\gamma_{\mathrm{p}} = (2m_{\mathrm{p}} + nm_{\pi})\beta\gamma.$$

Squaring and subtracting, we get

$$(m_{\mathrm{p}}\gamma_{\mathrm{p}} + m_{\mathrm{p}})^2 - m_{\mathrm{p}}^2\beta_{\mathrm{p}}^2\gamma_{\mathrm{p}}^2 = (2m_{\mathrm{p}} + nm_{\pi})^2\gamma^2(1-\beta^2).$$

Expanding and using the fact that $\gamma^2(1-\beta^2) = 1$, we have

$$m_{\mathrm{p}}^2\gamma_{\mathrm{p}}^2 + 2m_{\mathrm{p}}^2\gamma_{\mathrm{p}} + m_{\mathrm{p}}^2 - m_{\mathrm{p}}^2\beta_{\mathrm{p}}^2\gamma_{\mathrm{p}}^2 = 4m_{\mathrm{p}}^2 + 4nm_{\mathrm{p}}m_{\pi} + n^2m_{\pi}^2$$
$$\text{or} \quad m_{\mathrm{p}}^2 + 2m_{\mathrm{p}}^2\gamma_{\mathrm{p}} + m_{\mathrm{p}}^2 = 4m_{\mathrm{p}}^2 + 4nm_{\mathrm{p}}m_{\pi} + n^2m_{\pi}^2$$

from which it is found that

$$\gamma_{\mathrm{p}} = \frac{1}{2m_{\mathrm{p}}^2}\left(2m_{\mathrm{p}}^2 + 4nm_{\mathrm{p}}m_{\pi} + n^2m_{\pi}^2\right)$$

or

$$\gamma_{\mathrm{p}} = 1 + 2n\frac{m_{\pi}}{m_{\mathrm{p}}} + \frac{n^2}{2}\left(\frac{m_{\pi}}{m_{\mathrm{p}}}\right)^2.$$

The threshold energy is the kinetic energy of the incident proton, i.e.

$$K_{\mathrm{p}} = m_{\mathrm{p}}c^2(\gamma_{\mathrm{p}} - 1) = m_{\mathrm{p}}c^2\left[2n\frac{m_{\pi}}{m_{\mathrm{p}}} + \frac{n^2}{2}\left(\frac{m_{\pi}}{m_{\mathrm{p}}}\right)^2\right] \qquad \text{or}$$
$$K_{\mathrm{p}} = 2nm_{\pi}c^2\left(1 + \frac{nm_{\pi}}{4m_{\mathrm{p}}}\right).$$

In the case $\mathrm{p} + \mathrm{p} \to \mathrm{p} + \mathrm{p} + \pi$, it is $n = 1$ and, therefore, the threshold energy is

$$K_{\mathrm{p1}} = 2 \times 140 \times \left(1 + \frac{140}{4 \times 938}\right) = 290 \text{ MeV}.$$

In the case $\mathrm{p} + \mathrm{p} \to \mathrm{p} + \mathrm{p} + 2\pi$, it is $n = 2$ and, therefore, the threshold energy is

$$K_{\mathrm{p2}} = 2 \times 2 \times 140 \times \left(1 + \frac{2 \times 140}{4 \times 938}\right) = 602 \text{ MeV}.$$

7.16 Evaluate the threshold energy for the reaction $\pi^- + \mathrm{p} \rightarrow \Xi^- + \mathrm{K}^+ + \mathrm{K}^0$, during the incidence of a pion on a proton at rest. The rest energies of the particles involved, π^-, p, Ξ^-, K^+ and K^0, are 140, 938, 1321, 494 and 498 MeV, respectively.

Solution

For the available energy to be equal to the threshold energy, the particles produced will be stationary in the zero-momentum frame of reference. In the laboratory frame of reference, therefore, all the particles produced will move with the same velocity, of magnitude βc, say.

The laws of conservation give:

Conservation of energy: $$m_\pi c^2 \gamma_\pi + m_\mathrm{p} c^2 = (m_{\Xi^-} + m_{\mathrm{K}^+} + m_{\mathrm{K}^0}) c^2 \gamma\,.$$

Conservation of momentum: $$m_\pi c \beta_\pi \gamma_\pi = (m_{\Xi^-} + m_{\mathrm{K}^+} + m_{\mathrm{K}^0}) c \beta \gamma.$$

Having divided these equations by c^2 and by c, respectively, we square and subtract, to get

$$(m_\pi \gamma_\pi + m_\mathrm{p})^2 - (m_\pi \beta_\pi \gamma_\pi)^2 = (m_{\Xi^-} + m_{\mathrm{K}^+} + m_{\mathrm{K}^0})^2 \gamma^2 (1 - \beta^2).$$

Therefore,

$$m_\pi^2 \gamma_\pi^2 + 2 m_\pi m_\mathrm{p} \gamma_\pi + m_\mathrm{p}^2 - m_\pi^2 \gamma_\pi^2 \beta_\pi^2 = (m_{\Xi^-} + m_{\mathrm{K}^+} + m_{\mathrm{K}^0})^2$$

and

$$m_\pi^2 + 2 m_\pi m_\mathrm{p} \gamma_\pi + m_\mathrm{p}^2 = (m_{\Xi^-} + m_{\mathrm{K}^+} + m_{\mathrm{K}^0})^2.$$

Finally,

$$\gamma_\pi = \frac{(m_{\Xi^-} + m_{\mathrm{K}^+} + m_{\mathrm{K}^0})^2 - m_\pi^2 - m_\mathrm{p}^2}{2 m_\pi m_\mathrm{p}}.$$

The kinetic energy of π^- is

$$K_\pi = m_\pi c^2 (\gamma_\pi - 1) = \left[\frac{(m_{\Xi^-} + m_{\mathrm{K}^+} + m_{\mathrm{K}^0})^2 - m_\pi^2 - m_\mathrm{p}^2}{2 m_\mathrm{p}} - m_\pi \right] c^2 =$$

$$= \frac{(m_{\Xi^-} + m_{\mathrm{K}^+} + m_{\mathrm{K}^0})^2 - m_\pi^2 - m_\mathrm{p}^2 - 2 m_\pi m_\mathrm{p}}{2 m_\mathrm{p}} c^2$$

or $$K_\pi = \frac{(m_{\Xi^-} + m_{K^+} + m_{K^0})^2 - (m_\pi + m_p)^2}{2m_p} c^2,$$

which is also the threshold energy. Substituting,

$$K_\pi = \frac{(1321 + 494 + 498)^2 - (140 + 938)^2}{2 \times 938} = 2232 \text{ MeV} = 2.23 \text{ GeV}.$$

7.17 Find the threshold energy for the general case $A + B \to X_1 + X_2 + \ldots + X_i + \ldots + X_N$, in which a particle A falls on a particle B at rest, producing a number N of different particles X_i. The rest masses of the particles are $M_A,\ M_B,\ M_1, \ldots,\ M_i, \ldots,\ M_N$, respectively.

Solution

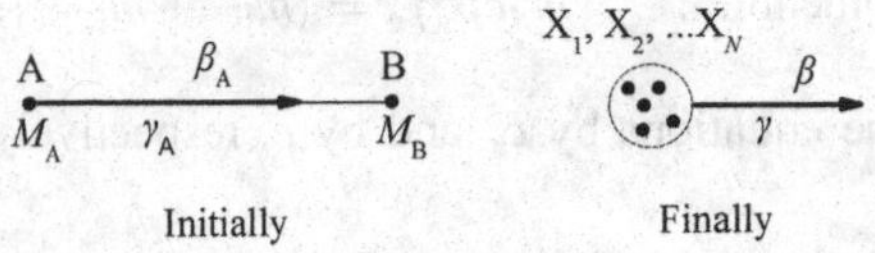

For the available energy to be equal to the threshold energy, the particles produced will be stationary in the zero-momentum frame of reference. In the laboratory frame of reference, therefore, all the particles produced will move with the same velocity, of magnitude βc, say (see figure). Let the reduced speed of A be β_A. We denote the sum of all the rest masses of the particles produced by $M = \sum_i M_i$.

The laws of conservation give:

Conservation of energy:

$$M_A c^2 \gamma_A + M_B c^2 = M c^2 \gamma.$$

Conservation of momentum:

$$M_A c \beta_A \gamma_A = M c \beta \gamma.$$

Having divided these equations by c^2 and by c, respectively, we square and subtract, to get

$$(M_A \gamma_A + M_B)^2 - (M_A \beta_A \gamma_A)^2 = M^2 \gamma^2 (1 - \beta^2).$$

Therefore,

$$M_A^2 \gamma_A^2 + 2 M_A M_B \gamma_A + M_B^2 - M_A^2 \beta_A^2 \gamma_A^2 = M^2$$

and

$$M_A^2 + M_B^2 + 2M_A M_B \gamma_A = M^2.$$

Solving,

$$\gamma_A = \frac{M^2 - M_A^2 - M_B^2}{2M_A M_B}.$$

The threshold energy is

$$E_T = M_A c^2(\gamma_A - 1) = M_A\left(\frac{M^2 - M_A^2 - M_B^2}{2M_A M_B} - 1\right)c^2$$

and, finally,

$$E_T = \frac{M^2 - (M_A + M_B)^2}{2M_B}c^2.$$

7.18 For the case of the rocket which expels mass in its own frame of reference at a constant rate equal to $dM/d\tau = -k$, find the rocket's acceleration in the Earth's frame of reference. Express the result in terms of the variable $\mu = M/M_0$, where M and M_0 are the instantaneous and the initial rest mass of the rocket, respectively.

Solution

We know (first of Eqs. 7.47) that the reduced speed of the rocket, $\beta = V/c$, is given, as a function of the variable $\mu = M/M_0$, by $\beta = \dfrac{\mu^{-\alpha} - \mu^{\alpha}}{\mu^{-\alpha} + \mu^{\alpha}}$.

Differentiating with respect to t,

$$\frac{d\beta}{dt} = \frac{d}{dt}\left(\frac{\mu^{-\alpha} - \mu^{\alpha}}{\mu^{-\alpha} + \mu^{\alpha}}\right) = \frac{d}{d\mu}\left(\frac{\mu^{-\alpha} - \mu^{\alpha}}{\mu^{-\alpha} + \mu^{\alpha}}\right)\frac{d\mu}{dt} = -\frac{4\alpha}{\mu(\mu^{-\alpha} + \mu^{\alpha})^2}\frac{d\mu}{dt}.$$

The proper time τ of the rocket and the time t, as measured by an observer on Earth, satisfy the relation $dt = \gamma d\tau$, where $\gamma = \frac{1}{2}(\mu^{-\alpha} + \mu^{\alpha})$. Substituting in the last equation, we find

$$\frac{d\beta}{dt} = \frac{8\alpha}{\mu(\mu^{\alpha} + \mu^{-\alpha})^3}\left(-\frac{d\mu}{d\tau}\right) \quad \text{or} \quad \frac{d(V/c)}{dt} = \frac{8\alpha}{\mu(\mu^{\alpha} + \mu^{-\alpha})^3}\left(-\frac{1}{M_0}\frac{dM}{d\tau}\right).$$

If the rate of change of the mass of the rocket in its own frame of reference is $\frac{dM}{d\tau} = -k$ and we define the time $\tau_0 = \frac{M_0}{k}$, we may write this equation as

$$\frac{d(V/c)}{d(t/\tau_0)} = \frac{8\alpha}{\mu(\mu^{\alpha} + \mu^{-\alpha})^3}.$$

The reduced acceleration as a function of the reduced rest mass of the rocket is plotted in the figure that follows, for various values of the parameter $\alpha = V_0/c$.

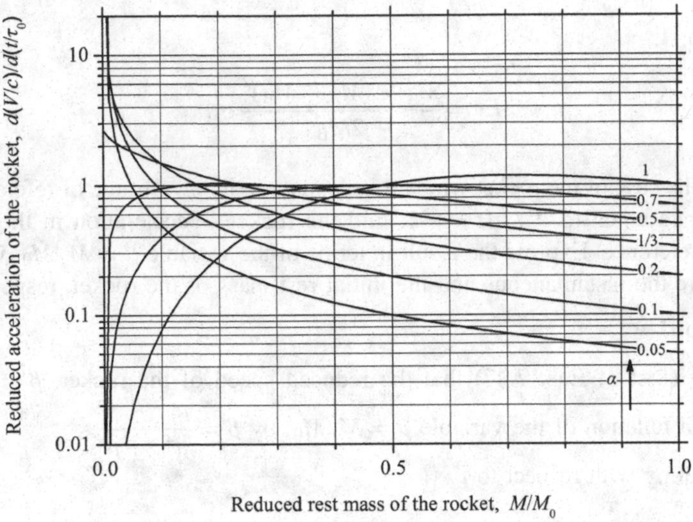

We notice that for $M/M_0 = 1$ the reduced acceleration has a value equal to the corresponding α. The curve for $\alpha = 1/3$ constitutes the separation line between the cases with $\alpha > 1/3$, for which the acceleration goes to zero with the rest mass, and the cases with $\alpha < 1/3$, for which the acceleration tends to infinity as the rest mass tends to zero.

7.19 Examine the motion of a rocket which moves with an exponential decrease of mass in its own frame of reference.

Solution

If the rest mass of the rocket is given as a function of its proper time by the relation $M = M_0 e^{-\kappa\tau}$, where κ is a constant and τ the proper time of the rocket, its rate of change is $dM/d\tau = -\kappa M$.

From Eq. (7.45), which is generally valid, we have

$$\beta = \frac{1-(M/M_0)^{2\alpha}}{1+(M/M_0)^{2\alpha}} = \frac{1-\mathrm{e}^{-2\alpha\kappa\tau}}{1+\mathrm{e}^{-2\alpha\kappa\tau}} = \tanh(\alpha\kappa\tau). \tag{1}$$

Also, from Eq. (7.46) we have

$$\gamma = \frac{1}{2}\left[\left(\frac{M}{M_0}\right)^{-\alpha} + \left(\frac{M}{M_0}\right)^{\alpha}\right] = \frac{\mathrm{e}^{\alpha\kappa\tau}+\mathrm{e}^{-\alpha\kappa\tau}}{2} = \cosh(\alpha\kappa\tau). \tag{2}$$

The times in the two frames of reference, of the rocket, τ, and of the Earth, t, are related through the equation $dt = \gamma d\tau = \cosh(\alpha\kappa\tau)d\tau$, which may be integrated, with $t = 0$ when $\tau = 0$, to give

$$t = \frac{1}{\alpha\kappa}\sinh(\alpha\kappa\tau). \tag{3}$$

Inverting this relationship, we get

$$\tau = \frac{1}{\alpha\kappa}\ln\left(\alpha\kappa t + \sqrt{1+(\alpha\kappa t)^2}\right). \tag{4}$$

The displacement of the rocket. From the relationship $dx = c\beta dt = c\beta\gamma d\tau$ and Eqs. (1) and (2), we have

$$dx = c\sinh(\alpha\kappa\tau)d\tau. \tag{5}$$

Integrating, with the condition $x = 0$ for $\tau = 0$, we have

$$x = \frac{c}{\alpha\kappa}[\cosh(\alpha\kappa\tau) - 1]. \tag{6}$$

Substituting from Eqs. (3) and (6) in the identity $\cosh^2(\alpha\kappa\tau) - \sinh^2(\alpha\kappa\tau) = 1$, we find

$$\left(1+\frac{\alpha\kappa}{c}x\right)^2 - (\alpha\kappa t)^2 = 1 \tag{7}$$

and, finally,

$$x = c\sqrt{\left(\frac{1}{\alpha\kappa}\right)^2 + t^2} - \frac{c}{\alpha\kappa}. \quad (8)$$

We notice that the motion of the rocket is hyperbolic.

The acceleration of the rocket. In the frame of reference of the Earth, it is

$$\frac{dV}{dt} = \frac{dV}{d\tau}\frac{d\tau}{dt} = \frac{1}{\gamma}\frac{dV}{d\tau} = \frac{c}{\gamma}\frac{d\beta}{d\tau} = \frac{c}{\gamma}\frac{\alpha\kappa}{\cosh^2(\alpha\kappa\tau)} = \frac{c\alpha\kappa}{\gamma^3} \quad (9)$$

or

$$\frac{dV}{dt} = \frac{c\alpha\kappa}{\cosh^3(\alpha\kappa\tau)}, \quad (10)$$

which tends to zero with time.

We have already proved that, for rectilinear motion, the acceleration of a body in a frame of reference is γ^3 times smaller than its proper acceleration (Sect. 3.3.1). From Eq. (9), therefore, there follows that the proper acceleration of the rocket, in this case, is constant and equal to $a_0 = c\alpha\kappa$. The case of a body moving with constant proper acceleration is examined in Sect. 4.7.

Chapter 8
Minkowski Spacetime and Four-Vectors

8.1 If $\underline{\mathbf{A}}$ and $\underline{\mathbf{B}}$ are two four-vectors, show that: (a) $\underline{\mathbf{A}} \cdot \underline{\mathbf{B}} = \underline{\mathbf{B}} \cdot \underline{\mathbf{A}}$, (b) $\underline{\mathbf{A}} \cdot (\underline{\mathbf{B}} + \underline{\mathbf{C}}) = \underline{\mathbf{A}} \cdot \underline{\mathbf{B}} + \underline{\mathbf{A}} \cdot \underline{\mathbf{C}}$, (c) $\delta(\underline{\mathbf{A}} \cdot \underline{\mathbf{B}}) = \delta\underline{\mathbf{A}} \cdot \underline{\mathbf{B}} + \underline{\mathbf{A}} \cdot \delta\underline{\mathbf{B}}$ and, in the limit, $d(\underline{\mathbf{A}} \cdot \underline{\mathbf{B}}) = d\underline{\mathbf{A}} \cdot \underline{\mathbf{B}} + \underline{\mathbf{A}} \cdot d\underline{\mathbf{B}}$.

Solution

(a) The inner product of the two four-vectors, $\underline{\mathbf{A}} = (A^0, \mathbf{a})$ and $\underline{\mathbf{B}} = (B^0, \mathbf{b})$, is, according to Eq. (8.23), $\underline{\mathbf{A}} \cdot \underline{\mathbf{B}} = A^0B^0 - \mathbf{a} \cdot \mathbf{b}$. Therefore,

$$\underline{\mathbf{A}} \cdot \underline{\mathbf{B}} = A^0B^0 - \mathbf{a} \cdot \mathbf{b} = B^0A^0 - \mathbf{b} \cdot \mathbf{a} = \underline{\mathbf{B}} \cdot \underline{\mathbf{A}}.$$

(b) If $\underline{\mathbf{A}} = (A^0, \mathbf{a})$, $\underline{\mathbf{B}} = (B^0, \mathbf{b})$ and $\underline{\mathbf{C}} = (C^0, \mathbf{c})$, then

$$\begin{aligned}\underline{\mathbf{A}} \cdot (\underline{\mathbf{B}} + \underline{\mathbf{C}}) &= (A^0, \mathbf{a}) \cdot \left[(B^0, \mathbf{b}) + (C^0, \mathbf{c})\right] = (A^0, \mathbf{a}) \cdot \left[(B^0 + C^0), (\mathbf{b} + \mathbf{c})\right] \\ &= A^0(B^0 + C^0) - \mathbf{a} \cdot (\mathbf{b} + \mathbf{c}) = A^0B^0 + A^0C^0 - \mathbf{a} \cdot \mathbf{b} - \mathbf{a} \cdot \mathbf{c} \\ &= (A^0B^0 - \mathbf{a} \cdot \mathbf{b}) + (A^0C^0 - \mathbf{a} \cdot \mathbf{c}) = \underline{\mathbf{A}} \cdot \underline{\mathbf{B}} + \underline{\mathbf{A}} \cdot \underline{\mathbf{C}}\end{aligned}$$

(c) If $\underline{\mathbf{A}} = (A^0, \mathbf{a})$ and $\underline{\mathbf{B}} = (B^0, \mathbf{b})$, it follows that, for very small variations,

$$\delta(\underline{\mathbf{A}}\cdot\underline{\mathbf{B}}) = \delta\left[A^0B^0 - \mathbf{a}\cdot\mathbf{b}\right] = \delta A^0B^0 + A^0\delta B^0 - \delta\mathbf{a}\cdot\mathbf{b} - \mathbf{a}\cdot\delta\mathbf{b}$$
$$= \left(\delta A^0B^0 - \delta\mathbf{a}\cdot\mathbf{b}\right) + \left(A^0\delta B^0 - \mathbf{a}\cdot\delta\mathbf{b}\right)$$
$$= \left(\delta A^0, \delta\mathbf{a}\right)\cdot\left(B^0, \mathbf{b}\right) + \left(A^0, \mathbf{a}\right)\cdot\left(\delta B^0, \delta\mathbf{b}\right) = \delta\underline{\mathbf{A}}\cdot\underline{\mathbf{B}} + \underline{\mathbf{A}}\cdot\delta\underline{\mathbf{B}}$$

In the limit, this becomes $d(\underline{\mathbf{A}}\cdot\underline{\mathbf{B}}) = d\underline{\mathbf{A}}\cdot\underline{\mathbf{B}} + \underline{\mathbf{A}}\cdot d\underline{\mathbf{B}}$.

8.2 If $\underline{\mathbf{U}}$ is the four-vector of the velocity of a particle and $\underline{\mathbf{A}}$ the four-vector of its acceleration, show that the two four-vectors are orthogonal, i.e. that it is $\underline{\mathbf{U}}\cdot\underline{\mathbf{A}} = 0$.

Solution

From Eqs. (8.53) and (8.63), the four-vectors of velocity and acceleration are, respectively,

$$\underline{\mathbf{U}} = (\gamma_P c,\ \gamma_P \boldsymbol{\upsilon}) \qquad \text{and} \qquad \underline{\mathbf{A}} = \left(\gamma_P \frac{d(\gamma_P c)}{dt},\ \gamma_P \frac{d(\gamma_P \boldsymbol{\upsilon})}{dt}\right).$$

Therefore, it is

$$\underline{\mathbf{U}}\cdot\underline{\mathbf{A}} = (\gamma_P c,\ \gamma_P \boldsymbol{\upsilon})\cdot\left(\gamma_P \frac{d(\gamma_P c)}{dt},\ \gamma_P \frac{d(\gamma_P \boldsymbol{\upsilon})}{dt}\right) = \gamma_P(c\gamma_P)\frac{d(\gamma_P c)}{dt} - \gamma_P(\gamma_P\boldsymbol{\upsilon})\cdot\frac{d(\gamma_P\boldsymbol{\upsilon})}{dt} =$$
$$= \frac{\gamma_P}{2}\left[\frac{d(c\gamma_P)^2}{dt} - \frac{d(\gamma_P\boldsymbol{\upsilon})^2}{dt}\right] = \frac{\gamma_P}{2}\frac{d}{dt}\left[\gamma_P^2(c^2 - \upsilon^2)\right] = \frac{\gamma_P}{2}\frac{d}{dt}(c^2) = 0$$

8.3 Show that the four-vector of acceleration is $\underline{\mathbf{A}} = \gamma_P(\dot{\gamma}_P c,\ \dot{\gamma}_P\boldsymbol{\upsilon} + \gamma_P\boldsymbol{a})$, where $\dot{\gamma}_P = d\gamma_P/dt$ and $\boldsymbol{a} = d\boldsymbol{\upsilon}/dt$ is the acceleration three-vector.

Solution

The four-vector of velocity is $\underline{\mathbf{U}} = (\gamma_P c,\ \gamma_P\boldsymbol{\upsilon})$.
Therefore,

$$\underline{\mathbf{A}} = \frac{d\underline{\mathbf{U}}}{d\tau} = \frac{d\underline{\mathbf{U}}}{dt}\frac{dt}{d\tau} = \gamma_P\frac{d\underline{\mathbf{U}}}{dt} = \gamma_P\frac{d}{dt}(\gamma_P c,\ \gamma_P\boldsymbol{\upsilon}) = \gamma_P(\dot{\gamma}_P c,\ \dot{\gamma}_P\boldsymbol{\upsilon} + \gamma_P\boldsymbol{a}).$$

8.4 Use the result of Problem 8.3 in order to find the four-acceleration of a particle in its own frame of reference, i.e. in the inertial frame of reference in which the particle is momentarily at rest, and show that this is equal to zero only if the particle's proper acceleration, $\boldsymbol{\alpha}$, is zero.

Solution

The result of Problem 8.3 is: $\underline{\mathbf{A}} = \gamma_P(\dot{\gamma}_P c,\ \dot{\gamma}_P\boldsymbol{\upsilon} + \gamma_P\boldsymbol{a})$.

In the inertial frame of reference in which the particle is momentarily at rest, it is $\boldsymbol{\upsilon} = 0$, in which case it is also $\gamma_P = 1$ and $\dot{\gamma}_P = 0$. The three-acceleration of the particle is then its vector proper acceleration, $\boldsymbol{\alpha}$. Therefore, $\underline{\mathbf{A}} = (0,\ \boldsymbol{\alpha})$.

Since it is $\underline{\mathbf{A}}^2 = -\boldsymbol{\alpha}^2 = -\alpha^2$, the four-acceleration of the particle vanishes only if it is $\boldsymbol{\alpha} = 0$.

8.5 Let $\underline{\mathbf{U}}$ and $\underline{\mathbf{V}}$ be the velocity four-vectors of two particles. Evaluating their inner product in the frame of reference of the particle with four-velocity $\underline{\mathbf{U}}$, show that it is $\underline{\mathbf{U}} \cdot \underline{\mathbf{V}} = c^2\gamma(\upsilon)$, where $\gamma(\upsilon)$ is the Lorentz factor corresponding to the relative speed υ of the particles.

Solution

In the frame of reference of the particle to which corresponds the four-velocity $\underline{\mathbf{U}}$, it is $\underline{\mathbf{U}} = (c,\ 0)$ and the other particle has a velocity equal to $\boldsymbol{\upsilon}$. The other particle's four-velocity is, therefore, $\underline{\mathbf{V}} = (\gamma(\upsilon)c,\ \gamma(\upsilon)\boldsymbol{\upsilon})$, in which case $\underline{\mathbf{U}} \cdot \underline{\mathbf{V}} = (c,\ 0) \cdot (\gamma(\upsilon)c,\ \gamma(\upsilon)\boldsymbol{\upsilon}) = c^2\gamma(\upsilon)$.

8.6 A photon and a particle have four-momenta $\underline{\mathbf{P}}$ and $\underline{\mathbf{Q}}$, respectively. Show that, in the frame of reference of the particle, it is $\underline{\mathbf{P}} \cdot \underline{\mathbf{Q}} = hfm_0$, where f is the frequency of the photon and m_0 the rest mass of the particle.

Solution

If $\hat{\mathbf{k}}$ is the unit vector in the direction of propagation of the photon, it is $\underline{\mathbf{P}} = \frac{h}{c}f(1,\ \hat{\mathbf{k}})$.

For the particle, with energy E_P and momentum $\mathbf{p}_P$, it is $\underline{\mathbf{Q}} = \left(\frac{E_P}{c},\ \mathbf{p}_P\right)$.

Therefore,

$$\underline{\mathbf{P}} \cdot \underline{\mathbf{Q}} = \frac{h}{c}f\left(\frac{E_P}{c} - \mathbf{p}_P \cdot \hat{\mathbf{k}}\right).$$

In the frame of reference of the particle it is $\mathbf{p}_P = 0$ and $E_P = m_0c^2$. Then, it is $\underline{\mathbf{P}} \cdot \underline{\mathbf{Q}} = hfm_0$.

8.7 Two particles have four-momenta $\underline{\mathbf{P}}_1$ and $\underline{\mathbf{P}}_2$, respectively, and relative speed between them υ. If m_{10} and $m_{20}$are the rest masses of the particles, m_1 is the mass of the first particle in the frame of reference of the second and m_2 is the mass of the second particle in the frame of reference of the first, show that $\underline{\mathbf{P}}_1 \cdot \underline{\mathbf{P}}_2 = m_{10}m_{20}c^2\gamma(\upsilon) = m_{10}m_2c^2 = m_{20}m_1c^2$.

Solution

If $\boldsymbol{\upsilon}_1$ and $\boldsymbol{\upsilon}_2$ are the velocities of the two particles and γ_1 and γ_2 are the Lorentz factors associated with them, it is $\underline{\mathbf{P}}_1 = \gamma_1 m_{10}(c,\ \boldsymbol{\upsilon}_1)$ and $\underline{\mathbf{P}}_2 = \gamma_2 m_{20}(c,\ \boldsymbol{\upsilon}_2)$. Then,

$$\underline{\mathbf{P}}_1 \cdot \underline{\mathbf{P}}_2 = \gamma_1\gamma_2 m_{10}m_{20}(c^2 - \boldsymbol{\upsilon}_1 \cdot \boldsymbol{\upsilon}_1) = m_{10}m_{20}c^2\gamma_1\gamma_2\left(1 - \frac{\boldsymbol{\upsilon}_1 \cdot \boldsymbol{\upsilon}_1}{c^2}\right).$$

The transformation of the Lorentz factor [Eq. (3.48)] gives $\gamma(\upsilon) = \gamma_1\gamma_2\left(1 - \frac{\boldsymbol{\upsilon}_1 \cdot \boldsymbol{\upsilon}_1}{c^2}\right)$.

Then, $\underline{\mathbf{P}}_1 \cdot \underline{\mathbf{P}}_2 = m_{10}m_{20}c^2\gamma(\upsilon)$. This also follows from the result of Problem 8.5.

The masses of each particle in the frame of reference of the other are given by $m_1 = m_{10}\gamma(\upsilon)$ and $m_2 = m_{20}\gamma(\upsilon)$. It follows that it is also $\underline{\mathbf{P}}_1 \cdot \underline{\mathbf{P}}_2 = m_{10}m_2c^2 = m_{20}m_1c^2$.

8.8 Consider an elastic collision between two particles. If the four-momenta of the two particles are, respectively, $\underline{\mathbf{P}}_1$ and $\underline{\mathbf{P}}_2$ before, and $\underline{\mathbf{P}}'_1$ and $\underline{\mathbf{P}}'_2$ after the collision, show that $\underline{\mathbf{P}}_1 \cdot \underline{\mathbf{P}}_2 = \underline{\mathbf{P}}'_1 \cdot \underline{\mathbf{P}}'_2$.

Solution

Conservation of momentum gives $\underline{\mathbf{P}}_1 + \underline{\mathbf{P}}_2 = \underline{\mathbf{P}}'_1 + \underline{\mathbf{P}}'_2$.

Squaring both sides, $\underline{\mathbf{P}}_1^2 + \underline{\mathbf{P}}_2^2 + 2\underline{\mathbf{P}}_1 \cdot \underline{\mathbf{P}}_2 = \underline{\mathbf{P}}_1'^2 + \underline{\mathbf{P}}_2'^2 + 2\underline{\mathbf{P}}'_1 \cdot \underline{\mathbf{P}}'_2$.

Since in an elastic collision the rest masses of the particles do not change, it is $\underline{\mathbf{P}}_1^2 = \underline{\mathbf{P}}_1'^2$ and $\underline{\mathbf{P}}_2^2 = \underline{\mathbf{P}}_2'^2$. Therefore, $\underline{\mathbf{P}}_1 \cdot \underline{\mathbf{P}}_2 = \underline{\mathbf{P}}'_1 \cdot \underline{\mathbf{P}}'_2$.

From the result of Problem 8.7, this equation means that the relative speed of the two particles remains the same before and after the collision.

8.9 Show that for two photons with energies E_1 and E_2 and four-momenta $\underline{\mathbf{P}}_1$ and $\underline{\mathbf{P}}_2$, it is $\underline{\mathbf{P}}_1 \cdot \underline{\mathbf{P}}_2 = (E_1E_2/c^2)(1 - \cos\theta)$, where θ is the angle between the directions of motion of the two photons.

Solution

If $\mathbf{p}_1$ and $\mathbf{p}_2$ are the three-momenta of the two photons, then their momentum four-vectors are $\underline{\mathbf{P}}_1 = (E_1/c,\ \mathbf{p}_1)$ and $\underline{\mathbf{P}}_2 = (E_2/c,\ \mathbf{p}_2)$, respectively. It follows that $\underline{\mathbf{P}}_1 \cdot \underline{\mathbf{P}}_2 = (E_1E_2/c^2 - \mathbf{p}_1 \cdot \mathbf{p}_2)$. However, it is $\mathbf{p}_1 \cdot \mathbf{p}_2 = (E_1/c)(E_2/c)\cos\theta$ and so $\underline{\mathbf{P}}_1 \cdot \underline{\mathbf{P}}_2 = (E_1E_2/c^2)(1 - \cos\theta)$.

8.10 Use four-vectors to prove Eq. (7.10) for the Compton effect.

Solution

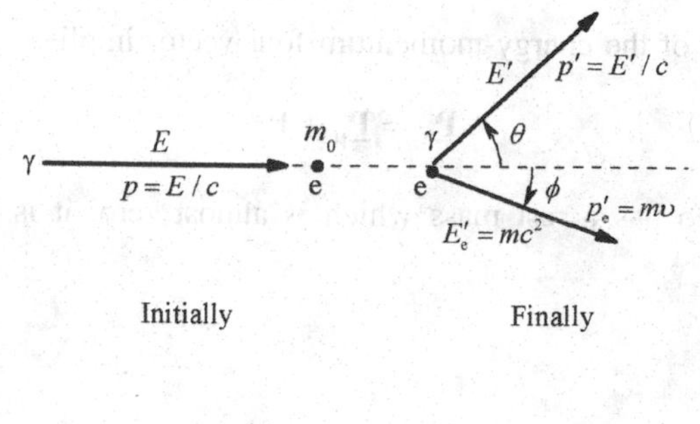

The energy-momentum four-vectors of the photon and the electron involved, are, respectively, before the scattering,

$$\underline{\mathbf{P}}_{\gamma} = \frac{1}{c}(E,\, E,\, 0,\, 0), \quad \underline{\mathbf{P}}_{\mathrm{e}} = (m_0 c,\, 0,\, 0,\, 0)$$

and after the scattering,

$$\underline{\mathbf{P}}'_{\gamma} = \frac{1}{c}(E',\, E' \cos\theta,\, E' \sin\theta,\, 0),$$

$$\underline{\mathbf{P}}'_{\mathrm{e}} = (\gamma m_0 c,\, \gamma m_0 \upsilon \cos\phi,\, -\gamma m_0 \upsilon \sin\phi,\, 0).$$

The conservation of four-momentum gives $\underline{\mathbf{P}}_{\gamma} + \underline{\mathbf{P}}_{\mathrm{e}} = \underline{\mathbf{P}}'_{\gamma} + \underline{\mathbf{P}}'_{\mathrm{e}}$.
We will use the fact that $\underline{\mathbf{P}}'^2_{\mathrm{e}} = m_0^2 c^2$ is a Lorentz invariant. Therefore,

$$\left(\underline{\mathbf{P}}_{\gamma} + \underline{\mathbf{P}}_{\mathrm{e}} - \underline{\mathbf{P}}'_{\gamma}\right)^2 = m_0^2 c^2.$$

It is $\underline{\mathbf{P}}^2_{\mathrm{e}} = m_0^2 c^2$ and, for the photons, $\underline{\mathbf{P}}^2_{\gamma} = 0$, $\underline{\mathbf{P}}'^2_{\gamma} = 0$. Expanding the square,

$$\underline{\mathbf{P}}^2_{\gamma} + \underline{\mathbf{P}}^2_{\mathrm{e}} + \underline{\mathbf{P}}'^2_{\gamma} + 2\underline{\mathbf{P}}_{\gamma} \cdot \underline{\mathbf{P}}_{\mathrm{e}} - 2\underline{\mathbf{P}}'_{\gamma} \cdot (\underline{\mathbf{P}}_{\gamma} + \underline{\mathbf{P}}_{\mathrm{e}}) = m_0^2 c^2.$$

This gives $0 + m_0^2 c^2 + 0 + 2E m_0 - 2E' m_0 - 0 - 2E'E(1 - \cos\theta)/c^2 = m_0^2 c^2$
or $2E m_0 - 2E' m_0 - 2E'E(1 - \cos\theta)/c^2 = 0$
and, finally, $\dfrac{1}{E'} = \dfrac{1}{E} + \dfrac{1}{m_0 c^2}(1 - \cos\theta)$.

It is obvious that using four-vectors we have done nothing more than apply the laws of conservation. This, however, was done in a compact way as far as notation is concerned.

8.11 A pion decays into a muon and a neutrino: $\pi^- \to \mu^- + \bar{\nu}_{\mu}$. Taking into account the fact that the neutrino has a negligible rest mass, show that $\underline{\mathbf{P}}_{\pi} \cdot \underline{\mathbf{P}}_{\mu} = \frac{1}{2}\left(m_{\pi}^2 + m_{\mu}^2\right)c^2$, where $\underline{\mathbf{P}}_{\pi}$ and $\underline{\mathbf{P}}_{\mu}$ are the energy-momentum four-vectors of the pion and the muon, respectively.

Solution

The conservation of the energy-momentum four-vector implies

$$\underline{\mathbf{P}}_{\pi} = \underline{\mathbf{P}}_{\mu} + \underline{\mathbf{P}}_{\nu}.$$

Since the neutrino has a rest mass which is almost zero, it is $\underline{\mathbf{P}}_{\nu} \cdot \underline{\mathbf{P}}_{\nu} = 0$. This gives

$$(\underline{\mathbf{P}}_\pi - \underline{\mathbf{P}}_\mu)^2 = \underline{\mathbf{P}}_\nu^2 = 0 \qquad \text{or} \qquad \underline{\mathbf{P}}_\pi^2 + \underline{\mathbf{P}}_\mu^2 - 2\underline{\mathbf{P}}_\pi \cdot \underline{\mathbf{P}}_\mu = 0.$$

It is $\underline{\mathbf{P}}_\pi^2 = m_\pi^2 c^2$ and $\underline{\mathbf{P}}_\mu^2 = m_\mu^2 c^2$.

Therefore, $\underline{\mathbf{P}}_\pi \cdot \underline{\mathbf{P}}_\mu = \frac{1}{2}\left(\underline{\mathbf{P}}_\pi^2 + \underline{\mathbf{P}}_\mu^2\right) = \frac{1}{2}\left(m_\pi^2 + m_\mu^2\right)c^2.$

8.12 A rocket is propelled by emitting photons backwards. The rocket is initially at rest and has a rest mass equal to M_{0i}. At time t, the rocket has a speed of υ and its rest mass has been reduced to M_0. Find the ratio M_{0i}/M_0 as a function of υ.

Solution

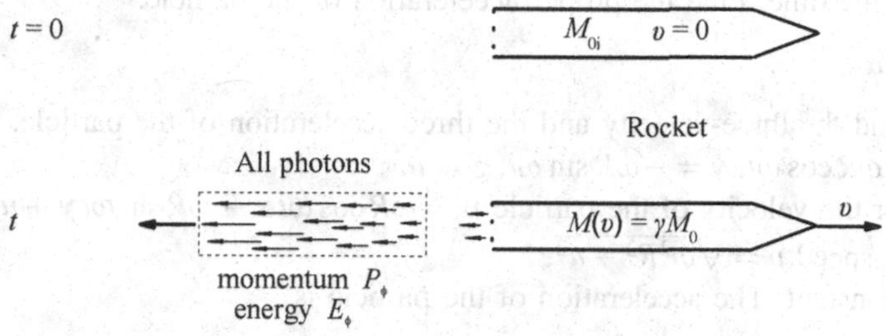

The initial rest mass of the rocket is equal to M_{0i} and its initial speed is zero. At time t, the rocket has a speed υ, to which there corresponds a Lorentz factor γ. Its mass is $M = \gamma M_0$, where M_0 is the rocket's rest mass at that moment. The photons that have already been emitted have a total energy equal to E_ϕ and a total momentum P_ϕ, in the direction opposite to that of the rocket's velocity.

The energy-momentum four-vectors are:

Rocket, initially:	$\underline{\mathbf{P}}_0 = (M_{0i}c,\ 0,\ 0,\ 0)$
Rocket at time t :	$\underline{\mathbf{P}} = (Mc,\ M\upsilon,\ 0,\ 0)$
Photons at time t :	$\underline{\mathbf{P}}_\phi = (E_\phi,\ -E_\phi,\ 0,\ 0)$

The law of conservation of the energy-momentum four-vector gives

$$\underline{\mathbf{P}}_0 = \underline{\mathbf{P}} + \underline{\mathbf{P}}_\phi.$$

Since for the photons it is $\underline{\mathbf{P}}_\phi \cdot \underline{\mathbf{P}}_\phi = 0$, it follows that $(\underline{\mathbf{P}}_0 - \underline{\mathbf{P}})^2 = 0$.

It is $\underline{\mathbf{P}}_0 - \underline{\mathbf{P}} = ((M_{0i} - M)c, -M\upsilon,\ 0,\ 0)$

so, therefore, $(M_{0i} - M)^2 c^2 - M^2\upsilon^2 = 0,$

$$\left(\frac{M_{0i}}{M}-1\right)^2=\frac{\upsilon^2}{c^2} \qquad \Rightarrow \qquad \frac{M_{0i}}{M}=1+\frac{\upsilon}{c}$$

where the negative root has been rejected as it would imply $M_{0i}<M$.

Since it is $M=\gamma M_0$, the final result is: $\dfrac{M_{0i}}{M_0}=\gamma(1+\beta)$ or $\dfrac{M_{0i}}{M_0}=\sqrt{\dfrac{1+\beta}{1-\beta}}$, in agreement with the result found in Example 6.3.

8.13 A particle in a magnetic field moves on a helical orbit given in parametric form by

$$x=R\sin\omega t, \quad y=R\cos\omega t, \quad z=ut,$$

where t is the time. Find the proper acceleration of the particle.

Solution

We first find the three-velocity and the three-acceleration of the particle.
From $\dot{x}=\omega R\cos\omega t$, $\dot{y}=-\omega R\sin\omega t$, $\dot{z}=u$,
we find for the velocity of the particle $\boldsymbol{\upsilon}=\omega R\cos\omega t\,\hat{\mathbf{x}}-\omega R\sin\omega t\,\hat{\mathbf{y}}+u\,\hat{\mathbf{z}}$
and for its speed $\upsilon=\sqrt{\omega^2R^2+u^2}$,
which is constant. The acceleration of the particle is

$$\boldsymbol{a}=-\omega^2R\sin\omega t\,\hat{\mathbf{x}}-\omega^2R\cos\omega t\,\hat{\mathbf{y}},$$

whose magnitude is $a=\omega^2R$.

Equation (8.68) gives for the proper acceleration: $\alpha=\gamma_{\mathrm{P}}^3\sqrt{a^2-(\boldsymbol{\upsilon}\times\boldsymbol{a})^2/c^2}$.
Here, it is $\gamma_{\mathrm{P}}=\left[1-(\omega^2R^2+u^2)/c^2\right]^{-1/2}$ and

$$\begin{aligned}\boldsymbol{\upsilon}\times\boldsymbol{a}&=(\omega R\cos\omega t\,\hat{\mathbf{x}}-\omega R\sin\omega t\,\hat{\mathbf{y}}+u\,\hat{\mathbf{z}})\times\left(-\omega^2R\sin\omega t\,\hat{\mathbf{x}}-\omega^2R\cos\omega t\,\hat{\mathbf{y}}\right)\\&=\omega^2Ru\cos\omega t\,\hat{\mathbf{x}}+\omega^2Ru\sin\omega t\,\hat{\mathbf{y}}-\omega^3R^2\hat{\mathbf{z}}=\omega^2R(u\cos\omega t\,\hat{\mathbf{x}}+u\sin\omega t\,\hat{\mathbf{y}}-\omega R\hat{\mathbf{z}})\end{aligned}$$

from which it is

$$\frac{(\boldsymbol{\upsilon}\times\boldsymbol{a})^2}{c^2}=\frac{\omega^4R^2}{c^2}(\omega^2R^2+u^2)=\omega^4R^2\left(1-\frac{1}{\gamma_{\mathrm{P}}^2}\right).$$

Substituting, we get $\alpha=\gamma_{\mathrm{P}}^3\sqrt{\omega^4R^2-\omega^4R^2+\dfrac{\omega^4R^2}{\gamma_{\mathrm{P}}^2}}$ or

$$\alpha=\gamma_{\mathrm{P}}^2\omega^2R=\frac{c^2\omega^2R}{c^2-\omega^2R^2-u^2}.$$

8.14 A particle P has four-momentum $\underline{\mathbf{P}}$ and velocity $\boldsymbol{\upsilon}_{\mathrm{P}}$ as observed in the frame of reference S. In the same frame of reference, a second observer, S′, has velocity $\boldsymbol{\upsilon}_{\mathrm{S}'}$ and four-velocity $\underline{\mathbf{U}}_{\mathrm{S}'}$. Show that $\underline{\mathbf{P}}\cdot\underline{\mathbf{U}}_{\mathrm{S}'}$ is the energy of the particle in the frame of reference S′.

Solution

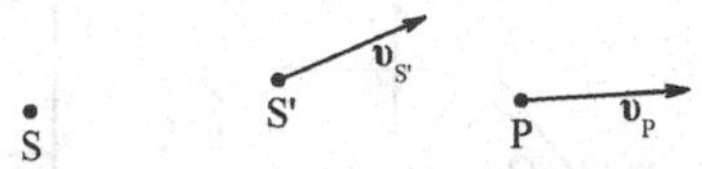

If the rest mass of the particle is m_0 and $\gamma_P = 1\big/\sqrt{1 - \upsilon_P^2/c^2}$ and $\gamma_{S'} = 1\big/\sqrt{1 - \upsilon_{S'}^2/c^2}$, then

$$\underline{\mathbf{P}} = (\gamma_P m_0 c,\ \gamma_P m_0 \boldsymbol{\upsilon}_P) \text{ and } \underline{\mathbf{U}}_{S'} = (\gamma_{S'} c,\ \gamma_{S'} \boldsymbol{\upsilon}_{S'}).$$

Their inner product is

$$\underline{\mathbf{P}} \cdot \underline{\mathbf{U}}_{S'} = \gamma_P \gamma_{S'} m_0 (c^2 - \boldsymbol{\upsilon}_P \cdot \boldsymbol{\upsilon}_{S'}) = m_0 c^2 \gamma_P \gamma_{S'} \left(1 - \frac{\boldsymbol{\upsilon}_P \cdot \boldsymbol{\upsilon}_{S'}}{c^2}\right).$$

If $\boldsymbol{\upsilon}'_P$ is the velocity of particle P in the frame of reference S′ and $\gamma'_P = 1\big/\sqrt{1 - \upsilon_P'^2/c^2}$, the transformation of the Lorentz factor [Eq. (3.48)] gives $\gamma'_P = \gamma_P \gamma_{S'} \left(1 - \frac{\boldsymbol{\upsilon}_P \cdot \boldsymbol{\upsilon}_{S'}}{c^2}\right)$.

Then it is $\underline{\mathbf{P}} \cdot \underline{\mathbf{U}}_{S'} = \gamma'_P m_0 c^2 = E'_P$,
where E'_P is the energy of the particle in the frame of reference S′.

Chapter 9
Electromagnetism

9.1 A capacitor with parallel plane plates is at rest, in vacuum, in a frame of reference S. The capacitor's plates are parallel to the *xz* plane and are rectangular, with sides *a* and *b* which are parallel to the *x* and *z* axes, respectively. The distance between the plates is *d*. The two plates have charges +*Q* and –*Q*, respectively. The potential difference across the plates is ΔU and the electric field in the region between them has magnitude *E*.

Given the invariance of the electric charge, find the capacitance of the capacitor, C', the strength of the electric field in the space between its plates, E', the potential difference across the plates of the capacitor, $\Delta U'$, the total electrostatic energy stored in the capacitor, W', and the density of electrostatic energy per unit volume between the plates of the capacitor, w', as these are measured by an observer in another frame of reference, S′, which is moving relative to frame S with a speed *V* in the direction of positive values of *x*. Express the results in terms of *V* and the corresponding values in frame S.

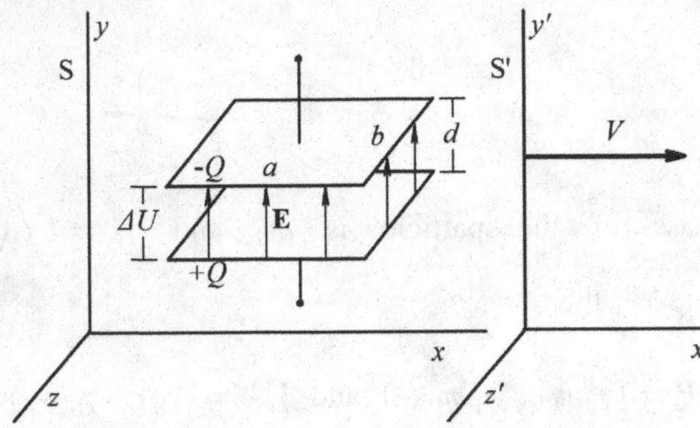

Solution

The capacitance of the capacitor in the frame of reference S is

$$C = \varepsilon_0 \frac{ab}{d}. \tag{1}$$

Since the charges on the plates are $\pm Q$, the potential difference across the capacitor is

$$\Delta U = \frac{Q}{C} = \frac{Q}{\varepsilon_0}\frac{d}{ab} \tag{2}$$

and the (uniform) electric field in the space between the plates is

$$E = \frac{\Delta U}{d} = \frac{Q}{\varepsilon_0}\frac{1}{ab}. \tag{3}$$

In the frame of reference S′, the charges on the capacitor's plates remain $\pm Q$, and the dimensions of the capacitor are the same, apart from the length a in the direction of the relative motion of the two frames of reference, which contracts to a/γ. From Eqs. (1)–(3), therefore, it follows that in frame S′

$$C' = \varepsilon_0 \frac{ab}{\gamma d} = \frac{C}{\gamma}, \quad \Delta U' = \frac{Q}{C'} = \frac{Q}{C/\gamma} = \gamma\,\Delta U, \quad E' = \frac{\Delta U'}{d} = \frac{\gamma \Delta U}{d} = \gamma E. \tag{4}$$

The total electrostatic energy stored in the capacitor is $W = \frac{1}{2} C \Delta U^2$ in frame S and $W' = \frac{1}{2} C' \Delta U'^2$ in frame S′. From Eq. (4),

$$W' = \frac{1}{2} C' \Delta U'^2 = \frac{1}{2}\frac{C}{\gamma}(\gamma \Delta U)^2 = \gamma W. \tag{5}$$

The density of electrostatic energy is $w = \dfrac{W}{abd}$ in frame S and $w' = \dfrac{W'}{a'b'd'}$ in frame S′. Therefore,

$$w' = \frac{W'}{a'b'd'} = \frac{\gamma W}{(a/\gamma)bd} = \gamma^2 \frac{W}{abd} = \gamma^2 w,$$

a result which also follows from $w = \dfrac{1}{2}\varepsilon_0 E^2$.

9.2 Using the transformations of the fields, find the electric field and the magnetic field in the region between the plates of the capacitor of Problem 9.1, as measured by an observer in the frame of reference S′. Express the results in terms of the components of the fields in frame S and the relative speed of the two frames of reference.

Solution

In the frame of reference S, the components of the fields in the space between the plates of the capacitor are:

$$E_x = 0, \quad E_y = E, \quad E_z = 0, \qquad B_x = 0, \quad B_y = 0, \quad B_z = 0.$$

The transformations of the fields give for frame S′:

$$E'_x = E_x = 0, \quad E'_y = \gamma(E_y - VB_z) = \gamma E, \quad E'_z = \gamma(E_z + VB_y) = 0$$
$$B'_x = B_x = 0, \quad B'_y = \gamma(B_y + VE_z/c^2) = 0, \quad B'_z = \gamma(B_z - VE_y/c^2) = -\gamma VE/c^2.$$

9.3 A capacitor with parallel plane plates is at rest in the frame of reference S. The capacitor's plates are rectangular with sides a and b, have a distance d between them, and are oriented so that they are parallel to the yz plane. The charges on the plates are $\pm Q$, which create an electric field $\mathbf{E} = E\hat{\mathbf{x}}$ in the space between the plates. The magnetic field in this region is zero. The potential difference across the plates is ΔU. Another frame of reference, S′, moves with a velocity $V\hat{\mathbf{x}}$ relative to S. The axes of the two frames are parallel and coincided at $t = t' = 0$.

(a) Without using the transformations of the electromagnetic field, find, in terms of C, $\mathbf{E}$, ΔU and the speed V, the capacitance of the capacitor, C', the strength of the electric field in the space between the plates of the capacitor, $\mathbf{E}'$, and the potential difference across the plates of the capacitor, $\Delta U'$, as measured by an observer in the frame of reference S′.

(b) Use now the transformations of the fields, in order to find the electric field again and the magnetic field in the space between the plates of the capacitor, in the frame of reference S′.

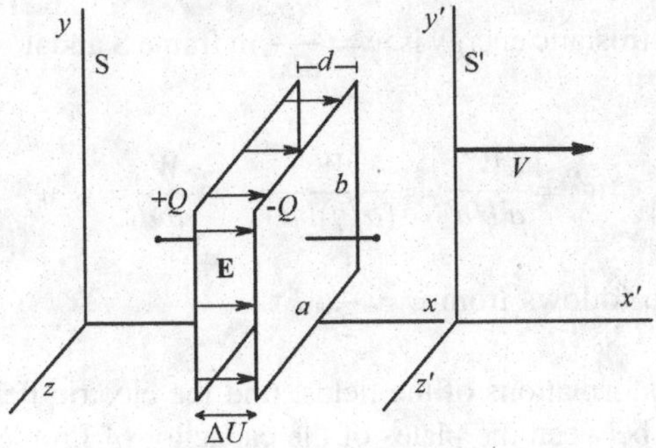

Solution

(a) In frame S′, the capacitor moves with a speed V towards the negative x's. The charge is invariant and remains equal to $\pm Q$ on the capacitor's plates. The distance between the plates will be $d' = d/\gamma$ and the capacitance of the capacitor is

$$C' = \varepsilon_0 \frac{a'b'}{d'} = \varepsilon_0 \frac{ab}{d/\gamma} = \varepsilon_0 \frac{ab}{d}\gamma = \gamma C.$$

The potential difference between the plates is

$$\Delta U' = \frac{Q}{C'} = \frac{Q}{\gamma C} = \frac{\Delta U}{\gamma}.$$

The electric field is

$$\mathbf{E}' = E'\hat{\mathbf{x}} = \frac{\Delta U'}{d'}\hat{\mathbf{x}} = \frac{\Delta U/\gamma}{d/\gamma}\hat{\mathbf{x}} = \frac{\Delta U}{d}\hat{\mathbf{x}} = E\hat{\mathbf{x}} = \mathbf{E}$$

(b) The transformations of the fields are

$$E'_x = E_x, \quad E'_y = \gamma(E_y - VB_z), \quad E'_z = \gamma(E_z + VB_y),$$
$$B'_x = B_x \quad B'_y = \gamma(B_y + VE_z/c^2), \quad B'_z = \gamma(B_z - VE_y/c^2).$$

In frame S, the fields have components $E_x = E$, $E_y = 0$, $E_z = 0$, $B_x = 0$, $B_y = 0$, $B_z = 0$.
Substituting in the transformation equations, we have:

$$E'_x = E_x = E, \quad E'_y = \gamma(E_y - VB_z) = 0, \quad E'_z = \gamma(E_z + VB_y) = 0,$$
$$B'_x = B_x = 0 \quad B'_y = \gamma(B_y + VE_z/c^2) = 0, \quad B'_z = \gamma(B_z - VE_y/c^2) = 0$$

Therefore, $\mathbf{E}' = E\hat{\mathbf{x}}$ and $\mathbf{B}' = 0$.

9.4 Show that the magnitudes $\mathbf{E}\cdot\mathbf{B}$ and $E^2 - c^2B^2$ remain invariant under the Lorentz transformations from one inertial system of reference S to another, S′, which is moving with a velocity $V\,\hat{\mathbf{x}}$ relative to the frame S.

Solution

The transformations for the components of the fields are

$$E_x = E'_x,\quad E_y = \gamma\left(E'_y + VB'_z\right),\quad E_z = \gamma\left(E'_z - VB'_y\right),$$
$$B_x = B'_x,\quad B_y = \gamma\left(B'_y - VE'_z/c^2\right),\quad B_z = \gamma\left(B'_z + VE'_y/c^2\right).$$

Therefore,

$$\begin{aligned}\mathbf{E}\cdot\mathbf{B} &= E_xB_x + E_yB_y + E_zB_z\\ &= E'_xB'_x + \gamma\left(E'_y + VB'_z\right)\gamma\left(B'_y - VE'_z/c^2\right) + \gamma\left(E'_z - VB'_y\right)\gamma\left(B'_z + VE'_y/c^2\right)\\ &= E'_xB'_x + \gamma^2\left(E'_yB'_y - \cancel{VE'_yE'_z/c^2} + \cancel{VB'_zB'_y} - V^2B'_zE'_z/c^2 + E'_zB'_z + \cancel{VE'_zE'_y/c^2} - \cancel{VB'_yB'_z} - V^2B'_yE'_y/c^2\right)\\ &= E'_xB'_x + \gamma^2\left(1 - V^2/c^2\right)\left(E'_yB'_y + E'_zB'_z\right) = E'_xB'_x + E'_yB'_y + E'_zB'_z = \mathbf{E}'\cdot\mathbf{B}'\end{aligned}$$

Also,

$$\begin{aligned}E^2 - c^2B^2 &= E_x^2 + E_y^2 + E_z^2 - c^2B_x^2 - c^2B_y^2 - c^2B_z^2\\ &= E_x'^2 + \gamma^2\left(E'_y + VB'_z\right)^2 + \gamma^2\left(E'_z - VB'_y\right)^2 - c^2B_x'^2 - c^2\gamma^2\left(B'_y - VE'_z/c^2\right)^2 - c^2\gamma^2\left(B'_z + VE'_y/c^2\right)^2\\ &= E_x'^2 - c^2B_x'^2 + \gamma^2\{(E'_y)^2 + \cancel{2E'_yVB'_z} + (VB'_z)^2 + (E'_z)^2 - \cancel{2E'_zVB'_y} + (VB'_y)^2 -\\ &\quad - c^2(B'_y)^2 + \cancel{2B'_yVE'_z} - c^2(VE'_z/c^2)^2 - c^2(B'_z)^2 - \cancel{2B'_zVE'_y} - c^2(VE'_y/c^2)^2\}\\ &= E_x'^2 - c^2B_x'^2 + \gamma^2\left\{\left(1 - V^2/c^2\right)\left(E_y'^2 + E_z'^2\right) - \left(c^2 - V^2\right)\left(B_y'^2 + B_z'^2\right)\right\}\\ &= E_x'^2 + E_y'^2 + E_z'^2 - c^2B_x'^2 - c^2B_y'^2 - c^2B_z'^2 = E'^2 - c^2B'^2\end{aligned}$$

The magnitudes $\mathbf{E}\cdot\mathbf{B}$ and $E^2 - c^2B^2$ are, therefore, invariant under the Lorentz transformations.

9.5 A rectilinear charge distribution of linear charge density λ', extending to infinity on both sides, lies on the x'-axis of an inertial frame of reference S′. The strength of the electric field it produces is given by the well known equations

$$E'_x = 0,\qquad E'_y = \frac{\lambda' y'}{2\pi\varepsilon_0(y'^2 + z'^2)},\qquad E'_z = \frac{\lambda' z'}{2\pi\varepsilon_0(y'^2 + z'^2)}.$$

In another frame of reference, S, in which frame S′ moves with a velocity $\mathbf{V} = V\hat{\mathbf{x}}$ (V positive), the charge distribution appears as a current I in the direction of positive x's. What are the magnetic and the electric fields in frame S? Compare the magnitudes of the electric and the magnetic forces exerted on a charge q which moves with velocity $\mathbf{V}$ in frame S and is at point (x, y, z).

Solution

The transformations of the fields' components are

$$E_x = E'_x, \quad E_y = \gamma\left(E'_y + VB'_z\right), \quad E_z = \gamma\left(E'_z - VB'_y\right),$$
$$B_x = B'_x, \quad B_y = \gamma\left(B'_y - VE'_z/c^2\right), \quad B_z = \gamma\left(B'_z + VE'_y/c^2\right).$$

Equating to zero those components that vanish in frame S′, we find:

$$E_x = 0, \quad E_y = \gamma E'_y, \quad E_z = \gamma E'_z,$$
$$B_x = 0, \quad B_y = -\gamma VE'_z/c^2, \quad B_z = \gamma VE'_y/c^2.$$

The coordinates transform simply as $y' = y$ and $z' = z$. Due to length contraction, the linear charge density will appear greater in frame S by a factor γ, $\lambda = \gamma\lambda'$. Substituting and using the relationship $c^2 = 1/\varepsilon_0\mu_0$,

$$E_x = 0, \quad E_y = \gamma\frac{\lambda' y'}{2\pi\varepsilon_0(y'^2 + z'^2)} = \frac{\lambda}{2\pi\varepsilon_0}\frac{y}{y^2 + z^2},$$
$$E_z = \gamma\frac{\lambda' z'}{2\pi\varepsilon_0(y'^2 + z'^2)} = \frac{\lambda}{2\pi\varepsilon_0}\frac{z}{y^2 + z^2},$$
$$B_x = 0, \quad B_y = -\frac{\gamma V}{c^2}\frac{\lambda' z'}{2\pi\varepsilon_0(y'^2 + z'^2)} = -\frac{\mu_0}{2\pi}(V\lambda)\frac{z}{y^2 + z^2},$$
$$B_z = \frac{\gamma V}{c^2}\frac{\lambda' y'}{2\pi\varepsilon_0(y'^2 + z'^2)} = \frac{\mu_0}{2\pi}(V\lambda)\frac{y}{y^2 + z^2}.$$

In vector form,

$$\mathbf{E} = \frac{\lambda}{2\pi\varepsilon_0}\frac{y\hat{\mathbf{y}} + z\hat{\mathbf{z}}}{y^2 + z^2}, \qquad \mathbf{B} = \frac{\mu_0}{2\pi}I\frac{z\hat{\mathbf{y}} - y\hat{\mathbf{z}}}{y^2 + z^2},$$

where $\lambda = \gamma\lambda'$ is the linear charge density and $I = -\lambda V$ is the electric current, as these are observed in the frame of reference S.
Because it is

$$\left|\frac{y\hat{\mathbf{y}}+z\hat{\mathbf{z}}}{y^2+z^2}\right| = \frac{1}{\sqrt{y^2+z^2}} = \frac{1}{r} \quad \text{and} \quad \left|\frac{z\hat{\mathbf{y}}-y\hat{\mathbf{z}}}{y^2+z^2}\right| = \frac{1}{\sqrt{y^2+z^2}} = \frac{1}{r},$$

where r is the distance from the x-axis, the magnitudes of the fields are

$$E = \frac{\lambda}{2\pi\varepsilon_0 r} \quad \text{are} \quad B = \frac{\mu_0 I}{2\pi r}.$$

The force exerted on the charge q, which is moving with velocity $\mathbf{V} = V\hat{\mathbf{x}}$ at the point (x, y, z), is

$$\mathbf{F} = q\mathbf{E} + qV\hat{\mathbf{x}} \times \mathbf{B} = \frac{\lambda q}{2\pi\varepsilon_0}\frac{y\hat{\mathbf{y}}+z\hat{\mathbf{z}}}{y^2+z^2} + qV\frac{\mu_0}{2\pi}I\,\hat{\mathbf{x}} \times \frac{z\hat{\mathbf{y}}-y\hat{\mathbf{z}}}{y^2+z^2}$$

$$\mathbf{F} = \left(\frac{\lambda q}{2\pi\varepsilon_0} + qV\frac{\mu_0}{2\pi}I\right)\frac{y\hat{\mathbf{y}}+z\hat{\mathbf{z}}}{y^2+z^2}.$$

The ratio

$$\frac{\text{Electric Force}}{\text{Magnetic force}} = \frac{\lambda q}{2\pi\varepsilon_0}\frac{2\pi}{\mu_0 qVI} = \frac{\lambda}{\varepsilon_0\mu_0 VI} = -\frac{c^2}{V^2}.$$

9.6 A constant current I flows in a thin conductor in the shape of a circle of radius ρ. Assume that all charges to which the current is due move with the same speed βc. Using the relativistic version of the differential form of the Biot-Savart law, Eq. (9.54), find the magnetic field at the center of the circle.

Solution

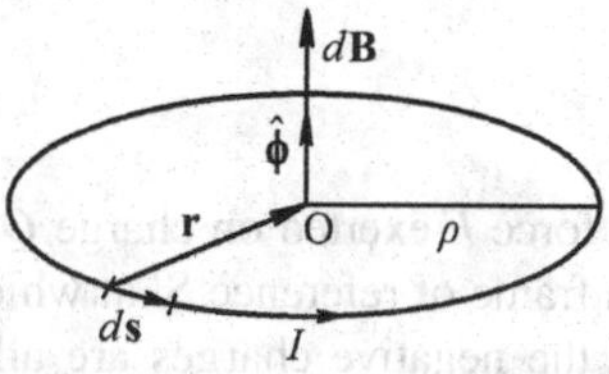

According to Eq. (9.54), $d\mathbf{B} = \dfrac{\mu_0 I}{4\pi r^2}\dfrac{1-\beta^2}{(1-\beta^2\sin^2\theta)^{3/2}}d\mathbf{s}\times\hat{\mathbf{r}}$.

In this case, it is $\theta = \pi/2$. Also, $d\mathbf{s}\times\hat{\mathbf{r}} = ds\,\widehat{\boldsymbol{\phi}}$. Therefore,

$$d\mathbf{B} = \frac{\mu_0 I}{4\pi\rho^2}\frac{\widehat{\boldsymbol{\phi}}}{\sqrt{1-\beta^2}}ds$$

is the magnetic field at the center of the circle, which is due to the current element Ids. Integrating ds over the whole circle, gives $2\pi\rho$. The total magnetic field at the center of the circle is, therefore,

$$d\mathbf{B} = \frac{\mu_0 I}{2\rho}\frac{\hat{\boldsymbol{\phi}}}{\sqrt{1-\beta^2}}.$$

The difference from the non-relativistic result is the presence of $\sqrt{1-\beta^2}$. Naturally, in a realistic case, this quantity must be thought of as being the mean value over all the charges to which the current is due. In practical cases it is negligible and the non-relativistic result follows.

9.7 A positive charge Q is at rest in the inertial frame of reference S (Figure a). In the same frame there is a rectilinear conductor, extending to infinity in both directions, which contains stationary positive ions, all with charge $+e$, with a linear charge density λ_+, and electrons, with charge $-e$ and linear charge density λ_-. All the electrons move with a common speed υ_- in frame S. In the same frame, the distances between the positive ions are all equal to d_+ and those between the electrons are all equal to d_-. It is $d_- > d_+$. The distance of the charge Q from the conductor is r.

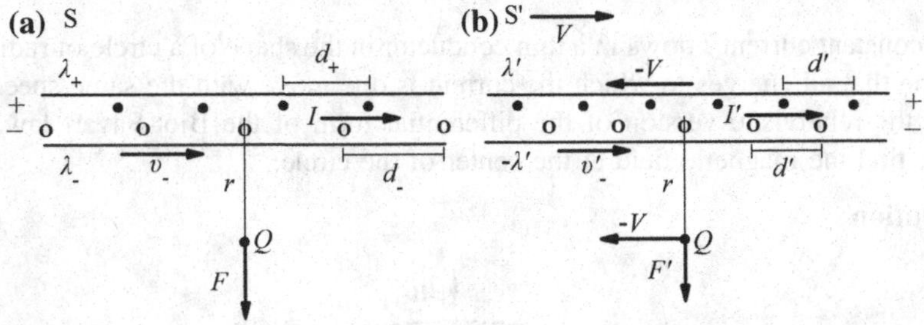

(a) Find the electrostatic force F exerted on charge Q in the frame of reference S.
(b) Find the speed V of a frame of reference S′ at which the distances between the positive charges and the negative charges are all equal to d' and, therefore, both the linear charge densities are equal to λ' in absolute values (Figure b).
(c) With the value of V found in (b), find the force exerted on the charge Q in frame S′. Since the conductor is neutral in frame S′, this force is a purely magnetic force. Express the force in terms of the electric current in the conductor and the distance r.

Solution

(a) The linear charge densities are: $\lambda_+ = e/d_+, \quad \lambda_- = -e/d_-$.

The intensity of the electric field at the charge Q is $E = \dfrac{\lambda_+ + \lambda_-}{2\pi\varepsilon_0 r} = \dfrac{e/d_+ - e/d_-}{2\pi\varepsilon_0 r}$

and the electrostatic force on the charge Q is

$$F = \frac{Qe}{2\pi\varepsilon_0 r}\left(\frac{1}{d_+} - \frac{1}{d_-}\right). \tag{1}$$

(b) In the frame of reference S, the positive charges are at a distance of d_- from each other and all move with a speed $-V$, to which there corresponds a Lorentz factor $\gamma = 1/\sqrt{1 - V^2/c^2}$.

In frame S, the negative charges are at a distance of d_- from each other and all move with a speed υ_-. The distance between these charges in their own frame of reference is $d_0 = \gamma_- d_-$, where $\gamma_- = 1/\sqrt{1 - \upsilon_-^2/c^2}$. In the frame of reference S′, which moves with a speed V relative to the frame S, the speed of the negative charges is $\upsilon'_- = \dfrac{\upsilon_- - V}{1 - \upsilon_- V/c^2}$. The Lorentz factor corresponding to υ'_- is $\gamma'_- = 1/\sqrt{1 - {\upsilon'_-}^2/c^2}$. The distance between the positive charges in frame S′ is

$$d' = \frac{d_+}{\gamma}. \tag{2}$$

The negative charges, in their own frame of reference, have a distance between them equal to $\gamma_- d_-$. In the frame of reference S′, therefore, they have a distance between them equal to

$$d' = d_- \frac{\gamma_-}{\gamma'_-}. \tag{3}$$

Equating Eqs. (2) and (3), we have

$$d' = \frac{d_+}{\gamma} = d_- \frac{\gamma_-}{\gamma'_-} \tag{4}$$

and,

$$d_+\sqrt{1-\frac{V^2}{c^2}}=d_-\frac{\sqrt{1-\frac{{\upsilon'_-}^2}{c^2}}}{\sqrt{1-\frac{\upsilon_-^2}{c^2}}}=d_-\frac{\sqrt{1-\frac{1}{c^2}\left(\frac{\upsilon_--V}{1-\upsilon_-V/c^2}\right)^2}}{\sqrt{1-\frac{\upsilon_-^2}{c^2}}}=d_-\frac{\sqrt{1-\frac{V^2}{c^2}}}{1-\frac{\upsilon_-V}{c^2}}. \quad (5)$$

Therefore,

$$1-\frac{\upsilon_-V}{c^2}=\frac{d_-}{d_+} \quad \text{or} \quad V=-\frac{c^2}{\upsilon_-}\left(\frac{d_-}{d_+}-1\right). \quad (6)$$

(c) To evaluate d_-, we need the factor γ'_-. It is

$$\gamma'_-=\frac{1}{\sqrt{1-\frac{{\upsilon'_-}^2}{c^2}}}=\frac{1}{\sqrt{1-\left(\frac{\upsilon_--V}{c^2-\upsilon_-V}\right)^2}}=\frac{1-\frac{\upsilon_-V}{c^2}}{\sqrt{1-\frac{V^2}{c^2}}\sqrt{1-\frac{\upsilon_-^2}{c^2}}}. \quad (7)$$

Substituting in Eq. (4), we have

$$d_-=d'\frac{\gamma'}{\gamma_-}=d'\frac{1-\frac{\upsilon_-V}{c^2}}{\sqrt{1-\frac{V^2}{c^2}}}=\gamma d'\left(1-\frac{\upsilon_-V}{c^2}\right). \quad (8)$$

The current in the conductor, in frame S′, is $I'=e\frac{(-V)}{d'}+(-e)\frac{\upsilon'_-}{d'}=-\frac{e}{d'}(V+\upsilon'_-)$ or

$$I'=-\frac{e}{d'}\left(V+\frac{\upsilon_--V}{1-\upsilon_-V/c^2}\right)=-\frac{e\upsilon_-}{d'\gamma^2}\frac{1}{1-\upsilon_-V/c^2}. \quad (9)$$

In the frame of reference S the electrostatic force on charge Q is $F=\frac{Q}{2\pi\varepsilon_0 r}\left(\frac{e}{d_+}-\frac{e}{d_-}\right)$.

Substituting for d_+ and d_- from Eqs. (2) and (8), it follows that

$$F=\frac{Q}{2\pi\varepsilon_0 r}\frac{e}{\gamma d'}\left(1-\frac{1}{1-\upsilon_-V/c^2}\right)=-\frac{Q}{2\pi\varepsilon_0 r}\frac{e}{\gamma d'}\frac{\upsilon_-V/c^2}{1-\upsilon_-V/c^2}.$$

Making use of Eq. (9), we find

$$F = -\frac{Q}{2\pi\varepsilon_0 r}\frac{e}{\gamma d'}\left(\frac{\upsilon_- V}{c^2}\right)\left(-\frac{d'\gamma^2}{e\upsilon_-}\right)I' = \frac{Q}{2\pi\varepsilon_0 r}\gamma\frac{VI'}{c^2}.$$

Because it is $c^2 = 1/\varepsilon_0\mu_0$, we finally have for the force in frame S,

$$F = \frac{\mu_0}{2\pi}\frac{QVI'}{r}\gamma.$$

In the frame of reference S′, the force exerted on the charge Q is $F' = F/\gamma$. Therefore, and since $r' = r$,

$$F' = \frac{\mu_0}{2\pi}\frac{I'}{r'}QV.$$

The force is perpendicular to the conductor. It may be considered as the magnetic force exerted on the charge Q, which is moving with speed V, by a magnetic field equal to

$$B' = \frac{\mu_0}{2\pi}\frac{I'}{r'},$$

the latter being normal to the plane defined by the conductor and the position at which the charge is situated.

Appendices

A1.1 Draw two Minkowski diagrams for the room and rod paradox: one with main frame of reference that of the room and another with main frame of reference that of the rod.

Solution

Diagram (a) refers to the frame of reference of the room, S. Diagram (b) refers to the frame of reference of the rod S′. The lengths and times in the two frames are also shown. The thick lines show the lengths of the x-dimension of the rod in diagram (a) and of the room in diagram (b).

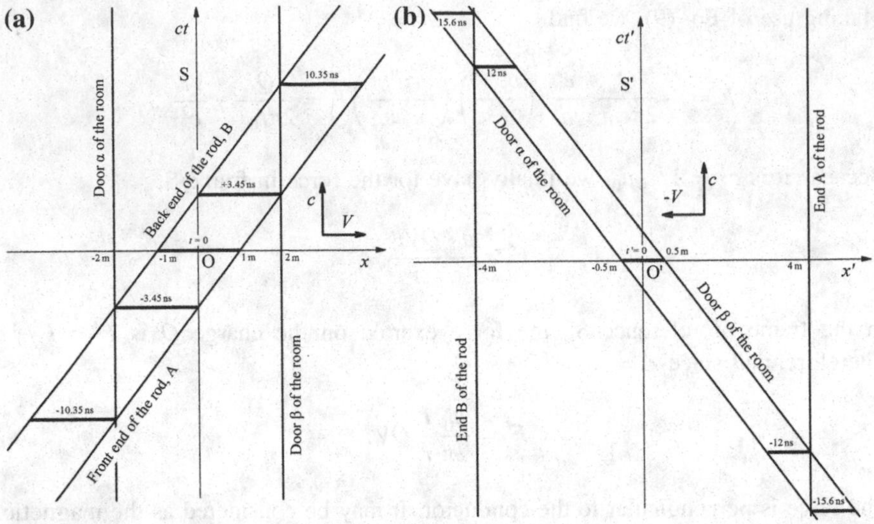

A4.1 A region of a space is characterized by a charge density and a current density which are the functions of position, $\rho(\mathbf{r})$ and $\mathbf{J}(\mathbf{r})$, respectively, as measured in a frame of reference S. Consider an element of volume $d\tau$, centered around a point at $\mathbf{r}$. At this point, let the values of the charge density and the current density be ρ and $\mathbf{J}$, respectively, and that the choice of the axes of frame S is such that $\mathbf{J} = J\hat{\mathbf{x}}$. The volume element and the charge it contains move, momentarily, with a velocity equal to $\mathbf{v} = \upsilon\hat{\mathbf{x}}$ in frame S. If the number density of the charge carriers in the element of volume is n and each charge carrier has a charge e, it is $\rho = ne$, $\mathbf{J} = ne\,\mathbf{v} = \rho\mathbf{v} = \rho\upsilon\hat{\mathbf{x}}$ and $J = \rho\upsilon$. Now view the element of volume from another frame of reference, S′, which is moving with a velocity $\mathbf{v} = \upsilon\hat{\mathbf{x}}$ with respect to frame S.

(a) In the frame of reference S′, find the volume $d\tau'$ of the element, and the charge density ρ' and the current density $\mathbf{J}'$ at the point considered.

(b) Prove, from first principles, that the magnitude $c^2\rho^2 - J^2$ is invariant under the Lorentz transformation.

(c) Use the transformation formulas for the magnitudes ρ and $\mathbf{J}$ in order to show that $c^2\rho^2 - J^2$ is invariant under the Lorentz transformation.

Solution

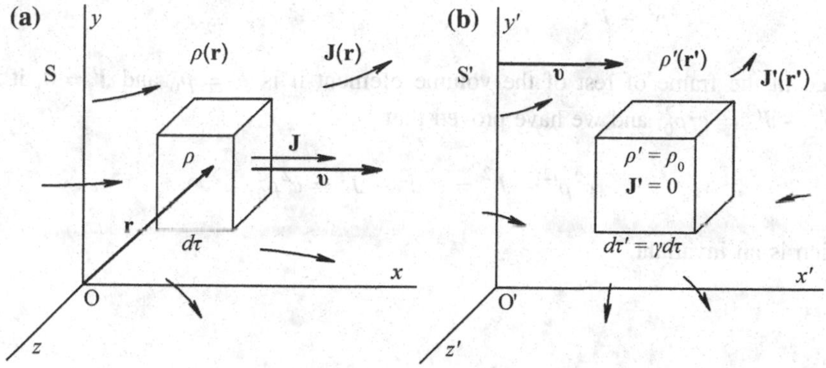

(a) Since both the volume element and the frame of reference S′ move with velocity $\mathbf{v} = \upsilon\hat{\mathbf{x}}$ in the frame of reference S, the volume element is at rest in S′. If $d\tau$ is the volume of the element in S, its volume in S′ will be $d\tau' = \gamma d\tau$. Since the charge enclosed by the volume element is an invariant, the charge densities are related by $\rho' = \rho/\gamma$. The charges are at rest in the volume element in S′, so $\mathbf{J}' = 0$. We denote ρ' by ρ_0. We have found that $\rho_0 = \rho/\gamma$ and $\mathbf{J}' = 0$.

(b) It is $c^2\rho^2 - J^2 = c^2\rho^2 - \rho^2\upsilon^2 = c^2\rho^2(1 - \upsilon^2/c^2) = c^2\rho^2/\gamma^2 = c^2\rho_0^2$.
Also, $c^2\rho'^2 - J'^2 = c^2\rho_0^2 - 0 = c^2\rho_0^2$.
It follows that $c^2\rho^2 - J^2 = c^2\rho'^2 - J'^2 = c^2\rho_0^2$
which is an invariant.

(c) The transformation formulas for the magnitudes ρ and $\mathbf{J}$ are, [Eq. (A4.96)],

$$J'_x = \gamma(J_x - V\rho), \quad J'_y = J_y, \quad J'_z = J_z, \quad \rho' = \gamma\left(\rho - \frac{V}{c^2}J_x\right).$$

It follows that

$$c^2\rho'^2 - J'^2 = c^2\rho'^2 - J'^2_x - J'^2_y - J'^2_z = c^2\gamma^2\left(\rho - \frac{\upsilon}{c^2}J_x\right)^2 - \gamma^2(J_x - \upsilon\rho)^2 - J_y^2 - J_z^2.$$

Expanding,

$$c^2\rho'^2 - J'^2 = c^2\gamma^2\rho^2 - \cancel{2\gamma^2\rho\upsilon J_x} + c^2\gamma^2\frac{\upsilon^2}{c^4}J_x^2 - \gamma^2 J_x^2 + \cancel{2\gamma^2\upsilon\rho J_x} - \gamma^2\upsilon^2\rho^2 - J_y^2 - J_z^2$$

or

$$c^2\rho'^2 - J'^2 = \gamma^2\rho^2(c^2 - \upsilon^2) - \gamma^2 J_x^2\left(1 - \frac{\upsilon^2}{c^2}\right) - J_y^2 - J_z^2 = c^2\rho^2 - J_x^2 - J_y^2 - J_z^2$$
$$= c^2\rho^2 - J^2.$$

Since in the frame of rest of the volume element it is $\rho' = \rho_0$ and $\mathbf{J}' = 0$, it is $c^2\rho'^2 - J'^2 = c^2\rho_0^2$, and we have proved that

$$c^2\rho'^2 - J'^2 = c^2\rho^2 - J^2 = c^2\rho_0^2,$$

which is an invariant.

Index

© Springer International Publishing Switzerland 2016
C. Christodoulides, *The Special Theory of Relativity*,
Undergraduate Lecture Notes in Physics, DOI 10.1007/978-3-319-25274-2

Printed in the United States
By Bookmasters